Die Grundlehren der mathematischen Wissenschaften

in Einzeldarstellungen
mit besonderer Berücksichtigung
der Anwendungsgebiete

Band 201

Herausgegeben von

S. S. Chern   J. L. Doob   J. Douglas, jr.
A. Grothendieck   E. Heinz   F. Hirzebruch
E. Hopf   W. Maak   S. MacLane
W. Magnus   M. M. Postnikov   F. K. Schmidt
D. S. Scott   K. Stein

Geschäftsführende Herausgeber

B. Eckmann   B. L. van der Waerden

Anatole Beck

Continuous Flows in the Plane

With the Assistance of
Jonathan and Mirit Lewin

Springer-Verlag
Berlin Heidelberg New York 1974

Professor Anatole Beck
University of Wisconsin, Madison, Wisconsin, USA

Dr. Jonathan and Dr. Mirit Lewin
Ben Gurion University of the Negev, Beer Sheva, Israel

With 47 Figures

AMS Subject Classifications (1970)
Primary 54 H 20, 34 C 35, 54 H 15, 57 E 25
Secondary 54 H 25, 57 E 05, 57 E 20, 58 F 99, 70 G 99

ISBN-13: 978-3-642-65550-0   e-ISBN-13: 978-3-642-65548-7
DOI: 10.1007/978-3-642-65548-7

Dedicated to the memory
of my mother

MINNIE BECK
BORNSTEIN

March 4 1904   January 20 1967

She made of her life a gift to those she
loved. Without her many sacrifices,
this book might never have been
written

אשת חיל מי ימצא   ורחק מפנינים מכרה.
◆ ◆ ◆
תנו לה מפרי ידיה   ויהללוה בשערים מעשיה.

PROLOGUE

Der junge Alexander eroberte Indien

Er allein?

The time of publishing a book is a time to remember and to give thanks. A time to remember all those who by their help have made the book possible and to give thanks for that help. In this case, where the book represents sixteen years of research built on an education of twenty years, the list of those who by their efforts and kindnesses have fostered the education, the research, and the writing of the book runs to many hundreds. Of these, a few dozen have contributed so much that I could not allow the book to go to press without explicit acknowledgement of their assistance.

I begin with my mother, whose contribution over the years was the greatest, and to whose memory this book is dedicated. Widowed at an early age with two young sons, she labored long hours at difficult and unrewarding work to make our educations possible. The price of those educations, to which she contributed unstintingly from her meager earnings, was high, for the accumulated damage to her health led her to an untimely death. Without her support, both financial and psychological, it is questionable whether I would have completed that education, without which this book would have been impossible.

I am deeply conscious of my debt to my teachers, both in school and in the various universities I have attended. Of these, I note especially Walter Prenowitz and Samuel Borofsky of Brooklyn College and Henry Helson, Nelson Dunford, Jacob Schwartz, and Shizuo Kakutani at Yale University. Professor Kakutani, who was my doctoral advisor, never stinted of his time and effort; his kind assistance and demanding discipline initiated me into the mathematical profession.

To Paul Mostert, I give my thanks for introducing me to the field of flows in the plane, and also to Professor Kakutani, Gustav Hedlund, Deane Montgomery, R. H. Bing, and Aryeh Dvoretsky for encouraging my efforts in this area. I thank Professor Hedlund for the administrative initiative and Doctor Harry Bakwin for the financial substance which together created a fellowship which enabled me to travel and study in Europe at a critical point in my career, and which contributed to my development as an independent mathematician.

My grateful thanks go to my wife, Evelyn, and to my children, Nina and Micah, who gamely endured at second hand many of the frustrations and difficulties of writing this book. Their contribution cannot be overstated.

I cannot thank all the friends whose encouragement fostered my work, but I must give special mention to my brother Bernard, to Donald Newman, Aryeh Dvoretsky, and Konrad Jacobs, and to all my close friends who are my colleagues at Madison.

More important than any in the actual task of writing this book were my students and assistants, Mirit and Jonathan Lewin. Their aid was invaluable in creating this work from the vast pile of notes, published and unpublished articles, jottings, ideas, and results, some correct and some incorrect, some raw and some polished, which represented the material of this book when they joined me in working on it. Reversing the roles of teacher and student, they more than repaid me for my work with them on their doctoral theses by correcting, criticizing, polishing, writing, re-writing, and editing portions of this text. Chapters 4, 5, 8, and 9 are the stuff of those theses, which I am honored to include in this book and proud to exhibit before the mathematical community.

Finally, I wish to give my genuine thanks to the many sources from which I have received financial assistance in my education and my research career: the City and State of New York, Yale University, the National Science Foundation, the Office of Naval Research, Doctor Harry Bakwin, the Wisconsin Alumni Research Foundation, the U.S. Army Mathematics Research Center, the Air Force Office of Scientific Research, the German Academic Exchange Service (DAAD), and the National Research Council.

I do not mean to compare this book to the conquest of India, but it does share this one aspect, that in each case, it is the product of many, many people's work. Although the protocols of the academy will call it my book, I wish here to note and to thank the many, many people who have helped to create it.

ANATOLE BECK

Madison, Wisconsin
9 February 1973

CONTENTS

INDEX OF SYMBOLS

INTRODUCTION

Topological Dynamics has its roots deep in the theory of differential equations, specifically in that portion called the "qualitative theory". The most notable early work was that of Poincaré and Bendixson, regarding stability of solutions of differential equations, and the subject has grown around this nucleus. It has developed now to a point where it is fully capable of standing on its own feet as a branch of Mathematics studied for its intrinsic interest and beauty, and since the publication of *Topological Dynamics* by Gottschalk and Hedlund, it has been the subject of widespread study in its own right, as well as for the light it sheds on differential equations. The *Bibliography for Topological Dynamics* by Gottschalk contains 1634 entries in the 1969 edition, and progress in the field since then has been even more prodigious.

The study of dynamical systems is an idealization of the physical studies bearing such names as aerodynamics, hydrodynamics, electrodynamics, *etc*. We begin with some space (call it X) and we imagine in this space some sort of idealized particles which change position as time passes. This change of position is in accordance with some rule or principle or formula φ. Without inquiring into the nature of the idealized particles, we simply state that after a time t, the particle which was at x will be transposed to the position $\varphi(t, x)$. Both x and $\varphi(t, x)$, being positions, or "space variables" are elements of the underlying space X. If our study were abstract aerodynamics, then certain physical considerations, such as compressability, pressure gradients, inertia, *etc*. would be reflected in specific laws concerning $\varphi(t, x)$ as a function of t and x. If it were hydrodynamics, different restraints on φ would apply. In the study of abstract dynamical systems, we replace these physical laws with mathematical considerations.

The most important mathematical constraint is the *flow equation*, also known as the *group property*: $\varphi\big(t_1, \varphi(t_2, x)\big) = \varphi(t_1 + t_2, x)$. What this says basically is that as the particles move, the behavior at each point does not change with the passage of time. Various special disciplines have their own names for this phenomenon. Stochastic processes which obey the flow equation are called *stationary*; differential equations which obey it are called *autonomous*. But they are special cases of this general principle of consistency of action at each location as time passes. If the flow equation is satisfied for all real values of t_1 and t_2, then φ is called a *flow*. If the equation holds only for positive t_1 and t_2, then it is a *semi-flow*.

In addition to the flow equation itself, we usually impose another condition on φ, and the nature of this other condition defines for us which of the several categories of dynamics we are studying. In this book, we will study continuous flows, *i.e.* those in which φ is a continuous function of t and x together. To do this, we must suppose that X is a topological space. Alternatively, we might have taken X to be a probability space and φ to be measurable. In that case, our study would have become that of stationary stochastic processes. We might have gone further and required that φ be measure-preserving, or ergodic, and each of these conditions would put us in one of the special categories of dynamics, many of which are venerable studies with their own vocabularies and histories. Especially important are the cases in which X is a topological linear space or an analytic manifold and in which the function

$$\dot{\varphi}(x) = \lim_{t \to 0} \frac{\varphi(t,x) - x}{t}.$$

exists at each point. These are the autonomous differential equations, which we see from the vantage point of dynamical theory as *differential flows*. There are important subclasses of the differential flows, in which the function $\dot{\varphi}(x)$ is continuous, or Lipschitzian, or differentiable in x, or even multiply differentiable in x. These are called continuous, or Lipschitzian, or C_1 differential equations, or, in the other cases, C_2, C_∞ or even analytic differential equations.

The study in this book concerns itself with continuous flows in $\mathbb{R}^2$, the Euclidean plane. The plane is an especially simple space, fully available to our visual intuition, and abundantly studied. Among its special properties is the Jordan Curve Theorem. Together with its corollary, the Theta Curve Theorem, it enables us to prove a basic result, the Gate Theorem, which simplifies the analysis of the orbits of any flow in the plane. It is interesting to see how the orbits of flows fit into the classical concepts of the theory of differential equations.

We might note that the differential equation $\dfrac{dy}{dx} = f(x, y)$ can be thought of as assigning to each point $\langle x, y \rangle$ of the plane a specific direction. Some treatments of differential equations do in fact think of the plane as being filled with tiny "lineal elements", one at each point, and directed according to the required slope. This concept is not far from the autonomous differential equation $\dot{v} = f(v)$, where v and $\dot{v}$ are two-dimentional vectors. In this latter situation, the lineal element is replaced by a vector. In the first case, a *solution* of the differential equation is a curve which has at each point the slope specified by the equation. In the second case, a solution is such a curve, given in parametric form with time as a parameter. If we imagine a particle moving so that its position at time t is $v(t)$, then the velocity of that particle at that time will be $f(v(t))$. In a certain sense, we could imagine the whole *set* of solution curves to be the solution in the first case. In the same way, we could think of the whole set of motions as constituting the solution of the second equation. This last construction is a motion of the whole plane, and it is something very like a flow. Conversely, if a flow φ is a differential flow and the velocity $\dot{\varphi}(v)$ satisfies $\dot{\varphi}(v) = f(v)$, then we can think of φ as a solution of the differential equation $\dot{v} = f(v)$.

However, the theory of differential equations usually focuses on the individual *solution curves*, raising the question of which curve passes through a given (initial) point. In the corresponding flow, the place of this curve is taken by the locus of points traversed by a particle which at some time is located at this initial point. If the system is a semi-flow, we are only interested in its path for positive time; if a flow, for positive and negative time. These curves are called the positive semi-orbit and the orbit respectively, and "trajectory" is sometimes used as a synonym for "orbit". The collection of solution curves is often called the *phase diagram* of a differential equation; we speak of the *orbit diagram* of a flow.

One of our major efforts will be to delineate, as well as we can, the possible orbit diagrams of continuous flows in the plane. In keeping with a topological approach to this problem, we will consider our question answered when we have obtained a solution up to a homeomorphic equivalence. By this we mean that if φ and ψ are flows in $\mathbb{R}^2$ (or in any topological space) which are related by a homeomorphism h in such a way that $\varphi(t, h(x)) = h(\psi(t,x))$, for all t and x, then for most purposes, we will not distinguish between φ and ψ. In such a situation, for example, the orbit diagrams will be homeomorphic.

As with any study, we will attack first the easy cases, in the hope that the tools acquired there will aid us in the more difficult ones. This raises immediately the question of which added conditions will yield a simple and understandable structure. It is especially in this area that our work brings forth a new insight. The theory of differential equations has classically considered that complexity of structure attends those points where the derivative vanishes, and has approached such points with elaborate care, using for them frightening names, such as *critical point* and *singular point*, which suggest that order and regularity end wherever such points are found. Indeed, if we stay away from such points, we find that the orbit diagrams consist locally of more or less "parallel" curves, and it is from such considerations that much of the qualitative theory of differential equations derives. These critical points are represented in the theory of flows as fixed points, and indeed if a flow has a Lipschitzian derivative, then every critical point is indeed a fixed point. But not all fixed points are equally difficult to manage; some are rather easy and some are quite complex. We have learned to be especially wary of what we call *stagnation points*, fixed points which are limit points of other orbits. Whenever a flow divides to avoid an obstacle, or changes direction, or shows any other signs of what we intuitively understand as "turbulence", stagnation points must come into the picture. And when a flow proceeds without stagnation points, things are rather simple, even though other fixed points might abound. It is thus these stagnation points, and other points closely associated with them (details of this must be sought in the body of the book), that we designate as *singular points*, and they include both fixed and moving points.

As an example of the difference created by this insight, we might consider the classic paper *Global structure of ordinary differential equations in the plane*, by Lawrence Markus [1] in which he discusses continuous differential equations (*i.e.* differential flows) in $\mathbb{R}^2$ with no critical (*i.e.* fixed) points. The analysis is quite difficult and is based in turn on some very subtle work of Wilfred Kaplan [1], [2]. This difficulty and complexity arises, from our point of view, from the fact that ∞ is a stagnation point, and a limit point of every orbit. Thus, in our concept, every point in the plane is a singular (moving) point, and it is not surprising that the results are so hard to obtain. This singularity of the flow is made manifest only in the neighborhood of ∞; in any bounded portion of the plane, it is not possible to distinguish it from a translation flow (up to a homeomorphism). Such observations as these convince us that any fruitful approach to flows in $\mathbb{R}^2$, which will reveal the true complexities, must consider them rather as flows in $\mathbb{S}^2$, the two-dimensional sphere, considered as the one-point compactification of $\mathbb{R}^2$. Every flow in the plane can be considered as a flow in the sphere which has ∞ as a fixed point; every flow in $\mathbb{S}^2$ having a fixed point at ∞ can be thought of as a flow in $\mathbb{R}^2$. Since every (continuous) flow in $\mathbb{S}^2$ has at least one fixed point, the theory of flows in the plane and that of flows in the sphere are equivalent.

Returning now to our concentration on stagnation points, we take for our simplest flows those with no stagnation points in $\mathbb{S}^2$ (not even at ∞). There is actually a simpler category, namely flows having only periodic orbits (including fixed points in this class), and we deal with these as well. But the flows without any stagnation points are the largest class for which we know essentially everything: not only the orbit diagrams but even the time structures on those diagrams. In the case of flows with only finitely many stagnation points, we have discovered much, but not the same sort of coverage as the "no-stagnation points" case. For flows whose stagnation points have countable closure, we are almost equally well informed. In each of these cases, for example, we can characterize completely the set $\mathscr{F}(\varphi)$ of fixed points by intrinsic topological properties without reference to flows, over and above the condition satisfied by the fixed set of any flow (*i.e.* that $\mathscr{F}(\varphi)$ be closed).

The most useful general tool in all these characterizations is this: that in the case of each flow φ, we can describe completely the structure of $\mathscr{G}_r(\varphi)$, the set of regular moving points. (Regular points are those which are cut off, in a way made formal in the body of the book, from the effects of the stagnation points.) The structure of $\mathscr{G}_r(\varphi)$ is revealed as a result of a major theorem called the Gate Theorem, which tells us roughly that if a moving point y belongs to the closure of another orbit $\mathcal{O}(x)$, then $\mathcal{O}(x)$ divides $\mathbf{S}^2$, in some sense, just as a periodic orbit (Jordan curve) does. Thus, the Gate Theorem is a generalization, in a sense, of the Jordan Curve Theorem, and it is this generalization which is the clue to the structure of continuous flows in the plane.

In addition to the principal thrust outlined above, this book contains studies based on some of the major techniques used to obtain the central theorems. *Inter alia*, it contains the entire texts of the doctoral dissertations of Jonathan Lewin and Mirit Lewin (cf. Prologue) on, respectively, reparametrization of flows and algebraic combinations of flows. Together, we have presented an introduction to a vital area of study, somewhat neglected in recent years, which offers to be a gateway to the study of Topological Dynamics and of Ordinary Differential Equations. The book presupposes an undergraduate education in Mathematics, including some Analytic Function Theory and Lebesgue Integration. Beyond this, and some elementary Topology, the reader is presumed to be a bright graduate student ready to take on a demanding area of study with only his wits for equipment, and some sophistication in Analysis for his nourishment. Everything else will be provided.

CHAPTER ONE

ELEMENTARY PROPERTIES OF FLOWS

Algebraic Flows

1.1. Definition. Let X be a set. A function φ from $\mathbb{R} \times X$ onto X will be called an *algebraic flow* in X if it satisfies the following group property:
$\varphi(s + t, x) = \varphi\big(s, \varphi(t, x)\big)$ for every $s, t \in \mathbb{R}$, and $x \in X$.

1.2. Theorem. *If φ is an algebraic flow in a set X, then for every $x \in X$, we have $\varphi(0, x) = x$.*

PROOF: Suppose $x \in X$. Choose $t \in \mathbb{R}$ and $y \in X$ such that $x = \varphi(t, y)$. Then

$$\varphi(0, x) = \varphi\big(0, \varphi(t, y)\big) = \varphi(0 + t, y) = \varphi(t, y) = x. \qquad \textbf{QED}$$

1.3. Definition. Let φ be an algebraic flow in a set X, and let $x \in X$. Then the set $\{\varphi(t, x) \mid t \in \mathbb{R}\}$ is called the *φ-orbit of x*, and is denoted by $\mathcal{O}_\varphi(x)$.

We denote the family of all φ-orbits, by $\mathbf{O}(\varphi)$.

1.4. Theorem. *Let φ be an algebraic flow in a set X. Then any two φ-orbits are either mutually disjoint, or equal.*

PROOF: Let $x, y \in X$, and suppose $\mathcal{O}_\varphi(x) \cap \mathcal{O}_\varphi(y) \neq \square$. Then we can choose $s_1, s_2 \in \mathbb{R}$ such that $\varphi(s_1, x) = \varphi(s_2, y)$.

Now, for any $t \in \mathbb{R}$, we have

$$\varphi(t, x) = \varphi(t - s_1 + s_1, x) = \varphi\big(t - s_1, \varphi(s_1, x)\big)$$
$$= \varphi\big(t - s_1, \varphi(s_2, y)\big) = \varphi(t - s_1 + s_2, y),$$

and it follows that $\mathcal{O}_\varphi(x) \subseteq \mathcal{O}_\varphi(y)$.

By symmetry, we have $\mathcal{O}_\varphi(y) \subseteq \mathcal{O}_\varphi(x)$, and the proof is complete. **QED**

1.5. Theorem. *Let φ be an algebraic flow in a set X, and for each $t \in \mathbb{R}$ let the function $\varphi_t : X \to X$ be defined by $\varphi_t(x) = \varphi(t, x)$, $\forall\; x \in X$.*

(a) *For each $t \in \mathbb{R}$, φ_t is a one-one mapping of X onto itself, i.e., φ_t is a permutation of X.*

(b) *$\{\varphi_t | t \in \mathbb{R}\}$ is an Abelian subgroup of the group of all permutations of X, and furthermore, for every $s, t \in \mathbb{R}$, we have $\varphi_s \cdot \varphi_t = \varphi_{s+t}$.*

PROOF: To prove (a), let $t \in \mathbb{R}$. φ_t is onto, for if $y \in X$, $y = \varphi(t, x)$ where $x = \varphi(-t, y)$. φ_t is one-one, for if $\varphi_t(x) = \varphi_t(y)$, then

$$x = \varphi\big(-t, \varphi(t, x)\big) = \varphi\big(-t, \varphi(t, y)\big) = y.$$

To prove (b), it is clearly sufficient to show that, for $s, t \in \mathbb{R}$, we have $\varphi_s \cdot \varphi_t = \varphi_{s+t}$.
But for any $x \in X$, and $s, t \in \mathbb{R}$,

$$(\varphi_s \cdot \varphi_t)(x) = \varphi_t\big(\varphi_s(x)\big) = \varphi\big(t, \varphi(s, x)\big) = \varphi(s + t, x) = \varphi_{s+t}(x).$$

Thus the proof is complete. **QED**

1.6. Theorem. *Let φ be an algebraic flow in a set X.*

(a) *For each $x \in X$, $\{t \in \mathbb{R} \mid \varphi(t, x) = x\}$ is an additive subgroup of $\mathbb{R}$.*

(b) *If $x, y \in X$ and $\mathcal{O}_\varphi(x) = \mathcal{O}_\varphi(y)$, then*

$$\{t \in \mathbb{R} \mid \varphi(t, x) = x\} \;=\; \{t \in \mathbb{R} \mid \varphi(t, y) = y\}.$$

PROOF: To prove (a), we note that, if $x \in X$, then $\varphi(0, x) = x$, and that whenever $\varphi(s, x) = \varphi(t, x) = x$, we have

$$\varphi(s - t, x) = \varphi\big(s, \varphi(-t, x)\big) = \varphi\big(s, \varphi(-t, \varphi(t, x))\big)$$
$$= \varphi\big(s, \varphi(t - t, x)\big) = \varphi(s, x) = x.$$

To prove (b), let $x, y \in X$, and suppose $y = \varphi(s, x)$. Then if $\varphi(t, x) = x$, we have $\varphi(t, y) = \varphi(t + s, x) = \varphi\big(s, \varphi(t, x)\big) = \varphi(s, x) = y$. Similarly, we have $\varphi(t, x) = x$ whenever $\varphi(t, y) = y$. This completes the proof of the theorem. **QED**

1.7. Fixed and Moving Points. Let φ be an algebraic flow in a set X. A point $x \in X$ is said to be a *φ-fixed point* if $\mathcal{O}_\varphi(x) = \{x\}$. The set of φ-fixed points is denoted by $\mathscr{F}(\varphi)$. If a point $x \in X$ is not a φ-fixed point, it is called a *φ-moving point*. The set of φ-moving points is denoted by $\mathscr{G}(\varphi)$. If $x \in \mathscr{F}(\varphi)$, we call $\mathcal{O}_\varphi(x)$ a *trivial φ-orbit*, and if $x \in \mathscr{G}(\varphi)$, we call $\mathcal{O}_\varphi(x)$ a *non-trivial φ-orbit*. A subset of X which is a union of a family of φ-orbits, is said to be *invariant of φ*.

We note that if a subset A of X is invariant of φ, then the restriction of φ to $\mathbb{R} \times A$ is an algebraic flow in A. We shall call this restriction, the *restriction of φ to A*.

Quasi-Flows

1.8. Definition. Let φ be an algebraic flow in a Hausdorff space X.

(a) If $x \in X$, then φ is said to be *quasi-continuous at the point x* if the function $t \to \varphi(t, x)$ is a continuous function from $\mathbb{R}$ into X.

(b) If $A \subseteq X$, then φ is said to be *quasi-continuous in A* if φ is quasi-continuous at every point $x \in A$.

(c) If φ is quasi-continuous in X, then φ is said to be a *quasi-flow*.

1.9. Theorem. *Let φ be a quasi-flow in a Hausdorff space X, and let $x \in X$. Then exactly one of the following occurs*:

(a) $\{t \in \mathbb{R} \mid \varphi(t, x) = x\} = \mathbb{R}$, *i.e.*, $x \in \mathscr{F}(\varphi)$.

(b) *There exists a least positive number t_0 such that $\varphi(t_0, x) = x$, and $\{t \in \mathbb{R} \mid \varphi(t, x) = x\}$ is the cyclic subgroup $\{nt_0 \mid n = 0, \pm 1, \pm 2, \ldots\}$ of $\mathbb{R}$ which is generated by t_0.*

(c) $\{t \in \mathbb{R} \mid \varphi(t, x) = x\} = \{0\}$, *and the function $t \to \varphi(t, x)$ is one-one.*

PROOF: Since $\{x\}$ is a closed subset of X, the continuity of the function $t \to \varphi(t, x)$ implies that $\{t \in \mathbb{R} \mid \varphi(t, x) = x\}$ is a closed subset of $\mathbb{R}$.

If $\{t > 0 \mid \varphi(t, x) = x\}$ is non-empty, but has no least member, then (a) holds, since the fact that $\{t \in \mathbb{R} \mid \varphi(t, x) = x\}$ is a group then implies that $\{t \in \mathbb{R} \mid \varphi(t, x) = x\}$ is dense in $\mathbb{R}$.

If $\{t > 0 \mid \varphi(t, x) = x\}$ has a least member, then it is easy to see that (b) holds.

Finally, if $\{t > 0 \mid \varphi(t, x) = x\}$ is empty, then (c) must clearly hold. This completes the proof.   **QED**

1.10. Period. Let φ be a quasi-flow in a Hausdorff space X. The *φ-period* is the function p_φ from X into $[0, \infty]$ which is defined by:

$$\mathrm{p}_\varphi(x) = \inf \{t > 0 \mid \varphi(t, x) = x\}, \qquad \forall\, x \in X,$$

where $\inf \{t > 0 \mid \varphi(t, x) = x\}$ is taken to be ∞ in case

$$\{t > 0 \mid \varphi(t, x) = x\}$$

is empty.

For each $x \in X$, $\mathrm{p}_\varphi(x)$ is called the *φ-period* of the point x.

1.11. Theorem. *Let φ be a quasi-flow in a Hausdorff space X. Then for any points x and y in X, if $y \in \mathcal{O}_\varphi(x)$, we have $\mathrm{p}_\varphi(y) = \mathrm{p}_\varphi(x)$.*

PROOF: The result follows at once from Theorem 1.5 (b). **QED**

1.12. Definition. Let φ be a quasi-flow in a Hausdorff space X, and let $x \in \mathscr{G}(\varphi)$.

If $\mathrm{p}_\varphi(x) < \infty$, we say that $\mathcal{O}_\varphi(x)$ is a *periodic* φ-orbit, and if $\mathrm{p}_\varphi(x) = \infty$, we say that $\mathcal{O}_\varphi(x)$ is an *aperiodic* φ-orbit.

1.13. Theorem. *Let φ be a quasi-flow in a Hausdorff space X, and suppose $x \in X$.*

(a) $\mathrm{p}_\varphi(x) = 0$ *iff* $x \in \mathscr{F}(\varphi)$.

(b) $0 < \mathrm{p}_\varphi(x) < \infty$ *iff* $\mathcal{O}_\varphi(x)$ *is a Jordan curve.*[1]

(c) $\mathrm{p}_\varphi(x) = \infty$ *iff the function* $t \to \varphi(t, x)$ *is one-one.*

PROOF: (a) and (c) are already known. It remains to prove (b). Suppose that $0 < \mathrm{p}_\varphi(x) < \infty$. Then the function

$$e^{\frac{2\pi i t}{\mathrm{p}_\varphi(x)}} \to \varphi(t, x)$$

can easily be seen to be a homeomorphism from the unit circle $\mathbf{S}^1$ onto $\mathcal{O}_\varphi(x)$, from which it follows that $\mathcal{O}_\varphi(x)$ is a Jordan curve.

Suppose, now, that $\mathcal{O}_\varphi(x)$ is a Jordan curve. Then clearly $\mathrm{p}_\varphi(x) \neq 0$. But we also have $\mathrm{p}_\varphi(x) \neq \infty$, for otherwise, the function $t \to \varphi(t, x)$ would be a one-one continuous function from $\mathbb{R}$ onto a Jordan curve, and this is impossible, by the following lemma. **QED**

1.14. Lemma. *There does not exist a continuous, one-one function from $\mathbb{R}$ onto $\mathbf{S}^1$.*

PROOF: Let f be a continuous, one-one function from $\mathbb{R}$ into $\mathbf{S}^1$. Since, for every $n = 1, 2, \ldots$, the set $f([-n, n])$ is a compact, connected, proper subset of $\mathbf{S}^1$, it is easy to see that for every $n = 1, 2, \ldots$, the set $f((-n, n))$ is an open, proper subset of $\mathbf{S}^1$. It follows that $\{\mathbf{S}^1 \setminus f((-n, n)) \mid n = 1, 2, \ldots\}$, being a nested sequence of non-empty, compact sets, has a non-empty intersection. Clearly, no point in this intersection can lie in the range of f, and therefore f cannot map $\mathbb{R}$ onto $\mathbf{S}^1$. **QED**

---

[1] If $0 < \mathrm{p}_\varphi(x) < \infty$, we can also regard $\mathcal{O}_\varphi(x)$ as an *oriented* Jordan curve, by identifying $\mathcal{O}_\varphi(x)$ with the oriented Jordan curve γ which is defined by

$$\gamma(t) = \varphi(t\mathrm{p}_\varphi(x), x), \quad \forall\, t \in [0, 1].$$

1.15. Examples

(a) Let φ be the algebraic flow in $\mathbb{R}$ which is defined by:

$$\varphi(t, x) = t + x, \qquad \forall\, t \in \mathbb{R} \qquad \forall\, x \in \mathbb{R}.$$

It is easy to see that φ is a quasi-flow in $\mathbb{R}$, and that $p_\varphi(x) = \infty$, $\forall\, x \in \mathbb{R}$. For any $x \in \mathbb{R}$, x "moves" under the influence of φ, from left to right with constant speed 1, and for this reason, we shall always denote this flow in the future by **1**.

(b) Let V be a real vector space, and let $y \in V$. Let φ be the algebraic flow in V defined by:

$$\varphi(t, x) = x + ty, \qquad \forall\, t \in \mathbb{R} \qquad \forall\, x \in V.$$

(c) Let φ be the algebraic flow in $\mathbf{S}^1$ which is defined by:

$$\varphi(t, x) = x \cdot e^{\frac{i\,2\pi t}{p}}, \qquad \forall\, t \in \mathbb{R} \qquad \forall\, x \in \mathbf{S}^1,$$

where p is a positive real number. It is clear that φ is a quasi-flow, and that for every $x \in \mathbf{S}^1$, we have $p_\varphi(x) = p$.

(d) Let Q be the set of rational numbers, and let X be the set of cosets of Q regarded as an additive subgroup of $\mathbb{R}$; *i.e.*, $X = \mathbb{R}/Q$. Let φ be the algebraic flow in X which is defined by:

$$\varphi(t, [x]) = [x + t], \qquad \forall\, x \in \mathbb{R} \qquad \forall\, t \in \mathbb{R}$$

where, for any $x \in \mathbb{R}$, $[x] = \{x + r \mid r \in Q\}$. It is clear that for any point $[x]$ of X, we have

$$\{t \in \mathbb{R} \mid \varphi(t, [x]) = [x]\} = Q.$$

Consequently, it is not possible to define a topology on X in such a way as to make φ a quasi-flow.

Flows

1.16. Definition. Let φ be an algebraic flow in a Hausdorff space X.

(a) If $x \in X$, then φ is said to be *continuous at the point* x if, for every $t \in \mathbb{R}$, φ, as a function from $\mathbb{R} \times X$ into X, is continuous at the point $\langle t, x \rangle$.

(b) If $A \subseteq X$, then φ is said to be *continuous in A* if φ is continuous at every point $x \in A$.

(c) If φ is continuous in X, then φ is said to be a *continuous flow* in X; *i.e.*, φ is a continous flow in X if φ is an algebraic flow in X, and is a *continuous* function from $\mathbb{R} \times X$ into X.

1.17. Note. In this book, the word *flow*, used without any modifying adjective or prefix, will always mean *continuous flow*.

1.18. Theorem. *Let φ be an algebraic flow in a Hausdorff space X.*

(a) *If φ is a flow in X, then φ is a quasi-flow in X.*

(b) *φ is a flow in X iff it is continuous at every point of X.*

PROOF: The proof is elementary, and is omitted.   **QED**

1.19. Theorem. *Let φ be a quasi-flow in a Hausdorff space X, and let $x \in X$. Suppose that either*

(i) $p_\varphi(x) < \infty$,

or

(ii) $\mathcal{O}_\varphi(x)$ *is homeomorphic to* $\mathbb{R}$.

Then the restriction of φ to $\mathcal{O}_\varphi(x)$ is a flow in $\mathcal{O}_\varphi(x)$, and when (ii) holds, the function $t \to \varphi(t, x)$ is a homeomorphism from $\mathbb{R}$ onto $\mathcal{O}_\varphi(x)$.

PROOF: Suppose (ii) holds. We shall show first that the map $t \to \varphi(t, x)$ is a homeomorphism from $\mathbb{R}$ onto $\mathcal{O}_\varphi(x)$. To do this, choose a homeomorphism h of $\mathcal{O}_\varphi(x)$ onto $\mathbb{R}$. Now Theorem 1.13, together with the fact that $\mathcal{O}_\varphi(x)$ is neither a point, nor a Jordan curve, implies that $p_\varphi(x) = \infty$, *i.e.*, that the function $t \to \varphi(t, x)$ is one-one. It follows that the function $t \to h\big(\varphi(t, x)\big)$ is a one-one, continuous function from $\mathbb{R}$ onto $\mathbb{R}$, and is therefore a homeomorphism. It is now clear that the function $t \to \varphi(t, x)$ is a homeomorphism from $\mathbb{R}$ onto $\mathcal{O}_\varphi(x)$.

Now since $\mathcal{O}_\varphi(x)$ is a metric space, to show that the restriction of φ to $\mathcal{O}_\varphi(x)$ is a flow, it is sufficient to show that whenever $t_n \to t \in \mathbb{R}$ and $y_n \to y \in \mathcal{O}_\varphi(x)$, $\big($where each $t_n \in \mathbb{R}$ and each $y_n \in \mathcal{O}_\varphi(x)\big)$, we have

$$\varphi(t_n, y_n) \to \varphi(t, y).$$

Suppose, then, that $t_n \to t \in \mathbb{R}$ and $y_n \to y \in \mathcal{O}_\varphi(x)$, as above. For each $n = 1, 2, \ldots$, choose $s_n \in \mathbb{R}$ such that $y_n = \varphi(s_n, y)$, and choose $s \in \mathbb{R}$ such that $y = \varphi(s, x)$. Then since $y_n = \varphi(s_n + s, x)$ for every n, and $y_n \to y = \varphi(s, x)$, we must have $s_n + s \to s$, *i.e.*, $s_n \to 0$. It follows that

$$\varphi(t_n, y_n) = \varphi(s_n + t_n, y) \to \varphi(0 + t, y) = \varphi(t, y)$$

as required. We have therefore shown that the theorem is true whenever (ii) holds.

Suppose, now, that (i) holds. Since the theorem is trivial if $p_\varphi(x) = 0$, we shall assume that $0 < p_\varphi(x) < \infty$. Once again, $\mathcal{O}_\varphi(x)$ is a metric space. Suppose $t_n \to t \in \mathbb{R}$ and $y_n \to y \in \mathcal{O}_\varphi(x)$, where each $t_n \in \mathbb{R}$ and each $y_n \in \mathcal{O}_\varphi(x)$. For each $n = 1, 2, \ldots$, choose $s_n \in [-\frac{1}{2} p_\varphi(x), \frac{1}{2} p_\varphi(x))$ such that $y_n = \varphi(s_n, y)$. (It is clear that, for each n, exactly one such number s_n van be found.) We claim that $s_n \to 0$; for otherwise $\{s_n\}$ would have a non-zero cluster point $s \in [-\frac{1}{2} p_\varphi(x), \frac{1}{2} p_\varphi(x)]$, and this is impossible, because $\{\varphi(s_n, y)\}$ would then have $\varphi(s, y)$ as a cluster point, and the fact that $\varphi(s_n, y) \to y$ would imply that $y = \varphi(s, y)$, contradicting the fact that $s \neq 0$.

We now have

$$\varphi(t_n, y_n) = \varphi(s_n + t_n, y) \to \varphi(0 + t, y) = \varphi(t, y)$$

as required. This completes the proof of the theorem.   **QED**

1.20. Examples

(a) Let φ be the quasi-flow in $\mathbb{R}^2$ defined as follows:

$$\varphi(t, \langle x, y \rangle) = \langle x + t, y \rangle, \quad \forall\, x \in \mathbb{R} \quad \forall\, y \in (0, \infty) \quad \forall\, t \in \mathbb{R},$$

and

$$\varphi(t, \langle x, y \rangle) = \langle x - t, y \rangle, \quad \forall\, x \in \mathbb{R} \quad \forall\, y \in (-\infty, 0] \quad \forall\, t \in \mathbb{R}.$$

It is clear that φ is a quasi-flow in $\mathbb{R}^2$, but is not a flow. However, the restriction of φ to each φ-orbit is a flow.

(b) Let X be the subset of $\mathbb{R}^2$, that consists of the interval $[-1, 1]$, the intersection A of the circle center $-\frac{1}{2}$, radius $\frac{1}{2}$, with the open upper half plane; and the intersection B of the circle center $\frac{1}{2}$, radius $\frac{1}{2}$, with the open upper half plane. Choose a continuous, one-one function f, from $\mathbb{R}$ onto X, which has the following properties:

(i) $f(x) = x$ for $-1 \leq x \leq 1$.

(ii) The restriction of f to $(-\infty, -1]$ is a homeomorphism of $(-\infty, -1]$ onto $A \cup \{-1\}$.

(iii) The restriction of f to $[1, \infty)$ is a homeomorphism of $[1, \infty)$ onto $B \cup \{1\}$.

Let φ be the quasi-flow in X, defined by:

$$\varphi(t, x) = f(t + f^{-1}(x)), \quad \forall\, t \in \mathbb{R} \quad \forall\, x \in X.$$

φ is a quasi-flow with exactly one orbit, but is not a flow, because it is clearly not continuous at the point 0.

1.21. Theorem. *Let φ be a flow in a Hausdorff space X. Then $\mathscr{F}(\varphi)$ is a closed subset of X.*

PROOF: For any $t \in \mathbb{R}$, it is clear that $\{x \in X \mid \varphi(t, x) = x\}$ is closed. The result now follows at once from the fact that

$$\mathscr{F}(\varphi) = \bigcap_{t \in \mathbb{R}} \{x \in X \mid \varphi(t, x) = x\}.$$ **QED**

1.22. Theorem. *Let φ be a flow in a Hausdorff space X.*

Then for each $t \in \mathbb{R}$, the permutation φ_t of X defined as in Theorem **1.5** *is a homeomorphism of X onto itself.*

PROOF: Let $t \in \mathbb{R}$. The continuity of φ_t is clear, and the continuity of φ_t^{-1} follows from the fact that $\varphi_t^{-1} = \varphi_{-t}$. **QED**

1.23. Theorem. *Let φ be a flow in a Hausdorff space X, let $x \in X$, and suppose that $\mathcal{O}_\varphi(x)$ is compact. Then $p_\varphi(x) < \infty$.*

PROOF: There is no loss of generality in assuming that $X = \mathcal{O}_\varphi(x)$. We shall prove that $p_\varphi(x) < \infty$, by showing that the map $t \to \varphi(t, x)$ is not one-one. For each $n = 1, 2, \ldots$, let $A_n = \{\varphi(t, x) \mid -n \le t \le n\}$. Then A_n is compact for every $n = 1, 2, \ldots$, and

$$X = \bigcup_{n=1}^{\infty} A_n.$$

It therefore follows from the Baire Category Theorem that, for some integer N, int $(A_N) \ne \square$. Choose $y \in$ int (A_N), and let $s_1 \in \mathbb{R}$ be so chosen that $y = \varphi(s_1, x)$.

Since X is compact, the sequence $\{\varphi(n, x)\}$ has a cluster point. Choose $s_2 \in \mathbb{R}$ such that $\varphi(s_2, x)$ is a cluster point of $\{\varphi(n, x)\}$. It is easy to see that y is a cluster point of the sequence $\{\varphi(n + s_1 - s_2, x)\}$, and consequently, that $\varphi(n + s_1 - s_2, x) \in A_N$ for arbitrarily large integers n. This shows that the function $t \to \varphi(t, x)$ is not one-one, and the proof is complete. **QED**

1.24. Remark. It follows from Theorems 1.13 and 1.23 that a compact, non-trivial orbit of a flow is always a Jordan curve. Example 1.20(b) shows, however, that this result is not true in general for quasi-flows.

1.25. Theorem. *Let φ be a flow in a Hausdorff space X, let $x \in X$, and suppose $p_\varphi(x) = \infty$. Let C be a compact, connected subset of $\mathcal{O}_\varphi(x)$. Then $\{t \in \mathbb{R} \mid \varphi(t, x) \in C\}$ is a compact interval I, and the restriction to I of the function $t \to \varphi(t, x)$ is a homeomorphism of I onto C.*

PROOF: Let $I = \{t \in \mathbb{R} \mid \varphi(t, x) \in C\}$. We shall show first that there exists a sequence $\{t_n\} \subseteq \mathbb{R} \setminus I$ such that $t_n \to \infty$:

If this is false, then $\varphi(t, x) \in C$ for all sufficiently large t, and in particular, $\varphi(n, x) \in C$ for all sufficiently large integers n. Let y be a cluster point of $\{\varphi(n, x)\}$. Then $y \in C \subseteq \mathcal{O}_\varphi(x)$. Further, for any $t \in \mathbb{R}$, $\varphi(t, y)$, being a cluster point of $\{\varphi(t + n, x)\}$, must also be a member of C. It follows that

$$\mathcal{O}_\varphi(x) = \mathcal{O}_\varphi(y) \subseteq C,$$

and consequently, that $C = \mathcal{O}_\varphi(x)$. But this implies that $\mathcal{O}_\varphi(x)$ is compact, and therefore periodic, a contradiction.

We have therefore shown that there exists a sequence $\{t_n\} \subseteq \mathbb{R}$ such that $t_n \to \infty$, and $\varphi(t_n, x) \notin C$ for all $n = 1, 2, \dots$.

We can see, similarly, that there exists a sequence $\{s_n\} \subseteq \mathbb{R}$ such that $s_n \to -\infty$, and $\varphi(s_n, x) \notin C$ for all $n = 1, 2, \dots$.

Choose sequences $\{s_n\}$ and $\{t_n\}$ of real numbers such that

(i) $\varphi(s_n, x) \notin C$ and $\varphi(t_n, x) \notin C$, $\forall n = 1, 2, \dots$

(ii) $\{t_n\}$ is a strictly increasing sequence of positive numbers and $t_n \to \infty$

(iii) $\{s_n\}$ is a strictly decreasing sequence of negative numbers, and $s_n \to -\infty$.

Let $A_0 = \{\varphi(t, x) \mid s_1 \leq t \leq t_1\} \cap C$, and for each $n = 1, 2, \dots$, let $A_n = \{\varphi(t, x) \mid s_{n+1} \leq t \leq s_n \text{ or } t_n \leq t \leq t_{n+1}\} \cap C$.

Then $\{A_n \mid n = 0, 1, 2, \dots\}$ is a pairwise-disjoint family of compact subsets of C, and $C = \bigcup_{n=0}^{\infty} A_n$.

It therefore follows from Theorem A. 11 that $C = A_n$ for some integer n, and the fact that the function $t \to \varphi(t, x)$ is one-one, consequently implies that I is a bounded subset of $\mathbb{R}$. But I is clearly closed, and is therefore compact. It follows that the restriction to I of the function $t \to \varphi(t, x)$, being a one-one, continuous function from a compact space into a Hausdorff space, must be a homeomorphism of I onto C. Therefore, I, being a continuous image of the connected set C, must be connected, and the proof is complete.   **QED**

1.26. Remark. Example 1.20(b) shows that Theorem 1.25 does not, in general, hold for quasi-flows. This is not surprising, in view of the remark made in Section 1.24, and the elementary observation that Theorem 1.23 is a weaker form of Theorem 1.25. Theorem 1.23 can be interpreted as the statement, that an aperiodic orbit of a flow (which must, of course, be connected) is never compact; while Theorem 1.25 gives a description of *all* compact, connected subsets of such an orbit.

Example 1.20(b) also shows that an aperiodic orbit of a quasiflow need not be homeomorphic to $\mathbb{R}$. In view of Theorems 1.19 and 1.25, one might expect that an aperiodic orbit of a *flow* must necessarily be homeomorphic to $\mathbb{R}$. The following example shows that this is false.

1.27. Example. Let X be the torus $\mathbf{S}^1 \times \mathbf{S}^1$, where $\mathbf{S}^1 = \{e^{i\theta} \mid \theta \in \mathbb{R}\}$ is, as usual, the unit circle. Let α be any irrational number, and let φ be the flow in X, which is defined by:

$$\varphi\left(t, \langle e^{i\theta_1}, e^{i\theta_2}\rangle\right) = \langle e^{i(\theta_1+t)}, e^{i(\theta_2+\alpha t)}\rangle, \quad \bigvee t \in \mathbb{R}, \quad \bigvee \langle e^{i\theta_1}, e^{i\theta_2}\rangle \in X.$$

It is easy to see that φ is a well-defined algebraic flow in X, and further, that φ is continuous.

Since α is irrational, it is clear that every φ-orbit is aperiodic. We claim that, for any point $x = \langle e^{i\theta_1}, e^{i\theta_2}\rangle$ of X, $\mathcal{O}_\varphi(x)$ is dense in X. We can see this by noting that the following lemma implies that, for every $t \in \mathbb{R}$,

$$\{e^{i(\theta_2+\alpha(t+2n\pi))}\mid n \text{ is an integer}\}$$

is a dense subset of $\mathbf{S}^1$.

In fact, this also shows that, for any $x \in X$, at least one of the sequences $\{\varphi(2\pi n, x)\}$, $\{\varphi(-2\pi n, x)\}$, must have x as a cluster point. Consequently, if $x \in X$, the function $t \to \varphi(t, x)$ is never a homeomorphism, and it follows from Theorem 1.19 that no φ-orbit is homeomorphic to $\mathbb{R}$.

1.28. Lemma. *If $\beta \in \mathbb{R}$ and $\beta/2\pi$ is irrational, then $\{e^{in\beta} \mid n$ is an integer$\}$ is a dense subgroup of $\mathbf{S}^1$.*

FIRST PROOF: (For readers familiar with compact Abelian groups.) Let

$$H = \overline{\{e^{in\beta} \mid n \text{ is an integer}\}}.$$

H is a compact subgroup of $\mathbf{S}^1$. If $H \neq \mathbf{S}^1$, then $\mathbf{S}^1/H$, being a nontrivial compact Abelian group, has a non-trivial character-group. It follows that there exists a member k of the character-group of $\mathbf{S}^1$, (*i.e.*, an integer k) such that $k \neq 0$, but

$$e^{in\beta k} = 1, \quad \bigvee \text{ integers } n.$$

In particular, $e^{i\beta k} = 1$, *i.e.*, $\beta k/2\pi$ is an integer, which contradicts the assumption that $\beta/2\pi$ is irrational. Thus, $H = \mathbf{S}^1$, as required.

QED

SECOND PROOF: It is clear that the map $n \to e^{in\beta}$ is one-one. Therefore, the set

$$K = \{e^{in\beta} \mid n \text{ is an integer}\},$$

being an infinite subset of $\mathbb{S}^1$, must have an accumulation-point. It follows that we can choose for each $j = 1, 2, \ldots,$ distinct integers m_j and n_j, such that

$$|e^{in_j\beta} - e^{im_j\beta}| \to 0 \text{ as } j \to \infty$$

i.e.,

$$e^{i(n_j-m_j)\beta} \to 1 \text{ as } j \to \infty.$$

It can be seen without difficulty that

$$\{e^{in(n_j-m_j)\beta} \mid j = 1, 2, \ldots, \text{ and } n \text{ is an integer}\}$$

must be a dense subset of $\mathbb{S}^1$.

It therefore follows that K is dense in $\mathbb{S}^1$, as required.   **QED**

1.29. Theorem. *Let φ be a flow in a locally compact Hausdorff space X, and let $X \cup \{\infty\}$ be the one-point compactification of X. Then φ has a unique extension to a flow in $X \cup \{\infty\}$, namely by making ∞ a fixed point.*

PROOF: To prove the theorem, we need only show that, if $t \in \mathbb{R}$, and $\{t_\alpha \mid \alpha \in A\}$ is a net of real numbers, and $\{x_\alpha \mid \alpha \in A\}$ is a net of points in X, and $t_\alpha \to t$ and $x_\alpha \to \infty$, then $\varphi(t_\alpha, x_\alpha) \to \infty$.

To prove this, let K be any compact subset of X. If $\varphi(t_\alpha, x_\alpha) \in K$ for arbitrarily large $\alpha \in A$, then $\{\varphi(t_\alpha, x_\alpha) \mid \alpha \in A\}$ has a cluster-point $x \in K$, and we may assume, without loss of generality, that $\varphi(t_\alpha, x_\alpha) \to x$. But this is impossible, for then

$$x_\alpha = \varphi(t_\alpha - t_\alpha, x) = \varphi(-t_\alpha, \varphi(t_\alpha, x_\alpha)) \to \varphi(-t, x) \neq \infty.$$

Thus $\varphi(t_\alpha, x_\alpha) \notin K$ for all sufficiently large α, and this shows that $\varphi(t_\alpha, x_\alpha) \to \infty$, as required.   **QED**

1.30. Theorem. *Let φ be a flow in a Hausdorff space X, let $x \in X$, let $\{x_\alpha \mid \alpha \in A\}$ be a net in X, converging to x, and suppose $p_\varphi(x) > 0$.*

Then each finite cluster point of the net $\{p_\varphi(x_\alpha) \mid \alpha \in A\}$ must lie in the set, $\{k \cdot p_\varphi(x) \mid k = 1, 2, \ldots\}$.

PROOF: We note first that since $x \in \mathscr{G}(\varphi)$, and $\mathscr{G}(\varphi)$ is open, we must have $x_\alpha \in \mathscr{G}(\varphi)$ for all sufficiently large α. There is consequently no loss of generality in assuming that $p_\varphi(x_\alpha) \neq 0$ for every $\alpha \in A$.

Now suppose δ is any finite cluster-point of the net $\{p_\varphi(x_\alpha) \mid \alpha \in A\}$. Since a subnet of $\{p_\varphi(x_\alpha) \mid \alpha \in A\}$ converges to δ, for the purpose of proving that $\delta \in \{k \cdot p_\varphi(x) \mid k = 1, 2, \ldots\}$, we do not lose generality in assuming for convenience that $p_\varphi(x_\alpha) \to \delta$.

Now for each $\alpha \epsilon A$, $x_\alpha = \varphi\big(p_\varphi(x_\alpha), x_\alpha\big)$, and therefore, since $x_\alpha \to x$, and $\varphi\big(p_\varphi(x_\alpha), x_\alpha\big) \to \varphi(\delta, x)$, we must have $\varphi(\delta, x) = x$, i.e., $\delta = k \cdot p_\varphi(x)$ for some integer $k \geq 0$.

The proof will therefore be complete once we have shown that $k \neq 0$, i.e., $\delta \neq 0$.

But if $\delta = 0$, we can choose, for each $\alpha \epsilon A$, a positive integer k_α, such that the net $\{k_\alpha \cdot p_\varphi(x_\alpha) \mid \alpha \epsilon A\}$ converges to a number ε which satisfies $0 < \varepsilon < p_\varphi(x)$. Clearly, $\varphi\big(k_\alpha \cdot p_\varphi(x_\alpha), x_\alpha\big) \to \varphi(\varepsilon, x) \neq x$, which is impossible, because $x_\alpha = \varphi\big(k_\alpha \cdot p_\varphi(x_\alpha), x_\alpha\big)$ for every $\alpha \epsilon A$. It follows that $\delta \neq 0$, and the proof is complete. **QED**

1.31. Corollary. *Let φ be a flow in a Hausdorff space X. Then p_φ is a lower-semi-continuous function from X into $[0, \infty]$.*

1.32. Theorem. *Let X be a compact Hausdorff space with the property that no fixed-point free homeomorphism of X onto itself is isotopic to the identity map on X.*

Then for any flow φ in X, we have $\mathscr{F}(\varphi) \neq \square$.

PROOF: Suppose we can find a flow φ in X such that $\mathscr{F}(\varphi) = \square$. Since p_φ is lower-semi-continuous and never zero, and X is compact, it is easy to see that there exists a number $\delta > 0$, such that

$$p_\varphi(x) > \delta, \quad \forall\, x \epsilon X.$$

It is clear that the homeomorphism φ_δ of X onto itself, defined by

$$\varphi_\delta(x) = \varphi(\delta, x), \quad \forall\, x \epsilon X$$

is fixed-point free, and isotopic to the identity map on X under the isotopy $H : [0, 1] \times X \to X$, where

$$H(t, x) = \varphi(\delta t, x), \quad \forall\, t \epsilon [0, 1] \ \forall\, x \epsilon X.$$

This contradiction completes the proof of the theorem. **QED**

1.33. Corollary

(a) *If n is a positive integer, and φ is a flow in $\mathbf{S}^{2n}$, then $\mathscr{F}(\varphi) \neq \square$.*

(b) *If φ is a flow in $\mathbf{S}^2$, $x \epsilon \mathbf{S}^2$, and $\mathcal{O}_\varphi(x)$ is periodic, then $\mathscr{F}(\varphi)$ contains points in each side of $\mathcal{O}_\varphi(x)$.*

PROOF: (a) is immediate from the fact that no fixed-point-free homeomorphism of $\mathbf{S}^{2n}$ onto itself is homotopic to the identity map on $\mathbf{S}^{2n}$.

(b) follows at once from the fact that the union of $\mathcal{O}_\varphi(x)$ with each of its sides, being homeomorphic to the closed unit disc in $\mathbb{R}^2$, has the fixed-point property. **QED**

Image of a Flow

1.34. Definition. Let X and Y be sets, and f a function from X onto Y. Let an algebraic flow φ in X satisfy the condition that for any points x_1, x_2 of X, $\{t \in \mathbb{R} \mid f(\varphi(t, x_1)) = f(\varphi(t, x_2))\}$ is either empty, or the whole of $\mathbb{R}$. Then the algebraic flow ψ in Y, defined by:

$$\psi(t, f(x)) = f(\varphi(t, x)), \quad \forall \, t \in \mathbb{R} \ \ \forall \, x \in X$$

is called the *image* of φ under f, and is denoted by $f[\varphi]$. In the special case in which X and Y are Hausdorff spaces and f is a homeomorphism of X onto Y, we say that φ and $f[\varphi]$ are *homeomorphic*.

1.35. Remark. If f is a one-one mapping of a set X onto a set Y, and φ is an algebraic flow in X, then $f[\varphi]$ is always defined.

1.36. Theorem. *Let f be a continuous function from a Hausdorff space X onto a Hausdorff space Y, let φ be a quasi-flow in X, and suppose $f[\varphi]$ is defined.*
Then $f[\varphi]$ is a quasi-flow in Y.

PROOF: Trivial.   **QED**

1.37. Theorem. *Let f be a continuous, open function from a Hausdorff space X onto a Hausdorff space Y, let φ be a flow in X, and suppose $f[\varphi]$ is defined.*
Then $f[\varphi]$ is a flow in Y.

PROOF: The proof of this theorem is elementary, and will be left to the reader.   **QED**

1.38. Example. Let ψ be the flow in $\mathbb{S}^1$, defined by

$$\psi(t, y) = y e^{it}, \quad \forall \, t \in \mathbb{R} \ \ \forall \, y \in \mathbb{S}^1.$$

Let $f: \mathbb{R} \to \mathbb{S}^1$ be defined by

$$f(x) = e^{ix} \ \ \forall \, x \in \mathbb{R}.$$

Then $\psi = f[1]$.

Endpoints

1.39. Definition. Let φ be a quasi-flow in a Hausdorff space X, and let $x \in X$ satisfy $p_\varphi(x) = \infty$. The set of φ-α-*endpoints of x*, or more simply, the set of φ-α-*points of x*, is denoted by $\alpha_\varphi(x)$, and is defined by:

$$\alpha_\varphi(x) = \bigcap_{t \in \mathbb{R}} \overline{\{\varphi(s, x) \mid s \leq t\}}.$$

The set of φ-ω-*endpoints of* x, or more simply, the set of φ-ω-*points of* x, is denoted by $\omega_\varphi(x)$, and is defined by:

$$\omega_\varphi(x) = \bigcap_{t \in \mathbf{R}} \overline{\{\varphi(s, x) \mid s \geq t\}}.$$

The set of φ-*endpoints of* x is defined to be $\alpha_\varphi(x) \cup \omega_\varphi(x)$.

It is clear that, whenever $y \in \mathcal{O}_\varphi(x)$, we have $\alpha_\varphi(y) = \alpha_\varphi(x)$, and $\omega_\varphi(y) = \omega_\varphi(x)$. In view of this, we sometimes call $\alpha_\varphi(x)$, $\omega_\varphi(x)$, and $\alpha_\varphi(x) \cup \omega_\varphi(x)$, respectively, the sets of φ-α-points, φ-ω-points, and φ-endpoints of $\mathcal{O}_\varphi(x)$, and write

$$\alpha_\varphi(x) = \alpha_\varphi\big(\mathcal{O}_\varphi(x)\big), \text{ and } \omega_\varphi(x) = \omega_\varphi\big(\mathcal{O}_\varphi(x)\big).$$

1.40. Theorem. *Let φ be a quasi-flow in a Hausdorff space X, and suppose the point $x \in X$ satisfies* $p_\varphi(x) = \infty$.

(a) $\alpha_\varphi(x)$ *and* $\omega_\varphi(x)$ *are closed subsets of X.*

(b) *If X is compact, then $\alpha_\varphi(x)$ and $\omega_\varphi(x)$ are non-empty, compact and connected.*

(c) $\alpha_\varphi(x)$ *is the set of all cluster-points of all nets* $\{\varphi(t_\alpha, x) \mid \alpha \in A\}$ *such that* $t_\alpha \to -\infty$; *and* $\omega_\varphi(x)$ *is the set of all cluster-points of all nets* $\{\varphi(t_\alpha, x) \mid \alpha \in A\}$ *such that* $t_\alpha \to \infty$.

(d) *If X satisfies the first axiom of countability, (in particular, if X is a metric space) then $\alpha_\varphi(x)$ is the set of all cluster-points of all sequences* $\{\varphi(t_n, x)\}$ *such that* $t_n \to -\infty$, *and* $\omega_\varphi(x)$ *is the set of all cluster-points of all sequences* $\{\varphi(t_n, x)\}$ *such that* $t_n \to \infty$.

PROOF: (a) is trivial, (b) follows at once from Theorem A. 5, and (c) and (d) are elementary, and are left to the reader.   **QED**

1.41. Theorem. *Let φ be a flow in a Hausdorff space X, let $x \in X$, and suppose* $p_\varphi(x) = \infty$.
Then we have

(a) $\alpha_\varphi(x)$ *and* $\omega_\varphi(x)$ *are invariant of φ.*

(b) *If* $y \in \alpha_\varphi(x)$ *and* $p_\varphi(y) = \infty$, *then* $\alpha_\varphi(y) \cup \omega_\varphi(y) \subseteq \alpha_\varphi(x)$.

(c) *If* $y \in \omega_\varphi(x)$, *and* $p_\varphi(y) = \infty$, *then* $\alpha_\varphi(y) \cup \omega_\varphi(y) \subseteq \omega_\varphi(x)$.

PROOF:

(a) Let $y \in \alpha_\varphi(x)$ and $t_0 \in \mathbf{R}$. Since the function $\varphi_{t_0} : X \to X$ defined by

$$\varphi_{t_0}(z) = \varphi(t_0, z), \qquad \forall\, z \in X$$

is a homeomorphism of X onto itself, it is clear that

$$\varphi(t_0, y) \in \cap_{t \in \mathbb{R}} \overline{\{\varphi(s, x) \mid s \leq t + t_0\}} = \alpha_\varphi(x).$$

A similar argument for $\omega_\varphi(x)$ completes the proof of (a).

(b) Since $\mathcal{O}_\varphi(y) \subseteq \alpha_\varphi(x)$, we have

$$\alpha_\varphi(y) \cup \omega_\varphi(y) \subseteq \overline{\mathcal{O}_\varphi(y)} \subseteq \overline{\alpha_\varphi(x)} = \alpha_\varphi(x).$$

This completes the proof of (b); the proof of (c) is similar.   **QED**

1.42. Theorem. *Let φ be a quasi-flow in a Hausdorff space X, let $x \in X$, and suppose $\mathrm{p}_\varphi(x) = \infty$.*
Then $\mathcal{O}_\varphi(x)$ is homeomorphic to $\mathbb{R}$ iff $\mathcal{O}_\varphi(x)$ contains no φ-endpoint of x.

PROOF: Suppose $\mathcal{O}_\varphi(x)$ is homeomorphic to $\mathbb{R}$. Then the function $t \to \varphi(t, x)$ is a homeomorphism. Therefore, whenever $t \in \mathbb{R}$, and $\{t_\alpha \mid \alpha \in A\}$ is a net in $\mathbb{R}$, and $\varphi(t_\alpha, x) \to \varphi(t, x)$, we must have $t_\alpha \to t$. It follows that, if $t \in \mathbb{R}$, and $\{t_\alpha \mid \alpha \in A\}$ is a net in $\mathbb{R}$, and $|t_\alpha| \to \infty$, we cannot have $\varphi(t_\alpha, x) \to \varphi(t, x)$. This shows that $\mathcal{O}_\varphi(x)$ contains no φ-endpoint of x.

Suppose, conversely, that $\mathcal{O}_\varphi(x)$ contains no φ-endpoint of x. To show that the map $t \to \varphi(t, x)$ is a homeomorphism, it is sufficient to show that whenever $t \in \mathbb{R}$, and $\{t_\alpha \mid \alpha \in A\}$ is a net in $\mathbb{R}$, and $\varphi(t_\alpha, x) \to \varphi(t, x)$, we have $t_\alpha \to t$. Suppose, then, that $\varphi(t_\alpha, x) \to \varphi(t, x)$. We see that $\{t_\alpha \mid \alpha \in A\}$ cannot have $-\infty$, nor ∞, as a cluster-point, for otherwise, $\varphi(t, x)$ would be a φ-endpoint of x. Therefore, to show that $t_\alpha \to t$, we need only show that t is the only finite cluster-point of $\{t_\alpha \mid \alpha \in A\}$. Suppose t' is any finite cluster-point of $\{t_\alpha \mid \alpha \in A\}$; then $\varphi(t', x)$ is a cluster-point of $\{\varphi(t_\alpha, x) \mid \alpha \in A\}$, and since $\varphi(t_\alpha, x) \to \varphi(t, x)$, we must therefore have $\varphi(t', x) = \varphi(t, x)$, *i.e.*, $t' = t$, as required.

This completes the proof of the theorem.   **QED**

1.43. Definition. Let φ be a quasi-flow in a Hausdorff space X.

(a) If $y \in \mathscr{F}(\varphi)$, and y is a φ-endpoint of some point of X, then y is said to be a φ-*stagnation point*. The set of all φ-stagnation points is denoted by $\mathscr{S}(\varphi)$.

(b) If $y \in \mathscr{F}(\varphi)$, and y is a φ-endpoint of a point x of X, then y is said to be a φ-*stagnation point of x*, and is also said to be a φ-*stagnation point of* $\mathcal{O}_\varphi(x)$.

(c) If $y \in \mathscr{F}(\varphi)$, and y is a φ-α-point (respectively a φ-ω-point) of a point x of X, then y is said to be a φ-α-*stagnation point of x*, (respectively a φ-ω-*stagnation point of x*).

1.44. Definition. Let φ be a quasi-flow in a Hausdorff space X.

(a) Let $x \in X$ satisfy $p_\varphi(x) = \infty$. A φ-α-point, y, of x, is said to be a *simple φ-α-point of x* if $\varphi(t, x) \to y$ as $t \to -\infty$. A φ-α-point of x which is not a simple φ-α-point of x, is called a *complex φ-α-point of x*. Simple and complex φ-ω-points of x are defined analogously.

(b) Let $x \in X$ satisfy $p_\varphi(x) = \infty$. A φ-endpoint y of x is said to be a *complex φ-endpoint of x* if it is either a complex φ-α-point of x, or a complex φ-ω-point of x. A φ-endpoint of x which is not a complex φ-endpoint of x is called a *simple φ-endpoint of x*.

(c) φ is said to be *simple* if, whenever $x \in X$ and $p_\varphi(x) = \infty$, x has a simple φ-α-point and a simple φ-ω-point.

(d) φ is said to be *semi-simple* if, whenever $x \in X$ and $p_\varphi(x) = \infty$, we have:

(i) either $\alpha_\varphi(x) \cap \mathscr{G}(\varphi) \neq \square$, or x has a simple φ-α-point,

and (ii) either $\omega_\varphi(x) \cap \mathscr{G}(\varphi) \neq \square$, or x has a simple φ-ω-point.

1.45. Theorem. *Let φ be a quasi-flow in a Hausdorff space X, let $x \in X$, and suppose $p_\varphi(x) = \infty$.*

(a) *If y is a simple φ-α-point of x, then $\{y\} = \alpha_\varphi(x)$, and if φ is a flow, (not merely a quasi-flow), we have $y \in \mathscr{F}(\varphi)$, i.e., $y \in \mathscr{S}(\varphi)$.*

(b) *If y is a simple φ-ω-point of x, then $\{y\} = \omega_\varphi(x)$, and if φ is a flow, we have $y \in \mathscr{F}(\varphi)$, i.e., $y \in \mathscr{S}(\varphi)$.*

PROOF: The result follows at once from Theorems 1.40(c) and 1.41(a).
QED

1.46. Theorem. *Let φ be a quasi-flow in a compact Hausdorff space X, and suppose $\mathscr{S}(\varphi)$ is totally disconnected.*

Then φ is semi-simple.

PROOF: Suppose $x \in X$, $p_\varphi(x) = \infty$, and $\alpha_\varphi(x) \cap \mathscr{G}(\varphi) = \square$. Then by Theorem 1.40(b), $\alpha_\varphi(x)$ is a connected, non-empty subset of $\mathscr{S}(\varphi)$, and therefore consists of a single point. It is easy to see that this point is a simple φ-α-point of x. An analogous argument for any point $x \in X$ which satisfies $\omega_\varphi(x) \cap \mathscr{G}(\varphi) = \square$ completes the proof of the theorem.
QED

Velocity, Speed, and Arc Length

1.47. Definition. Let φ be a quasi-flow in a real, normed linear space X. For each $x \in X$, if

$$\lim_{s \to 0} \frac{\varphi(s, x) - x}{s}$$

exists, then φ is said to have a velocity at x, and the above limit, which is called the *velocity of φ at x*, is denoted by $\dot\varphi(x)$.

1.48. Note. If φ is a quasi-flow in a real, normed linear space X, $x \in X$ and $t \in \mathbb{R}$, we have

$$\dot\varphi\big(\varphi(t, x)\big) = \lim_{s \to t} \frac{\varphi(s, x) - \varphi(t, x)}{s - t}$$

whenever $\dot\varphi\big(\varphi(t, x)\big)$ exists. For simplicity of notation, we shall write $\dot\varphi\big(\varphi(t, x)\big)$ as $\dot\varphi(t, x)$.

1.49. Definition. Let φ be a quasi-flow in a metric space X. For each $x \in X$, if

$$\lim_{s \to 0} \frac{d(\varphi(s, x), x)}{|s|}$$

exists (finite or infinite), then φ is said to have a speed at x, and the above limit, which is called the *speed of φ at x*, is denoted by $|\dot\varphi|(x)$.

1.50. Note. If φ is a quasi-flow in a metric space X, $x \in X$, and $t \in \mathbb{R}$, we have

$$|\dot\varphi|\big(\varphi(t, x)\big) = \lim_{s \to t} \frac{d\big(\varphi(s, x), \varphi(t, x)\big)}{|s - t|}$$

whenever $|\dot\varphi|\big(\varphi(t, x)\big)$ exists. For simplicity of notation, we shall write $|\dot\varphi|\big(\varphi(t, x)\big)$ as $|\dot\varphi|(t, x)$.

1.51. Theorem. *Let φ be a quasi-flow in a real, normed linear space X, let $x \in X$, and suppose φ has a velocity at x.*

Then φ has a speed at x, and $|\dot\varphi|(x) = |\dot\varphi(x)|$, where, for each $y \in X$, $|y|$ denotes the norm of y.

PROOF: Trivial.   **QED**

In order to simplify notation, we shall now give a slightly wider definition of speed:

1.52. Definition. Let $-\infty \le a < b \le \infty$, and let h be a continuous function from (a, b) into a metric space X. The *lower* and *upper speeds*, $\underline{D}h$ and $\overline{D}h$, of h, are the functions from (a, b) into $[0, \infty]$, defined respectively, by:

$$(\underline{D}h)(t) = \liminf_{s \to t} \frac{d\big(h(s), h(t)\big)}{|s - t|}, \qquad \forall\, t \in (a, b),$$

and

$$(\bar{D}h)(t) = \limsup_{s \to t} \frac{d\big(h(s), h(t)\big)}{|s-t|}, \qquad \forall \, t \in (a, b).$$

For each $t \in (a, b)$, if $(\underline{D}h)(t) = (\bar{D}h)(t)$, then h is said to have a *speed* at t, which we denote by $(Dh)(t)$, and define by

$$(Dh)(t) = (\underline{D}h)(t) = (\bar{D}h)(t).$$

Where there is no ambiguity, $(Dh)(t)$ will also be called the speed of h at the point $h(t)$ in its range.

1.53. Note. If φ is a quasi-flow in a metric space X, $x \in X$, and $h(t) = \varphi(t, x)$, $\forall \, t \in \mathbb{R}$, then for each $t \in \mathbb{R}$, the expressions $|\dot{\varphi}|(t, x)$ and $(Dh)(t)$ have the same meaning. It is easy to see that with a function h of this type, there is no ambiguity in calling $(Dh)(t)$ the speed of h at the point $h(t)$.

1.54. Theorem. *Let* $-\infty \leq a < b \leq \infty$, *and let h be a continuous function from (a, b) into a metric space X.*

Then $\underline{D}h$ and $\bar{D}h$ are Borel functions.

PROOF: We shall show that $\underline{D}h$ is a Borel function; an analogous argument holds for $\bar{D}h$.

For each non-zero real number s, the function

$$t \to \frac{d\big(h(t+s), h(t)\big)}{|s|}$$

is continuous.

Therefore, for each positive number δ, the function $\varDelta_\delta h$ from (a, b) into $[0, \infty]$, defined by

$$(\varDelta_\delta h)(t) = \inf \left\{ \frac{d\big(h(t+s), h(t)\big)}{|s|} \,\middle|\, |s| \leq \delta \right\}, \qquad \forall \, t \in (a, b),$$

being the infimum of a family of continuous functions, is upper semi-continuous, and is therefore a Borel function.

But for each $t \in (a, b)$, we have

$$(\underline{D}h)(t) = \lim_{n \to \infty} \left(\varDelta_{\frac{1}{n}} h\right)(t),$$

and therefore $\underline{D}h$, being a pointwise-limit of a sequence of Borel functions, is itself a Borel function. **QED**

1.55. Definition. Let h be a function from a compact interval $[a, b]$ into a metric space X. The *variation of h over* $[a, b]$ is denoted by $V_h([a,b])$, and is defined by

$$V_h([a, b]) = \sup \sum_{j=1}^{n} d\big(h(t_{j-1}), h(t_j)\big),$$

where d is the metric in X and the supremum is taken over all finite families $\{t_0, t_1, \ldots, t_n\}$ with n a positive integer and

$$a = t_0 < t_1 < \ldots < t_n = b.$$

1.56. Theorem. *Let h be a function from a compact interval $[a, b]$ into a metric space X, and suppose $c \in [a, b]$. Then*

$$V_h([a, b]) = V_h([a, c]) + V_h([c, b]).$$

PROOF: The proof is elementary and will be left to the reader.   **QED**

1.57. Theorem. *Let h be a homeomorphism from a compact interval $[a, b]$ onto a subset A of a metric space X, and suppose we have another homeomorphism $\tilde{h}$ from a compact interval $[\tilde{a}, \tilde{b}]$ onto A.*
Then

$$V_{\tilde{h}}([\tilde{a}, \tilde{b}]) = V_h([a, b]).$$

PROOF: Since $\tilde{h}^{-1}h$ is a homeomorphism from $[a, b]$ onto $[\tilde{a}, \tilde{b}]$, it carries a family $\{t_0, t_1, \ldots, t_n\}$ with $a = t_0 < t_1 < \ldots < t_n = b$, onto a family of the same type in $[\tilde{a}, \tilde{b}]$, except that the order may be reversed. In view of this, the theorem is obvious.   **QED**

1.58. Definition. Let a subset A of a metric space be homeomorphic to a compact interval. Then the *arc length* of A is denoted by $\mathscr{L}(A)$, and is defined to be $V_h([0, 1])$, where h is any homeomorphism from $[0, 1]$ onto A.

1.59. Theorem. *Let $-\infty \le a < b \le \infty$, and let h be a continuous function from (a, b) into a metric space X. Then*

(a) *For every compact subinterval $[\alpha, \beta]$ of (a, b), we have*

$$V_h([\alpha, \beta]) \ge \int_\alpha^\beta (\overline{D}h)\, dm,$$

and

(b) *if $\overline{D}h$ is finite everywhere in (a, b), then for each compact subinterval $[\alpha, \beta]$ of (a, b), we have*

$$V_h([\alpha, \beta]) = \int_\alpha^\beta (\overline{D}h)\, dm.$$

PROOF: This proof relies heavily on the discussion of the Kurzweil integral that appears in Appendix B.

(a): Let $[\alpha, \beta]$ be a compact subinterval of (a, b), and let v be the function from $[\alpha, \beta]$ into $[0, \infty]$ defined by

$$v(t) = V_h([\alpha, t]), \qquad \forall\, t \in [\alpha, \beta].$$

If $v(\beta) = \infty$, we have nothing to prove. Suppose that $v(\beta) < \infty$. Then v is an increasing function from $[\alpha, \beta]$ into $[0, v(\beta)]$; in fact v is continuous and therefore maps $[\alpha, \beta]$ *onto* $[0, v(\beta)]$.

It is clear that, for any $s, t \in [\alpha, \beta]$,

$$d\big(h(s), h(t)\big) \leq |v(t) - v(s)|,$$

and we deduce from this that

$$(\bar{D}h)(t) \leq (\bar{D}v)(t), \qquad \forall\, t \in [\alpha, \beta].$$

Therefore

$$V_h([\alpha, \beta]) = v(\beta) - v(\alpha) \geq \int_\alpha^\beta v'(t)\, dt$$

$$= \int_\alpha^\beta (\bar{D}v)(t)\, dt \geq \int_\alpha^\beta (\bar{D}h)(t)\, dt.$$

This proves (a).

(b): Assume that $\bar{D}h$ is finite everywhere in (a, b). We shall show first that whenever $a < \alpha < \beta < b$, we have

$$d\big(h(\alpha), h(\beta)\big) \leq \int_\alpha^\beta (\bar{D}h)\, dm.$$

This inequality is trivial if $\int_\alpha^\beta (\bar{D}h)\, dm = \infty$. Suppose that $\int_\alpha^\beta (\bar{D}h)\, dm < \infty$. Let $\varepsilon > 0$. Choose a function δ_1 from $[\alpha, \beta]$ into $(0, \infty)$ such that whenever $t \in [\alpha, \beta]$ and $|s| \leq \delta_1(t)$, we have

$$\frac{d\big(h(t+s), h(t)\big)}{|s|} \leq (\bar{D}h)(t) + \varepsilon.$$

By Theorem B.7, we can choose a function δ_2 from $[\alpha, \beta]$ into $(0, \infty)$ such that, whenever a family $\{\tau_0, t_1, \tau_1, \ldots, t_n, \tau_n\}$ of points of $[\alpha, \beta]$, $\mathcal{K}$-conforms to δ_2, we have

$$\left| \sum_{j=1}^n (\bar{D}h)(t_j)(\tau_j - \tau_{j-1}) - \int_\alpha^\beta (\bar{D}h)\, dm \right| < \varepsilon.$$

Let $\delta = \delta_1 \wedge \delta_2$, and choose a family $\{\tau_0, t_1, \tau_1, \ldots, t_n, \tau_n\}$ of points of $[\alpha, \beta]$, which $\mathscr{X}$-conforms to δ. (See Theorem B. 4.) We now have:

$$\int_\alpha^\beta (\bar{D}h)\, dm \geq \sum_{j=1}^n (\bar{D}h)(t_j)(\tau_j - \tau_{j-1}) - \varepsilon$$

$$= \sum_{j=1}^n (\bar{D}h)(t_j)(\tau_j - t_j) + \sum_{j=1}^n (\bar{D}h)(t_j)(t_j - \tau_{j-1}) - \varepsilon$$

$$\geq \sum_{j=1}^n \left[\frac{d\big(h(\tau_j), h(t_j)\big)}{|\tau_j - t_j|} - \varepsilon \right](\tau_j - t_j)$$

$$+ \sum_{j=1}^n \left[\frac{d\big(h(\tau_{j-1}), h(t_j)\big)}{|\tau_{j-1} - t_j|} - \varepsilon \right](t_j - \tau_{j-1}) - \varepsilon$$

$$= \sum_{j=1}^n d\big(h(\tau_j), h(t_j)\big) + \sum_{j=1}^n d\big(h(\tau_{j-1}), h(t_j)\big) - \varepsilon(\beta - \alpha) - \varepsilon$$

$$\geq d\big(h(\alpha), h(\beta)\big) - \varepsilon(\beta - \alpha) - \varepsilon.$$

This shows that, whenever $a < \alpha < \beta < b$, we have

$$d\big(h(\alpha), h(\beta)\big) \leq \int_\alpha^\beta (\bar{D}h)\, dm.$$

Now suppose $a < \alpha < \beta < b$. Then for any family $\{t_0, t_1, \ldots, t_n\}$, where $\alpha = t_0 < t_1 < \ldots < t_n = \beta$, we have

$$\sum_{j=1}^n d\big(h(t_{j-1}), h(t_j)\big) \leq \sum_{j=1}^n \int_{t_{j-1}}^{t_j} (\bar{D}h)\, dm = \int_\alpha^\beta (\bar{D}h)\, dm.$$

It therefore follows at once that

$$\mathscr{L}\big(h([\alpha, \beta])\big) \leq \int_\alpha^\beta (\bar{D}h)\, dm.$$

This completes the proof of the theorem.   **QED**

1.60. Corollary. *Let φ be a quasi-flow in a metric space X, let $x \in X$, and suppose $|\dot{\varphi}|(t, x)$ exists and is finite for all $t \in \mathbb{R}$.*

Then whenever $0 < t < \mathrm{p}_\varphi(x)$, we have

$$\mathscr{L}\big(\{\varphi(s, x) \mid 0 \leq s \leq t\}\big) = \int_0^t |\dot{\varphi}|(s, x)\, ds.$$

1.61. Example. Let h be a strictly increasing, continuous function from $[0, 1]$ into $\mathbb{R}$, which has the property that $h'(t) = 0$ for almost all $t \in [0, 1]$. (Such functions h will be constructed in Example 8.45.) It is easy to see that Theorem 1.58(a) does not hold for h.

This shows that the requirement in Theorem 1.58(a) that. $\overline{D}h$ be finite-valued, is essential, and cannot even be replaced by the assumption that $\overline{D}h$ be finite-valued almost everywhere.

Bounded and Canonical Flows

1.62. Definition. Let φ be an algebraic flow in a subspace Y of a metric space X. φ is said to be *bounded with respect to* a subset K of X if there exists a monotone function f from $[0, \infty]$ into $[0, \infty]$ such that

(*i*) $f(s) \to 0$ as $s \to 0$,

and

(ii) whenever $x \in Y$ and $|t| \leq 1$, we have

$$d\big(x, \varphi(t, x)\big) \leq f\big(d(x, K)\big).$$

When this happens, φ is said to be *bounded by f with respect to K*.

If b is a positive real number, then the function f from $[0, \infty]$ into $[0, \infty]$, defined by $f(s) = b \cdot s$ for all $s \in [0, \infty]$, always satisfies (i), and the statement that φ is bounded by f with respect to K becomes:

$$d\big(x, \varphi(t, x)\big) \leq b \cdot d(x, K) \quad \text{whenever } x \in Y \text{ and } |t| \leq 1.$$

When this happens, φ is said to be *bounded by b with respect to K*. No confusion should arise out of this abuse of notation.

φ is said to be *canonical with respect to* a subset K of X, if for each $x \in Y$, φ has a speed at x, and we have

$$|\dot{\varphi}|(x) = d(x, K).$$

1.63. Theorem. *Let φ be an algebraic flow in a subspace Y of a metric space X, and let K be a subset of X.*

Then φ is bounded [respectively canonical] with respect to K iff φ is bounded [respectively canonical] with respect to $\bar{K}$.

PROOF: Trivial.   **QED**

1.64. Lemma. *Let φ be an algebraic flow in a subspace Y of a metric space X. Let K be a subset of X, and f a function from $[0, \infty]$ into $[0, \infty]$, and suppose φ is bounded by f with respect to K.*

Define a sequence $\{g_n\}$ of functions from $[0, \infty]$ into $[0, \infty]$, inductively as follows:

$$g_1(s) = s + f(s), \qquad \forall\, s \in [0, \infty],$$

and

$$g_{n+1}(s) = g_1\big(g_n(s)\big), \qquad \forall\; s \in [0, \infty],$$

and for each positive integer n, define f_n by:

$$f_n(s) = g_n(s) - s, \qquad \forall\; s \in [0, \infty].$$

Then for each positive integer n, we have:

(i) *f_n is monotone, $f_n(s) \to 0$ as $s \to 0$, and $f_n(0) = 0$*

and

(ii) *whenever $x \in Y$ and $|t| \leq n$, we have*

$$d\big(x, \varphi(t, x)\big) \leq f_n\big(d(x, K)\big).$$

PROOF: (i) can be proved easily, by induction. To prove (ii), we note first that (ii) is clear if $n = 1$. Suppose we have already shown, for a given integer $n \geq 1$, that whenever $x \in Y$ and $|t| \leq n$, we have

$$d\big(x, \varphi(t, x)\big) \leq f_n\big(d(x, K)\big),$$

and assume now that $x \in Y$ and $|t| \leq n + 1$. Choose $\tau \in \mathbb{R}$ such that $|\tau| \leq n$, and $|t - \tau| \leq 1$. Then

$$\begin{aligned}
d\big(x, \varphi(t, x)\big) &\leq d\big(x, \varphi(\tau, x)\big) + d\big(\varphi(\tau, x), \varphi(t, x)\big) \\
&\leq f_n\big(d(x, K)\big) + f_1\big(d(\varphi(\tau, x), K)\big) \\
&\leq f_n\big(d(x, K)\big) + f_1\big(d(\varphi(\tau, x), x) + d(x, K)\big) \\
&\leq f_n\big(d(x, K)\big) + f_1\big(f_n\big(d(x, K)\big) + d(x, K)\big) \\
&= f_{n+1}\big(d(x, K)\big).
\end{aligned}$$

Thus (ii) follows, by induction.   **QED**

1.65. Theorem. *Let φ be an algebraic flow in a subspace Y of a metric space X, and suppose φ is bounded with respect to a subset K of X.*
Then

(a) *$Y \cap K \subseteq \mathscr{F}(\varphi)$*

and

(b) *φ is continuous at each point of $Y \cap K$.*

PROOF: The result follows easily from Theorem 1.64.   **QED**

1.66. Theorem. *Let φ be a flow in a compact metric space X. Then φ is bounded with respect to $\mathscr{F}(\varphi)$.*

PROOF: If $\mathscr{F}(\varphi) = \square$, then $d\big(x, \mathscr{F}(\varphi)\big) = \infty$ for all $x \in X$, and therefore φ is bounded by f, where $f(s) = s$, $\qquad \forall\; s \in [0, \infty]$.

Suppose that $\mathscr{F}(\varphi) \neq \square$. Now for each real number $s \geq 0$, since

$$[-1, 1] \times \{x \in X \mid d(x, \mathscr{F}(\varphi)) \leq s\}$$

is a non-empty compact subset of $\mathbb{R} \times X$, it is clear that

$$\sup \{d(x, \varphi(t, x)) \mid d(x, \mathscr{F}(\varphi)) \leq s \text{ and } |t| \leq 1\} < \infty.$$

We define f from $[0, \infty]$ into $[0, \infty]$ as follows:

$$f(s) = \sup \{d(x, \varphi(t, x)) \mid d(x, \mathscr{F}(\varphi)) \leq s \text{ and } |t| \leq 1\}, \qquad \forall\, s \in [0, \infty],$$

and

$$f(\infty) = \infty.$$

The only non-trivial thing we need to prove, in order to show that φ is bounded by f with respect to $\mathscr{F}(\varphi)$, is that $f(s) \to 0$ as $s \to 0$. But if this is false, we can find a number $\delta > 0$, and a net $\{s_\alpha \mid \alpha \in A\}$ of points in $[0, \infty]$, such that $s_\alpha \to 0$, but $f(s_\alpha) > \delta$, $\forall\, \alpha \in A$. For each $\alpha \in A$, choose $t_\alpha \in [-1, 1]$, and x_α satisfying $d(x_\alpha, \mathscr{F}(\varphi)) \leq s_\alpha$, such that

$$d(x_\alpha, \varphi(t_\alpha, x_\alpha)) > \delta/2.$$

Now $[-1, 1] \times X$ is compact, and therefore $\{\langle t_\alpha, x_\alpha\rangle \mid \alpha \in A\}$ has a cluster-point $\langle t, x \rangle$ in $[-1, 1] \times X$. Clearly, x is a cluster-point of $\{x_\alpha \mid \alpha \in A\}$, t is a cluster-point of $\{t_\alpha \mid \alpha \in A\}$, and $d(x, \varphi(t, x))$ is a cluster-point of $\{d(x_\alpha, \varphi(t_\alpha, x_\alpha))\}$. It follows that

$$d(x, \varphi(t, x)) \geq \delta/2.$$

But since $s_\alpha \to 0$, it is clear that $x \in \mathscr{F}(\varphi)$, and therefore

$$d(x, \varphi(t, x)) = d(x, x) = 0.$$

This contradiction proves that $f(s) \to 0$ as $s \to 0$. The rest of the proof is left to the reader.   **QED**

The following theorem enables us to prove that a canonical quasi-flow is bounded, and to find a best possible bound.

1.67. Theorem. *Let* $-\infty \leq a < b \leq \infty$, *and let* h *be a continuous function from* (a, b) *into a metric space* X. *Let* K *be a non-empty, closed subset of* X, *and suppose that, for each* $t \in (a, b)$, *we have*

$$(\overline{D} h)(t) \leq d(h(t), K).$$

Then whenever $[\alpha, \beta]$ *is a compact subinterval of* (a, b), *and* $\alpha \leq \beta \leq \alpha + 1/n$, *we have*

$$d(h(\alpha), h(\beta)) \leq \frac{n}{n-1}(\beta - \alpha)\, d(h([\alpha, \beta]), K).$$

PROOF: Choose t_0 and t_1 in $[\alpha, \beta]$ such that $d\big(h(t_0), K\big) = d\big(h([\alpha, \beta]), K\big)$ and $\quad d\big(h(t_1), K\big) = \sup\{d\big(h(t), K\big) \mid \alpha \le t \le \beta\}$.   We claim that $d\big(h(t_1), K\big) \le \dfrac{n}{n-1} d\big(h(t_0), K\big)$. To see this, we note that

$$d\big(h(t_0), h(t_1)\big) \le \left| \int_{t_0}^{t_1} (\overline{D}h)(t)\, dt \right| \qquad \text{(by Theorem 1.59.)}$$

$$\le |t_1 - t_0| \sup\{(\overline{D}h)(t) \mid t \in [t_0, t_1]\}$$

$$\le \frac{1}{n} d\big(h(t_1), K\big)$$

and therefore

$$d\big(h(t_1), K\big) \le d\big(h(t_1), h(t_0)\big) + d\big(h(t_0), K\big)$$

$$\le \frac{1}{n} \cdot d\big(h(t_1), K\big) + d\big(h(t_0), K\big)$$

and therefore $d\big(h(t_1), K\big) \le \dfrac{n}{n-1} d\big(h(t_0), K\big)$, as promised.

Now

$$d\big(h(\alpha), h(\beta)\big) \le \int_{\alpha}^{\beta} (\overline{D}h)(t)\, dt \le (\beta - \alpha)\, d\big(h(t_1), K\big)$$

$$\le \frac{n}{n-1} (\beta - \alpha)\, d\big(h(t_0), K\big).$$

This completes the proof of the theorem.   **QED**

1.68. Corollary. *Let* $-\infty \le a < b \le \infty$ *and let* h *be a continuous function from* (a, b) *into a metric space* X. *Let* K *be a non-empty closed subset of* X, *and suppose that, for each* $t \in (a, b)$, *we have*

$$(\overline{D}h)(t) \le d\big(h(t), K\big).$$

Suppose α *and* β *are points in* (a, b), *and*

$$d\big(h(\alpha), K\big) > \frac{n}{n-1} d\big(h(\beta), K\big).$$

Then $|\alpha - \beta| > \dfrac{1}{n}$.

PROOF: This follows easily from Theorem 1.67.   **QED**

1.69. Corollary. *Let* φ *be a quasi-flow in a subspace* Y *of a metric space* X, *let* K *be a non-empty, closed subset of* X, *and suppose* φ *is canonical with respect to* K.

Then φ is bounded by $(e - 1)$ with respect to K. Further, the bound $(e - 1)$ is best possible.

PROOF: Let $x \in Y$ and suppose $|t| \leq 1$. Then

$$d\big(x, \varphi(t, x)\big) \leq \sum_{i=1}^{n} d\left(\varphi\left(\frac{i-1}{n} t, x\right), \varphi\left(\frac{i}{n} t, x\right)\right)$$

$$\leq \sum_{i=1}^{n} \left(\frac{n}{n-1}\right) \left|\frac{t}{n}\right| d\left(\varphi\left(\left[\frac{i-1}{n} t, \frac{i}{n} t\right], x\right), K\right)$$

$$\leq \sum_{i=1}^{n} \left(\frac{n}{n-1}\right) \left|\frac{t}{n}\right| d\left(\varphi\left(\frac{i-1}{n} t, x\right), K\right)$$

$$\leq \sum_{i=1}^{n} \left(\frac{n}{n-1}\right) \left|\frac{t}{n}\right| \left(\frac{n}{n-1}\right)^{i-1} d(x, K)$$

by Corollary 1.68

$$= d(x, K) |t| \sum_{i=1}^{n} \frac{1}{n} \left(\frac{n}{n-1}\right)^{i}.$$

Since $\displaystyle\sum_{i=1}^{n} \frac{1}{n} \left(\frac{n}{n-1}\right)^{i} \to e - 1$ as $n \to \infty$, we have

$$d\big(x, \varphi(t, x)\big) \leq (e - 1) |t| d(x, K) \leq (e - 1) d(x, K).$$

This shows that φ is bounded by $(e - 1)$ with respect to K.

In order to show that the bound $(e - 1)$ is best possible, we define a flow φ in $\mathbb{R}$ as follows:

$$\varphi(t, x) = xe^{t}, \qquad \forall\, t \in \mathbb{R} \qquad \forall\, x \in \mathbb{R}.$$

Clearly, φ is canonical with respect to $\{0\}$. But for every $x \in \mathbb{R}$, we have

$$\sup \{d\big(x, \varphi(t, x)\big) \mid |t| \leq 1\} = (e - 1) |x|$$

$$= (e - 1) d(x, \{0\}),$$

and it follows that $(e - 1)$ is the smallest possible bound.   **QED**

1.70. Uniform Continuity. Let φ be a flow in a metric space X. Then for any compact subinterval $[a, b]$ of $\mathbb{R}$ and compact subset C of X, the restriction of φ to the compact subset $[a, b] \times C$ of $\mathbb{R} \times X$ is uniformly continuous. The following theorems, with which we conclude the chapter, give two extensions of this result.

1.71. Theorem. *Let φ be a flow in a metric space X, $[a, b]$ a compact subinterval of $\mathbb{R}$, and C a compact subset of X.*

Then given $\varepsilon > 0$, there exists $\delta > 0$ such that, whenever $x \in C$, $x' \in X$, $t \in [a, b]$, $t' \in \mathbb{R}$, and $d(x, x') < \delta$ and $|t - t'| < \delta$, we have

$$d\big(\varphi(t, x), \dot{\varphi}(t', x')\big) < \varepsilon.$$

PROOF: This theorem may be proved by exactly the same techniques as are used to prove that a continuous, real function on a closed interval (or more generally, a continuous function from a compact metric space into a metric space), is uniformly continuous.   **QED**

1.72. Lemma. *Let φ be a flow in a Hausdorff space X, let $x \in X$, and let $[a, b]$ be a compact subinterval of $\mathbb{R}$.*

Then given a neighborhood U of $\{\varphi(t, x) \mid t \in [a, b]\}$, there exist neighborhoods V and W, of x and $[a, b]$ respectively, such that

$$y \in V \text{ and } t \in W \Rightarrow \varphi(t, y) \in U.$$

PROOF: For each $t \in [a, b]$, choose neighborhoods V_t and W_t of x and t respectively, such that, whenever $y \in V_t$ and $s \in W_t$, we have $\varphi(s, y) \in U$. Choose a finite subset $\{t_1, \ldots, t_n\}$ of $[a, b]$ such that

$$[a, b] \subseteq \bigcup_{j=1}^{n} W_{t_j},$$

and let $V = \bigcap_{j=1}^{n} V_{t_j}$ and $W = \bigcup_{j=1}^{n} W_{t_j}$.   **QED**

1.73. Theorem. *Let φ be a flow in a Hausdorff space X, let $[a, b]$ be a compact subinterval of $\mathbb{R}$ and let C be a compact subset of X.*

Then given a neighborhood U of $\{\varphi(t, x) \mid t \in [a, b]$ and $x \in C\}$, there exist neighborhoods V and W, of C and $[a, b]$ respectively, such that

$$x \in V \text{ and } t \in W \Rightarrow \varphi(t, x) \in U.$$

PROOF: For each $x \in C$, choose neighborhoods V_x and W_x, of x and $[a, b]$ respectively, such that

$$y \in V_x \text{ and } t \in W_x \Rightarrow \varphi(t, y) \in U.$$

Choose a finite subset $\{x_1, \ldots, x_n\}$ of C such that

$$C \subseteq \bigcup_{j=1}^{n} V_{x_j},$$

and let

$$V = \bigcup_{j=1}^{n} V_{x_j} \text{ and } W = \bigcap_{j=1}^{n} W_{x_j}. \qquad \textbf{QED}$$

Notes and Remarks to Chapter 1

Section 1.25. Theorem 1.25 was proved by the Lewins as a tool for use in the proof of Theorem 4.3 (e). Theorem 4.3 (e) is trivial for orbits which are homeomorphic to $\mathbb{R}$, and Theorem 1.25 may be interpreted as saying that, in a certain sense, aperiodic orbits of a flow are locally homeomorphic to $\mathbb{R}$. Further results of this type may be found in Beck, Lewin and Lewin [1], and in Ura [1] (in particular, Theorem 2, p. 113), and in Nadler and Quinn [1].

Sections 1.30, 1.31. Corollary 1.31 first appeared in Beck [2], (Lemma 3), and the proof contained there yields the slightly stronger assertion of Theorem 1.30.

Section 1.43. The concept of a stagnation point appeared in Beck [3] in order to generalize the results contained in Beck [2] concerning flows in the plane with compact or closed orbits. In Beck [3] it is shown that in the absence of stagnation points, the behavior of a flow in $\mathbb{R}^2$ resembles the behavior of a flow with compact orbits. (These results may be found in Chapter 3).

Section 1.59. Theorem 1.59 is an interesting generalization of the fundamental theorem of Calculus. It was proved jointly with the Lewins, and the proof is based on an idea communicated orally to us by J. Kurzweil. By drawing attention to the close relationship between the Riemann integral and the Lebesgue integral via the integral he had defined in Kurzweil [1], Kurzweil made it possible to generalize an earlier proof of this theorem by the Lewins, which held only for bounded Riemann-integrable functions.

Sections 1.62 to 1.69. The concepts of canonical and bounded flows first appeared in Beck [3] (Sections 38 ff), in order to provide conditions that would insure that a quasi-flow would be continuous and fixed in a given set K. The result that a canonical flow is bounded by 3 appears there, with a proof that does not rely on the Kurzweil Integral. (The applications of this material that are made in Beck [3] may be found in Chapter 6).

Theorem 1.67 and its Corollary 1.68 were first obtained by the Lewins for the case $n = 2$, as a tool for use in the proof of Theorem 5.12. Their proof was essentially the one that appears here. The author later observed that these results generalize to their present forms, and used them to replace the bound 3 of a canonical flow, by the sharp bound $e - 1$; (Corollary 1.69).

CHAPTER TWO

SPECIAL PROPERTIES OF PLANE FLOWS

2.1. Introduction. By a *plane flow*, we mean a flow in a subspace of $\mathbf{S}^2$, (where $\mathbf{S}^2$ is, as usual, the unit sphere in $\mathbb{R}^3$).

Chapter 1 provides us with some ready information about flows in $\mathbf{S}^2$; for example, since $\mathbf{S}^2$ is compact, we have that, for any flow φ in $\mathbf{S}^2$, $\alpha_\varphi(x)$ and $\omega_\varphi(x)$ are non-empty, compact and connected, whenever $p_\varphi(x) = \infty$. By Corollary 1.33 we also know that, if φ is a flow in $\mathbf{S}^2$, then $\mathscr{F}(\varphi) \neq \square$. If φ is a flow in $\mathbf{S}^2$, and a point $x \in \mathscr{F}(\varphi)$ is chosen to play the role of ∞, (see Appendix C) then since $\mathbf{S}^2 \setminus \{x\}$ is homeomorphic to $\mathbb{R}^2$, we can identify the restriction of φ to $\mathbf{S}^2 \setminus \{x\}$ with a flow in $\mathbb{R}^2$. On the other hand, given a flow φ in $\mathbb{R}^2$, Theorem 1.29 implies that φ has a unique extension to the one-point compactification $\mathbb{R}^2 \cup \{\infty\}$ of $\mathbb{R}^2$, obtained by making ∞ fixed. Since $\mathbb{R}^2 \cup \{\infty\}$ is homeomorphic to $\mathbf{S}^2$, this unique extension may be identified with a flow in $\mathbf{S}^2$.

We can go further: If φ is a flow in a subspace Y of $\mathbf{S}^2$, then φ may be identified with a flow in a subspace of $\mathbb{R}^2$ as follows:

(i) If $Y = \mathbf{S}^2$, follow the procedure described above.

(ii) If $Y \neq \mathbf{S}^2$, choose a point $x \in \mathbf{S}^2 \setminus Y$ to play the role of ∞, and note that Y is a subspace of the homeomorphic image $\mathbf{S}^2 \setminus \{x\}$ of $\mathbb{R}^2$.

It is clear that this method identifies a flow in an open [respectively connected open] subset of $\mathbf{S}^2$, with a flow in an open [respectively connected open] subset of $\mathbb{R}^2$.

Given a plane flow φ, we shall use this method to regard φ as a flow in a subspace of $\mathbf{S}^2$ whenever we wish to make use of the chordal metric (see Appendix C), the compactness of $\mathbf{S}^2$, or any other properties of $\mathbf{S}^2$ which are not enjoyed directly by $\mathbb{R}^2$. On the other hand, we shall regard φ as a flow in a subspace of $\mathbb{R}^2$ when we wish to make use of the Euclidean metric, the identification of $\mathbb{R}^2$ with the complex field $\mathbb{C}$, or other properties of $\mathbb{R}^2$ which are not enjoyed directly by $\mathbf{S}^2$. The value of this ability to switch will become evident in the theorems which follow.

The Gate Theorem

2.2. Introduction. Let φ be a flow in $\mathbf{S}^2$, $y \in \mathscr{G}(\varphi)$, and let a point x_1 of $\mathbf{S}^2$ be "very close" to y. Then as t increases from 0, $\varphi(t, x_1)$ and $\varphi(t, y)$ move off together, away from y. Suppose that, at some later time t_2, the point $x_2 = \varphi(t_2, x_1)$ is again "very close" to y, and that t_2 is, in some sense, substantially less than $p_\varphi(x_1)$. Then as t increases beyond t_2, $\varphi(t, x_1)$ again moves away from y, before $\mathcal{O}_\varphi(x_1)$ can make a complete period, and return to x_1. This part of $\mathcal{O}_\varphi(x_1)$ begins a "spiral", either towards, or away from, y. (See Figs. 2.1 and 2.2.)

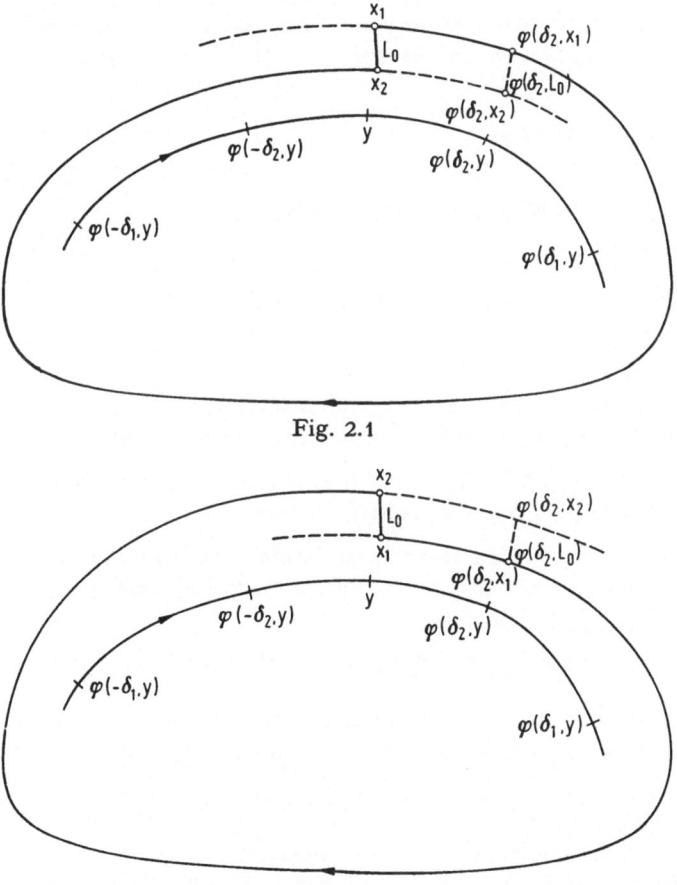

Fig. 2.1

Fig. 2.2

The Gate Theorem provides the technical machinery that enables us to prove that once this spiral has started, it must continue. Roughly speaking, the Gate Theorem states that $\{\varphi(t, x_1) \mid 0 < t < t_2\}$ and the line segment $[x_1, x_2]$, together constitute a Jordan curve J_0, of which $[x_1, x_2]$ is called the gate, and denoted by L_0, and that $\varphi(t, x_1)$ lies on one side of J_0 for all $t < 0$, and on the other side for all $t > t_2$. Further, no other φ-orbit can cross J_0, except through its gate.

Unfortunately, the precise statement of the Gate Theorem is rather complicated, but the reader will find, that with the help of the figures provided, the statement will become progressively clearer as he reads the proof.

2.3. The Gate Theorem, I. *Let φ be a flow in an open subset Ω of $\mathbb{R}^2$. Let $y \in \mathcal{G}(\varphi)$, and suppose $0 < \delta_1 < \frac{1}{2} \mathrm{p}_\varphi(y)$.*

Then there exists a number $\delta_2 > 0$ such that $\delta_1 = k \cdot \delta_2$ for some integer $k \geq 2$, and such that, for any neighborhood V of y, a number $\varepsilon > 0$ can be so chosen that, given any point $x \in \Omega$ which satisfies

either

(a) *y is a φ-endpoint of x*

or

(b) *$d(x, y) < \varepsilon$ and $\mathrm{p}_\varphi(y) + 2\delta_1 < \mathrm{p}_\varphi(x)$,*

we can find numbers t_1 and t_2 for which the following statements hold:

(i) *Defining $x_1 = \varphi(t_1, x)$ and $x_2 = \varphi(t_2, x)$, we have $x_1 \in V$ and $x_2 \in V$.*

(ii) *$t_1 + \delta_2 < t_2$.*

(iii) *The line segment $L_0 = [x_1, x_2]$ and $\{\varphi(t, x) \mid t_1 < t < t_2\}$ together constitute a Jordan curve J_0 which is contained in Ω.*

(iv) *$[\varphi(\delta_2, L_0) \cup \{\varphi(t, x) \mid t_2 < t < t_2 + \delta_2\}] \smallsetminus \{\varphi(\delta_2, x_1)\}$ is an arc which joins x_2 to $\varphi(\delta_2, x_1)$, and is disjoint from J_0.*

(v) *$\varphi(\delta_2, J_0)$ is a Jordan curve whose inside either contains, or is contained in, ins (J_0). Further, the symmetric difference between ins (J_0) and ins $(\varphi(\delta_2, J_0))$, is a subset of Ω.*

(vi) *If $z \in L_0$ and $-\delta_2 \leq t \leq \delta_2$, then $\varphi(t, z) \notin \varphi(-\delta_1, L_0) \cup \varphi(\delta_1, L_0)$.*

PROOF: Let M be the least of the distances between any two of the three points $\varphi(-\delta_1, y), y, \varphi(\delta_1, y)$. Choose $\delta_2 > 0$, so small that

$$d\big(y, \varphi(t, y)\big) < M/4 \text{ whenever } -\delta_2 \leq t \leq \delta_2.$$

By making δ_2 smaller, if necessary, make it satisfy the further requirement that $\delta_1 = k \cdot \delta_2$, for some integer $k \geq 2$. Choose $\varepsilon_1 > 0$ such that, firstly, $\varepsilon_1 < M/16$, secondly, no two of the three ε_1-neighborhoods of the points $\varphi(-\delta_2, y)$, y, $\varphi(\delta_2, y)$, intersect, thirdly, $N(y, \varepsilon_1) \subseteq V \cap \Omega$, and fourthly, there exists a simply-connected region U which is contained in Ω, and which contains the ε_1-neighborhood of the set $\{\varphi(t, y) \mid 0 \leq t \leq \delta_2\}$. (It is easy to see how to choose ε_1 small enough to satisfy the first three of these four requirements. To see that the fourth requirement can also be met, we need only prove that there exists a simply-connected region U such that $U \subseteq \Omega$, and

$$\{\varphi(t, y) \mid 0 \leq t \leq \delta_2\} \subseteq U.$$

But the existence of such a set U is easy to prove in the special case $\{\varphi(t, y) \mid 0 \leq t \leq \delta_2\} = [0, 1]$, and therefore since the Schoenflies Theorem (see Theorem C. 22) always provides a homeomorphism of $\mathbb{R}^2$ onto itself that carries $\{\varphi(t, y) \mid 0 \leq t \leq \delta_2\}$ onto $[0, 1]$, the existence of U is always assured.) By Theorem 1.71, we can choose $\varepsilon_2 > 0$ such that $\varepsilon_2 < \varepsilon_1$, and $d\big(\varphi(t, y), \varphi(t, z)\big) < \varepsilon_1$ whenever $d(y, z) < \varepsilon_2$ and $-\delta_1 \leq t \leq \delta_1$. If $\mathrm{p}_\varphi(y) < \infty$, choose $\varepsilon > 0$ such that $\varepsilon < \varepsilon_2$, and $d\big(\varphi(t, y), \varphi(t, z)\big) < \varepsilon_2$ whenever $d(y, z) < \varepsilon$ and $-\delta_1 \leq t \leq \mathrm{p}_\varphi(y) + \delta_1$. If $\mathrm{p}_\varphi(y) = \infty$, the statement of the theorem makes no requirement of the number ε (because (b) is always false). Therefore, if $\mathrm{p}_\varphi(y) = \infty$, we shall assign to ε the value 1.

Now let x be a point in Ω which satisfies either (a) or (b).

If x satisfies (a), let $\tilde{x}_1$ and $\tilde{x}_2$ be chosen in $\mathcal{O}_\varphi(x) \cap N(y, \varepsilon_2)$ such that

$$\tilde{x}_2 = \varphi(\tilde{l}, \tilde{x}_1) \text{ for some } \tilde{l} > 2\delta_1.$$

Since every point of $\mathcal{O}_\varphi(x)$ also satisfies (a), and since the conclusions of the theorem are not altered if we substitute for x another point in $\mathcal{O}_\varphi(x)$, we can assume for simplicity of notation that $\tilde{x}_1 = x$.

If x satisfies (b), but does not satisfy (a), we define $\tilde{x}_1 = x$, and $\tilde{x}_2 = \varphi\big(\mathrm{p}_\varphi(y), x\big)$, and we let $\tilde{l} = \mathrm{p}_\varphi(y)$.

So whether x satisfies (a) or (b), we have defined $\tilde{x}_1$ and $\tilde{x}_2$ in $N(y, \varepsilon_2)$, we have a number $\tilde{l} > 2\delta_1$ such that $\tilde{x}_2 = \varphi(\tilde{l}, \tilde{x}_1)$, and we have $\tilde{x}_1 = x$. Also, since $\tilde{l} + 2\delta_1 < \mathrm{p}_\varphi(x)$, we see that the restriction to $[-\delta_1, \tilde{l} + \delta_1]$ of the function $t \to \varphi(t, x)$, is one-one.

Let $\Lambda = [\tilde{x}_1, \tilde{x}_2]$. (See Fig. 2.3) Let x_1 be that point on

$$\Lambda \cap \{\varphi(t, x) \mid -\delta_1 \leq t \leq \tilde{l} - \delta_1\}$$

that lies closest to $\tilde{x}_2$. Now let x_2 be that point on

$$[x_1, \tilde{x}_2] \cap \{\varphi(t, x) \mid \tilde{l} - \delta_1 \leq t \leq \tilde{l} + \delta_1\}$$

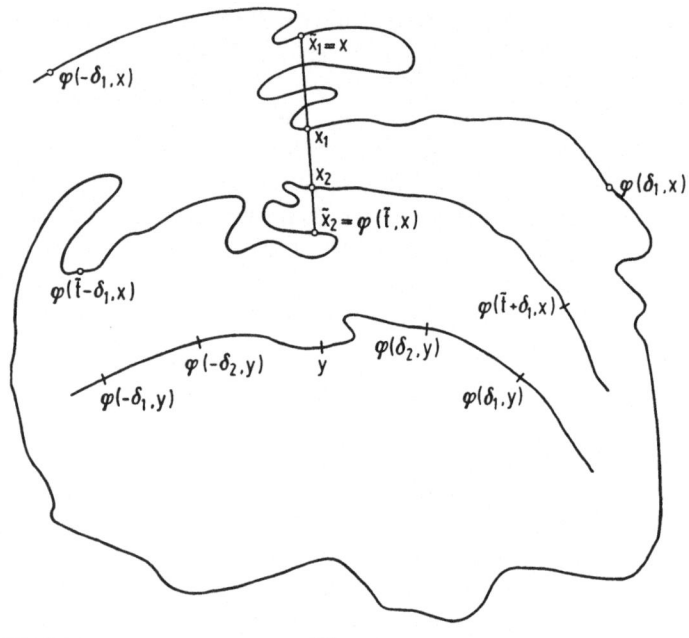

Fig. 2.3

that lies closest to x_1. Let t_1 and t_2 be, respectively, the unique numbers in $[-\delta_1, \bar{t} - \delta_1]$ and $[\bar{t} - \delta_1, \bar{t} + \delta_1]$ that satisfy

$$x_1 = \varphi(t_1, x) \quad \text{and} \quad x_2 = \varphi(t_2, x).$$

Clearly, both x_1 and x_2 lie in $N(y, \varepsilon_2) \subseteq V$.

We claim that $t_1 + \delta_2 < t_2$: for suppose $t_2 \leq t_1 + \delta_2$. Then we have

$$t_1 \leq \bar{t} - \delta_1 \leq t_2 \leq t_1 + \delta_2.$$

But on the one hand, since $\tilde{x}_2 \in N(y, \varepsilon_2)$, we have

$$d\big(\varphi(-\delta_1, \tilde{x}_2), \varphi(-\delta_1, y)\big) < \varepsilon_1 < M/4,$$

and therefore $d\big(\varphi(-\delta_1, \tilde{x}_2), y\big) > 3M/4$, by the choice of M.

Therefore $d\big(\varphi(\bar{t} - \delta_1, x), y\big) > 3M/4.$

And on the other hand, whenever $t_1 \leq t \leq t_1 + \delta_2$, the conditions $0 \leq t - t_1 \leq \delta_2$ and $x_1 \in N(y, \varepsilon_2)$ give us

$$d\big(\varphi(t, x), y\big) = d\big(\varphi(t - t_1, x_1), y\big)$$
$$\leq d\big(\varphi(t - t_1, x_1), \varphi(t - t_1, y)\big) + d\big(\varphi(t - t_1, y), y\big)$$
$$< \varepsilon_1 + M/4 < 3M/4.$$

This contradiction establishes that $t_1 + \delta_2 < t_2$, as promised.

It follows from this that $x_1 \neq x_2$. Let $L_0 = [x_1, x_2]$. Clearly, L_0 does not meet the arc $\{\varphi(t, x) \mid t_1 < t < t_2\}$. In fact, L_0 does not even meet $\{\varphi(t, x) \mid -\delta_1 \leq t \leq \bar{t} + \delta_1\}$ except in the points x_1 and x_2. Therefore, if we define

$$J_0 = L_0 \cup \{\varphi(t, x) \mid t_1 < t < t_2\},$$

we see that J_0 is a Jordan curve.

At this point of the proof, we have established parts (i), (ii) and (iii) of the theorem.

Define Λ_1, Λ_2 and Λ_3 as follows: (the reader should consult Figs. 2.1 and 2.2)

$$\Lambda_1 = \{\varphi(t, x) \mid t_1 + \delta_2 < t < t_2\}$$

$$\Lambda_2 = \left[L_0 \cup \{\varphi(t, x) \mid t_1 \leq t < t_1 + \delta_2\} \right] \setminus \{x_2\}$$

$$\Lambda_3 = \left[\varphi(\delta_2, L_0) \cup \{\varphi(t, x) \mid t_2 < t \leq t_2 + \delta_2\} \right] \setminus \{\varphi(\delta_2, x_1)\}.$$

It is clear that Λ_1 and Λ_2 are arcs which join x_2 to $\varphi(\delta_2, x_1)$. To see that Λ_3 is also an arc which joins x_2 to $\varphi(\delta_2, x_1)$, we need only note that, since L_0 does not meet $\{\varphi(t, x) \mid t_1 < t < t_2\}$, $\varphi(\delta_2, L_0)$ does not meet $\{\varphi(t, x) \mid t_2 < t < t_2 + \delta_2\}$.

Before showing that no two of the arcs Λ_1, Λ_2, Λ_3, can intersect, we need to show first that $t_1 - \delta_2 > -\delta_1$ and $t_2 + \delta_2 < \bar{t} + \delta_1$. Suppose $t_1 - \delta_2 \leq -\delta_1$. The we have $-\delta_2 \leq -t_1 - \delta_1 \leq 0$. But on the one hand, since $x \in N(y, \varepsilon_2)$, we have $d\big(\varphi(-\delta_1, x), \varphi(-\delta_1, y)\big) < \varepsilon_1$, and therefore $d\big(\varphi(-\delta_1, x), y\big) > 3M/4$. And on the other hand, since $x_1 \in N(y, \varepsilon_2)$,

$$d\big(\varphi(-t_1 - \delta_1, x_1), y\big) \leq d\big(\varphi(-t_1 - \delta_1, x_1), \varphi(-t_1 - \delta_1, y)\big)$$
$$+ d\big(\varphi(-t_1 - \delta_1, y), y\big)$$
$$< \varepsilon_1 + M/4 < 3M/4.$$

This is impossible, since $\varphi(-\delta_1, x) = \varphi(-t_1 - \delta_1, x_1)$. We have therefore shown that $t_1 - \delta_2 > -\delta_1$; a similar argument shows that $t_2 + \delta_2 < \bar{t} + \delta_1$.

These two inequalities, together with the fact that L_0 does not meet $\{\varphi(t, x) \mid -\delta_1 \leq t \leq \bar{t} + \delta_1\}$ except in the points x_1 and x_2, and the fact that L_0 and $\varphi(\delta_2, L_0)$, being contained respectively in $N(y, \varepsilon_1)$ and $N\big(\varphi(\delta_2, y), \varepsilon_1\big)$, are mutually disjoint, can easily be used to show that no two of the arcs Λ_1, Λ_2, Λ_3, can intersect. Part (iv) of the theorem is now established.

Now $\Lambda_1 \cup \Lambda_2 \cup \{x_2, \varphi(\delta_2, x_1)\} = J_0$, and

$$\Lambda_1 \cup \Lambda_3 \cup \{x_2, \varphi(\delta_2, x_1)\} = \varphi(\delta_2, J_0).$$

The first half of part (v) of the theorem will now follow from the "Theta-Curve Theorem" (see Theorem C. 26) once we have shown that Λ_1 does not lie inside the Jordan curve

$$\Lambda_2 \cup \Lambda_3 \cup \{x_2, \varphi(\delta_2, x_1)\}.$$

But the Theta-Curve Theorem implies, that to show this, we need only show that

$$\text{diam}\,(\Lambda_1) > \text{diam}\,(\Lambda_2 \cup \Lambda_3).$$

Now on the one hand, both $\varphi(\bar{t} - \delta_1, x)$ and x_2 lie in $\overline{\Lambda}_1$, and therefore

$$\text{diam}\,(\Lambda_1) \geq d\big(x_2, \varphi(\bar{t} - \delta_1, x)\big)$$
$$> M - 2\varepsilon_1 > 7M/8.$$

And on the other hand, it is easy to see that every point of $\Lambda_2 \cup \Lambda_3$ lies in the ε_1-neighborhood of a point of $\{\varphi(t, y) \mid 0 \leq t \leq \delta_2\}$, and therefore lies in $N(y, 5M/16)$, from which it follows that

$$\text{diam}\,(\Lambda_2 \cup \Lambda_3) \leq 5M/8 < \text{diam}\,(\Lambda_1).$$

This establishes the first half of part (v) of the theorem. The second half of part (v), follows at once from the elementary observation that the Jordan curve

$$\Lambda_2 \cup \Lambda_3 \cup \{x_2, \varphi(\delta_2, x_1)\}$$

lies in U, together with the fact that U is simply-connected.

Finally, to prove (vi), suppose $z \in L_0$ and $-\delta_2 \leq t \leq \delta_2$. Then since $z \in N(y, \varepsilon_2)$, we have

$$d\big(y, \varphi(t, z)\big) \leq d\big(\varphi(t, y), \varphi(t, z)\big) + d\big(\varphi(t, y), y\big)$$
$$< \varepsilon_1 + M/4 < M/2.$$

But for any point $w \in \varphi(-\delta_1, L_0) \cup \varphi(\delta_1, L_0)$, it is easy to see that

$$d(w, y) > M - \varepsilon_1 > M/2.$$

Therefore $\varphi(t, z) \notin \varphi(-\delta_1, L_0) \cup \varphi(\delta_1, L_0)$, and this completes the proof of the theorem. **QED**

2.4. The Gate Theorem, II. *Let φ be a flow in a subregion Ω of $\mathbb{R}^2$. Let $y \in \mathscr{G}(\varphi)$, and suppose $0 < \delta_1 < \frac{1}{2}\,\mathrm{p}_\varphi(y)$. Let $\delta_2 > 0$ be chosen* *satisfying the conditions of Theorem 2.3, and let* $k \geq 2$ *be the integer for which* $\delta_1 = k \cdot \delta_2$. *Let* V *be a neighborhood of* y. *Let* $\varepsilon > 0$ *be chosen satisfying the conditions of Theorem 2.3, and let the point* $x \in \Omega$ *satisfy either condition* (a) *or condition* (b) *of Theorem 2.3. Let* t_1 *and* t_2 *be chosen satisfying the requirements* (i),..., (vi) *of Theorem 2.3, and let* x_1, x_2, L_0, *and* J_0, *be defined accordingly, as in Theorem 2.3. Suppose also that the sense in which requirement* (v) *of Theorem 2.3 is satisfied, is that*

$$\text{ins}\left(\varphi(\delta_2, J_0)\right) \subseteq \text{ins}\left(J_0\right).$$

(a) *The family* $\{\varphi(n\delta_2, J_0) \mid n \text{ is an integer}\}$ *is a family of Jordan curves, all of which lie in* Ω, *and which has the property that, for each* n,

$$\text{ins}\left(\varphi((n+1)\delta_2, J_0)\right) \subseteq \text{ins}\left(\varphi(n\delta_2, J_0)\right).$$

(b) $\varphi(t, x_1) \in \text{outs}(J_0)$, $\forall\, t < 0$, *and* $\varphi(t, x_2) \in \text{ins}(J_0)$, $\forall\, t > 0$.

(c) *Let an integer* m *be chosen such that* $m > k$, *and* $m\delta_2 > t_2 - t_1$. *For every integer* n, *define*

$$J_n = \varphi(nm\delta_2, J_0).$$

Then the family $\{J_n \mid n \text{ is an integer}\}$ *is a pairwise disjoint family of Jordan curves, and for every* n, *we have*

$$J_{n+1} \subseteq \text{ins}(J_n).$$

(d) *For every integer* n, *define*

$$L_n = \varphi(nm\delta_2, L_0).$$

Then if n *is an integer and* $z \in L_n$, *we have*

$$\varphi(t, z) \in \text{outs}(J_{n+1}), \quad \forall\, t < 0 \quad and \quad \varphi(t, z) \in \text{ins}(J_{n-1}), \quad \forall\, t > 0.$$

(e) *For every integer* n, *let* T_n *be the annulus between* J_n *and* J_{n+1}. *Let*

$$T = \bigcup_{n=-\infty}^{\infty} \overline{T}_n.$$

Then T *is an annulus, is invariant of* φ, *and contains* $\mathcal{O}_\varphi(x)$. *In fact, for each integer* n,

$$T_n = \varphi(nm\delta_2, T_0),$$

and furthermore,

$$T = \bigcup_{z \in L_0} \mathcal{O}_\varphi(z).$$

(f) *If n is an integer, and $z \in \overline{T}_n$, we have*

$$\varphi(t, z) \in \text{outs}(J_n), \qquad \forall\, t < -2m\delta_2,$$

and

$$\varphi(t, z) \in \text{ins}(J_{n+1}), \qquad \forall\, t > 2m\delta_2.$$

(g) $p_\varphi(x) = \infty$; *in fact*, $p_\varphi(z) = \infty, \qquad \forall\, z \in T.$

(h) *No member of T can be a φ-endpoint*, i.e., *if $z \in \Omega$, then $\alpha_\varphi(z) \cup \omega_\varphi(z)$ does not meet T.*

PROOF:

(a) For each $t \in \mathbb{R}$, let φ_t be, as usual, the homeomorphism of Ω onto itself defined by

$$\varphi_t(z) = \varphi(t, z), \qquad \forall\, z \in \Omega.$$

It is clear that, for every integer n, since $\varphi(n\delta_2, J_0) = \varphi_{n\delta_2}(J_0)$, $\varphi(n\delta_2, J_0)$ is a Jordan curve which lies in Ω.

Now for each integer n, since

$$\text{ins}\big(\varphi(\delta_2, J_0)\big) \subseteq \text{ins}(J_0),$$

we have

$$\varphi_{n\delta_2}\big(\text{ins}\big(\varphi(\delta_2, J_0)\big)\big) \subseteq \varphi_{n\delta_2}\big(\text{ins}(J_0)\big),$$

and therefore, since $\varphi_{n\delta_2}$ satisfies the hypotheses of Theorem C. 67, we have

$$\text{ins}\big(\varphi_{n\delta_2}\big(\varphi(\delta_2, J_0)\big)\big) \subseteq \text{ins}\big(\varphi_{n\delta_2}(J_0)\big),$$

i.e.,

$$\text{ins}\big(\varphi((n+1)\,\delta_2, J_0)\big) \subseteq \text{ins}\big(\varphi(n\delta_2, J_0)\big).$$

This proves (a).

(b) Let $t > 0$, and write $t = n\delta_2 + t^*$, where n is an integer and $0 < t^* \leq \delta_2$. Since

$$\varphi(t^*, x_2) \in \varphi(\delta_2, J_0) \subseteq \overline{\text{ins}\big(\varphi(\delta_2, J_0)\big)}$$

$$\subseteq \overline{\text{ins}(J_0)} = J_0 \cup \text{ins}(J_0),$$

and since $\varphi(t^*, x_2) \notin J_0$, we have $\varphi(t^*, x_2) \in \text{ins}(J_0)$. Therefore, since $\varphi_{n\delta_2}$ satisfies the hypotheses of Theorem C. 67, we have

$$\varphi(t, x_2) = \varphi\big(n\delta_2, \varphi(t^*, x_2)\big) \in \text{ins}\big(\varphi_{n\delta_2}(J_0)\big)$$

$$\subseteq \text{ins}(J_0) \qquad\qquad \text{by (a)}.$$

This proves the second part of (b).

Now let $t < 0$, and write $t = n\delta_2 + t^*$, where $0 \leq t^* < \delta_2$ and n is an integer. Clearly, $n \leq -1$. Since $\varphi(t^*, x_1) \in J_0 \setminus \varphi(\delta_2, J_0)$, it is clear that $\varphi(t^*, x_1) \in \text{outs}\left(\varphi(\delta_2, J_0)\right)$. Therefore, since $\varphi_{n\delta_2}$ satisfies the hypotheses of Theorem C. 67, we have

$$\varphi(t, x_1) = \varphi\left(n\delta_2, \varphi(t^*, x_1)\right) \in \text{outs}\left(\varphi\left((n+1)\delta_2, J_0\right)\right).$$

Therefore since (a) implies that

$$\overline{\text{ins}\,(J_0)} \subseteq \overline{\text{ins}\left(\varphi(n+1)\delta_2, J_0\right)},$$

we must have

$$\varphi(t, x_1) \in \text{outs}\,(J_0).$$

This completes the proof of (b).

(c) We note first that, for each integer n, $J_n = \varphi_{nm\delta_2}(J_0)$ is clearly a Jordan curve. The proof of (c) will be complete once we have shown that $J_1 \subseteq \text{ins}\,(J_0)$. But since $m\delta_2 > t_2 - t_1$, (b) implies that

$$J_1 \cap \mathcal{O}_\varphi(x) \subseteq \text{ins}\,(J_0).$$

But $J_1 \setminus \mathcal{O}_\varphi(x) = \varphi\left(m\delta_2, (x_1, x_2)\right) = \varphi\left((m-1)\delta_2, \varphi\left(\delta_2, (x_1, x_2)\right)\right)$

$$\subseteq \varphi\left((m-1)\delta_2, \text{ins}\,(J_0)\right) = \text{ins}\left(\varphi\left((m-1)\delta_2, J_0\right)\right)$$

$$\subseteq \text{ins}\,(J_0) \qquad\qquad \text{by part (a).}$$

This completes the proof of (c).

(d) It is clearly sufficient to prove the result for $n = 0$. Let $z \in L_0$. Suppose $t > 0$, and write $t = t^* + q\delta_2$ where q is an integer and $0 \leq t^* < \delta_2$. We shall show first that $\varphi(t^*, z) \in \text{ins}\,(J_{-1})$. This is easy to see if $z \in \mathcal{O}_\varphi(x)$, for then either $z = x_1$ or $z = x_2$. Suppose $z \in L_0 \setminus \mathcal{O}_\varphi(x)$. Then since $\{\varphi(s, z) \mid 0 \leq s \leq t^*\}$ is connected, and by Theorem 2.3 (vi) does not meet $\varphi(-\delta_1, L_0)$, it is clear that $\{\varphi(s, z) \mid 0 \leq s \leq t^*\}$ does not meet $\varphi(-\delta_1, J_0)$. But since $\delta_1 = k\delta_2$, we have $z \in \text{ins}\left(\varphi(-\delta_1, J_0)\right)$. It follows that

$$\varphi(t^*, z) \in \text{ins}\left(\varphi(-\delta_1, J_0)\right) \subseteq \text{ins}\,(J_{-1}).$$

Therefore

$$\varphi(t, z) = \varphi\left(q\delta_2, \varphi(t^*, z)\right) \in \text{ins}\left(\varphi(q\delta_2, J_{-1})\right) \subseteq \text{ins}\,(J_{-1}).$$

A similar proof shows that if $z \in L_0$ and $t < 0$, then $\varphi(t, z) \in \text{outs}\,(J_1)$. This completes the proof of (d).

e) It is easy to see that T is connected. Now

$$\mathbf{S}^2 \setminus T = \left(\bigcap_{n=-\infty}^{\infty} \text{ins} \, (J_n) \right) \cup \left(\bigcap_{n=-\infty}^{\infty} \text{outs} \, (J_n) \cup \{\infty\} \right)$$

$$= \left(\bigcap_{n=-\infty}^{\infty} \overline{\text{ins} \, (J_n)} \right) \cup \left(\bigcap_{n=-\infty}^{\infty} \overline{\text{outs} \, (J_n) \cup \{\infty\}} \right),$$

and therefore, since $\mathbf{S}^2 \setminus T$ is compact, and has exactly two components (see Theorem A. 5), T must be an annulus.

We shall now show that $T \subseteq \Omega$. Suppose $z \in T$. If for some integer n, $z \in \varphi(n\delta_2, J_0)$, it is clear that $z \in \Omega$. Otherwise, we can choose an integer n such that

$$z \in \text{ins} \left(\varphi(n\delta_2, J_0) \right) \cap \text{outs} \left(\varphi((n+1)\delta_2, J_0) \right).$$

Let $\Lambda_1, \Lambda_2, \Lambda_3$ be defined as in the proof of Theorem 2.3, and let the Jordan curve

$$\Lambda_2 \cup \Lambda_3 \cup \{x_2, \varphi(\delta_2, x_1)\}$$

be denoted by H_0. Then the homeomorphism $\varphi_{n\delta_2}$ carries Λ_1, Λ_2 and Λ_3 onto three non-intersecting arcs, each of which joins $\varphi(n\delta_2, x_2)$ to $\varphi((n+1)\delta_2, x_1)$. Furthermore, by Theorem C. 67,

$$\varphi(n\delta_2, \Lambda_3) \subseteq \text{ins} \left(\varphi(n\delta_2, \Lambda_1) \cup \varphi(n\delta_2, \Lambda_2) \cup \{\varphi(n\delta_2, x_2), \varphi((n+1)\delta_2, x_1)\} \right).$$

It is now easy to see that $z \in \text{ins} \left(\varphi(n\delta_2, H_0) \right)$. It therefore remains to show that $\text{ins} \left(\varphi(n\delta_2, H_0) \right) \subseteq \Omega$. But, by Theorem 2.3 (v), $\text{ins} \, (H_0) \subseteq \Omega$. Therefore, since

$$\varphi_{n\delta_2}\left(\text{ins} \, (H_0)\right) = \Omega \cap \text{ins} \left(\varphi(n\delta_2, H_0) \right),$$

and $\varphi_{n\delta_2}\left(\text{ins} \, (H_0)\right)$ is a simply-connected region which contains all points of $\text{ins} \left(\varphi(n\delta_2, H_0) \right)$ which lie sufficiently close to $\varphi(n\delta_2, H_0)$, we must have

$$\varphi_{n\delta_2}\left(\text{ins} \, (H_0)\right) = \text{ins}\left(\varphi(n\delta_2, H_0) \right),$$

and therefore $\text{ins}\left(\varphi(n\delta_2, H_0) \right) \subseteq \Omega$.

This proves that $T \subseteq \Omega$. We complete the proof of (e) by showing, firstly, that for each integer n,

$$T_n = \varphi(nm\delta_2, T_0)$$

and secondly, that

$$T = \bigcup_{z \in L_0} \mathcal{O}_\varphi(z).$$

Now, for any integer n, the homeomorphism $\varphi_{nm\delta_2}$ carries J_0 onto J_n and J_1 onto J_{n+1}. It now follows by Theorem C. 67 that since $T \subseteq \Omega$, we have $T_n = \varphi(nm\delta_2, T_0)$. Suppose now that $w \in T$. If $\mathcal{O}_\varphi(w)$ does not meet L_0, then $\mathcal{O}_\varphi(w)$ does not meet $\bigcup\limits_{n=-\infty}^{\infty} L_n \cup \mathcal{O}_\varphi(x)$. Choose n such that $w \in \overline{T}_n$. Then, clearly, $w \in T_n$, and therefore $\varphi(m\delta_2, w) \in T_{n+1}$. It follows that $\mathcal{O}_\varphi(w)$, being a connected set which contains points on both sides of J_{n+1}, must meet J_{n+1}, a contradiction. This shows that

$$T \subseteq \bigcup_{z \in L_0} \mathcal{O}_\varphi(z).$$

Now suppose $z \in L_0$ and $t \in \mathbb{R}$. Choose an integer n such that $|t| < nm\delta_2$. It is a simple consequence of (d) that

$$\varphi(t, z) \in \text{ins}\,(J_{-(n+1)}) \cap \text{outs}\,(J_{n+1}) \subseteq T.$$

This completes the proof of (e).

(f) Let n be an integer, $z \in \overline{T}_n$, and $t < -2m\delta_2$. Choose $t^* \in [0, m\delta_2]$ such that $\varphi(-t^*, z) \in J_n$. Clearly, $\varphi(-m\delta_2 - t^*, z) \in J_{n-1}$. But $t + m\delta_2 + t^* < 0$, and it therefore follows from (d) that

$$\varphi(t, z) = \varphi\big(t + m\delta_2 + t^*, \varphi(-m\delta_2 - t^*, z)\big) \in \text{outs}\,(J_n).$$

A similar argument shows that if $z \in \overline{T}_n$ and $t > 2m\delta_2$, then $\varphi(t, z) \in \text{ins}\,(J_{n+1})$.

(g) Let $z \in L_0$. Then since (d) implies that

$$\varphi(t, z) \in \text{outs}\,(J_0), \qquad \forall\, t < -m\delta_2, \text{ and}$$

$$\varphi(t, z) \in \text{ins}\,(J_0), \qquad \forall\, t > m\delta_2,$$

we must have $\mathrm{p}_\varphi(z) = \infty$.

(h) Let $z \in \Omega$, and suppose $\mathrm{p}_\varphi(z) = \infty$. If z has a φ-endpoint $w \in T$, then since T is open and invariant, we must have $z \in T$. Assume without loss of generality that $z \in L_0$, and choose n such that $w \in \overline{T}_n$. By (e), we have

$$\varphi(t, z) \in \text{ins}\,(J_{n+2}), \qquad \forall\, t > (n + 3)\, m\delta_2,$$

and

$$\varphi(t, z) \in \text{outs}\,(J_{n-1}), \qquad \forall\, t < (n - 2)\, m\delta_2.$$

This contradicts the assumption that w is a φ-endpoint of z, and we have therefore shown that no member of T can be a φ-endpoint.

This completes the proof of the theorem. **QED**

2.5. Remark. A theorem dual to Theorem 2.4 can be obtained by replacing "ins" by "outs", and "outs" by "ins" wherever they appear there, *i.e.*, reversing the sense in which (v) of Theorem 2.3 is assumed to be satisfied.

2.6. Definition. Let φ be a flow in a subregion Ω of $\mathbb{R}^2$. Let $y \in \mathscr{G}(\varphi)$, and suppose $0 < \delta_1 < \frac{1}{2} p_\varphi(y)$. Let $\delta_2 > 0$ be chosen satisfying the conditions of Theorem 2.3. Let V be a neighborhood of y. Let $\varepsilon > 0$ be chosen satisfying the conditions of Theorem 2.3, and let the point $x \in \Omega$ satisfy either condition (a) or condition (b) of Theorem 2.3. Let $t_1, t_2, x_1, x_2, J_0, L_0$, and T be defined as in Theorems 2.3 and 2.4. Then we call the 7-tuple $\langle t_1, t_2, x_1, x_2, J_0, L_0, T \rangle$ a *φ-gate construction for the pair* $\langle x, y \rangle$. Usually, we shall not need all these symbols, and we shall refer to the φ-gate construction as $\langle J_0, L_0, T \rangle$.

2.7. Definition. Let φ be a flow in an open subset Ω of $\mathbf{S}^2$, let $x \in \Omega$, and suppose $p_\varphi(x) = \infty$. If $\alpha_\varphi(x) \cup \omega_\varphi(x)$ meets $\mathscr{G}(\varphi)$, we say $\mathcal{O}_\varphi(x)$ *spirals*. If $\alpha_\varphi(x)$ meets $\mathscr{G}(\varphi)$, we say $\mathcal{O}_\varphi(x)$ *α-spirals*, and if $\omega_\varphi(x)$ meets $\mathscr{G}(\varphi)$, we say $\mathcal{O}_\varphi(x)$ *ω-spirals*. If $y \in \alpha_\varphi(x) \cap \mathscr{G}(\varphi)$ [respectively $\omega_\varphi(x) \cap \mathscr{G}(\varphi)$], then we say $\mathcal{O}_\varphi(x)$ *α-spirals* [respectively *ω-spirals*] *to* $\mathcal{O}_\varphi(y)$. In either case, we say $\mathcal{O}_\varphi(x)$ *spirals to* $\mathcal{O}_\varphi(y)$.

2.8. Definition. Let φ be a quasi-flow in a subspace Y of $\mathbf{S}^2$, let $x \in Y$, and suppose $p_\varphi(x) = \infty$. The *complete set $\bar{\alpha}_\varphi(x)$ of φ-α-points of x* is defined to be

$$\bigcap_{t \in \mathbb{R}} \overline{\{\varphi(s, x) \mid s \leq t\}},$$

(where, as usual, the closure operation is taken in $\mathbf{S}^2$, not in Y). The *complete set $\bar{\omega}_\varphi(x)$ of φ-ω-points of x* is analogously defined to be

$$\bigcap_{t \in \mathbb{R}} \overline{\{\varphi(s, x) \mid s \geq t\}}.$$

2.9. Note. If φ is a quasi-flow in a subspace Y of $\mathbf{S}^2$, $x \in Y$, and $p_\varphi(x) = \infty$, then while $\alpha_\varphi(x)$ and $\omega_\varphi(x)$ can fail to be non-empty or compact or connected, it is nevertheless clear that $\bar{\alpha}_\varphi(x)$ and $\bar{\omega}_\varphi(x)$ are always non-empty, compact, and connected. Furthermore, $\alpha_\varphi(x) = Y \cap \bar{\alpha}_\varphi(x)$, and $\omega_\varphi(x) = Y \cap \bar{\omega}_\varphi(x)$.

2.10. Theorem. *Let φ be a flow in a subregion Ω of $\mathbb{R}^2$, let $x, y \in \Omega$, and suppose $\mathcal{O}_\varphi(x)$ spirals to $\mathcal{O}_\varphi(y)$.*

Then there exists a φ-gate construction $\langle J_0, L_0, T \rangle$ for the pair $\langle x, y \rangle$.

PROOF: Since y is a φ-endpoint of x, (condition (a) of Theorem 2.3) any choice of δ_1 satisfying $0 < \delta_1 < \frac{1}{2} p_\varphi(y)$ will give us this result as a consequence of Theorem 2.3. **QED**

2.11. Theorem. *Let φ be a flow in a subregion Ω of $\mathbb{R}^2$, let $x, y \in \Omega$, and suppose $\mathcal{O}_\varphi(x)$ spirals to $\mathcal{O}_\varphi(y)$. Let $\langle J_0, L_0, T \rangle$ be a φ-gate construction for the pair $\langle x, y \rangle$.*

Then $\bigcup\limits_{z \in T} \bar{\alpha}_\varphi(z)$ is contained in one component of $\partial(T)$, and $\bigcup\limits_{z \in T} \bar{\omega}_\varphi(z)$ is contained in the other.[1]

PROOF: It is obvious that

$$\bigcup_{z \in T} \bar{\alpha}_\varphi(z) \cup \bigcup_{z \in T} \bar{\omega}_\varphi(z) \subseteq \bar{T}$$

and therefore, by Theorem 2.4 (h), we have

$$\bigcup_{z \in T} \bar{\alpha}_\varphi(z) \cup \bigcup_{z \in T} \bar{\omega}_\varphi(z) \subseteq \partial(T).$$

Let m be an integer satisfying the condition of Theorem 2.4 (c), and for each integer n, let J_n, L_n and T_n be defined as in Theorem 2.4. The result now follows at once from Theorem 2.4 (d). **QED**

2.12. Corollary. *Let φ be a flow in an open subset Ω of $\mathbf{S}^2$, let $x \in \Omega$, and suppose $\mathcal{O}_\varphi(x)$ spirals.*

Then $\bar{\alpha}_\varphi(x) \cap \bar{\omega}_\varphi(x) = \square$.

PROOF: Assume without loss of generality that Ω is connected. Now use the methods of 2.1. to regard φ as a flow in a subregion of $\mathbb{R}^2$, and apply Theorem 2.11. **QED**

2.13. Theorem. *Let φ be a flow in an open subset Ω of $\mathbf{S}^2$, let $x, y \in \Omega$ and suppose y is a φ-endpoint of x.*

Then $\mathcal{O}_\varphi(y)$ does not spiral.

PROOF: We may clearly assume that φ is a flow in a subregion of $\mathbb{R}^2$. If $\mathcal{O}_\varphi(y)$ spirals, we can choose $z \in \Omega$ such that z is a moving φ-endpoint of y, and we can choose a φ-gate construction $\langle J_0, L_0, T \rangle$ for the pair $\langle y, z \rangle$. Then since $y \in T$, Theorem 2.4 (h) gives us a contradiction. **QED**

2.14. Theorem.[2] *Let φ be a flow in an open subset Ω of $\mathbf{S}^2$, let $x \in \Omega$, and suppose $\mathrm{p}_\varphi(x) = \infty$.*

Then $\mathcal{O}_\varphi(x)$ is homeomorphic to $\mathbb{R}$.

---

[1] $\partial(T)$ refers to the boundary of T, where T is regarded as a subset of $\mathbf{S}^2$. Unless otherwise stated, closures, boundaries, neighborhoods, *etc.*, will always be taken in the topology of $\mathbf{S}^2$, rather than in the relative topology of a subspace of $\mathbf{S}^2$. Thus, since T is an annulus, $\partial(T)$ has two components (see Appendix C).

[2] Compare with Sections 1.26 and 1.27.

PROOF: Since, by Theorem 2.13, $\mathcal{O}_\varphi(x)$ cannot contain any φ-endpoints of x, the result follows at once from Theorem 1.42.   **QED**

2.15. Theorem. *Let φ be a flow in an open subset Ω of $\mathbf{S}^2$, let $y \in \Omega$, and suppose $0 < p_\varphi(y) < \infty$.*

Then for any $\eta > 0$, there exists a neighborhood W of y, such that, for any $x \in W$, we have

either

(a) $|p_\varphi(x) - p_\varphi(y)| < \eta$

or

(b) $p_\varphi(x) = \infty$.

PROOF: Assume without loss of generality that Ω is connected, and in order to regard it as a plane region, choose a point of $\mathscr{F}(\varphi) \cup (\mathbf{S}^2 \setminus \Omega)$ to play the role of ∞. By Corollary 1.31, there exists a neighborhood V of y such that, for any $x \in V$, we have $p_\varphi(x) > p_\varphi(y) - \eta$. Choose $\delta_1 > 0$ such that $2\delta_1 < \eta$ and $2\delta_1 < p_\varphi(y)$. First choose $\delta_2 > 0$, and then choose $\varepsilon > 0$, to satisfy the conditions of Theorem 2.3. Let $W = V \cap N(y, \varepsilon)$. Now, if $x \in W$, then either $p_\varphi(x) < p_\varphi(y) + \eta$, or there exists a φ-gate construction for the pair $\langle x, y \rangle$, in which case $p_\varphi(x) = \infty$.

This completes the proof of the theorem.   **QED**

2.16. Theorem. *Let φ be a flow in a subregion Ω of $\mathbb{R}^2$, let $y \in \Omega$, and suppose $0 < p_\varphi(y) < \infty$. Let U be a neighborhood of $\mathcal{O}_\varphi(y)$.*

Then there exists a neighborhood W of y, such that for any $x \in W$ we have either

(a) $p_\varphi(x) < \infty$, $\mathcal{O}_\varphi(x)$ *and* $\mathcal{O}_\varphi(y)$ *are similarly oriented (see Definition C. 68), and the annulus between $\mathcal{O}_\varphi(x)$ and $\mathcal{O}_\varphi(y)$ is contained in U,*

or

(b) $p_\varphi(x) = \infty$, *and there exists a φ-gate construction $\langle J_0, L_0, T \rangle$ for the pair $\langle x, y \rangle$ such that the annulus between J_0 and $\mathcal{O}_\varphi(y)$ is contained in U.*

PROOF: There is no loss of generality in assuming that ins $\big(\mathcal{O}_\varphi(y)\big)$ is not a subset of U, and that $U \subseteq \mathscr{G}(\varphi)$. Let $\varepsilon = d\big(\mathcal{O}_\varphi(y), \mathbb{R}^2 \setminus U\big) > 0$. Choose $\delta > 0$ such that

$$|t_1 - t_2| < \delta \Rightarrow d\big(\varphi(t_1, y), \varphi(t_2, y)\big) < \varepsilon/4.$$

Choose a neighborhood W_1 of y such that, whenever $x \in W_1$, we have

(i) $d\big(\varphi(t, x), \varphi(t, y)\big) < \varepsilon/4$ whenever $0 \le t \le p_\varphi(y) + \delta$,

and

(ii) $p_\varphi(x) = \infty$ or $|p_\varphi(y) - p_\varphi(x)| < \delta$.

We claim that, if $x \in W_1$ and $p_\varphi(x) < \infty$, then the annulus between $\mathcal{O}_\varphi(x)$ and $\mathcal{O}_\varphi(y)$ is contained in U, for suppose $x \in W_1$ and $p_\varphi(x) < \infty$. Let γ_1 and γ_2 be the (oriented) Jordan curves defined as follows:

$$\gamma_1(t) = \varphi\big(t p_\varphi(y), y\big), \qquad \forall\, t \in [0, 1],$$

and

$$\gamma_2(t) = \varphi\big(t p_\varphi(x), x\big), \qquad \forall\, t \in [0, 1].$$

Then, for each $t \in [0, 1]$, we have

$$|\gamma_1(t) - \gamma_2(t)| = |\varphi\big(t p_\varphi(y), y\big) - \varphi\big(t p_\varphi(x), x\big)|$$
$$\leq |\varphi\big(t p_\varphi(y), y\big) - \varphi\big(t p_\varphi(x), y\big)|$$
$$+ |\varphi\big(t p_\varphi(x), y\big) - \varphi\big(t p_\varphi(x), x\big)|$$
$$< \varepsilon/2.$$

Consequently, if $z \in \mathbb{R}^2 \setminus U$, we have

$$|\gamma_1(t) - \gamma_2(t)| < |\gamma_1(t) - z|, \qquad \forall\, t \in [0, 1],$$

and therefore, by Theorem C. 62 and Remark C. 64 (c),

$$\mathrm{ind}_{\gamma_1}(z) = \mathrm{ind}_{\gamma_2}(z).$$

It follows that, since $\mathrm{ins}\big(\mathcal{O}_\varphi(y)\big)$ is not a subset of U, we have

$$\mathrm{ins}\big(\mathcal{O}_\varphi(y)\big) \cap \mathrm{ins}\big(\mathcal{O}_\varphi(x)\big) \neq \square.$$

This shows that $\mathcal{O}_\varphi(x)$ and $\mathcal{O}_\varphi(y)$ are mutually disjoint Jordan curves, and one of the two sets $\mathrm{ins}\big(\mathcal{O}_\varphi(x)\big)$, $\mathrm{ins}\big(\mathcal{O}_\varphi(y)\big)$, contains the other. Theorem C. 66 now implies that if z is any point in the annulus between $\mathcal{O}_\varphi(x)$ and $\mathcal{O}_\varphi(y)$, then

$$\mathrm{ind}_{\gamma_1}(z) \neq \mathrm{ind}_{\gamma_2}(z)$$

and from this we deduce that U contains the annulus between $\mathcal{O}_\varphi(x)$ and $\mathcal{O}_\varphi(y)$.

We shall now define a neighborhood W_2 of y which will have the property that, for any $x \in W_2$, if $p_\varphi(x) = \infty$ and $x \in \mathrm{ins}\big(\mathcal{O}_\varphi(y)\big)$, then there exists a φ-gate construction $\langle J_0, L_0, T \rangle$ for the pair $\langle x, y \rangle$ such that the annulus between J_0 and $\mathcal{O}_\varphi(y)$ is contained in U. In order to define W_2, we proceed as follows:

Choose $\eta > 0$ such that $N\big(\mathcal{O}_\varphi(y), \eta\big) \subseteq U$, and such that, whenever $z \in N\big(\mathcal{O}_\varphi(y), \eta\big)$, $\mathcal{O}_\varphi(z)$ meets W_1. It is easy to see, using the properties of W_1 established above, that whenever $z \in N\big(\mathcal{O}_\varphi(y), \eta\big)$ and $p_\varphi(z) < \infty$, we have

$$\mathrm{ins}\big(\mathcal{O}_\varphi(y)\big) \setminus U \subseteq \mathrm{ins}\big(\mathcal{O}_\varphi(z)\big).$$

Now use the techniques that were used to define W_1 to find a neighborhood $\widetilde{W}_1$ of y such that, whenever $x \in \widetilde{W}_1$ and $\mathrm{p}_\varphi(x) < \infty$, the annulus between $\mathcal{O}_\varphi(x)$ and $\mathcal{O}_\varphi(y)$ is contained in $N\big(\mathcal{O}_\varphi(y), \eta\big)$. Choose $\bar\eta > 0$ such that $\bar\eta < \eta$, and such that, whenever $z \in N\big(\mathcal{O}_\varphi(y), \bar\eta\big)$, $\mathcal{O}_\varphi(z)$ meets $\widetilde{W}_1$.

If $\widetilde{W}_1$ contains a point inside $\mathcal{O}_\varphi(y)$ with finite period, let z be such a point, and let $J = \mathcal{O}_\varphi(z)$.

If $\widetilde{W}_1$ contains no point inside $\mathcal{O}_\varphi(y)$ with finite period, choose a Jordan curve J inside $\mathcal{O}_\varphi(y)$, but so close to $\mathcal{O}_\varphi(y)$ that the annulus between J and $\mathcal{O}_\varphi(y)$ is contained in $N\big(\mathcal{O}_\varphi(y), \bar\eta\big)$ (see Theorem C. 30).

In either case, let A be the annulus between J and $\mathcal{O}_\varphi(y)$.

Choose $\delta_1 > 0$ such that $2\delta_1 < \mathrm{p}_\varphi(y)$, and let δ_2 be chosen accordingly, satisfying the conditions of Theorem 2.3. Choose a convex neighborhood V of y, such that, for any $x \in V$, we have

$$\varphi(t, x) \in \mathrm{outs}\,(J) \quad \text{whenever} \quad |t| \le \mathrm{p}_\varphi(y) + 2\delta_1.$$

Accordingly, choose $\varepsilon > 0$ satisfying the conditions of Theorem 2.3. We can now define W_2 by

$$W_2 = N(y, \varepsilon) \cap V.$$

We shall now show that W_2 has the promised property. Suppose x is any point in $W_2 \cap \mathrm{ins}\big(\mathcal{O}_\varphi(y)\big)$, and $\mathrm{p}_\varphi(x) = \infty$. By Theorem 2.3 there exists a φ-gate construction $\langle t_1, t_2, x_1, x_2, J_0, L_0, T \rangle$ for the pair $\langle x, y \rangle$ with the property that both x_1 and x_2 lie in V. Because of the way in which t_1 and t_2 are defined, we have

$$|t_1 - t_2| \le \mathrm{p}_\varphi(y) + 2\delta_1,$$

and consequently, by the choice of V, we have $J_0 \subseteq A$. What we need to show is that the annulus between J_0 and $\mathcal{O}_\varphi(y)$ is contained in A. (This is because $A \subseteq U$.) To show this, we need only show that J_0 separates $\partial(A)$.

In the case $J = \mathcal{O}_\varphi(z)$, it is clear that both $\bar\alpha_\varphi(x)$ and $\bar\omega_\varphi(x)$ are contained in $\bar{A}$, and therefore do not meet $\mathscr{F}(\varphi) \cup (\mathbf{S}^2 \setminus \varOmega)$. It follows from Theorem 2.13 and 1.41 that no point in $\bar\alpha_\varphi(x) \cup \bar\omega_\varphi(x)$ can have infinite period, either. This shows that $\bar\alpha_\varphi(x) \cup \bar\omega_\varphi(x)$ is a union of periodic φ-orbits. (We shall show later that it is a union of exactly two periodic φ-orbits, but this information is not needed here; see Theorem 2.18.) But each of these periodic φ-orbits, being contained in $\bar{A}$, is also contained in $N\big(\mathcal{O}_\varphi(y), \eta\big)$, and consequently must either coincide with $\mathcal{O}_\varphi(y)$, or coincide with $\mathcal{O}_\varphi(z)$, or separate $\partial(\mathrm{A})$. But $\bar\alpha_\varphi(x)$ and $\bar\omega_\varphi(x)$ lie in opposite sides of J_0, and it therefore follows at once that J_0 separates $\partial(A)$.

In the case $\widetilde{W}_1$ contains no point inside $\mathcal{O}_\varphi(y)$ with finite period, we note once again that $\bar{\alpha}_\varphi(x)$ and $\bar{\omega}_\varphi(x)$ lie in opposite sides of J_0, and therefore, that if J_0 does not separate $\partial(A)$, then one of the sets $\bar{\alpha}_\varphi(x)$, $\bar{\omega}_\varphi(x)$, must be contained in A. But if $\bar{\alpha}_\varphi(x)$ (say) is contained in A, then every point in $\bar{\alpha}_\varphi(x)$ has infinite φ-period, which, by Theorems 2.13 and 1.41, is impossible. Therefore, in this case too, J_0 separates $\partial(A)$.

Now using the same techniques that were used to define W_2, choose a neighborhood W_3 of y, which has the property that, for any $x \in W_3$, if $p_\varphi(x) = \infty$ and $x \in \text{outs}\big(\mathcal{O}_\varphi(y)\big)$, then there exists a φ-gate construction $\langle J_0, L_0, T \rangle$ for the pair $\langle x, y \rangle$ such that the annulus between J_0 and $\mathcal{O}_\varphi(y)$ is contained in U.

Define $W = W_1 \cap W_2 \cap W_3$. It is clear that W has the required properties.   **QED**

2.17. Theorem. *Let φ be a flow in a subregion Ω of $\mathbb{R}^2$, let $x, y \in \Omega$, and suppose $\mathcal{O}_\varphi(x)$ α-spirals [respectively ω-spirals] to $\mathcal{O}_\varphi(y)$. Let U be a neighborhood of $\bar{\alpha}_\varphi(x)$ [respectively $\bar{\omega}_\varphi(x)$].*

Then there exists a φ-gate construction $\langle J_0, L_0, T \rangle$ for the pair $\langle x, y \rangle$ such that the annulus between J_0 and $\bar{\alpha}_\varphi(x)$ [respectively $\bar{\omega}_\varphi(x)$] is contained in U.

PROOF: Suppose, to be precise, that $\mathcal{O}_\varphi(x)$ α-spirals to $\mathcal{O}_\varphi(y)$. Using Theorem C.30, choose a Jordan curve J in the disc $[x \text{ side} : \bar{\alpha}_\varphi(x)]$, such that the annulus A between J and $\bar{\alpha}_\varphi(x)$ is contained in U, and does not meet $\bar{\omega}_\varphi(x)$. It is easy to see that we can choose t_0 such that

$$\varphi(t, x) \in A \quad \text{whenever } t \leq t_0.$$

Let V be a convex neighborhood of y which does not meet J, and choose a φ-gate construction $\langle t_1, t_2, x_1, x_2, J_0, L_0, T \rangle$ for the pair $\langle x, y \rangle$ in such a way that $x_1 \in V$, $x_2 \in V$, and $t_1 < t_2 < t_0$. Clearly $J_0 \subseteq A$. To see that the annulus between J_0 and $\bar{\alpha}_\varphi(x)$ is contained in U, we need only show that J_0 separates $\partial(A)$. But this is obvious because, on the one hand, $\bar{\alpha}_\varphi(x)$ and $\bar{\omega}_\varphi(x)$ lie in opposite sides of J_0, and on the other hand, J and $\bar{\omega}_\varphi(x)$ lie in the same side of J_0.   **QED**

2.18. Theorem. *Let φ be a flow in an open subset Ω of $\mathbf{S}^2$, let $x \in \Omega$, and suppose $p_\varphi(x) = \infty$.*

(a) *Either*

   (i) *there exists $y \in \Omega$ such that $0 < p_\varphi(y) < \infty$, and*

   $$\bar{\alpha}_\varphi(x) = \alpha_\varphi(x) = \mathcal{O}_\varphi(y),$$

   or

   (ii) *$\bar{\alpha}_\varphi(x)$ contains no periodic φ-orbit, and $\bar{\alpha}_\varphi(x)$ meets*

   $$\mathcal{F}(\varphi) \cup (\mathbf{S}^2 \setminus \Omega).$$

(b) *Either*

   (i) *there exists* $y \in \Omega$ *such that* $0 < p_\varphi(y) < \infty$, *and*

   $$\bar{\omega}_\varphi(x) = \omega_\varphi(x) = \mathcal{O}_\varphi(y),$$

   or

   (ii) $\bar{\omega}_\varphi(x)$ *contains no periodic* φ-*orbit, and* $\bar{\omega}_\varphi(x)$ *meets*

   $$\mathscr{F}(\varphi) \cup (\mathbf{S}^2 \setminus \Omega).$$

PROOF: We shall prove (a); the proof of (b) is analogous. Suppose there exists $y \in \Omega$ such that $0 < p_\varphi(y) < \infty$, and $\mathcal{O}_\varphi(y) \subseteq \bar{\alpha}_\varphi(x)$. Then $\bar{\alpha}_\varphi(x) = \mathcal{O}_\varphi(y)$, for otherwise, we can choose $z \in \bar{\alpha}_\varphi(x) \setminus \mathcal{O}_\varphi(y)$. Now since z and x lie in the same side of $\mathcal{O}_\varphi(y)$, we can apply Theorem 2.15 to find a φ-gate construction $\langle J_0, L_0, T \rangle$ for the pair $\langle x, y \rangle$ such that J_0 lies so close to $\mathcal{O}_\varphi(y)$, that y and z lie in opposite sides of J_0. But this is impossible, since J_0 does not meet $\bar{\alpha}_\varphi(x)$, and $\bar{\alpha}_\varphi(x)$ is connected.

To complete the proof, we therefore need only show that if $\bar{\alpha}_\varphi(x)$ contains no periodic φ-orbits, then $\bar{\alpha}_\varphi(x)$ meets $\mathscr{F}(\varphi) \cup (\mathbf{S}^2 \setminus \Omega)$. But this is obvious, for if $z \in \bar{\alpha}_\varphi(x) \setminus (\mathscr{F}(\varphi) \cup (\mathbf{S}^2 \setminus \Omega))$, we have $z \in \alpha_\varphi(x) \cap \mathscr{G}(\varphi)$, and therefore, firstly, $\square \neq \bar{\alpha}_\varphi(z) \subseteq \overline{\mathcal{O}_\varphi(z)} \subseteq \bar{\alpha}_\varphi(x)$, and secondly, Theorem 2.13 implies that $\alpha_\varphi(z)$ cannot meet $\mathscr{G}(\varphi)$, and therefore that

$$\bar{\alpha}_\varphi(z) \subseteq \mathscr{F}(\varphi) \cup (\mathbf{S}^2 \setminus \Omega). \qquad \textbf{QED}$$

2.19. Theorem. *Let* φ *be a flow in a subregion* Ω *of* $\mathbb{R}^2$, *let* $x, y \in \Omega$, *and suppose* $\mathcal{O}_\varphi(x)$ ω-*spirals to* $\mathcal{O}_\varphi(y)$. *Let* $\langle J_0, L_0, T \rangle$ *be a* φ-*gate construction for the pair* $\langle x, y \rangle$.

(a) *For any* $z \in T$, *we have either* $\bar{\omega}_\varphi(z) = \bar{\omega}_\varphi(x)$ *or* $\bar{\omega}_\varphi(z) \cap \bar{\omega}_\varphi(x) = \square$.

(b) *If there exists* $z \in T$ *such that* $\bar{\omega}_\varphi(z) \cap \bar{\omega}_\varphi(x) = \square$, *then there exists* $z' \in T$ *such that* $\mathcal{O}_\varphi(z')$ *does not* ω-*spiral*.

(c) *The following are equivalent:*

   (i) $\bar{\omega}_\varphi(z) = \bar{\omega}_\varphi(x) \ \bigvee \ z \in T$.

   (ii) $\bar{\omega}_\varphi(x)$ *is a component of* $\partial(T)$.

   (iii) $T = \{z \mid z \in [x \text{ side} : \bar{\omega}_\varphi(x)] \text{ and } \bar{\omega}_\varphi(z) = \bar{\omega}_\varphi(x)\}$.

   (iv) $[\bar{\omega}_\varphi(x) \text{ side} : J_0] \cap [x \text{ side} : \bar{\omega}_\varphi(x)]$ *does not meet* $\mathscr{F}(\varphi) \cup (\mathbf{S}^2 \setminus \Omega)$.

   (v) $[\bar{\omega}_\varphi(x) \text{ side} : J_0] \cap [x \text{ side} : \bar{\omega}_\varphi(x)]$ *does not meet* $\mathscr{S}(\varphi) \cup (\mathbf{S}^2 \setminus \Omega)$.

PROOF: To prove (a), we shall first show that if $z \in L_0$, then either $\bar{\omega}_\varphi(z) \subseteq \bar{\omega}_\varphi(x)$, or $\bar{\omega}_\varphi(z) \cap \bar{\omega}_\varphi(x) = \square$. Suppose $z \in L_0$, and $\bar{\omega}_\varphi(z) \setminus \bar{\omega}_\varphi(x) \neq \square$. Choose $u \in \bar{\omega}_\varphi(z) \setminus \bar{\omega}_\varphi(x)$. Using Theorem 2.17, choose a φ-gate construction $\langle \tilde{J}_0, \tilde{L}_0, \tilde{T} \rangle$ for the pair $\langle x, y \rangle$ such that

$\overline{\omega}_\varphi(x)$ and u lie in opposite sides of $\tilde{J}_0$. Now since $\overline{\omega}_\varphi(z)$, being connected and disjoint from $\tilde{J}_0$, must lie wholly in one side of $\tilde{J}_0$, we see that $\overline{\omega}_\varphi(x)$ and $\overline{\omega}_\varphi(z)$ lie in opposite sides of $\tilde{J}_0$, and consequently

$$\overline{\omega}_\varphi(x) \cap \overline{\omega}_\varphi(z) = \square.$$

Next, we shall show that if $z \in L_0$, and $\overline{\omega}_\varphi(z) \subseteq \overline{\omega}_\varphi(x)$, then $y \in \overline{\omega}_\varphi(z)$. Suppose $z \in L_0$, and $\overline{\omega}_\varphi(z) \subseteq \overline{\omega}_\varphi(x)$, and let V be any neighborhood of y. Choose a φ-gate construction $\langle \tilde{J}_0, \tilde{L}_0, \tilde{T} \rangle$ for the pair $\langle x, y \rangle$ such that $\tilde{L}_0 \subseteq V$, and the sets J_0 and $\overline{\omega}_\varphi(x)$ lie in opposite sides of $\tilde{J}_0$. Since $\mathcal{O}_\varphi(z)$ is connected, and must contain points in both sides of $\tilde{J}_0$, we have

$$\mathcal{O}_\varphi(z) \cap \tilde{J}_0 \neq \square, \qquad i.e., \qquad \mathcal{O}_\varphi(z) \cap \tilde{L}_0 \neq \square.$$

This shows that $\mathcal{O}_\varphi(z)$ contains points in each neighborhood of y, and therefore, since $\overline{\alpha}_\varphi(z)$ and y lie in opposite sides of J_0, we must have $y \in \overline{\omega}_\varphi(z)$.

Now to complete the proof of (a), suppose $z \in T$ and $\overline{\omega}_\varphi(z) \cap \overline{\omega}_\varphi(x) \neq \square$. In order to show that $\overline{\omega}_\varphi(z) = \overline{\omega}_\varphi(x)$, it is sufficient to assume that $z \in L_0$. By what we have proved above,

$$y \in \overline{\omega}_\varphi(z) \subseteq \overline{\omega}_\varphi(x).$$

Choose a φ-gate construction $\langle \tilde{J}_0, \tilde{L}_0, \tilde{T} \rangle$ for the pair $\langle z, y \rangle$ in such a way that J_0 and $\overline{\omega}_\varphi(z)$ lie in opposite sides of $\tilde{J}_0$. Since $\mathcal{O}_\varphi(x)$ is connected, and must contain points in both sides of $\tilde{J}_0$, we must have $\mathcal{O}_\varphi(x) \cap \tilde{J}_0 \neq \square$, i.e., $x \in \tilde{T}$. Therefore, by the above arguments, with the roles of x and z interchanged, we have, since $\overline{\omega}_\varphi(x) \cap \overline{\omega}_\varphi(z) \neq \square$, that $\overline{\omega}_\varphi(x) \subseteq \overline{\omega}_\varphi(z)$. This shows that $\overline{\omega}_\varphi(x) = \overline{\omega}_\varphi(z)$, and (a) is proved.

To prove (b), choose $z \in T$ such that $\overline{\omega}_\varphi(z) \cap \overline{\omega}_\varphi(x) = \square$, and choose a φ-gate construction $\langle \tilde{J}_0, \tilde{L}_0, \tilde{T} \rangle$ for the pair $\langle x, y \rangle$ such that $\overline{\omega}_\varphi(x)$ and $J_0 \cup \overline{\omega}_\varphi(z)$ lie in opposite sides of $\tilde{J}_0$. Then since $x \in \tilde{T}$, $z \notin \tilde{T}$, and T is connected, T must contain a point $z' \in \partial(\tilde{T})$. Now since $\tilde{J}_0 \cup T$ is connected, and does not meet $\overline{\omega}_\varphi(z')$, it is clear that both J_0 and $\tilde{J}_0$ are contained in $[z' \text{ side} : \overline{\omega}_\varphi(z')]$. Therefore if $\mathcal{O}_\varphi(z')$ ω-spirals, and y' is chosen in $\omega_\varphi(z') \cap \mathcal{G}(\varphi)$, we can choose a φ-gate construction $\langle J_0' L_0' T' \rangle$ for the pair $\langle z', y' \rangle$ such that $J_0 \cup \tilde{J}_0$ and $\overline{\omega}_\varphi(z')$ lie in opposite sides of J_0'. In particular, $\overline{\omega}_\varphi(z')$ and $\tilde{J}_0$ lie in opposite sides of J_0'. Now since J_0 and $\overline{\omega}_\varphi(x)$ lie in opposite sides of $\tilde{J}_0$, and $\mathcal{O}_\varphi(z')$ meets J_0, and $\mathcal{O}_\varphi(z')$ does not meet $\tilde{J}_0$, it is clear that J_0' and $\overline{\omega}_\varphi(x)$ lie in opposite sides of $\tilde{J}_0$. From this, it follows that the discs $[\overline{\omega}_\varphi(x) \text{ side} : \tilde{J}_0]$ and $[\overline{\omega}_\varphi(z') \text{ side} : J_0']$ are mutually disjoint. We therefore have $\tilde{T} \cap T' = \square$, for if $u \in \tilde{T} \cap T'$, then

$$\overline{\omega}_\varphi(u) \in [\overline{\omega}_\varphi(x) \text{ side} : \tilde{J}_0] \cap [\overline{\omega}_\varphi(z') \text{ side} : J_0'],$$

which is impossible. But T' is open, and $z' \in T' \cap \partial(\tilde{T})$, and we therefore have a contradiction to the assumption that $\mathcal{O}_\varphi(z')$ ω-spirals. This completes the proof of (b).

We shall prove (c) by showing that, firstly, (i) $\Rightarrow$ (ii) $\Rightarrow$ (iv) $\Rightarrow$ (v) $\Rightarrow$ (i), secondly, (iii) $\Rightarrow$ (i), and thirdly, (i) and (ii) $\Rightarrow$ (iii).

(c) : (i) $\Rightarrow$ (ii).

Let B be the component of $\partial(T)$ that contains y. We know that $\overline{\omega}_\varphi(x) \subseteq \partial(T)$, and thus that $\overline{\omega}_\varphi(x) \subseteq B$. Suppose that $\overline{\omega}_\varphi(z) = \overline{\omega}_\varphi(x)$, $\forall z \in T$, and suppose we can find $u \in B \setminus \overline{\omega}_\varphi(x)$. Choose a sequence $\{t_n\}$ of real numbers, and a sequence $\{z_n\}$ of points of L_0, such that $u = \lim_{n\to\infty} \varphi(t_n, z_n)$. By taking a subsequence, if necessary, (using the compactness of L_0), we may assume that $z_n \to z \in L_0$. It is clear that $t_n \to \infty$, for by Theorem 2.4 (d), $\{t_n\}$ cannot have $-\infty$ as a cluster-point, and since $u \notin T$, $\{t_n\}$ cannot have a finite cluster-point. Choose a φ-gate construction $\langle \tilde{J}_0, \tilde{L}_0, \tilde{T} \rangle$ for the pair $\langle z, y \rangle$ such that $\overline{\omega}_\varphi(z)$ and $J_0 \cup \{u\}$ lie in opposite sides of $\tilde{J}_0$. It is easy to see, by Theorem 2.4 (f), that there exists a neighborhood U of z, and a number t, such that

$$\varphi(t', z') \in [\overline{\omega}_\varphi(z) \text{ side} : \tilde{J}_0] \text{ whenever } t' \geq t \text{ and } z' \in U.$$

But this implies that $\varphi(t_n, z_n) \in [\overline{\omega}_\varphi(z) \text{ side} : \tilde{J}_0]$ for n sufficiently large, which is clearly impossible.

(c) : (ii) $\Rightarrow$ (iv):

Since (ii) implies that the annulus $[\overline{\omega}_\varphi(x) \text{ side} : J_0] \cap [x \text{ side} : \overline{\omega}_\varphi(x)]$ is a subset of T, (iv) follows at once from (ii).

(c) : (iv) $\Rightarrow$ (v): Trival.

(c) : (v) $\Rightarrow$ (i):

If (i) is false, then we can choose $z' \in T$ such that $\mathcal{O}_\varphi(z')$ does not ω-spiral. But

$$\overline{\omega}_\varphi(z') \subseteq \mathcal{S}(\varphi) \cup (\mathbf{S}^2 \setminus \Omega),$$

and also

$$\overline{\omega}_\varphi(z') \subseteq [\overline{\omega}_\varphi(x) \text{ side} : J_0] \cap [x \text{ side} : \overline{\omega}_\varphi(x)],$$

and therefore (v) is also false.

(c) : (iii) $\Rightarrow$ (i). Trivial.

Finally, we show that (i) and (ii) together imply (iii). Assume that (i) and (ii) hold. Then clearly

$$T \subseteq \{z \mid z \in [x \text{ side} : \overline{\omega}_\varphi(x)] \text{ and } \overline{\omega}_\varphi(z) = \overline{\omega}_\varphi(x)\},$$

and on the other hand, if $z \in [x \text{ side} : \overline{\omega}_\varphi(x)]$, and $\overline{\omega}_\varphi(z) = \overline{\omega}_\varphi(x)$, then choosing an integer m satisfying the condition in Theorem 2.4(c), and defining J_n, for each integer n, accordingly, it follows easily from (ii), that for n sufficiently large, z and y lie in opposite sides of J_n. Therefore, for n sufficiently large, $\mathcal{O}_\varphi(z)$ contains points in both sides of J_n, and consequently meets J_n, from which it follows that $z \in T$.

This completes the proof of the theorem.   **QED**

2.20. Theorem. *Let φ be a flow in a subregion Ω of $\mathbb{R}^2$, let $x, y \in \Omega$ and suppose $\mathcal{O}_\varphi(x)$ α-spirals to $\mathcal{O}_\varphi(y)$. Let $\langle J_0, L_0, T \rangle$ be a φ-gate construction for the pair $\langle x, y \rangle$.*

Then results analogous to (a), (b) *and* (c) *of Theorem 2.19 hold with the symbol $\overline{\omega}_\varphi$ replaced by $\overline{\alpha}_\varphi$.*

PROOF: The proof is similar to that of 2.19.   **QED**

2.21. Definition. Let φ be a flow in a subregion Ω of $\mathbb{R}^2$, let $x, y \in \Omega$, and suppose $\mathcal{O}_\varphi(x)$ α-spirals [respectively ω-spirals] to $\mathcal{O}_\varphi(y)$. A φ-gate construction $\langle J_0, L_0, T \rangle$ for the pair $\langle x, y \rangle$ is said to be *unobstructed* if for all $z \in L_0$, we have $\overline{\alpha}_\varphi(z) = \overline{\alpha}_\varphi(x)$ [respectively $\overline{\omega}_\varphi(z) = \overline{\omega}_\varphi(x)$], and is otherwise said to be *obstructed*.

2.22. Illustrative Figure. Fig. 2.4 illustrates an obstructed gate construction.

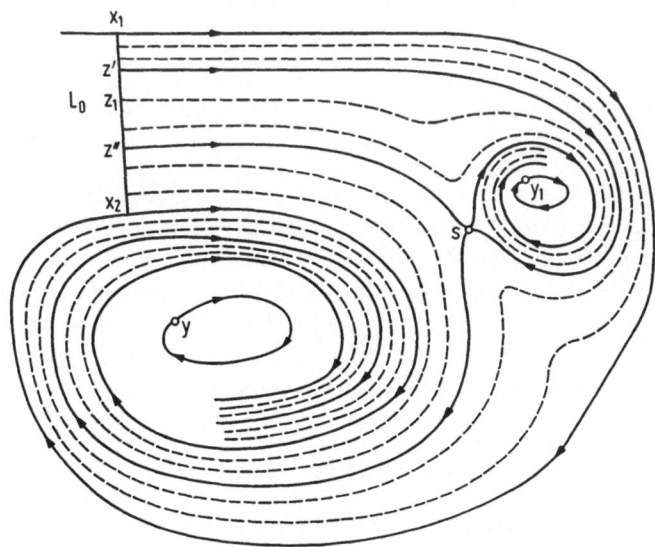

Fig. 2.4

2.23. Theorem. *Let φ be a flow in an open subset Ω of $\mathbf{S}^2$, let $y \in \Omega$, and suppose $0 < \mathrm{p}_\varphi(y) < \infty$.*

Then

$$\{x \in \Omega \mid \overline{\omega}_\varphi(x) = \mathcal{O}_\varphi(y)\}$$

and

$$\{x \in \Omega \mid \overline{\alpha}_\varphi(x) = \mathcal{O}_\varphi(y)\}$$

are open.

PROOF: We shall prove that

$$\{x \in \Omega \mid \overline{\omega}_\varphi(x) = \mathcal{O}_\varphi(y)\}$$

is open. The second result is similar.

The result is trivial if this set is empty. Otherwise, choose $x \in \Omega$ such that $\overline{\omega}_\varphi(x) = \mathcal{O}_\varphi(y)$. Suppose without loss of generality that Ω is connected, and in order to regard φ as a flow in a subregion of $\mathbb{R}^2$, choose a point in $\mathscr{F}(\varphi) \cup (\mathbf{S}^2 \setminus \Omega)$ to play the role of ∞. Using Theorem 2.15, choose a φ-gate construction for the pair $\langle x, y \rangle$ with J_0 so close to $\mathcal{O}_\varphi(y)$, that the annulus between J_0 and $\mathcal{O}_\varphi(y)$ is contained in $\mathscr{G}(\varphi)$. It follows from Theorem 2.19(c) that $\langle J_0, L_0, T \rangle$ is unobstructed, and consequently, that

$$\{z \mid z \in [x \text{ side} : \mathcal{O}_\varphi(y)] \text{ and } \overline{\omega}_\varphi(z) = \mathcal{O}_\varphi(y)\} = T,$$

which is open.

We have therefore shown that the intersection of

$$\{x \in \Omega \mid \overline{\omega}_\varphi(x) = \mathcal{O}_\varphi(y)\}$$

with each side of $\mathcal{O}_\varphi(y)$, is open, and this completes the proof of the theorem.   **QED**

2.24. Theorem. *Let φ be a flow in a subregion Ω of $\mathbb{R}^2$, let $x, y \in \Omega$, and suppose $\mathcal{O}_\varphi(x)$ α-spirals [respectively ω-spirals] to $\mathcal{O}_\varphi(y)$. Let $\langle J_0, L_0, T \rangle$ be a φ-gate construction for the pair $\langle x, y \rangle$, and let the point $x' \in T$ satisfy $\overline{\alpha}_\varphi(x') = \overline{\alpha}_\varphi(x)$ [respectively $\overline{\omega}_\varphi(x') = \overline{\omega}_\varphi(x)$]. Let y' be a moving φ-α-point [respectively φ-ω-point] of x', and let $\langle J_0', L_0', T' \rangle$ be a φ-gate construction for the pair $\langle x' y' \rangle$. Let B be the component of $\partial(T)$ which contains y, and let B' be the component of $\partial(T')$ which contains y'.*

Then we have

$$B = B' \text{ iff } T = T'.$$

PROOF: It is obvious that, if $T = T'$, then $B = B'$. Suppose that $B = B'$. Choose an integer m satisfying the requirements of Theorem

2.4 (c) with respect to $\langle J_0, L_0, T \rangle$, and define $\{ J_n \mid n$ is an integer$\}$ accordingly. Similarly, choose an integer m' satisfying the requirements of Theorem 2.4 (c) with respect to $\langle J_0', L_0', T' \rangle$, and define $\{ J_n' \mid n$ is an integer$\}$ accordingly.

Now since $[B$ side$: J_0]$ is a neighborhood of B', we can choose an integer n_1 such that $J_{n_1}' \subseteq [B$ side$: J_0]$. Since each of the Jordan curves J_0 and J_{n_1}' satisfies that $\bar{\alpha}_\varphi (x)$ lies in one side, and $\bar{\omega}_\varphi (x)$ lies in the other, it is easy to see that J_0 and B lie in opposite sides of J_{n_1}'. Since $[B$ side $: J_{n_1}']$ is a neighborhood of B, we can choose an integer n_2 such that $J_{n_2} \subseteq [B$ side$: J_{n_1}']$. Since J_0 and J_{n_2} lie in opposite sides of J_{n_1}', it follows that $T \subseteq T'$.

We can see similarly, by choosing an integer n_3 such that $J_{n_3}' \subseteq [B$ side$: J_{n_2}]$, that $T' \subseteq T$. This completes the proof of the theorem.   **QED**

2.25. Corollary. *Let φ be a flow in a subregion Ω of $\mathbb{R}^2$, let $x, y \in \Omega$, and suppose $\mathcal{O}_\varphi (x)$ spirals to $\mathcal{O}_\varphi (y)$. Let $\langle J_0, L_0, T \rangle$ and $\langle J_0', L_0', T' \rangle$ be two φ-gate constructions for the pair $\langle x, y \rangle$, and let B and B' be, respectively, the components of $\partial (T)$ and $\partial (T')$ which contain y.*

Then we have

$$B = B' \; iff \; T = T',$$

a fortiori, if $\langle J_0, L_0, T \rangle$ and $\langle J_0', L_0', T' \rangle$ are both unobstructed, we have $T = T'$.

PROOF: This follows at once from Theorem 2.24.   **QED**

Notes and Remarks to Chapter 2

The principal result of Chapter 2 is, of course, the Gate Theorem, which is an extension and sharpening of certain "folk theorems" concerning the nature of a plane flow in the vicinity of a periodic orbit. Many "proofs" of such theorems have been offered, including some actual proofs, but statements of these theorems are even harder to find than proofs, and the best known sources rely heavily on intuition and have gaps of rather large size.

The reader should notice that the Gate Theorem really consists of two theorems, one for condition (a) of Theorem 2.3, and one for condition (b). The Gate Theorem in Beck [2] (Lemma 4) is of the (b)-type, and corresponds to Theorem 2.15. Beck [3] contains an (a)-type Gate Theorem which is used to deduce much of the material stated here in Theorems 2.4, 2.10, 2.11, 2.12, 2.13 and 2.14.

The formulation of the Gate Theorem given here, which is due to the Lewins, combines the Gate Theorem in Beck [2] and Beck [3], states them precisely, and increases their generality.

Sections 2.19 to 2.25 are joint work, developed together with the Lewins, and of a later vintage than the rest of the chapter. These sections arise out of the observation made by the Lewins that if $\langle J_0, L_0, T \rangle$ and $\langle J_0', L_0', T' \rangle$ are gate constructions for a pair $\langle x, y \rangle$, then we do not always have $T = T'$. In view of this observation, it was necessary to make a detailed study of the conditions under which the annulus T is unique.

CHAPTER THREE

REGULAR AND SINGULAR POINTS

3.1. Introduction. Suppose φ is a flow in $\mathbf{S}^2$, and $\mathscr{S}(\varphi) = \square$. As we shall show later, the behavior of φ is very simple. In fact, if Ω is any component of $\mathscr{G}(\varphi)$, then Ω is an annulus, and if $x \in \Omega$, then either

(i) $\mathscr{O}_\varphi(x)$ is a Jordan curve which separates $\partial(\Omega)$,

or

(ii) $p_\varphi(x) = \infty$, and there exist z_1 and z_2 in Ω, such that $\mathscr{O}_\varphi(z_1)$ and $\mathscr{O}_\varphi(z_2)$ are Jordan curves, x belongs to the annulus A between $\mathscr{O}_\varphi(z_1)$ and $\mathscr{O}_\varphi(z_2)$, and each point $x' \in A$ satisfies that $\mathscr{O}_\varphi(x')$ α-spirals to $\mathscr{O}_\varphi(z_1)$ and ω-spirals to $\mathscr{O}_\varphi(z_2)$.

More generally, every flow φ in $\mathbf{S}^2$ exhibits similar simple behavior in that part of $\mathbf{S}^2$ (which we call the φ-regular part of $\mathbf{S}^2$) which is somehow divorced from $\mathscr{S}(\varphi)$. In the rest of $\mathbf{S}^2$ (which we call the φ-singular part of $\mathbf{S}^2$) the behavior of φ is quite different, and much more complicated.

In this chapter, we shall study some of the behavior of φ in its regular and singular parts, and the way these parts fit together.

3.2. Definition. Let φ be a flow in $\mathbf{S}^2$. We define the φ-singular part $\mathscr{F}_s(\varphi)$ of $\mathscr{F}(\varphi)$ by

$$\mathscr{F}_s(\varphi) = \mathscr{C}\left(\mathscr{F}(\varphi), \overline{\mathscr{S}(\varphi)}\right),$$

i.e., $\mathscr{F}_s(\varphi)$ is the union of those components of $\mathscr{F}(\varphi)$ that meet $\overline{\mathscr{S}(\varphi)}$. We define the φ-regular part $\mathscr{F}_r(\varphi)$ of $\mathscr{F}(\varphi)$ to be $\mathscr{F}(\varphi) \setminus \mathscr{F}_s(\varphi)$.

3.3. Definition. Let φ be a flow in $\mathbf{S}^2$. A periodic φ-orbit, $\mathscr{O}_\varphi(x)$, is said to be φ-singular on a certain side if, in that side, there are points aribtrarily close to $\mathscr{O}_\varphi(x)$ whose orbits have φ-stagnation points. If a periodic φ-orbit is not φ-singular on a certain side, it is said to be φ-regular on that side. A periodic φ-orbit is said to be φ-regular if it is φ-regular on both sides; otherwise it is said to be φ-singular.

An aperiodic φ-orbit, $\mathcal{O}_\varphi(x)$, is said to be φ-*regular* if both $\alpha_\varphi(x)$ and $\omega_\varphi(x)$ are periodic φ-orbits, and each of them is φ-regular on the side which contains $\mathcal{O}_\varphi(x)$. If $\mathcal{O}_\varphi(x)$ is not φ-regular, it is said to be φ-*singular*.

We define the φ-*singular part* $\mathcal{G}_s(\varphi)$ of $\mathcal{G}(\varphi)$ by

$$\mathcal{G}_s(\varphi) = \{x \in \mathcal{G}(\varphi) \mid \mathcal{O}_\varphi(x) \text{ is } \varphi\text{-singular}\},$$

and we define the φ-*regular part* $\mathcal{G}_r(\varphi)$ of $\mathcal{G}(\varphi)$ to be $\mathcal{G}(\varphi) \setminus \mathcal{G}_s(\varphi)$.

3.4. Definition. Let φ be a flow in $\mathbf{S}^2$. $\mathscr{F}_s(\varphi) \cup \mathscr{G}_s(\varphi)$ is called the φ-*singular part* of $\mathbf{S}^2$, and a member of $\mathscr{F}_s(\varphi) \cup \mathscr{G}_s(\varphi)$ is called a φ-*singular point*. $\mathscr{F}_r(\varphi) \cup \mathscr{G}_r(\varphi)$ is called the φ-*regular part* of $\mathbf{S}^2$, and a member of $\mathscr{F}_r(\varphi) \cup \mathscr{G}_r(\varphi)$ is called a φ-*regular point*.

3.5. Theorem. *Let φ be a flow in $\mathbf{S}^2$, let $x \in \mathbf{S}^2$, and suppose $p_\varphi(x) = \infty$. Then $x \in \mathcal{G}_s(\varphi)$ iff either*

(i) *$\mathcal{O}_\varphi(x)$ has a φ-stagnation point,*

or

(ii) *at least one of the sets $\alpha_\varphi(x)$, $\omega_\varphi(x)$ is a periodic φ-orbit which is φ-singular on the side containing $\mathcal{O}_\varphi(x)$.*

PROOF: The result follows at once from Theorem 2.18.   **QED**

3.6. Theorem. *Let φ be a flow in $\mathbf{S}^2$. Then $\mathscr{F}_s(\varphi)$ is compact.*

PROOF: The result follows at once from Theorem A.10.   **QED**

Structure of $\mathcal{G}_r(\varphi)$

3.7. Theorem. *Let φ be a flow in $\mathbf{S}^2$, let $x, y \in \mathbf{S}^2$, suppose $\mathcal{O}_\varphi(x)$ spirals to $\mathcal{O}_\varphi(y)$, and suppose $\mathcal{O}_\varphi(y)$ is a periodic φ-orbit, which is φ-regular on the side containing $\mathcal{O}_\varphi(x)$.*

(a) *Both $\alpha_\varphi(x)$ and $\omega_\varphi(x)$ are periodic φ-orbits, which are φ-regular on the side containing $\mathcal{O}_\varphi(x)$.*

(b) *Let A be the annulus between $\alpha_\varphi(x)$ and $\omega_\varphi(x)$. Then if $z \in A$, we have*

$$\alpha_\varphi(z) = \alpha_\varphi(x) \text{ and } \omega_\varphi(z) =: \omega_\varphi(x).$$

(c) *If $\bar{y} \in \alpha_\varphi(x) \cup \omega_\varphi(x)$, and $\langle \bar{J}_0, \bar{L}_0, \bar{T} \rangle$ is a φ-gate construction for the pair $\langle x, \bar{y} \rangle$, then $\langle \bar{J}_0, \bar{L}_0, \bar{T} \rangle$ is unobstructed, and $\bar{T} = A$.*

PROOF: In order to regard φ as a flow in $\mathbb{R}^2$, choose a point in $\mathscr{F}(\varphi)$ to play the role of ∞.

We may suppose that $\mathcal{O}_\varphi(x)$ ω-spirals to $\mathcal{O}_\varphi(y)$; the other case is dual.

Let $\langle J_0, L_0, T \rangle$ be an unobstructed φ-gate construction for the pair $\langle x, y \rangle$. Then $T \subseteq A$, and

$$T = \{z \in [x \text{ side} : \mathcal{O}_\varphi(y)] \mid \omega_\varphi(z) = \mathcal{O}_\varphi(y)\},$$

and since $\mathcal{O}_\varphi(y)$ is φ-regular on the side containing T, we see that, for each $z \in T$, $\alpha_\varphi(z)$ is a periodic φ-orbit. For each $z \in T$, define

$$U_z = \{z' \in T \mid \alpha_\varphi(z') = \alpha_\varphi(z)\}.$$

By Theorem 2.23, U_z is open for each $z \in T$, and it is therefore easy to see that $\{U_z \mid z \in T\}$ is a pairwise-disjoint family of non-empty open subsets of T. It follows, since T is connected, that

$$\alpha_\varphi(z) = \alpha_\varphi(x), \quad \forall z \in T.$$

Now choose $y' \in \alpha_\varphi(x)$, and let $\langle J_0', L_0', T' \rangle$ be an unobstructed φ-gate construction for the pair $\langle x, y' \rangle$. Then $T' \subseteq A$,

$$T' = \{z \in [x \text{ side} : \alpha_\varphi(x)] \mid \alpha_\varphi(z) = \alpha_\varphi(x)\},$$

and $\alpha_\varphi(x)$ is a component of $\partial(T')$. Furthermore, the fact that $\alpha_\varphi(z) = \alpha_\varphi(x)$ for all $z \in T$ implies that $T \subseteq T'$.

It is clear, by Theorem 2.11, that $\omega_\varphi(x) \subseteq \partial(T')$. But the annulus between J_0 and $\omega_\varphi(x)$ is contained in T, and is therefore contained in T'. It follows that $\omega_\varphi(x)$ must be a whole component of $\partial(T')$, and from this we deduce that $T' = A$. Again, by Theorem 2.11, we have

$$\omega_\varphi(z) = \omega_\varphi(x), \quad \forall z \in T',$$

and from this it follows that $T' \subseteq T$, i.e., $T' = T = A$. The theorem now follows easily. **QED**

3.8. Corollary. *Let φ be a flow in $\mathbf{S}^2$, let $x \in \mathbf{S}^2$, and suppose $\mathcal{O}_\varphi(x)$ spirals to a periodic φ-orbit which is φ-regular on the side containing $\mathcal{O}_\varphi(x)$.*

Then there exists an annulus A, which is invariant of φ, whose boundary consists of two periodic φ-orbits, and which satisfies

$$x \in A \subseteq \mathscr{G}_r(\varphi).$$

3.9. Theorem. *Let φ be a flow in $\mathbf{S}^2$, let $y \in \mathbf{S}^2$, and suppose $\mathcal{O}_\varphi(y)$ is periodic. In order to refer to a certain side of $\mathcal{O}_\varphi(y)$ as $\mathrm{ins}\big(\mathcal{O}_\varphi(y)\big)$, let a φ-fixed point in the other side be chosen to play the role of ∞. (Corollary 1.33 shows that such a choice is possible.)*

(a) *If there exists a neighborhood U of $\mathcal{O}_\varphi(y)$ such that $p_\varphi(z) = \infty$, $\forall \, z \in U \cap \text{ins}\big(\mathcal{O}_\varphi(y)\big)$, then there exists a neighborhood V of $\mathcal{O}_\varphi(y)$ such that $\mathcal{O}_\varphi(z)$ spirals to $\mathcal{O}_\varphi(y)$ whenever $z \in V \cap \text{ins}\big(\mathcal{O}_\varphi(y)\big)$.*

(b) *If, for each neighborhood U of $\mathcal{O}_\varphi(y)$, there exist points with finite φ-period in $U \cap \text{ins}\big(\mathcal{O}_\varphi(y)\big)$, then $\mathcal{O}_\varphi(y)$ is regular on its inside.*

(c) *If $\mathcal{O}_\varphi(y)$ is regular on its inside, then there exists a periodic φ-orbit contained in $\text{ins}\big(\mathcal{O}_\varphi(y)\big)$ such that the annulus between this φ-orbit and $\mathcal{O}_\varphi(y)$ is contained in $\mathcal{G}_r(\varphi)$.*

PROOF: To prove (a), suppose there exists a neighborhood U of $\mathcal{O}_\varphi(y)$ such that

$$p_\varphi(z) = \infty, \qquad \forall \, z \in U \cap \text{ins}\big(\mathcal{O}_\varphi(y)\big).$$

Using Theorem 2.16, choose $x \in \text{ins}\big(\mathcal{O}_\varphi(y)\big)$, so close to y that there exists a φ-gate construction $\langle J_0, L_0, T\rangle$ for the pair $\langle x, y\rangle$, with the property that the annulus A between J_0 and $\mathcal{O}_\varphi(y)$ is contained in U. Suppose, without loss of generality, that $\alpha_\varphi(x) \subseteq \text{outs}(J_0)$; the other case is dual. Now since A does not meet $\mathscr{F}(\varphi)$, Theorem 2.18 implies that $\alpha_\varphi(x)$ is a periodic φ-orbit, and we must therefore have $\alpha_\varphi(x) = \mathcal{O}_\varphi(y)$. By Theorem 2.19, $\langle J_0, L_0, T\rangle$ is unobstructed, and therefore $\mathcal{O}_\varphi(y)$ is a component of $\partial(T)$. Let $V = T \cup \mathcal{O}_\varphi(y) \cup \text{outs}\big(\mathcal{O}_\varphi(y)\big)$. Clearly, V has the required properties, and (a) is proved.

To prove (b), use Theorem 2.16 to choose $x \in \text{ins}\big(\mathcal{O}_\varphi(y)\big)$, such that $p_\varphi(x) < \infty$, and x is so close to y that the annulus A between $\mathcal{O}_\varphi(x)$ and $\mathcal{O}_\varphi(y)$ is contained in $\mathcal{G}(\varphi)$. It is clear that $A \subseteq \mathcal{G}_r(\varphi)$, and (b) is proved.

To prove (c), suppose $\mathcal{O}_\varphi(y)$ is regular on its inside. The proof of (b) shows that (c) holds if there are points of finite φ-period inside $\mathcal{O}_\varphi(y)$, arbitrarily close to y. If all points sufficiently close to y, inside $\mathcal{O}_\varphi(y)$, have infinite φ-period, then by (a), we can choose $x \in \text{ins}\big(\mathcal{O}_\varphi(y)\big)$ such that $\mathcal{O}_\varphi(x)$ spirals to $\mathcal{O}_\varphi(y)$, and in this case, (c) follows from Theorem 3.7.

QED

3.10. Corollary. *Let φ be a flow in $\mathbf{S}^2$, let $x \in \mathbf{S}^2$, suppose $p_\varphi(x) < \infty$, and suppose $\mathcal{O}_\varphi(x)$ is φ-regular.*

Then there exists an annulus A, which is invariant of φ, whose boundary is a union of two periodic φ-orbits, and which satisfies

$$x \in A \subseteq \mathcal{G}_r(\varphi).$$

PROOF: Immediate from Theorem 3.9.   **QED**

3.11. Corollary. *Let φ be a flow in $\mathbf{S}^2$. Then $\mathcal{G}_r(\varphi)$ is open.*

PROOF: Combine Corollaries 3.8 and 3.10.   **QED**

3.12. Theorem. *Let φ be a flow in $\mathbf{S}^2$, and let U be an annulus, which is invariant of φ, and is contained in $\mathscr{G}_r(\varphi)$.*

(a) *If $x \in U$ and $\mathrm{p}_\varphi(x) < \infty$, then there exist two periodic φ-orbits, one in each side of $\mathcal{O}_\varphi(x)$, with the property that the annulus between them is contained in U.*

(b) *If $x \in U$ and $\mathrm{p}_\varphi(x) = \infty$, then $\alpha_\varphi(x)$ and $\omega_\varphi(x)$ are periodic φ-orbits, and the annulus between them is contained in U.*

(c) *If B is a component of $\partial(U)$, then B is either a periodic φ-orbit, or B consists solely of φ-fixed points and aperiodic φ-orbits which do not spiral.*

PROOF: To prove (a), suppose $x \in U$ and $\mathrm{p}_\varphi(x) < \infty$. In each side of $\mathcal{O}_\varphi(x)$ there are two possibilities:

Either

There exist points with finite φ-period, lying arbitrarily close to x, in which case we can use Theorem 2.16 to choose z in this side of $\mathcal{O}_\varphi(x)$ satisfying $\mathrm{p}_\varphi(z) < \infty$, and in such a way that the annulus between $\mathcal{O}_\varphi(z)$ and $\mathcal{O}_\varphi(x)$ is contained in U.

Or

There exists a neighborhood V of $\mathcal{O}_\varphi(x)$ such that, in this side of $\mathcal{O}_\varphi(x)$, V contains no points of finite φ-period. In this case, using Theorem 2.16, choose z in this side of $\mathcal{O}_\varphi(x)$, and so close to $\mathcal{O}_\varphi(x)$ that we can find a φ-gate construction $\langle J_0, L_0, T \rangle$ for the pair $\langle z, x \rangle$ such that the annulus between J_0 and $\mathcal{O}_\varphi(x)$ is contained in V. Clearly the φ-orbit of z (and also of every other member of T) spirals to $\mathcal{O}_\varphi(x)$, and consequently, $T \subseteq U$, and by Theorem 3.7, T is the annulus between the two periodic φ-orbits, $\alpha_\varphi(z)$ and $\omega_\varphi(z)$.

(a) follows immediately from the above two observations.

To prove (b), suppose $x \in U$ and $\mathrm{p}_\varphi(x) = \infty$. It is already known that $\alpha_\varphi(x)$ and $\omega_\varphi(x)$ are periodic φ-orbits, and we need only show that the annulus A between them is contained in U. Suppose to the contrary that there exists a point $z \in A \setminus U$. Then since $\mathcal{O}_\varphi(z)$ α-spirals to $\alpha_\varphi(x)$, and ω-spirals to $\omega_\varphi(x)$, and U is open, it is clear that neither $\alpha_\varphi(x)$ nor $\omega_\varphi(x)$ is contained in U. Consequently, since U is connected, $U \subseteq A$. But if z' is any point in $A \setminus U$, then $\mathcal{O}_\varphi(z')$ connects z', in $\mathbf{S}^2 \setminus U$, to both $\alpha_\varphi(x)$ and $\omega_\varphi(x)$, and therefore $(A \setminus U) \cup \alpha_\varphi(x) \cup \omega_\varphi(x)$ is connected. This implies that U is simply-connected; contradiction. This completes the proof of (b).

To prove (c) we note first that B is compact, connected, and invariant of φ. Suppose B contains a periodic φ-orbit, $\mathcal{O}_\varphi(y)$. Clearly, all points of $B \setminus \mathcal{O}_\varphi(y)$ (if any) must lie in $[U$ side: $\mathcal{O}_\varphi(y)]$. In order to prove that $B = \mathcal{O}_\varphi(y)$, suppose to the contrary that there exists a point $w \in B \setminus \mathcal{O}_\varphi(y)$. We shall consider two cases:

CASE 1: There exist points in $[U$ side $: \mathcal{O}_\varphi(y)]$, arbitrarily close to y, and with finite φ-period. In this case, choose $z \in [U$ side $: \mathcal{O}_\varphi(y)]$, such that $p_\varphi(z) < \infty$, and such that z lies so close to y that the annulus between $\mathcal{O}_\varphi(z)$ and $\mathcal{O}_\varphi(y)$ does not contain w. Since both sides of $\mathcal{O}_\varphi(z)$ contain points of U, and since U is connected and invariant of φ, it is clear that $\mathcal{O}_\varphi(z) \subseteq U$, and from this it follows that w and $\mathcal{O}_\varphi(y)$ cannot be contained in the same component of $\partial(U)$; contradiction.

CASE 2: There exists a neighborhood V of $\mathcal{O}_\varphi(y)$ such that $V \cap [U$ side $: \mathcal{O}_\varphi(y)]$ contains no points of finite φ-period. In this case, choose $z \in U$, so close to y that we can find a φ-gate construction $\langle J_0, L_0, T \rangle$ for the pair $\langle z, y \rangle$ such that the annulus between J_0 and $\mathcal{O}_\varphi(y)$ is contained in V, and does not contain w. Clearly, $\mathcal{O}_\varphi(z)$ must spiral to $\mathcal{O}_\varphi(y)$, and (b) implies that the annulus T between $\alpha_\varphi(z)$ and $\omega_\varphi(z)$ is contained in U. Since $\mathcal{O}_\varphi(y)$ and w lie in opposite sides of J_0, and $w \notin U$, it is clear that $\mathcal{O}_\varphi(y)$ and w lie in opposite sides of T, and cannot therefore be contained in the same component of $\partial(U)$; contradiction.

We have therefore shown that either B is a periodic φ-orbit, or B does not contain a periodic φ-orbit.

Suppose, now, that B is not a periodic φ-orbit, *i.e.*, B contains no periodic φ-orbits. To prove that B contains no φ-orbits which spiral, suppose to the contrary, that there exists $x \in B$ such that $\mathcal{O}_\varphi(x)$ spirals. Choose $y \in \mathbf{S}^2$ such that $\mathcal{O}_\varphi(x)$ spirals to $\mathcal{O}_\varphi(y)$, and let $\langle J_0, L_0, T \rangle$ be a φ-gate construction for the pair $\langle x, y \rangle$. Since T is open and $x \in \partial(U)$, we can find a point $z \in T \cap U$. Choose an arc $\Lambda \subseteq T$ which joins z to x. Since Λ meets no periodic φ-orbit, $p_\varphi(x) = p_\varphi(z) = \infty$, and $z \in U$, we must have that x belongs to the annulus between the periodic φ-orbits $\alpha_\varphi(z)$ and $\omega_\varphi(z)$, which, by (b), is contained in U. Thus $x \in U$; contradiction.

This completes the proof of the theorem.   **QED**

3.13. Theorem. *Let φ be a flow in $\mathbf{S}^2$, and let A be a component of $\mathcal{G}_r(\varphi)$.*

(a) *A is an annulus, and is invariant of φ.*

(b) *Each component of $\partial(A)$ is either a periodic φ-orbit which is singular on exactly one side, or consists solely of φ-fixed points and aperiodic φ-orbits which do not spiral.*

PROOF: Let $x \in A$, and using Corollary 3.10, choose an annulus W which is invariant of φ, and which satisfies $x \in W \subseteq \mathcal{G}_r(\varphi)$. Let $\mathcal{U}$

be the family of all annuli which are invariant of φ, are contained in $\mathscr{G}_r(\varphi)$, and contain W. Then since $\mathscr{U}$ is non-empty, and is closed under unions of chains (nests), we can use Zorn's lemma to find a maximal member U of $\mathscr{U}$.

In order to show that $U = A$, it is sufficient to show that $\partial(U)$ does not meet $\mathscr{G}_r(\varphi)$. Suppose that B is a component of $\partial(U)$. We already know that B either consists entirely of φ-fixed points and aperiodic φ-orbits which do not spiral, or B is a periodic φ-orbit. In the first case, B certainly does not meet $\mathscr{G}_r(\varphi)$, and in the second, B must be singular on one side, for otherwise we could use Theorem 3.9 to contradict the maximality of U.

This completes the proof of the theorem. **QED**

3.14. Three Examples. In this section, we shall give examples to illustrate three simple situations that can occur in $\mathscr{G}_r(\varphi)$ (where φ is a flow in $\mathbf{S}^2$), and in the next theorem, we shall show that, in a sense, these are the building blocks of $\mathscr{G}_r(\varphi)$.

Let $0 < r_1 < r_2 < \infty$, and let Γ_{r_1} and Γ_{r_2} be, respectively, the circles center 0, with radii r_1 and r_2. Let A be the annulus between Γ_{r_1} and Γ_{r_2}; i.e., (identifying $\mathbb{R}^2$ with $\mathbb{C}$, as usual)

$$A = \{z \in \mathbb{R}^2 \mid r_1 < |z| < r_2\}.$$

(a) Let φ be the flow in $\bar{A}$ defined by:

$$\varphi(t, z) = z e^{it}, \qquad \forall \, z \in \bar{A} \qquad \forall \, t \in \mathbb{R}.$$

Every φ-orbit is periodic, and all the φ-orbits are oriented in the same sense.

Before we can describe examples (b) and (c) we need the following notation:

Let P be the vertical strip in the plane defined by:

$$P = \{(x + iy) \mid -\pi/2 \le x \le \pi/2 \text{ and } y \in \mathbb{R}\}.$$

Now choose real numbers c_1 and c_2 such that

$$e^{-\frac{\pi}{2}c_1 + c_2} = r_1 \qquad \text{and} \qquad e^{\frac{\pi}{2}c_1 + c_2} = r_2;$$

in fact,

$$c_1 = \frac{1}{\pi} \log\left(\frac{r_2}{r_1}\right) \qquad \text{and} \qquad c_2 = \frac{1}{2} \log\,(r_1 r_2).$$

Now let f be the function from P onto $\bar{A}$ defined by:

$$f(z) = e^{c_1 z + c_2}, \qquad \forall\, z \in P.$$

We note that, if z and z' belong to P, then we have

$$f(z) = f(z') \text{ iff } z - z' = \frac{2n\pi i}{c_1} \text{ for some integer } n.$$

(b) We first define a flow ψ in P as follows: If $x = -\pi/2$ or $x = \pi/2$, we define

$$\psi(t, x + iy) = x + i\left(y + \frac{t}{c_1}\right), \qquad \forall\, t \in \mathbb{R} \qquad \forall\, y \in \mathbb{R},$$

i.e., in the lines $x = -\pi/2$ and $x = \pi/2$, ψ moves upwards with constant speed $1/c_1$. Between the lines $x = -\pi/2$ and $x = \pi/2$, the orbits of ψ will be the curves with equations $y = \tan x + \alpha$. To see where a typical point $x_0 + iy_0$ moves under the influence of ψ in time t, we first select that α for which $x_0 + iy_0$ lies on the curve $y = \tan x + \alpha$, and then choose $\psi(t, x_0 + iy_0)$ to be the unique point on this curve with ordinate (i.e., imaginary part) $y_0 + \dfrac{t}{c_1}$. Explicitly, if $-\pi/2 < x < \pi/2$, $y \in \mathbb{R}$ and $t \in \mathbb{R}$,

$$\psi(t, x + iy) = \arctan\left(\frac{t}{c_1} + \tan x\right) + i\left(y + \frac{t}{c_1}\right).$$

It is easy to see that $f[\psi]$ is a well-defined flow in $\bar{A}$. (See Definition 1.34 and Theorem 1.37.)

Define $\varphi = f[\psi]$. φ has exactly two periodic orbits, namely Γ_{r_1} and Γ_{r_2}, and these have period 2π, and are both oriented counter-clockwise. Every other φ-orbit is aperiodic, α-spirals to Γ_{r_1} and ω-spirals to Γ_{r_2}.

(c) We first define a flow ψ in P as follows:

$$\psi\left(t, -\frac{\pi}{2} + iy\right) = -\frac{\pi}{2} + i\left(y - \frac{t}{c_1}\right), \qquad \forall\, t \in \mathbb{R} \qquad \forall\, y \in \mathbb{R},$$

$$\psi\left(t, \frac{\pi}{2} + iy\right) = \frac{\pi}{2} + i\left(y + \frac{t}{c_1}\right), \qquad \forall\, t \in \mathbb{R} \qquad \forall\, y \in \mathbb{R}.$$

Between the lines $x = -\pi/2$ and $x = \pi/2$, the orbits of ψ will be the curves with equations $y = \sec x + \alpha$. To see where a typical point $x_0 + iy_0$ moves under the influence of ψ in time t, we first select that α for which $x_0 + iy_0$ lies on the curve $y = \sec x + \alpha$, and then choose $\psi(t, x_0 + iy_0)$ to be the unique point $x_1 + iy_1$ on this curve which satisfies firstly, that the arc-length along this curve from $x_0 + iy_0$ to $x_1 + iy_1$ is t/c_1, and secondly, that $x_1 - x_0$ and t have the same sign.

It is clear that ψ is a well-defined flow in P, and that further, $f[\psi]$ is a well-defined flow in $\bar{A}$.

Define $\varphi = f[\psi]$. φ has exactly two periodic orbits, namely Γ_{r_1} and Γ_{r_2}, but although these have the same period 2π, they are oppositely oriented; Γ_{r_1} is oriented clockwise, while Γ_{r_2} is oriented counter-clockwise. Every other φ-orbit is aperiodic, α-spirals to Γ_{r_1}, and ω-spirals to Γ_{r_2}.

3.15. Definition. Suppose $0 < r_1 < r_2 < \infty$, and let Γ_{r_1}, Γ_{r_2}, and A be as in Section 3.14. The flow φ defined in Section 3.14(b) is called the *standard orientation-preserving spiral between* Γ_{r_1} *and* Γ_{r_2}. The flow φ defined in Section 3.14(c) is called the *standard orientation-reversing spiral between* Γ_{r_1} *and* Γ_{r_2}.

3.16. Theorem. *Let* φ *be a flow in* $\mathbb{S}^2$, *and by choosing a point in* $\mathscr{F}(\varphi)$ *to play the role of* ∞, *let* φ *also be regarded as a flow in* $\mathbb{R}^2$. *Let* A *be a component of* $\mathscr{G}_r(\varphi)$.

(a) *The union of the periodic* φ-*orbits in* A *is relatively closed in* A.

(b) *If* $y \in A$ *and* $\mathrm{p}_\varphi(y) < \infty$, *then* $\mathcal{O}_\varphi(y)$ *separates* $\partial(A)$.

(c) *If* U *is the annulus between any two distinct periodic* φ-*orbits,* $\mathcal{O}_\varphi(y)$ *and* $\mathcal{O}_\varphi(z)$, *in* $\bar{A}$, *then* $U \subseteq A$, *and either*

(i) *There exist points in* U, *with finite* φ-*period,*

or

(ii) *Every point* $x \in U$ *satisfies* $\mathrm{p}_\varphi(x) = \infty$, $\mathcal{O}_\varphi(x)$ *spirals to* $\mathcal{O}_\varphi(y)$, *and* $\mathcal{O}_\varphi(x)$ *spirals to* $\mathcal{O}_\varphi(z)$.

(d) *There exist at most countably many pairs of periodic* φ-*orbits in* $\bar{A}$, *which have the property that the annulus between them contains no point whose* φ-*period is finite.*

(e) *Given any two periodic* φ-*orbits,* $\mathcal{O}_\varphi(y)$ *and* $\mathcal{O}_\varphi(z)$, *in* $\bar{A}$, *which are oppositely oriented, the annulus between* $\mathcal{O}_\varphi(y)$ *and* $\mathcal{O}_\varphi(z)$ *contains a point* x *which satisfies*

(i) $\mathrm{p}_\varphi(x) = \infty$

and

(ii) $\alpha_\varphi(x)$ *and* $\omega_\varphi(x)$ *are oppositely oriented periodic* φ-*orbits.*

(f) *Given any two periodic* φ-*orbits* $\mathcal{O}_\varphi(y)$ *and* $\mathcal{O}_\varphi(z)$ *in* $\bar{A}$, *the annulus between* $\mathcal{O}_\varphi(y)$ *and* $\mathcal{O}_\varphi(z)$ *contains at most finitely many pairs of oppositely oriented periodic* φ-*orbits with the property that the annulus between them contains no point whose* φ-*period is finite.*

PROOF: Theorem 3.7 implies that the union of the aperiodic φ-orbits in A, is open, and from this, (a) follows at once.

To prove (b), suppose $y \in A$ and $p_\varphi(y) < \infty$. Since each side of $\mathcal{O}_\varphi(y)$ meets $\mathscr{F}(\varphi)$, and $A \cap \mathscr{F}(\varphi) = \square$, it is clear that $\mathcal{O}_\varphi(y)$ separates $\partial(A)$.

To prove (c), suppose $y, z \in \bar{A}$, and that $\mathcal{O}_\varphi(y)$ and $\mathcal{O}_\varphi(z)$ are distinct and periodic. Let U be the annulus between $\mathcal{O}_\varphi(y)$ and $\mathcal{O}_\varphi(z)$. It follows easily from (b) that $U \subseteq A$, and the rest of (c) follows at once from Theorem 3.12 (b).

(d) is a trivial consequence of the fact that each such pair of orbits bounds an annulus, the family of these annuli is pairwise disjoint, and $\mathbf{S}^2$ is separable.

Now to prove (e) and (f), suppose $\mathcal{O}_\varphi(y)$ and $\mathcal{O}_\varphi(z)$ are two periodic φ-orbits in $\bar{A}$, and let U be the annulus between them. Let f be a continuous function from $[0, 1]$ into $\bar{U}$ such that $f(0) \in \mathcal{O}_\varphi(y)$ and $f(1) \in \mathcal{O}_\varphi(z)$, and whenever $0 < s < 1$, $f(s) \in U$. For each $s \in [0, 1]$ for which $p_\varphi\big(f(s)\big) < \infty$, let γ_s be the oriented Jordan curve defined by

$$\gamma_s(t) = \varphi\big(t \cdot p_\varphi\big(f(s)\big), f(s)\big), \qquad \forall\, t \in [0, 1].$$

Choose, for once and for all, a point w such that

$$w \in \text{ins}\big(\mathcal{O}_\varphi(y)\big) \cap \text{ins}\big(\mathcal{O}_\varphi(z)\big).$$

It is an elementary consequence of Theorem 2.16 (a) and the continuity of f, that, for each $s \in [0, 1]$ for which γ_s is defined, we have

$$\text{ind}_{\gamma_{s'}}(w) = \text{ind}_{\gamma_s}(w)$$

for all s' sufficiently close to s for which $\gamma_{s'}$ is defined.

To prove (e), suppose that $\mathcal{O}_\varphi(y)$ and $\mathcal{O}_\varphi(z)$ are oppositely oriented (where y and z are as above). Let s_1 be the largest member of $[0, 1]$ such that

$$\text{ind}_{\gamma_{s_1}}(w) = \text{ind}_{\gamma_s}(w).$$

(s_1 can clearly be found, by the abovementioned properties of the curves γ_s and part (a) of this theorem.) Let s_2 be the smallest member of $[s_1, 1]$ such that

$$\text{ind}_{\gamma_1}(w) = \text{ind}_{\gamma_{s_2}}(w).$$

Since

$$\text{ind}_{\gamma_{s_1}}(w) \neq \text{ind}_{\gamma_{s_2}}(w),$$

we have $s_1 < s_2$, and since $\{f(s) \mid s_1 < s < s_2\}$ meets every periodic φ-orbit in the annulus between $\mathcal{O}_\varphi(f(s_1))$ and $\mathcal{O}_\varphi(f(s_2))$, there are, therefore, no periodic φ-orbits in this annulus. This completes the proof of (e).

To prove (f), let $\mathcal{O}_\varphi(y)$ and $\mathcal{O}_\varphi(z)$ be periodic φ-orbits in $\bar{A}$, and let f, w, and the curves γ_s be defined as above. Let

$$F^- = \{s \in [0, 1] \mid \gamma_s \text{ is defined and } \mathrm{ind}_{\gamma_s}(w) = -1\}$$

and

$$F^+ = \{s \in [0, 1] \mid \gamma_s \text{ is defined and } \mathrm{ind}_{\gamma_s}(w) = 1\}.$$

Then F^- and F^+ are mutually disjoint closed subsets of $[0, 1]$. Let

$$\delta = d(F^-, F^+) > 0.$$

It is clear that, if (f) is false, we can find a sequence of distinct pairs $\langle s_n, s_n' \rangle$ such that, for each $n = 1, 2, \ldots$, $0 \le s_n < s_n' \le 1$, γ_s is not defined for any $s \in (s_n, s_n')$, and

$$\mathrm{ind}_{\gamma_{s_n}}(w) \ne \mathrm{ind}_{\gamma_{s_n'}}(w).$$

But it is clear that, for each n, $s_n' - s_n \ge \delta$, and that the family $\{(s_n, s_n') \mid n = 1, 2, \ldots\}$ is pairwise disjoint, and this is clearly impossible. This completes the proof of the theorem.   **QED**

Some Properties of $\mathcal{G}_s(\varphi)$

3.17. Theorem. *Let φ be a flow in $\mathbf{S}^2$, $x \in \mathcal{G}_s(\varphi)$, $y \in \mathcal{G}(\varphi)$, and suppose $\mathcal{O}_\varphi(x)$ spirals to $\mathcal{O}_\varphi(y)$. Let a point of $\mathcal{F}(\varphi)$ be chosen to play the role of ∞.*

Then given a φ-gate construction $\langle J_0, L_0, T \rangle$ for the pair $\langle x, y \rangle$, we have $T \subseteq \mathcal{G}_s(\varphi)$. A fortiori, $x \in \mathrm{int}\,(\mathcal{G}_s(\varphi))$.

PROOF: Let $\langle J_0, L_0, T \rangle$ be a φ-gate construction for the pair $\langle x, y \rangle$. If there exists a point $z \in T \cap \mathcal{G}_r(\varphi)$, then $\alpha_\varphi(z)$ and $\omega_\varphi(z)$ are periodic φ-orbits, and the annulus A between them is contained in $\mathcal{G}_r(\varphi)$. But clearly $T \subseteq A$, which is impossible, since $x \in \mathcal{G}_s(\varphi)$.

This completes the proof of the theorem.   **QED**

3.18. Theorem. *Let φ be a flow in $\mathbf{S}^2$.*

(a) $\mathcal{G}(\varphi) \cap \partial(\mathcal{G}_s(\varphi)) = \mathcal{G}(\varphi) \cap \partial(\mathcal{G}_r(\varphi)) \subseteq \mathcal{G}_s(\varphi).$

(b) *For every $x \in \mathcal{G}(\varphi) \cap \partial(\mathcal{G}_s(\varphi))$, $\mathcal{O}_\varphi(x)$ is either periodic and singular on exactly one side, or is aperiodic and does not spiral.*

PROOF: (a) is trivial.

To prove (b), suppose $x \in \mathcal{G}(\varphi) \cap \partial\big(\mathcal{G}_s(\varphi)\big)$. If $\mathrm{p}_\varphi(x) = \infty$, then $\mathcal{O}_\varphi(x)$ does not spiral, by Theorem 3.17. If $\mathrm{p}_\varphi(x) < \infty$, then since $x \in \mathcal{G}_s(\varphi)$, $\mathcal{O}_\varphi(x)$ is singular on at least one side. But $\mathcal{O}_\varphi(x)$ cannot be singular on both sides, for if it were, then Theorem 3.9 would imply that whenever z lies sufficiently close to $\mathcal{O}_\varphi(x)$, $\mathcal{O}_\varphi(z)$ spirals to $\mathcal{O}_\varphi(x)$, and consequently $z \in \mathcal{G}_s(\varphi)$. This completes the proof of the theorem.   **QED**

3.19. Theorem. *Let φ be a flow in $\mathbf{S}^2$, and let $x \in \mathcal{G}_s(\varphi)$.*

Then there exists an arc $\Lambda \subseteq \mathcal{G}_s(\varphi)$ and a point $z \in \mathcal{G}_s(\varphi)$ such that

(i) *Λ joins x to z*

and

(ii) *either $\mathcal{O}_\varphi(z)$ does not α-spiral or $\mathcal{O}_\varphi(z)$ does not ω-spiral; i.e., either $\alpha_\varphi(z) \subseteq \mathcal{S}(\varphi)$ or $\omega_\varphi(z) \subseteq \mathcal{S}(\varphi)$.*

PROOF:

CASE 1: $\mathrm{p}_\varphi(x) = \infty$ and $\mathcal{O}_\varphi(x)$ has a φ-stagnation point:

Suppose, without loss of generality, that $\mathcal{O}_\varphi(x)$ has an ω-stagnation point; the other case is dual. If $\mathcal{O}_\varphi(x)$ does not ω-spiral, there is nothing to prove. Suppose, then, that $\mathcal{O}_\varphi(x)$ ω-spirals to $\mathcal{O}_\varphi(y)$, and after choosing a point of $\mathcal{F}(\varphi)$ to play the role of ∞, choose a φ-gate construction $\langle J_0, L_0, T \rangle$ for the pair $\langle x, y \rangle$. Since $\mathcal{O}_\varphi(x) \subseteq \mathcal{G}_s(\varphi)$, we may assume, without loss of generality, that $x \in L_0$. Now if $\langle J_0, L_0, T \rangle$ is obstructed, then by Theorem 2.19, there is a point $z \in L_0$ such that $\mathcal{O}_\varphi(z)$ does not ω-spiral, and the choice of Λ is clear, since $T \subseteq \mathcal{G}_s(\varphi)$. On the other hand, if $\langle J_0, L_0, T \rangle$ is unobstructed, let Λ_1 be an arc contained in $[y$ side : $J_0] \cap \mathcal{G}(\varphi)$ which joins a point $z_1 \in T$ to y, and let Λ_2 be an arc, contained in T, which joins x to z_1. Let z be the first point in $\Lambda_1 \cap \partial(T)$. (See Section C. 19.) Since z and y lie in the same side of J_0, and $\langle J_0, L_0, T \rangle$ is unobstructed, we must have $z \in \omega_\varphi(x)$. Therefore, since $z \in \mathcal{G}(\varphi)$, and $\omega_\varphi(x)$ contains no periodic φ-orbit (by Theorem 2.18), we have $\mathrm{p}_\varphi(z) = \infty$, and $\mathcal{O}_\varphi(z)$ does not ω-spiral. The choice of Λ is now clear. This completes Case 1 of the proof.

CASE 2: $\mathrm{p}_\varphi(x) < \infty$:

Choose a φ-fixed point to play the role of ∞, in such a way that $\mathcal{O}_\varphi(x)$ is singular on its inside. Using Theorem 3.9, choose a connected neighborhood V of $\mathcal{O}_\varphi(x)$, such that $\mathcal{O}_\varphi(z)$ spirals to $\mathcal{O}_\varphi(x)$ whenever $z \in V \cap \mathrm{ins}\big(\mathcal{O}_\varphi(x)\big)$. Choose $z_1 \in V \cap \mathrm{ins}\big(\mathcal{O}_\varphi(x)\big)$ such that $\mathcal{O}_\varphi(z_1)$ has a φ-stagnation point. Let Λ_1 be an arc contained in V which joins x to z_1, and let x_1 be the last point in $\Lambda_1 \cap \mathcal{O}_\varphi(x)$. Since that part of Λ_1 that joins x_1 to z_1 is contained in $\mathcal{G}_s(\varphi)$, and since $\mathcal{O}_\varphi(x)$ is contained in

$\mathscr{G}_s(\varphi)$, it is easy to choose an arc $\varLambda_2 \subseteq \mathscr{G}_s(\varphi)$ such that $\varLambda_2$ joins x to z_1. But z_1 satisfies the conditions of Case 1, and the rest of Case 2 therefore follows easily.

CASE 3: $p_\varphi(x) = \infty$ and $\mathcal{O}_\varphi(x)$ has no φ-stagnation point:

$\omega_\varphi(x)$ is a periodic φ-orbit. Write $\omega_\varphi(x) = \mathcal{O}_\varphi(y)$. Then $\mathcal{O}_\varphi(y)$ is singular on $[x$ side : $\mathcal{O}_\varphi(y)]$. Using Theorem 3.9, choose a connected neighborhood V of $\mathcal{O}_\varphi(y)$ such that $\mathcal{O}_\varphi(z)$ spirals to $\mathcal{O}_\varphi(y)$ whenever $z \in V \cap [x$ side : $\mathcal{O}_\varphi(y)]$. Choose a point $x_1 \in V \cap \mathcal{O}_\varphi(x)$ and an arc $\varLambda_1$ in $V \cap [x$ side : $\mathcal{O}_\varphi(y)]$ which joins x_1 to a point $y_1 \in \mathcal{O}_\varphi(y)$. Since $\mathcal{O}_\varphi(x) \cup \varLambda_1 \subseteq \mathscr{G}_s(\varphi)$, it is easy to choose an arc $\varLambda_2 \subseteq \mathscr{G}_s(\varphi)$ which joins x to y_1. But y_1 satisfies the conditions of Case 2, and the rest of Case 3 therefore follows easily. **QED**

3.19. Corollary. *Let φ be a flow in $\mathbf{S}^2$, and let $x \in \mathscr{G}_s(\varphi)$.*

Then there exists an arc $\varLambda \subseteq \mathscr{G}_s(\varphi)$, which joins x to a subset of $\mathscr{S}(\varphi)$.

PROOF: Immediate from Theorem 3.18. **QED**

3.20. Corollary. *Let φ be a semi-simple flow in $\mathbf{S}^2$, and let $x \in \mathscr{G}_s(\varphi)$.*

Then there exists an arc $\varLambda \subseteq \mathscr{G}_s(\varphi)$, which joins x to a point of $\mathscr{S}(\varphi)$.

PROOF: Immediate from Theorem 3.18. **QED**

3.21. The Location of Boundary Points of $\mathscr{G}_s(\varphi)$. Given a flow φ in $\mathbf{S}^2$, we have already seen that $\mathscr{G}_s(\varphi)$ is relatively closed in $\mathscr{G}(\varphi)$, *i.e.*, no limit points of $\mathscr{G}_s(\varphi)$ are to be found in $\mathscr{G}_r(\varphi)$. But $\mathscr{G}_s(\varphi)$ need not be closed; on the contrary, $\mathscr{S}(\varphi) \subseteq \overline{\mathscr{G}_s(\varphi)}$, and therefore every component of $\mathscr{F}_s(\varphi)$ meets $\overline{\mathscr{G}_s(\varphi)}$. This poses the question as to whether any component of $\mathscr{F}_r(\varphi)$ can meet $\overline{\mathscr{G}_s(\varphi)}$. The answer is *no*, as we shall see shortly; in fact, we shall show later that $\mathscr{F}_r(\varphi)$ is "surrounded" by $\mathscr{G}_r(\varphi)$. From the fact that $\mathscr{F}_r(\varphi)$ does not meet $\overline{\mathscr{G}_s(\varphi)}$, we shall make the elementary observation that $\mathscr{F}_s(\varphi) \cup \mathscr{G}_s(\varphi)$ is closed.

3.22. Lemma. *Let φ be a flow in $\mathbf{S}^2$, let K be a component of $\mathscr{F}(\varphi)$, and let $x \in K$. Let $\{x_n\}$ be a sequence of points of $\mathbf{S}^2$ converging to x, let U be a neighborhood of K, and suppose that for each n, there exists $t < 0$ such that $\varphi(t, x_n) \notin U$.*

Then given any neighborhood V of K, there exists a point $y \in V$ such that
(i) $y \in \mathscr{G}(\varphi)$,
(ii) *there exists a subsequence $\{x_{n_j}\}$ of $\{x_n\}$, and a sequence $\{\tau_j\}$ such that*

$$\tau_j \to -\infty \quad and \quad \varphi(\tau_j, x_{n_j}) \to y \quad as \quad j \to \infty,$$

(iii) $\varphi(t, y) \in V, \qquad \forall\, t \geq 0.$

PROOF: Let V be any neighborhood of K, and choose a neighborhood W of K such that $\overline{W} \subseteq U \cap V$. There is no loss of generality in assuming that $x_n \in W$ for every $n = 1, 2, \dots$. Now for each n, let t_n be the largest member of the set

$$\{t < 0 \mid \varphi(t, x_n) \notin W\}.$$

Since $\varphi(t_n, x_n) \in \partial(W)$ for every $n = 1, 2, \dots$, $\{\varphi(t_n, x_n)\}$ has a cluster-point $x_0 \in \partial(W)$, and by reducing to a subsequence, if necessary, we may assume that

$$\varphi(t_n, x_n) \to x_0.$$

Let D be the set of all cluster-points of all sequences of the form $\{\varphi(s_n, x_n)\}$ which satisfy $t_n \leq s_n \leq 0$ for each n.

We claim that D is connected; for otherwise, we can choose two mutually disjoint open sets P and Q, both meeting D, which satisfy $D \subseteq P \cup Q$. Since $x \in D$, we may clearly assume that $x \in P$. Choose $x' \in D \cap Q$, and choose a sequence $\{s_n\}$ satisfying $t_n \leq s_n \leq 0$ for every $n = 1, 2, \dots$, and such that x' is a cluster point of $\{\varphi(s_n, x_n)\}$. For infinitely many integers n, we have $x_n \in P$ and $\varphi(s_n, x_n) \in Q$. For each such n, $\{\varphi(t, x_n) \mid s_n \leq t \leq 0\}$ being connected, we can find a number q_n between s_n and 0 such that $\varphi(q_n, x_n) \notin P \cup Q$. For each other n, let $q_n = 0$. It is clear that $t_n \leq q_n \leq 0$ for every $n = 1, 2, \dots$, and that $\{\varphi(q_n, x_n)\}$ has a cluster-point outside $P \cup Q$. This is impossible, since $D \subseteq P \cup Q$, and it follows that D must be connected.

It now follows that D must meet $\mathscr{G}(\varphi)$, for otherwise, the fact that D is connected and that both x and x_0 belong to D, would imply that $x_0 \in K$, which would contradict the fact that $K \subseteq W$.

Let y be any member of $D \cap \mathscr{G}(\varphi)$. Then y satisfies (i). Now since $y \in D$, we can choose a sequence $\{s_n\}$ satisfying $t_n \leq s_n \leq 0$ for every $n = 1, 2, \dots$, and such that y is a cluster-point of $\{\varphi(s_n, x_n)\}$. Some subsequence $\{\varphi(s_{n_j}, x_{n_j})\}$ of $\{\varphi(s_n, x_n)\}$ converges to y. For each $j = 1, 2, \dots$, let $\tau_j = s_{n_j}$. In order to show that y satisfies (ii), we need only show that $\tau_j \to -\infty$ as $j \to \infty$, and to do this, it is sufficient to show that $\{\tau_j\}$ has no finite cluster-point. But if $\{\tau_j\}$ had a finite cluster-point τ, then since $x_n \to x$, $\varphi(\tau, x)$ would be a cluster-point of $\{\varphi(\tau_j, x_{n_j})\}$, and we would have $\varphi(\tau, x) = y$. But $\varphi(\tau, x) = x$, because $x \in \mathscr{F}(\varphi)$, and we would therefore be led to the absurd conclusion that $x = y$. It follows that y satisfies (ii).

To show that y satisfies (iii), suppose that $t \geq 0$. For every $n = 1, 2, \dots$, let $q_n = (s_n + t) \wedge 0$. Then we clearly have $t_n \leq q_n \leq 0$ for every $n = 1, 2, \dots$. But we also have $q_{n_j} = s_{n_j} + t$ for all sufficiently large j

(because $s_{n_j} \to -\infty$ as $j \to \infty$), and therefore $\varphi(t, y)$ is a cluster-point of $\{\varphi(q_n, x_n)\}$. Thus $\varphi(t, y) \in D \subseteq \overline{W} \subseteq V$.

This completes the proof of the lemma.   **QED**

3.23. Theorem. *Let φ be a flow in $\mathbf{S}^2$, let K be a component of $\mathscr{F}(\varphi)$, and let $x \in K$. Let $\{x_n\}$ be a sequence of points of $\mathbf{S}^2$ converging to x, let U be a neighborhood of K, and suppose that, for each n, there exists $t < 0$ such that $\varphi(t, x_n) \notin U$.*

(a) *Given any neighborhood V of K, there exists a point $y \in V$ such that $p_\varphi(y) = \infty$, $\mathcal{O}_\varphi(y)$ does not ω-spiral, and $\omega_\varphi(y) \subseteq V$.*

(b) $K \subseteq \mathscr{F}_s(\varphi)$.

PROOF: To prove (a), let V be a neighborhood of K, and choose a neighborhood W of K such that $\overline{W} \subseteq U \cap V$. Using Lemma 3.22, choose a point $z \in W \cap \mathscr{G}(\varphi)$ such that $\varphi(t, z) \in W$, $\forall\, t \geq 0$, and z is the limit of a sequence $\{\varphi(\tau_j, x_{n_j})\}$ where $\tau_j \to -\infty$ as $j \to \infty$. Since the sequence $\{x_{n_j}\}$ converges to x, it satisfies the hypotheses of the theorem. Thus, we can simplify the notation by assuming, without loss of generality, that $\{x_{n_j}\} = \{x_n\}$, *i.e.*, that $\varphi(\tau_n, x_n) \to z$.

If $p_\varphi(z) = \infty$, and $\omega_\varphi(z) \subseteq \mathscr{F}(\varphi)$, let $y = z$.

If $p_\varphi(z) = \infty$, and $\omega_\varphi(z)$ meets $\mathscr{F}(\varphi)$, but $\mathcal{O}_\varphi(z)$ ω-spirals, choose y to be any member of $\mathscr{G}(\varphi) \cap \omega_\varphi(z)$.

It is easy to see that, in either of these two cases, the choice of y satisfies the requirements of the theorem.

Suppose now that z does not satisfy either of these two conditions. Then either $\mathcal{O}_\varphi(z)$ or $\omega_\varphi(z)$ is a periodic φ-orbit. Whichever it is, call this orbit γ. Since γ is compact and disjoint from K, we can choose two mutually disjoint open subsets, P and Q, of $U \cap V$ such that $K \subseteq P$ and $\gamma \subseteq Q$.

We note, firstly, that $p_\varphi(x_n) = \infty$ for all sufficiently large n; this is obvious if $\gamma = \omega_\varphi(z)$ (see Theorem 2.23), and if $\gamma = \mathcal{O}_\varphi(z)$, then for any sufficiently large n for which $p_\varphi(x_n) < \infty$, we have the impossible situation that $\mathcal{O}_\varphi(x_n) \subseteq Q \subseteq U$ (see Theorem 2.16).

Secondly, we assert that, for all sufficiently large n, we have $\omega_\varphi(x_n) \subseteq Q$; this follows at once from Theorem 2.23 if $\gamma = \omega_\varphi(z)$, and in case $\gamma = \mathcal{O}_\varphi(z)$, it follows from Theorem 2.16, the fact that $Q \subseteq U$, and the fact that, for each $n = 1, 2, \ldots$, $\{\varphi(t, x_n) \mid t < 0\}$ meets $\mathbf{S}^2 \setminus U$.

Choose a neighborhood P_1 of K such that $\overline{P}_1 \subseteq P \cap W$, and using Lemma 3.22 choose a point $z_1 \in P_1 \cap \mathscr{G}(\varphi)$ which satisfies $\varphi(t, z_1) \in P_1$ for all $t \geq 0$, and such that z_1 is the limit of a sequence $\{\varphi(\sigma_j, x_{n_j})\}$, where $\sigma_j \to -\infty$ as $j \to \infty$. Now $p_\varphi(z_1) = \infty$ and $\omega_\varphi(z_1)$ meets

$\mathscr{F}(\varphi)$, for otherwise, since both $\{\varphi(t, z_1) \mid t \geq 0\}$ and $\omega_\varphi(z_1)$ (if defined) are contained in P, we would have $\omega_\varphi(x_{n_j}) \subseteq P$ for all sufficiently large j, contradicting the fact that $\omega_\varphi(x_n) \subseteq Q$ for all sufficiently large n.

If $\omega_\varphi(z_1) \subseteq \mathscr{F}(\varphi)$, let $y = z_1$, and if $\omega_\varphi(z_1)$ meets $\mathscr{G}(\varphi)$, choose y_1 to be any member of $\omega_\varphi(z_1) \cap \mathscr{G}(\varphi)$.

It is clear that y satisfies the requirements of the theorem. This completes the proof of (a).

To prove (b), choose y_n, for each positive integer n, such that $p_\varphi(y_n) = \infty$, $\mathscr{O}_\varphi(y_n)$ does not ω-spiral, and $\omega_\varphi(y_n) \subseteq N_\varrho(K, \frac{1}{n})$. For each positive integer n, choose $z_n \in \omega_\varphi(y_n)$. Let z be a cluster-point of $\{z_n\}$. It is clear that $z \in K \cap \overline{\mathscr{S}(\varphi)}$, and from this, it follows at once that $K \subseteq \mathscr{F}_s(\varphi)$.

                                              QED

3.24. Theorem. *Let φ be a flow in $\mathbf{S}^2$, let K be a component of $\mathscr{F}(\varphi)$ and let $x \in K$. Let $\{x_n\}$ be a sequence of points of $\mathbf{S}^2$ converging to x, let U be a neighborhood of K, and suppose that for each n, there exists $t \in \mathbb{R}$ such that $\varphi(t, x_n) \notin U$.*

Then given any neighborhood V of K, there exists a point $y \in V$ such that $p_\varphi(y) = \infty$, and

either

(i) $\mathscr{O}_\varphi(y)$ *does not α-spiral, and $\alpha_\varphi(y) \subseteq V$*

or

(ii) $\mathscr{O}_\varphi(y)$ *does not ω-spiral, and $\omega_\varphi(y) \subseteq V$.*

In either case, $K \subseteq \mathscr{F}_s(\varphi)$.

PROOF: For each $n = 1, 2, \ldots$, choose t_n so that $\varphi(t_n, x_n) \notin U$. If $t_n < 0$ for infinitely many n, then the result follows immediately from Theorem 3.23. Otherwise, we have $t_n > 0$ for infinitely many n, and the result follows from an obvious analogue of Theorem 3.23.   **QED**

3.25. Theorem. *Let φ be a flow in $\mathbf{S}^2$, let K be a component of $\mathscr{F}(\varphi)$, and let $x \in K$. Let $\{x_n\}$ be a sequence of points of $\mathbf{S}^2$ converging to x, and suppose that for each $n = 1, 2, \ldots$, $\mathscr{O}_\varphi(x_n)$ has a φ-stagnation point.*

Then given any neighborhood V of K, there exists a point $y \in V$ such that $p_\varphi(y) = \infty$, and
either

(i) $\mathscr{O}_\varphi(y)$ *does not α-spiral, and $\alpha_\varphi(y) \subseteq V$,*

or

(ii) $\mathscr{O}_\varphi(y)$ *does not ω-spiral, and $\omega_\varphi(y) \subseteq V$.*

In either case, $K \subseteq \mathscr{F}_s(\varphi)$.

PROOF:

CASE 1: There exists a neighborhood W of K such that, for each n, $\mathcal{O}_\varphi(x_n)$ has a φ-endpoint in $\mathbf{S}^2 \setminus W$.

Choose a neighborhood U of K such that $\bar{U} \subseteq W$. Then since for $n = 1, 2, \ldots$, there exists $t \in \mathbb{R}$ such that $\varphi(t, x_n) \notin U$, the result follows at once from Theorem 3.24.

CASE 2: For every neighborhood W of K, we have $\alpha_\varphi(x_n) \cup \omega_\varphi(x_n) \subseteq W$ for some n.

Let V be any neighborhood of K, and choose n_0 such that

$$\alpha_\varphi(x_{n_0}) \cup \omega_\varphi(x_{n_0}) \subseteq V.$$

We may assume without loss of generality that $\mathcal{O}_\varphi(x_{n_0})$ has a φ-α-stagnation point. Now if $\alpha_\varphi(x_{n_0}) \subseteq \mathscr{F}(\varphi)$, choose y to be any member of $\mathcal{O}_\varphi(x_{n_0}) \cap V$. y clearly has the required properties. On the other hand, if $\alpha_\varphi(x_{n_0})$ meets $\mathscr{G}(\varphi)$, choose y to be any member of $\alpha_\varphi(x_{n_0}) \cap \mathscr{G}(\varphi)$. Again, y has the required properties, and the proof is complete.   **QED**

3.26. Theorem. *Let φ be a flow in $\mathbf{S}^2$, let K be a component of $\mathscr{F}(\varphi)$, and let $x \in K$. Let $\{x_n\}$ be a sequence of points of $\mathscr{G}_s(\varphi)$ converging to x.*

Then given any neighborhood V of K, there exists a point $y \in V$ such that $p_\varphi(y) = \infty$, and either

(i) $\mathcal{O}_\varphi(y)$ does not α-spiral, and $\alpha_\varphi(y) \subseteq V$,

or

(ii) $\mathcal{O}_\varphi(y)$ does not ω-spiral, and $\omega_\varphi(y) \subseteq V$.

A fortiori, $K \subseteq \mathscr{F}_s(\varphi)$.

PROOF: It is sufficient to demonstrate the existence of a sequence $\{z_n\}$ converging to a member of K, which has the property that, for each n, $\mathcal{O}_\varphi(z_n)$ has a φ-stagnation point.

CASE 1: For infinitely many integers n, $\mathcal{O}_\varphi(x_n)$ has a φ-stagnation point. In this case, the existence of $\{z_n\}$ is trivial.

CASE 2: For infinitely many integers n, $\mathcal{O}_\varphi(x_n)$ is a singular periodic φ-orbit.

In this case, choose a sequence $\{z_n\}$ in such a way that, for these infinitely many n, $\mathcal{O}_\varphi(z_n)$ has a φ-stagnation point, and $\varrho(z_n, x_n) < \frac{1}{n}$.[1]

---

[1] ϱ is the chordal metric; see Appendix C.

CASE 3: For infinitely many integers n, $\mathcal{O}_\varphi(x_n)$ spirals to a periodic φ-orbit.

If there is no sequence $\{z_n\}$ converging to a member of K which satisfies that for each n, $\mathcal{O}_\varphi(z_n)$ has a φ-stagnation point, then there is a neighborhood W of K, such that no member of W has a φ-stagnation point. It is clear that no member of $W \cap \mathcal{G}_s(\varphi)$ has finite period.

Choose a neighborhood U of K such that $\overline{U} \subseteq W$. Then since $\{t \in \mathbb{R} \mid \varphi(t, x_n) \notin U\}$ is unbounded for infinitely many n, the conclusion of the theorem follows from Theorem 3.24. This being so, the apparent nonexistence of $\{z_n\}$ does not worry us, and the proof is complete.

(The reader should note that the conclusion of the theorem assures the existence of $\{z_n\}$, so that really, the argument given above leads to a contradiction.)   **QED**

3.27. Corollary. *Let φ be a flow in $\mathbf{S}^2$. Then $\mathcal{F}_s(\varphi) \cup \mathcal{G}_s(\varphi)$ is closed.*

PROOF: Immediate from Corollary 3.11, Theorem 3.6, and Theorem 3.26.
                                                              QED

Topological Preliminaries to the Relationship between the sets $\mathcal{F}_r(\varphi)$, $\mathcal{G}_r(\varphi)$, $\mathcal{F}_s(\varphi)$, and $\mathcal{G}_s(\varphi)$

3.28. T-sets.[2] Let Ω be a connected open subset of $\mathbf{S}^2$. A subset F of Ω is said to be a *T-set of Ω* if

(i) every component of $\Omega \setminus F$ is an annulus

and

(ii) every component of F is compact.

3.29. Theorem. *Let F be a T-set of a connected, open subset Ω of $\mathbf{S}^2$, let H be the union of a family of components of $\mathbf{S}^2 \setminus \Omega$, and suppose that $\Omega \cup H$ is open.*
Then $F \cup H$ is a T-set of $\Omega \cup H$, and each component of $F \cup H$ is either a component of F or a component of H.

PROOF: We remark first that since Theorem C.5 implies that $\Omega \cup H$ is connected, it is meaningful to speak of a *T-set* of $\Omega \cup H$. Now since

$$(\Omega \cup H) \setminus (F \cup H) = \Omega \setminus F,$$

---

[2] The reason for the name "*T-set*" is that the removal of a *T-set* from Ω leaves annuli, and the symbol T is often used in this book to denote an annulus.

it is clear that every component of $(\Omega \cup H) \setminus (F \cup H)$ is an annulus. Therefore, to prove that $F \cup H$ is a T-set of $\Omega \cup H$, it suffices to show that every component of $F \cup H$ is compact, and therefore, since all components of F, and all components of H, are compact, we shall have completed the proof of the theorem when we have shown that each component of $F \cup H$ is either a component of F or a component of H.

But if there exists a component K of $F \cup H$, which is neither a component of F, nor of H, then K must meet both F and H. Choose $x \in K \cap F$. Then $\bar{K}$, being a continuum from x to $\mathbf{S}^2 \setminus \Omega$, must, by Theorem A. 7, contain a minimal continuum K_0 from x to $\mathbf{S}^2 \setminus \Omega$. By Theorem A. 9, $K_0 \cap \Omega = K_0 \setminus (\mathbf{S}^2 \setminus \Omega)$ is connected, and has limit points in $\mathbf{S}^2 \setminus \Omega$. But $K_0 \cap \Omega \subseteq F$, and from this it follows that F must have a non-compact component, a contradiction.

This completes the proof of the theorem. **QED**

3.30. Theorem. *Let F be a T-set of a connected, open subset Ω of $\mathbf{S}^2$, let B be a component of $\partial(\Omega)$, and let A be a component of $\Omega \setminus F$.*

Then we have either

i) $\bar{A} \cap B = \square$

or

(ii) *B is a component of $\partial(A)$, and*

$$[B \text{ side} : A] = [B \text{ side} : \Omega].$$

PROOF: Each component of $\partial(A)$ is a connected subset of $F \cup (\mathbf{S}^2 \setminus \Omega)$, and is therefore (by Theorem 3.29) contained in a component of F, or in a component of $\mathbf{S}^2 \setminus \Omega$. It follows that each component of $\partial(A)$ is either contained in a component of F, or in a component of $\partial(\Omega)$.

Therefore, if $\bar{A} \cap B \neq \square$, then B contains a component $\tilde{B}$ of $\partial(A)$. From Theorem C. 9, we have $\tilde{B} = B$ and $[B \text{ side} : \Omega] = [B \text{ side} : A]$.

This completes the proof. **QED**

3.31. Theorem. *Let F be a T-set of a connected, open subset Ω of $\mathbf{S}^2$, let A be a component of $\Omega \setminus F$, and suppose that $\bar{A}$ meets more than one component of $\partial(\Omega)$.*

Then $A = \Omega$, and $F = \square$.

PROOF: Immediate from Theorem 3.30. **QED**

3.32. Theorem. *Let F be a T-set of a connected, open subset Ω of $\mathbf{S}^2$, and let B be a component of $\partial(\Omega)$. Then either*

(i) *every point of B is a limit-point of F, and no component of $\Omega \setminus F$ has a limit-point in B,*

or

(ii) *no point of B is a limit-point of F, and there exists a component A of $\Omega \setminus F$ such that B is a component of $\partial(A)$, $A \cup [B$ side : $\Omega]$ is a neighborhood of $[B$ side : $\Omega]$, $A \cup [B$ side : $\Omega]$ does not meet F, and $[B$ side : $\Omega]$ is the only component of $\mathbf{S}^2 \setminus \Omega$ which meets $A \cup [B$ side : $\Omega]$.*

PROOF: We shall show that, if there exists a point of B which is not a limit-point of F, then some component A of $\Omega \setminus F$ has a limit-point in B. Once this has been established, the theorem will follow easily from Theorem 3.30.

Suppose, then, that some point $x \in B$ is not a limit-point of F, and choose a connected neighborhood U of x which does not meet F. Choose $y \in U \cap \Omega$, and let A be the component of $\Omega \setminus F$ which contains y. We shall show that A has a limit-point in B.

Suppose, to the contrary, that $\bar{A} \cap B = \square$, and let L be an arc in U which joins y to x. Let z be the last point in $L \cap \bar{A}$. Then z lies in a component B_1 of $\partial(\Omega)$. Since B_1 meets $\bar{A}$, we have $B_1 \neq B$, and B_1 is a component of $\partial(A)$. Further, $[B_1$ side : $\Omega] = [B_1$ side : $A]$. It follows that since that part Λ of L which joins z to x is a connected subset of $\mathbf{S}^2 \setminus A$, and meets B_1, we have

$$\Lambda \subseteq [B_1 \text{ side} : \Omega].$$

Therefore Λ is a connected subset of $\mathbf{S}^2 \setminus \Omega$ which meets both B and B_1, contradicting the fact that B and B_1 are different components of $\partial(\Omega)$.

Thus, A has a limit-point in B, and the proof is complete.   **QED**

3.33. Remark. It is a simple consequence of Theorem 3.30 that if F is a T-set of a connected, open subset Ω of $\mathbf{S}^2$, then only countably many components of $\partial(\Omega)$ can fail to contain a limit-point of F. If B is one of those components of $\partial(\Omega)$ which do not contain a limit-point of F, and if A is that component of $\Omega \setminus F$ for which B is a component of $\partial(A)$, then using Theorem C. 30, we can find in A analytic Jordan curves which separate $\partial(A)$, and furthermore we can find such curves as close as we please to either component of $\partial(A)$. Because of this, it is possible to "trim" away Ω, replacing B by one of these analytic Jordan curves, without removing any part of F.

3.34. Theorem. *Let F be a non-empty T-set of a connected, open subset Ω of $\mathbf{S}^2$, and let $\{B_n \mid n = 1, 2, ...\}$ be a family of components of $\partial(\Omega)$*

which contain no limit-points of F. For each n, let A_n be that component of $\Omega \setminus F$ for which B_n is a component of $\partial(A_n)$, and let J_n be a Jordan curve contained in A_n, which separates $\partial(A_n)$. Suppose further, that for all but finitely many positive integers n, the Jordan curve J_n lies close to B_n, relative to the width of the annulus A_n, in the following sense:

There exists a number $\gamma < 1$ such that, for all but finitely many integers n, writing δ_n for the ϱ-distance between the two components of $\partial(A_n)$, we have that the annulus between J_n and B_n is contained in $N_\varrho(B_n, \gamma \delta_n)$.

For each $n = 1, 2, ...,$ let $\tilde{A}_n$ be that part of A_n that remains after the removal of $J_n \cup [B_n$ side : $J_n]$, i.e.,

$$\tilde{A}_n = A_n \setminus (J_n \cup [B_n \text{ side} : J_n]).$$

Let

$$\tilde{\Omega} = \Omega \setminus \bigcup_n (J_n \cup [B_n \text{ side} : J_n]).$$

Then $\tilde{\Omega}$ is a region, F is a T-set of $\tilde{\Omega}$, and the components of $\partial(\tilde{\Omega})$ are the components of $\partial(\Omega)$ other than $B_1, B_2, ...,$ together with the Jordan curves $J_1, J_2, ...$

PROOF: We note first that, since $F \neq \square$, Theorem 3.31 implies that no component of $\Omega \setminus F$ can have limit-points in more than one component of $\partial(\Omega)$. Therefore the family $\{A_n \mid n = 1, 2, ...\}$ is pairwise-disjoint, and consequently, $\tilde{A}_n$ is well-defined for each $n = 1, 2, ...$ We note also that each $\tilde{A}_n$ is an annulus.

Now clearly $F \subseteq \tilde{\Omega}$, and the above remark shows that each component of $\tilde{\Omega} \setminus F$ is either a member of the family $\{\tilde{A}_n \mid n = 1, 2, ...\}$, or is a component of $\Omega \setminus F$, other than $A_1, A_2, ...$

$\tilde{\Omega}$ must be open, for certainly $\tilde{\Omega} \setminus F$ is open, and if there is a point $x \in \tilde{\Omega} \cap F$ which is not an interior point of $\tilde{\Omega}$, then we can obtain a contradiction as follows: choose a sequence $\{x_j\}$ of points in $\Omega \setminus \tilde{\Omega}$ such that $x_j \to x$ as $j \to \infty$. For each $j = 1, 2, ...,$ choose n_j such that

$$x_j \in A_{n_j} \setminus \tilde{A}_{n_j}.$$

It is clear that for each integer n, there are at most finitely many integers j for which $n_j = n$, for otherwise, $\{x_j\}$ would have a cluster point in $J_n \cup [B_n$ side : $J_n]$, contradicting the fact that $x_j \to x$ as $j \to \infty$, and $x \in F$.

Therefore, by reducing $\{x_j\}$ to a subsequence, if necessary, we may assume that for each j, the annulus between J_{n_j} and B_{n_j} is contained in $N_\varrho(B_{n_j}, \gamma \delta_{n_j})$.

For each $j = 1, 2, ...,$ choose $y_j \in B_{n_j}$ such that

$$\varrho(x_j, y_j) < \gamma \delta_{n_j}.$$

Now since $x \in F$, we have that for each j, x and y_j lie in opposite sides of A_{n_j}, and consequently,

$$\varrho(x, y_j) \geq \delta_{n_j}.$$

It follows that

$$\varrho(x, x_j) \geq (1 - \gamma) \delta_{n_j},$$

and therefore $\delta_{n_j} \to 0$ as $j \to \infty$. Therefore, since $\varrho(x_j, y_j) \to 0$ as $j \to \infty$, we have $y_j \to x$ as $j \to \infty$, which is a contradiction, since $y_j \in \mathbf{S}^2 \setminus \Omega$ for every $j = 1, 2, ...$

This shows that $\tilde{\Omega}$ is open, and the proof will therefore be complete when we have shown that $\tilde{\Omega}$ is connected. But if $\tilde{\Omega}$ is not connected, we can choose two mutually disjoint, non-empty open sets U and V such that $\tilde{\Omega} = U \cup V$. For each n, we note that A_n cannot meet both U and V: in fact, $\tilde{A}_n$, being connected, must be contained in either U or V. Let U_1 be the union of U and all the annuli A_n which meet U, and let V_1 be the union of V and all the annuli A_n which meet V. Then $\Omega = U_1 \cup V_1$, in spite of the fact that Ω is connected, U_1 and V_1 are mutually disjoint, and both U_1 and V_1 are open.

Thus, $\tilde{\Omega}$ is connected, and the proof is complete.   **QED**

3.35. Definition. Let F be a non-empty T-set of a connected, open subset Ω of $\mathbf{S}^2$, and let $\{B_n \mid n = 1, 2, ...\}$ be a family of components of $\partial(\Omega)$ which contain no limit-points of F. Let $0 < \gamma < 1$.

We say that we *trim Ω around F, at $\{B_n \mid n = 1, 2, ...\}$, moderated by γ*, when we carry out a procedure as outlined in the next paragraph, to replace Ω by a smaller region $\tilde{\Omega}$, of which F is also a T-set.

For each $n = 1, 2, ...,$ let A_n be that component of $\Omega \setminus F$ for which B_n is a component of $\partial(A_n)$, and for each $n = 1, 2, ...,$ choose an analytic Jordan curve $J_n \subseteq A_n$, such that J_n separates $\partial(A_n)$, and such that, for all but finitely many integers n, writing δ_n for the ϱ-distance between the two components of $\partial(A_n)$, we have that the annulus between J_n and B_n is contained in $N_\varrho(B_n, \gamma \delta_n)$. Let

$$\tilde{\Omega} = \Omega \setminus \bigcup_n (J_n \cup [B_n \text{ side} : J_n]).$$

3.36. Remark. A trivial analogue of Theorem 3.34 holds for empty T-sets. (Of course, if Ω is a connected, open subset of $\mathbf{S}^2$, then $\square$ is a T-set of Ω iff Ω is an annulus.) This analogue of Theorem 3.34 is:

Let $F = \square$ be a T-set of an annulus Ω, and let B_1 and B_2 be the two components of $\partial(\Omega)$. Let J_1 and J_2 be mutually disjoint Jordan curves contained in Ω, and suppose that both J_1 and J_2 separate $\partial(\Omega)$. Let $\tilde{\Omega}$ be defined in any one of the following four ways:

(i) $\tilde{\Omega} = \square$.

(ii) $\tilde{\Omega}$ is the annulus between B_1 and J_2.

(iii) $\tilde{\Omega}$ is the annulus between B_2 and J_1.

(iv) $\tilde{\Omega}$ is the annulus between J_1 and J_2.

Then F is a T-set of $\tilde{\Omega}$. Once again, we say that $\tilde{\Omega}$ is obtained by trimming Ω around F.

3.37. Settings and T_σ-sets. Let Ω be a connected open subset of $\mathbf{S}^2$, and let F be a relatively closed subset of Ω. A pairwise disjoint family $\{\Omega_j \mid j = 1, 2, \ldots\}$ of connected open subsets of Ω is said to be a *setting for F in Ω* if

(i) $F \subseteq \bigcup_j \Omega_j$.

(ii) For each $j = 1, 2, \ldots$, $F \cap \Omega_j$ is a T-set of Ω_j.

(iii) Every point of $\Omega \setminus \bigcup_j \Omega_j$ can be joined by an arc in $\Omega \setminus \bigcup_j \Omega_j$ to a subset of $\partial(\Omega)$.

Of course, a setting must necessarily be a countable family, but it can also be finite or even empty.

Let F be a relatively closed subset of a connected, open subset Ω of $\mathbf{S}^2$. A setting $\{\Omega_j \mid j = 1, 2, \ldots\}$ for F in Ω is said to be *trim* if:

(i) The family $\{\bar{\Omega}_j \mid j = 1, 2, \ldots\}$ is pairwise disjoint.

(ii) For every $j = 1, 2, \ldots$, and every component B of $\partial(\Omega_j)$, B is either a component of $\partial(\Omega)$, or B is an analytic Jordan curve contained in Ω and some neighborhood of B does not meet $\bigcup_{i \neq j} \bar{\Omega}_i$.

(iii) For every $j = 1, 2, \ldots$, we have

either $\Omega_j \cap F \neq \square$ or $\partial(\Omega_j) \cap \partial(\Omega) \neq \square$.

Roughly speaking, a setting is trim if it contains no unnecessary members, no member is unnecessarily large, and where possible, boundaries are smooth.

Let F be a relatively closed subset of a connected, open subset Ω of $\mathbf{S}^2$. F is said to be a *T_σ-set of Ω* if there exists a setting for F in Ω.

3.38. Remark. Let F be a T-set of a connected, open subset Ω of $\mathbf{S}^2$. Then since $\{\Omega\}$ is a trim setting for F in Ω, F is a T_σ-set of Ω.

3.39. Theorem. *Let* $\{\Omega_j \mid j = 1, 2, \ldots\}$ *be a setting for a set F in a connected open subset Ω of $\mathbf{S}^2$. Suppose that for each $j = 1, 2, \ldots$, $0 < \gamma_j < 1$, and that $\mathscr{B}_j$ is the family of those components of $\partial(\Omega_j)$ which meet Ω. For each $j = 1, 2, \ldots$, let $\varDelta_j$ be obtained by trimming Ω_j around F, at $\mathscr{B}_j$, moderated by γ_j.*

Then the family of those sets $\varDelta_j$ which satisfy that either $\varDelta_j \cap F \neq \square$ or $\partial(\varDelta_j) \cap \partial(\Omega) \neq \square$ is a trim setting for F in Ω.

Furthermore, if $\{\Omega_j \mid j = 1, 2, \ldots\}$ has the additional property that, for every $x \in \Omega \setminus \bigcup_j \Omega_j$, there exists an arc $\varLambda \subseteq \Omega \setminus \bigcup_j \Omega_j$, which joins x to a single point of $\partial(\Omega)$, then $\{\varDelta_j \mid j = 1, 2, \ldots\}$ has the same property.

PROOF: We shall show first, that for every $j_0 = 1, 2, \ldots$, and every $B \in \mathscr{B}_{j_0}$, B contains no limit-points of F: for suppose $B \in \mathscr{B}_{j_0}$ for some $j_0 = 1, 2, \ldots$ Choose $x \in B \cap \Omega$. Then since $x \notin \bigcup_j \Omega_j$, and F is relatively closed in Ω, and $F \subseteq \bigcup_j \Omega_j$, we see that x cannot be a limit-point of F, and consequently, by Theorem 3.32, B contains no limit-points of F.

Thus we see that, for each j, it is possible to define $\varDelta_j$.

Now suppose that, for some $j = 1, 2, \ldots$, B is a component of $\partial(\varDelta_j)$. Then either B is an analytic Jordan curve which is contained in Ω_j, or $B \subseteq \partial(\Omega)$. Therefore, either $B \subseteq \Omega$, or (using Theorem C. 9) B is a component of $\partial(\Omega)$, and furthermore, if $B \subseteq \Omega$, then Ω_j is a neighborhood of B which does not meet $\bigcup_{i \neq j} \bar{\varDelta}_i$.

Now to show that $\{\bar{\varDelta}_j \mid j = 1, 2, \ldots\}$ is pairwise disjoint, suppose $j_1 \neq j_2$, and $x \in \bar{\varDelta}_{j_1} \cap \bar{\varDelta}_{j_2}$. Then $x \in \partial(\varDelta_{j_1})$. Let B be that component of $\partial(\varDelta_{j_1})$ which contains x. Since we cannot have $B \subseteq \Omega_{j_1}$, we must have that B is a component of $\partial(\Omega)$. It follows from Theorem C. 9 that $[B \text{ side}: \Omega_{j_1}]$ does not meet Ω, and consequently that $[\Omega_{j_2} \text{ side}: \Omega_{j_1}] \neq [B \text{ side}: \Omega_{j_1}]$. Therefore $[\Omega_{j_2} \text{ side}: \Omega_{j_1}]$ is a compact set which contains Ω_{j_2}, but does not meet B, contradicting the fact that $x \in \bar{\varDelta}_{j_2}$. Thus $\{\bar{\varDelta}_j \mid j = 1, 2, \ldots\}$ is pairwise disjoint.

Let Q be the set of those positive integers j for which either $\varDelta_j \cap F \neq \square$ or $\partial(\varDelta_j) \cap \partial(\Omega) \neq \square$. Now since $F \subseteq \bigcup_{j \in Q} \varDelta_j$, and since $F \cap \varDelta_j$ is a T-set of $\varDelta_j$ for every $j = 1, 2, \ldots$, the proof will be completed once we have shown that every point of $\Omega \setminus \bigcup_{j \in Q} \varDelta_j$ can be joined by an arc in $\Omega \setminus \bigcup_{j \in Q} \varDelta_j$ to a subset of $\partial(\Omega)$, or a single point if necessary.

Suppose, then, that $x \in \Omega \setminus \bigcup_{j \in Q} \varDelta_j$.

Firstly, if $x \in \Omega \setminus \bigcup_j \Omega_j$, then the existence of the required arc is already known.

Secondly, if $x \in \Omega_j \setminus \Delta_j$ for some $j = 1, 2, \ldots$, then there exists a component B of $\partial(\Omega_j)$ and a component J of $\partial(\Delta_j)$ such that J is an analytic Jordan curve which is contained in Ω_j, and either $x \in J$ or x belongs to the annulus A between B and J. Choose an arc $\Lambda_1 \subseteq A \cup J$ such that Λ_1 joins x to a point $y \in B$. If $y \in \partial(\Omega)$, then Λ_1 satisfies our requirements. Otherwise, $y \in \Omega \setminus \bigcup_j \Omega_j$, and we obtain an arc satisfying our requirements by following Λ_1 by an arc joining y to a subset of $\partial(\Omega)$.

Thirdly, if $x \in \Delta_j$ for some $j \notin Q$, then choose an arc $\Lambda_1 \subseteq \Delta_j$, which joins x to a point $y \in \partial(\Delta_j)$. Now $y \in \Omega_j \setminus \Delta_j$, and we can therefore follow Λ_1 by an arc which joins y to a subset of $\partial(\Omega)$, thus obtaining an arc with the required properties.

This completes the proof of the theorem. **QED**

3.40. Definition. Let $\{\Omega_j \mid j = 1, 2, \ldots\}$ be a trim setting for a set F in a connected, open subset Ω of $\mathbf{S}^2$.

A component B of $\partial(\Omega)$ is said to be *sealed* by $\{\Omega_j \mid j = 1, 2, \ldots\}$ if $B \cap \bigcup_j \bar{\Omega}_j \neq \square$.

We remark that if a component B of $\partial(\Omega)$ is sealed by $\{\Omega_j \mid j = 1, 2, \ldots\}$, then for exactly one positive integer j, B is a component of $\partial(\Omega_j)$.

3.41. Theorem. *Let* $\{\Omega_j \mid j = 1, 2, \ldots\}$ *be a trim setting for a set F in a connected, open subset Ω of $\mathbf{S}^2$, let B be a component of $\partial(\Omega)$, and suppose that B is not sealed by* $\{\Omega_j \mid j = 1, 2, \ldots\}$. *For each* $j = 1, 2, \ldots$, *let*

$$\Omega_j^B = \mathbf{S}^2 \setminus [B \text{ side} : \Omega_j].$$

For each $j = 1, 2, \ldots$, *let*

$$\mathscr{C}_j^B = \{\Omega_i^B \mid \Omega_i^B \supseteq \Omega_j^B\}.$$

(a) *For every* $j = 1, 2, \ldots$, Ω_j^B *is a disc, and* $\partial(\Omega_j^B)$ *is an analytic Jordan curve which is contained in Ω.*

(b) *For any choice of j_1 and j_2, if $j_1 \neq j_2$, then either*

$$\overline{\Omega_{j_1}^B} \cap \overline{\Omega_{j_2}^B} = \square$$

or one of the two sets, $\Omega_{j_1}^B$, $\Omega_{j_2}^B$, contains the closure of the other.

(c) *For every* $j = 1, 2, \ldots$, $\mathscr{C}_j^B$ *is a chain (nest), and either $\mathscr{C}_j^B$ has a largest member, or its union is* $\mathbf{S}^2 \setminus [B \text{ side} : \Omega]$.

(d) *Suppose there exists j_0 such that $\mathscr{C}_{j_0}^B$ has no largest member. Then*

(i) *For every j, $\mathscr{C}_j^B$ has no largest member.*

(ii) *For any choice of j_1 and j_2, $\mathscr{C}_{j_1}^B$ and $\mathscr{C}_{j_2}^B$ have members in common.*

(iii) *Every point of B is a limit-point of F.*

(e) *Suppose that, for every j, $\mathscr{C}_j^B$ has a largest member. Then the family of these largest members is pairwise disjoint.*

PROOF: To prove (a), let j be a positive integer. It is obvious that Ω_j^B is a disc, and since $\partial(\Omega_j^B)$ is a component of $\partial(\Omega_j)$, $\partial(\Omega_j^B)$ must either be an analytic Jordan curve contained in Ω, or must be a component of $\partial(\Omega)$. But $\partial(\Omega_j^B)$ cannot be a component of $\partial(\Omega)$, for otherwise we would have $\partial(\Omega_j^B) = B$, contradicting the fact that B is not sealed by $\{\Omega_j \mid j = 1, 2, \ldots\}$. This proves (a).

To prove (b), suppose $j_1 \neq j_2$. Clearly, if B lies in the annulus between $\partial(\Omega_{j_1}^B)$ and $\partial(\Omega_{j_2}^B)$, then

$$\overline{\Omega_{j_1}^B} \cap \overline{\Omega_{j_2}^B} = \square,$$

and otherwise, one of the sets $\Omega_{j_1}^B, \Omega_{j_2}^B$, must contain the closure of the other. This proves (b).

To prove (c), we note first that every family $\mathscr{C}_j^B$ is obviously a chain. Now suppose, for some choice of j, that $\mathscr{C}_j^B$ has no largest member, and let Δ be the union of the chain $\mathscr{C}_j^B$. Since Δ is connected and open, and

$$\mathbf{S}^2 \setminus \Delta = \cap \{\mathbf{S}^2 \setminus \Omega_i^B \mid \Omega_i^B \in \mathscr{C}_j^B\},$$

it is clear that Δ is a disc.

We claim that every point of $\partial(\Delta)$ is a limit-point of F. To prove this, suppose $x \in \partial(\Delta)$, and let U be any connected neighborhood of x. Let y be any member of $U \cap \Delta$, and choose an arc $\Lambda \subseteq U$ which joins x to y. Choose j_1 such that $\Omega_{j_1}^B \in \mathscr{C}_j^B$, and $y \in \Omega_{j_1}^B$. Then x and y lie in opposite sides of $\partial(\Omega_{j_1}^B)$. Now choose j_2 such that $\Omega_{j_2}^B \in \mathscr{C}_j^B$, and $\overline{\Omega_{j_1}^B} \subseteq \Omega_{j_2}^B$. Then x and $\partial(\Omega_{j_1}^B)$ lie in opposite sides of $\partial(\Omega_{j_2}^B)$. Let A be that component of $\Omega_{j_2} \setminus F$ for which $\partial(\Omega_{j_2}^B)$ is a component of $\partial(A)$, and let C be the other component of $\partial(A)$. It is clear that C is not a component of $\partial(\Omega_{j_2})$, because the presence of Ω_{j_1} makes it impossible for C to be a component of $\partial(\Omega)$, and Theorem 3.31 and part (iii) of Definition 3.37, would make it impossible for us to have $C \subseteq \Omega$. Therefore $C \subseteq F$. Consequently, since x and y lie in opposite sides of C, we must have $\Lambda \cap F \neq \square$, and it follows that $U \cap F \neq \square$. Thus, every point of $\partial(\Delta)$ is a limit-point of F.

Now to prove that $\varDelta = \mathbf{S}^2 \setminus [B \text{ side}: \varOmega]$, suppose $x \in \partial(\varDelta)$. Then $x \notin \bigcup_i \varOmega_i^B$, for otherwise, we could choose i such that $x \in \varOmega_i^B \in \mathscr{C}_j^B$, from which it would follow that $x \in \varDelta$, contradicting the fact that $\varDelta$ is open. But x is a limit-point of F, $F \subseteq \bigcup_i \varOmega_i^B$, and F is relatively closed in $\varOmega$. Therefore $x \notin \varOmega$.

We have now shown that $\partial(\varDelta) \cap \varOmega = \square$, and since $\varOmega$ is connected and $\varOmega \cap \varDelta \neq \square$, it follows that $\varOmega \subseteq \varDelta$. Therefore $\mathbf{S}^2 \setminus \varDelta$ is a connected subset of $\mathbf{S}^2 \setminus \varOmega$, and therefore, since $B \subseteq \mathbf{S}^2 \setminus \varDelta$, we have

$$\mathbf{S}^2 \setminus \varDelta \subseteq [B \text{ side}: \varOmega].$$

But we also have

$$\varDelta \subseteq \mathbf{S}^2 \setminus [B \text{ side}: \varOmega],$$

i.e.,

$$\mathbf{S}^2 \setminus \varDelta \supseteq [B \text{ side}: \varOmega].$$

Thus, $\varDelta = \mathbf{S}^2 \setminus [B \text{ side}: \varOmega]$, and (c) is proved.

To prove (d) (i) and (ii), it is sufficient to show that if there exists j_0 such that $\mathscr{C}_{j_0}^B$ has no largest member, then for every j_1, the chains $\mathscr{C}_{j_0}^B$ and $\mathscr{C}_{j_1}^B$ have members in common. To show this, we note that since the closure of $\varOmega_{j_1}^B$ is a compact subset of $\mathbf{S}^2 \setminus [B \text{ side}: \varOmega]$, and therefore of the union of $\mathscr{C}_{j_0}^B$, some member $\varOmega_i^B$ of $\mathscr{C}_{j_0}^B$ must contain $\varOmega_{j_1}^B$. (d) (iii) was established in the course of proving (c). This proves (d).

Finally, we observe that (e) follows trivially from (b). This completes the proof of the theorem. **QED**

3.42. Definition. Let $\{\varOmega_j = 1, 2, \ldots\}$ be a trim setting for a set F in a connected, open subset $\varOmega$ of $\mathbf{S}^2$, and let B be a component of $\partial(\varOmega)$. For each $j = 1, 2, \ldots$, define

$$\varOmega_j^B = \mathbf{S}^2 \setminus [B \text{ side}: \varOmega_j],$$

and for each $j = 1, 2, \ldots$, define

$$\mathscr{C}_j^B = \{\varOmega_i^B \mid \varOmega_i^B \supseteq \varOmega_j^B\}.$$

B is said to be *wrapped* by $\{\varOmega_j \mid j = 1, 2, \ldots\}$, if either

(i) B is sealed by $\{\varOmega_j \mid j = 1, 2, \ldots\}$

or

(ii) For each $j = 1, 2, \ldots$, the chain $\mathscr{C}_j^B$ has no largest member.

3.43. Remark. It is easy to verify that if two trim settings $\{\varOmega_j \mid j = 1, 2, \ldots\}$, $\{\varDelta_j \mid j = 1, 2, \ldots\}$ for a set F in a connected, open subset $\varOmega$ of $\mathbf{S}^2$ are obtained by two (possibly different) applications of Theorem 3.39

to a given setting for F in Ω, then a component B of $\partial(\Omega)$ is sealed by $\{\Omega_j \mid j = 1, 2, \ldots\}$ iff it is sealed by $\{\Delta_j \mid j = 1, 2, \ldots\}$, and is wrapped by $\{\Omega_j \mid j = 1, 2, \ldots\}$ iff it is wrapped by $\{\Delta_j \mid j = 1, 2, \ldots\}$.

3.44. Theorem. *Let $\{\Omega_j \mid j = 1, 2, \ldots\}$ be a trim setting for a set F in a connected, open subset Ω of $\mathbf{S}^2$. For every component B of $\partial(\Omega)$, define*

$$\Omega_j^B = \mathbf{S}^2 \setminus [B \text{ side} : \Omega_j].$$

(a) If a component B of $\partial(\Omega)$ is wrapped by $\{\Omega_j \mid j = 1, 2, \ldots\}$ then $B = \bar{\Omega} \setminus \bigcup_j \Omega_j^B$, and B is a component of $\bar{\Omega} \setminus \bigcup_j \Omega_j$.

(b) If a component B of $\partial(\Omega)$ is not wrapped by $\{\Omega_j \mid j = 1, 2, \ldots\}$, then $\bar{\Omega} \setminus \bigcup_j \Omega_j^B$ is the component of $\bar{\Omega} \setminus \bigcup_j \Omega_j$ that contains B. Further, $\bar{\Omega} \setminus \bigcup_j \Omega_j^B$ contains B properly.

(c) $\{\bar{\Omega} \setminus \bigcup_j \Omega_j^B \mid B \text{ is a component of } \partial(\Omega)\}$ is the set of components of $\bar{\Omega} \setminus \bigcup_j \Omega_j$.

PROOF: The proof of (a) is straightforward, and will be left to the reader.

(b) will also be straightforward once we have shown that whenever a component B of $\partial(\Omega)$ is not wrapped by $\{\Omega_j \mid j = 1, 2, \ldots\}$, the set $\bar{\Omega} \setminus \bigcup_j \Omega_j^B$ is connected. To show this, suppose B is not wrapped by $\{\Omega_j \mid j = 1, 2, \ldots\}$. It is easy to see that for any positive integer n, the set $\Omega \setminus \bigcup_{j=1}^{n} \Omega_j^B$ is connected. Therefore, $\bar{\Omega} \setminus \bigcup_{j=1}^{n} \Omega_j^B$, being the closure of $\Omega \setminus \bigcup_{j=1}^{n} \Omega_j^B$, must also be connected, for every $n = 1, 2, \ldots$. It now follows from Theorem A.5 that $\bar{\Omega} \setminus \bigcup_j \Omega_j^B$ is connected, and (b) is proved.

To prove (c), we need only show that every component of $\bar{\Omega} \setminus \bigcup_j \Omega_j$ is of the form $\bar{\Omega} \setminus \bigcup_j \Omega_j^B$ for some component B of $\partial(\Omega)$.

Suppose then that C is a component of $\bar{\Omega} \setminus \bigcup_j \Omega_j$. It is clear that if $C \cap \Omega = \square$, then C is a wrapped component of $\partial(\Omega)$. Suppose now that $C \cap \Omega \neq \square$. Choose $x \in C \cap \Omega$, and using the fact that $\{\Omega_j \mid j = 1, 2, \ldots\}$ is a setting and $x \in \Omega \setminus \bigcup_j \Omega_j$, choose an arc $\Lambda \subseteq \Omega \setminus \bigcup_j \Omega_j$ which joins x to a subset P of $\partial(\Omega)$. Let B be that component of $\partial(\Omega)$ that contains P. Then it is clear that B is not wrapped and that $C = \bar{\Omega} \setminus \bigcup_j \Omega_j^B$.

This completes the proof of the theorem.   **QED**

3.45. Definition. Let F be a T_σ-set of a connected open subset Ω of $\mathbf{S}^2$ and let Y be a subset of $\mathbf{S}^2 \setminus \Omega$.

A setting $\{\Omega_j \,|\, j = 1, 2, \ldots\}$ for F in Ω is said to be *weakly Y-accessible* if

(i) The setting $\{\Omega_j \,|\, j = 1, 2, \ldots\}$ is trim.

(ii) Given any component B of $\partial(\Omega)$ which is not wrapped by $\{\Omega_j \,|\, j = 1, 2, \ldots\}$ and given any neighborhood U of B, if we define $\Omega_j^B = \mathbf{S}^2 \setminus [B \text{ side} : \Omega_j]$ for every $j = 1, 2, \ldots$, then there exists an arc in $\Omega \setminus \bigcup_j \overline{\Omega_j^B}$, which joins a point of $\Omega \setminus \bigcup_j \overline{\Omega_j^B}$ to a subset of $Y \cap U$.

If, in addition, the arcs in (ii) can always be so chosen that the corresponding subset of $Y \cap U$ (which must, of course, always be non-empty, compact and connected), consists of exactly one point, then we say that the setting $\{\Omega_j = 1, 2, \ldots\}$ for F in Ω, is Y-*accessible*.

3.46. Definition. Let F be a T_σ-set of a connected, open subset Ω of $\mathbf{S}^2$, and suppose $Y \subseteq \mathbf{S}^2 \setminus \Omega$. If there exists a [weakly] Y-accessible setting for F in Ω, then F is said to be a [*weakly*] Y-*accessible* T_σ-set of Ω.

3.47. Examples.

(a) Let F be a T-set of a connected, open subset Ω of $\mathbf{S}^2$. Then for every subset Y of $\mathbf{S}^2 \setminus \Omega$, F is a Y-accessible T_σ-set of Ω.

(b) Let $\Omega = \{re^{i\vartheta} \,|\, 1 < r < 2, \vartheta \in \mathbb{R}\}$. For each even positive integer n, let

$$F_n = \left\{ re^{i\vartheta} \,\middle|\, r = 1 + \frac{1}{n} \text{ and } \frac{\pi}{3} \le \vartheta \le \frac{5\pi}{3} \right\},$$

and for each odd integer $n \ge 3$, let

$$F_n = \left\{ re^{i\vartheta} \,\middle|\, r = 1 + \frac{1}{n} \text{ and } -\frac{2\pi}{3} \le \vartheta \le \frac{2\pi}{3} \right\}.$$

Let $F = \bigcup_{n=2}^{\infty} F_n$, and, as usual, let $\Gamma_r = \{re^{i\vartheta} \,|\, \vartheta \in \mathbb{R}\}$ for all positive real numbers r. It is easy to see that F is a T_σ-set of Ω. Let Y be any totally disconnected subset of $\Gamma_1 \cup \Gamma_2$. Then there exists no weakly Y-accesible setting for F in Ω, because there does not exist an arc in $\Omega \setminus F$ which joins a point of Ω to a point of Γ_1.

(c) Let Ω and F be as in (b), choose $x_0 \in \Gamma_2$, and let $Y = \Gamma_1 \cup \{x_0\}$. Then it is easy to find a weakly Y-accessible setting for F in Ω, but there exists no Y-accessible setting for F in Ω.

3.48. Definition. Let F be a compact subset of $\mathbf{S}^2$ and let $Y \subseteq F$.

(a) Y is said to be a WATI *subset of* F if, for each component Ω of $\mathbf{S}^2 \setminus \mathscr{C}(F, \overline{Y})$, $F \cap \Omega$ is a weakly Y-accessible T_σ-set of Ω. ("WATI" stands for "weakly-accessible T_σ-set inducing".)

(b) Y is said to be an ATI *subset of* F if, for each component Ω of $\mathbf{S}^2 \setminus \mathscr{C}(F, \overline{Y})$, $F \cap \Omega$ is a Y-accessible T_σ-set of Ω.

("ATI" stands for "accessible T_σ-set inducing".)

We remark that every ATI subset of F is a WATI subset of F.

The Relationship between the Sets $\mathscr{F}_r(\varphi)$, $\mathscr{G}_r(\varphi)$, $\mathscr{F}_s(\varphi)$, $\mathscr{G}_s(\varphi)$

3.49. Theorem. *Let φ be a flow in $\mathbf{S}^2$, and let Ω be a component of*

$\mathbf{S}^2 \setminus \big(\mathscr{F}_s(\varphi) \cup \mathscr{G}_s(\varphi)\big)$, *i.e., Ω is a component of $\mathscr{F}_r(\varphi) \cup \mathscr{G}_r(\varphi)$.*

Then $\mathscr{F}(\varphi) \cap \Omega$ is a T-set of Ω.

PROOF: Since every component of $\Omega \setminus \mathscr{F}(\varphi)$ is a component of $\mathscr{G}_r(\varphi)$, it follows from Theorem 3.13 that every component of $\Omega \setminus \mathscr{F}(\varphi)$ is an annulus. Furthermore, every component of $\mathscr{F}(\varphi) \cap \Omega$, being a component of $\mathscr{F}(\varphi)$, must be compact. This completes the proof of the theorem.
QED

3.50. Theorem. *Let φ be a flow in $\mathbf{S}^2$, let Ω be a component of $\mathbf{S}^2 \setminus \mathscr{F}_s(\varphi)$, and let $\{\Omega_j \mid j = 1, 2, \ldots\}$ be the family of components of $\Omega \setminus \mathscr{G}_s(\varphi)$.*

Then $\mathscr{F}(\varphi) \cap \Omega$ is a T_σ-set of Ω; in fact, $\{\Omega_j \mid j = 1, 2, \ldots\}$ is a setting for $\mathscr{F}(\varphi) \cap \Omega$ in Ω.

PROOF: It is clear that for every $j = 1, 2, \ldots$, Ω_j, being a component of $\mathscr{F}_r(\varphi) \cup \mathscr{G}_r(\varphi)$, is open. Further, it is trivial that $\mathscr{F}(\varphi) \cap \Omega$ is relatively closed in Ω, and that

$$\mathscr{F}(\varphi) \cap \Omega = \mathscr{F}_r(\varphi) \cap \Omega \subseteq \bigcup_j \Omega_j.$$

Theorem 3.49 implies that $\mathscr{F}(\varphi) \cap \Omega_j$ is a T-set of Ω_j for each $j = 1, 2, \ldots$. Finally, we note that for every $x \in \Omega \setminus \bigcup_j \Omega_j$, we have $x \in \mathscr{G}_s(\varphi)$, and Corollary 3.19 implies that there exists an arc $\varLambda \subseteq \mathscr{G}_s(\varphi)$, which joins x to a subset P of $\mathscr{S}(\varphi)$. Clearly, $\varLambda \subseteq \Omega \setminus \bigcup_j \Omega_j$, and $P \subseteq \partial(\Omega)$.

This completes the proof of the theorem. **QED**

3.51. Theorem. *Let φ be a flow in $\mathbf{S}^2$, let Ω be a component of $\mathbf{S}^2 \setminus \mathscr{F}_s(\varphi)$, let*

$$F = \mathscr{F}(\varphi) \cap \Omega,$$

and let $\{\Omega_j \mid j = 1, 2, \ldots\}$ *be the family of components of* $\Omega \setminus \mathscr{G}_s(\varphi)$. *Let* $\{\varDelta_j \mid j = 1, 2, \ldots\}$ *be a trim setting for F in* Ω, *obtained by an application of Theorem 3.39 to* $\{\Omega_j \mid j = 1, 2, \ldots\}$.[3]

Then $\{\varDelta_j \mid j = 1, 2, \ldots\}$ *is a weakly* $\mathscr{S}(\varphi)$-*accessible setting for F in* Ω.

Furthermore, if φ *is semi-simple, then* $\{\varDelta_j \mid j = 1, 2, \ldots\}$ *is an* $\mathscr{S}(\varphi)$-*accessible setting for F in* Ω.

PROOF: Let $\varPi$ be the set of positive integers, and let $\varPi_1$ be the set of those positive integers j for which $\varDelta_j$ is defined; *i.e.*, $\varPi_1$ is the set of those positive integers j for which either $F \cap \Omega_j \neq \square$, or some component of $\partial(\Omega_j)$ does not meet Ω.

Now suppose B is a component of $\partial(\Omega)$ which is not wrapped by $\{\varDelta_j \mid j \in \varPi_1\}$. For every $j \in \varPi$, let

$$\Omega_j^B = \mathbf{S}^2 \setminus [B \text{ side} : \Omega_j],$$

and for every $j \in \varPi_1$, let

$$\varDelta_j^B = \mathbf{S}^2 \setminus [B \text{ side} : \varDelta_j].$$

Let $\varPi_2$ be the set of those $j \in \varPi_1$ for which $\varDelta_j^B$ is maximal in $\{\varDelta_i^B \mid i \in \varPi_1\}$.

For every $j \in \varPi_2$, $\partial(\Omega_j^B)$ is one of the two components of the boundary of a component annulus T_j of $\Omega_j \setminus F$, and it is not hard to see that the other component of $\partial(T_j)$ is a subset of $\mathscr{F}(\varphi)$. For every $j \in \varPi_2$, define $F_j^B = \Omega_j^B \setminus T_j$. Then $\{F_j^B \mid j \in \varPi_2\}$ is a pairwise disjoint family of compact, connected sets, and for every $j \in \varPi_2$, we have $\partial(F_j^B) \subseteq \mathscr{F}(\varphi)$.

What we have to do to complete the proof of the theorem is to show that B satisfies condition (ii) of Definition 3.45. The rest of the proof can be motivated by the observation that had we been fortunate enough to have

$$\mathbf{S}^2 \setminus \Omega \subseteq \mathscr{F}(\varphi), \text{ and } F_j^B \subseteq \mathscr{F}(\varphi), \qquad \forall j \in \varPi_2,$$

then to show that B satisfies condition (ii) of Definition 3.45, it would have been sufficient to show that, given a neighborhood U of B, there exists $y \in U$ such that $\mathrm{p}_\varphi(y) = \infty$, and either $\alpha_\varphi(y) \subseteq U \cap \mathscr{S}(\varphi)$, or $\omega_\varphi(y) \subseteq U \cap \mathscr{S}(\varphi)$. But this latter condition will follow from Theorem 3.26, as soon as we show that there are points of $\mathscr{G}_s(\varphi)$ lying arbitrarily close to B.

Fortunately, it is possible to reduce to this special case, by redefining φ to be everywhere fixed in K, where

$$K = (\mathbf{S}^2 \setminus \Omega) \cup \bigcup_{j \in \varPi_2} F_j^B.$$

---

[3] Of course, $\varDelta_j$ need not be defined for every positive integer j.

In order to show that this is possible, let us examine the set K. It is easy to see that $\mathscr{F}(\varphi) \subseteq K$. Also, K is compact, for if $x \in \mathbf{S}^2 \setminus K$, then since $x \in \mathscr{G}(\varphi)$, we can find a connected neighborhood V of x such that $V \subseteq \mathscr{G}(\varphi)$, and since $\partial(\Omega) \subseteq \mathscr{F}(\varphi)$, V cannot meet $\mathbf{S}^2 \setminus \Omega$, and since $\partial(F_j^B) \subseteq \mathscr{F}(\varphi)$ for every $j \in \Pi_2$, V cannot meet F_j^B for any $j \in \Pi_2$. From the fact that K is compact, it follows that

$$\partial(K) \subseteq \partial(\Omega) \cup \bigcup_{j \in \Pi_2} \partial(F_j^B) \subseteq \mathscr{F}(\varphi).$$

It is clear that for each $j \in \Pi_2$, F_j^B is an isolated component of K, in fact, F_j^B and $K \setminus F_j^B$ lie in opposite sides of T_j. From this, it follows easily that the components of K are precisely the sets $F_j^B \ (j \in \Pi_2)$ together with those components of $\mathbf{S}^2 \setminus \Omega$ which do not meet $\bigcup_{j \in \Pi_2} F_j^B$.

In particular, $[B \text{ side} : \Omega]$ is a component of K. We remark finally that K is invariant of φ.

Using these properties of K, we now define a flow ψ in $\mathbf{S}^2$ as follows:

$$\psi(t, x) = x, \quad \forall\, x \in K \quad \forall\, t \in \mathbb{R},$$

and

$$\psi(t, x) = \varphi(t, x), \quad \forall\, x \in \mathbf{S}^2 \setminus K \quad \forall\, t \in \mathbb{R}.$$

It is clear that ψ is well defined and continuous, that $\mathscr{S}(\psi) \subseteq \mathscr{S}(\varphi)$, and that $\mathscr{F}(\psi) = K$. Further, it is easy to see that for every $j \in \Pi_2$, $T_j \subseteq \mathscr{G}_r(\psi)$, and therefore

$$\mathscr{G}_s(\psi) \subseteq \Omega \setminus \bigcup_{j \in \Pi_2} \Omega_j^B \subseteq \Omega \setminus \bigcup_{j \in \Pi_2} \overline{\Delta_j^B} = \Omega \setminus \bigcup_{j \in \Pi_1} \overline{\Delta_j^B}.$$

Now suppose we could prove that there are points of $\mathscr{G}_s(\psi)$, lying arbitrarily close to B, and suppose U is a neighborhood of B. Using Theorem 3.26, we could then choose $y \in U$ such that $p_\psi(y) = \infty$, and either $\alpha_\psi(y) \subseteq U \cap \mathscr{S}(\psi)$, or $\omega_\psi(y) \subseteq U \cap \mathscr{S}(\psi)$. This, together with the fact that

$$\mathscr{O}_\psi(y) \subseteq \mathscr{G}_s(\psi) \subseteq \Omega \setminus \bigcup_{j \in \Pi_1} \overline{\Delta_j^B}$$

would then show that $\{\Delta_j \mid j \in \Pi_1\}$ satisfies condition (ii) of Definition 3.45, thus completing the proof of the theorem.

We must therefore show that there are points of $\mathscr{G}_s(\psi)$ lying arbitrarily close to B. Suppose, to obtain a contradiction, that some neighborhood of B does not meet $\mathscr{G}_s(\psi)$. Using Theorem C. 30, choose a Jordan curve $J \subseteq \Omega$, so close to B that neither J, nor the annulus A between J and B, contains any point of $\mathscr{G}_s(\psi)$.

We claim that $(A \cup J) \cap \Omega$ is arc-connected, for suppose $x, y \in (A \cup J) \cap \Omega$, and choose an arc $\Lambda \subseteq \Omega$ which joins x to y. If $\Lambda \cap J = \square$, then $\Lambda \subseteq \Omega \cap A$. Otherwise, there is a first and last time that Λ meets J, and by replacing Λ between these two points by a part of J, we obtain an arc $\Lambda' \subseteq (A \cup J) \cap \Omega$ which joins x to y. Therefore $(A \cup J) \cap \Omega$ is connected. Now

$$(A \cup J) \cap \Omega \subseteq (\mathscr{F}(\psi) \cap \Omega) \cup \mathscr{G}_r(\psi)$$

$$\subseteq (K \cap \Omega) \cup \mathscr{G}_r(\varphi)$$

$$\subseteq \bigcup_{j \in \Pi_1} \Omega_j^B \cup \bigcup_{j \in \Pi} \Omega_j$$

$$= \bigcup_{j \in \Pi_2} \Omega_j^B \cup \bigcup_{j \in \Pi_3} \Omega_j$$

where Π_3 is the set of those positive integers $j \in \Pi \setminus \Pi_1$ for which

$$\Omega_j \subseteq [B \text{ side} : \Omega_i], \qquad \forall\, i \in \Pi_2.$$

It follows, since $(A \cup J) \cap \Omega$ is connected, and the family

$$\{\Omega_j^B \mid j \in \Pi_2\} \cup \{\Omega_j \mid j \in \Pi_3\}$$

is pairwise disjoint, that there exists $j_0 \in \Pi_2 \cup \Pi_3$ such that

$$(A \cup J) \cap \Omega \subseteq \Omega_{j_0}^B.$$

It is clear that $B \subseteq \partial(\Omega_{j_0}^B)$. But on the other hand, since $J \subseteq \Omega_{j_0}^B$, we must have

$$\partial(\Omega_{j_0}^B) \subseteq [B \text{ side} : J],$$

and therefore $\partial(\Omega_{j_0}^B)$ cannot meet Ω, i.e., $\partial(\Omega_{j_0}^B) \subseteq B$.

It follows that $B = \partial(\Omega_{j_0}^B)$, and therefore, that B is a component of $\partial(\Delta_{j_0})$. But this implies that B is sealed by $\{\Delta_j \mid j \in \Pi_1\}$, contrary to the assumption that B is not wrapped by $\{\Delta_j \mid j \in \Pi_1\}$. This contradiction completes the proof of the theorem.   **QED**

3.52. Theorem. *Let φ be a flow in $\mathbf{S}^2$. Then $\mathscr{S}(\varphi)$ is a WATI subset of $\mathscr{F}(\varphi)$. If, in addition, φ is semi-simple, then $\mathscr{S}(\varphi)$ is an ATI subset of $\mathscr{F}(\varphi)$.*

PROOF: Immediate from Theorem 3.51.   **QED**

3.53. Theorem. *Let F be a compact subset of $\mathbf{S}^2$, and let Y_1 be a WATI subset [respectively ATI subset] of F. Let Y_2 be a subset of F, and suppose $Y_1 \subseteq Y_2$.*
Then Y_2 is a WATI subset [respectively ATI subset] of F.

PROOF: We need to show that for each component $\varDelta$ of $\mathbf{S}^2 \setminus \mathscr{C}(F, \overline{Y_2})$, there exists a weakly Y_2-accessible [respectively, Y_2-accessible] setting for $F \cap \varDelta$ in $\varDelta$.

Suppose $\varDelta$ is a component of $\mathbf{S}^2 \setminus \mathscr{C}(F, \overline{Y_2})$, and let $\varOmega$ be that component of $\mathbf{S}^2 \setminus \mathscr{C}(F, \overline{Y_1})$ for which $\varDelta \subseteq \varOmega$. Let $\{\varOmega_j \mid j = 1, 2, \ldots\}$ be a weakly Y_1-accessible [respectively, Y_1-accessible] setting for $F \cap \varOmega$ in $\varOmega$, and for each $j = 1, 2, \ldots$, let $\varDelta_j = \varOmega_j \cap \varDelta$. Let $\varPi$ be the set of positive integers, and let $\varPi_1 = \{j \mid \varDelta_j \neq \square\}$.

The remainder of the proof shows that $\{\varDelta_j \mid j \in \varPi_1\}$ is a weakly Y_2-accessible [respectively, Y_2-accessible] setting for $F \cap \varDelta$ in $\varDelta$.

First, we shall look at the components of $\partial(\varDelta)$, and the components of the sets $\partial(\varDelta_j)$, $j = 1, 2, \ldots$. If B is a component of $\partial(\varDelta)$, then since B is contained in a component of $\mathscr{C}(F, \overline{Y_2})$, we see that depending on whether or not this component of $\mathscr{C}(F, \overline{Y_2})$ meets $\overline{Y_1}$, B is either contained in (and therefore, by Theorem C.9, coincides with) a component of $\partial(\varOmega)$, or B is contained in $F \cap \varOmega_j$ for some $j \in \varPi$. If, for any j, B is a component of $\partial(\varDelta_j)$, then since $\partial(\varDelta_j) \subseteq \partial(\varDelta) \cup \partial(\varOmega_j)$, it follows that either

(i) $B \cap \varDelta = \square$,

or

(ii) B is an analytic Jordan curve which is contained in $\varOmega$ and is a component of $\partial(\varOmega_j)$, and since $B \cap F = \square$, B must be contained in $\varDelta$, and $\varDelta_j$ must contain the component of $\varOmega_j \setminus F$ which has B in its boundary.

Now if (i) applies, B is clearly included in a component of $\partial(\varDelta)$, and therefore, if only we knew that $\varDelta_j$ were connected, we could invoke Theorem C.9, and conclude that if (i) applies, B is a component of $\partial(\varDelta)$.

We shall now show that for each $j \in \varPi$, $\varDelta_j$ is connected. It is trivial that $\varDelta_j$ is connected whenever $\varDelta_j = \square$. Suppose, then, that $\varDelta_j \neq \square$, and choose a point of $\varDelta_j$ to play the role of ∞. Let $\{B_\alpha \mid \alpha \in A\}$ be the family of those components of $\partial(\varDelta_j)$ which are subsets of $\varDelta$. It is easy to see that each component of $\partial(\varOmega_j)$ which meets $\varDelta$ must be one of the sets $B_\alpha (\alpha \in A)$. Because of this, we have

$$\varDelta_j = \varDelta \setminus \bigcup_{\alpha \in A} \overline{\text{ins} (B_\alpha)}$$
$$= [\varDelta \cup \bigcup_{\alpha \in A} \text{ins} (B_\alpha)] \setminus \bigcup_{\alpha \in A} \overline{\text{ins} (B_\alpha)},$$

and an application of Corollary C.12 to the connected, open set $\varDelta \cup \bigcup_{\alpha \in A} \text{ins} (B_\alpha)$ shows that $\varDelta_j$ is connected.

At this stage, it is easy to check that $\{\varDelta_j \mid j \in \Pi_1\}$ is a trim setting for $F \cap \varDelta$ in $\varDelta$. We leave this simple task to the reader.

Finally, to show that $\{\varDelta_j \mid j \in \Pi_1\}$ is a weakly Y_2-accessible [respectively Y_2-accessible] setting for $F \cap \varDelta$ in $\varDelta$, suppose B is a component of $\partial(\varDelta)$, and B is not wrapped by $\{\varDelta_j \mid j \in \Pi_1\}$. We clearly cannot have $B \subseteq \Omega_j$ for any $j \in \Pi$, and therefore B is a component of $\partial(\Omega)$. Let

$$\Omega_j^B = \mathbf{S}^2 \setminus [B \text{ side} : \Omega_j], \qquad \forall\, j \in \Pi,$$

and

$$\varDelta_j^B = \mathbf{S}^2 \setminus [B \text{ side} : \varDelta_j], \qquad \forall\, j \in \Pi_1.$$

We note that, if $j \in \Pi_1$, then $\varDelta_j^B = \Omega_j^B$, for $\partial(\varDelta_j^B)$, being a subset of $\varDelta$, must be a component of $\partial(\Omega_j)$, and both $\varDelta_j$ and Ω_j lie in one side of $\partial(\varDelta_j^B)$ while B lies in the other side.

Next, we note that if C is a component of $\mathbf{S}^2 \setminus \varDelta$, and $C \cap \Omega \neq \square$, then $C \subseteq \Omega_j^B$ for some $j \in \Pi_1$, for given such a component C, it is clear that $\partial(C)$ cannot be a component of $\partial(\Omega)$, and therefore since $\partial(C)$ is a component of $\partial(\varDelta)$, we see that

$$\partial(C) \subseteq F \cap \Omega_j \quad \text{for some } j \in \Pi.$$

Clearly, Ω_j, being a neighborhood of $\partial(C)$, must meet $\varDelta$, i.e., $j \in \Pi_1$. Since $\Omega_j^B = \varDelta_j^B$, we have $\partial(\Omega_j^B) \subseteq \varDelta$, and consequently

$$C \cap \partial(\Omega_j^B) = \square,$$

and therefore, since $\Omega_j \cap C \neq \square$, C and Ω_j lie in the same side of $\partial(\Omega_j^B)$, and therefore $C \subseteq \Omega_j^B$.

We note also, that if $j \in \Pi \setminus \Pi_1$, then there exists $j_1 \in \Pi_1$ such that $\Omega_j^B \subseteq \Omega_{j_1}^B$, for given $j \in \Pi \setminus \Pi_1$, let C be the component of $\mathbf{S}^2 \setminus \varDelta$ that contains Ω_j, and choose $j_1 \in \Pi_1$ such that $C \subseteq \Omega_{j_1}^B$. Clearly,

$$\Omega_j^B \subseteq C \subseteq \Omega_{j_1}^B.$$

It is now easy to see that B, regarded as a component of $\partial(\Omega)$, is not wrapped by $\{\Omega_j \mid j \in \Pi\}$, and therefore in every neighborhood U of B, there is an arc

$$\varLambda \subseteq \Omega \setminus \bigcup_{j \in \Pi} \overline{\Omega_j^B},$$

such that $\varLambda$ joins a point of $\Omega \setminus \bigcup_j \overline{\Omega_j^B}$ to a subset of $Y_1 \cap U$ [respectively to a member of $Y_1 \cap U$]. We complete the proof of the theorem by showing that

$$\varLambda \subseteq \varDelta \setminus \bigcup_{j \in \Pi_1} \overline{\varDelta_j^B}.$$

Now

$$\bigcup_{j\in\Pi_1} \overline{\mathit{\Delta}_j^B} = \bigcup_{j\in\Pi} \overline{\mathit{\Omega}_j^B},$$

and we therefore need only note that since every component of $\mathbf{S}^2 \setminus \mathit{\Delta}$ which meets $\mathit{\Omega}$ must be contained in one of the sets $\overline{\mathit{\Omega}_j^B}$, we have

$$\mathit{\Omega} \setminus \mathit{\Delta} \subseteq \bigcup_{j\in\Pi} \overline{\mathit{\Omega}_j^B}.$$

Thus,

$$\mathit{\Lambda} \subseteq \mathit{\Delta} \setminus \bigcup_{j\in\Pi_1} \overline{\mathit{\Delta}_j^B},$$

and the proof is complete.   **QED**

3.54. Theorem. *Let Y be a subset of a compact subset F of $\mathbf{S}^2$.*

Then a necessary condition for the existence of a flow [respectively semi-simple flow] φ in $\mathbf{S}^2$, such that

$$\mathscr{F}(\varphi) = F \text{ and } \mathscr{S}(\varphi) \subseteq Y,$$

is that Y be a WATI subset [respectively ATI subset] of F.

PROOF: Immediate from Theorems 3.52 and 3.53.   **QED**

3.55. Remark. Theorem 3.54 is important, because it suggests the following question: Given a subset Y of a compact subset F of $\mathbf{S}^2$, is the condition that Y be a WATI subset [respectively ATI subset] of F, *sufficient* for the existence of a flow [respectively semi-simple flow] φ in $\mathbf{S}^2$ such that

$$\mathscr{F}(\varphi) = F \text{ and } \mathscr{S}(\varphi) \subseteq Y?$$

In a very large class of special cases, most of which will be dealt with later in this book, the answer to this question is YES, but in general, the answer is unknown.[4]

We note that Theorem 3.53 implies that if F is a non-empty compact subset of $\mathbf{S}^2$, then large subsets of F are more likely to be WATI-subsets of F than are small ones. (In fact, the reader can easily verify that F is always an ATI subset of itself.) In view of this, it is of interest that the class of special cases mentioned above includes all cases in which Y is, in some sense, small. In fact, the answer to the above question is known to be *yes* whenever $\mathscr{C}(F, \overline{Y})$ has only countably many components.

---

[4] (Added in proof) The answer is YES in the general case; the proof will appear later.

Special Cases

3.56. Introduction. Broadly speaking, there are two types of special cases which concern us.

The first type occurs when $\mathscr{S}(\varphi)$ is totally disconnected (for instance, if $\mathscr{S}(\varphi)$ is countable), and some simplification of the theory results from the fact that φ is automatically semi-simple.

The second type of special case occurs when we assume that $\mathscr{F}_s(\varphi)$ does not have too many components, and in particular, if we assume that $\overline{\mathscr{S}(\varphi)}$ itself does not have too many components.

The rest of the chapter is concerned with this second type of special case.

3.57. Theorem. *Let φ be a flow in $\mathbf{S}^2$.*

(a) *The number of components of $\mathscr{F}_s(\varphi)$ does not exceed the number of components of $\overline{\mathscr{S}(\varphi)}$.*

(b) *For each component Ω of $\mathbf{S}^2 \setminus \mathscr{F}_s(\varphi)$, the number of components of $\mathbf{S}^2 \setminus \Omega$ does exceed not the number of components of $\mathscr{F}_s(\varphi)$, i.e., the connectivity of Ω does not exceed the number of components of $\mathscr{F}_s(\varphi)$.*

PROOF: Trivial.   **QED**

3.58. Theorem. *Let $\{\Omega_j \mid j = 1, 2, \ldots\}$ be a setting for a set F in a connected, open subset Ω of $\mathbf{S}^2$.*

Then for each j, the connectivity of Ω_j is no more than the connectivity of Ω. Further, if Ω is finitely connected, then all but finitely many of the sets $\Omega_j (j = 1, 2, \ldots)$ are discs.

PROOF: Suppose, for some j_0, that the connectivity of Ω_{j_0} is greater than the connectivity of Ω. Then since each component of $\mathbf{S}^2 \setminus \Omega$ is contained in a component of $\mathbf{S}^2 \setminus \Omega_{j_0}$, there exists a component C of $\mathbf{S}^2 \setminus \Omega_{j_0}$ such that $C \cap (\mathbf{S}^2 \setminus \Omega) = \square$, i.e., $C \subseteq \Omega$. Choose $x \in \partial(C)$. Then since $x \in \Omega \setminus \bigcup_j \Omega_j$, there exists an arc $\Lambda \subseteq \Omega \setminus \bigcup_j \Omega_j$, such that Λ joins x to a subset of $\partial(\Omega)$. But this is impossible, since $\Lambda \subseteq C$, and C is a compact subset of Ω.

It follows that, for each j, the connectivity of Ω_j is no more than the connectivity of Ω.

Now suppose Ω is finitely connected, and let n be the connectivity of Ω. For each j, Ω_j determines a partition of the set of components of $\mathbf{S}^2 \setminus \Omega$, into equivalence classes, in the following way: Two components of $\mathbf{S}^2 \setminus \Omega$ lie in the same class iff they lie in the same component of $\mathbf{S}^2 \setminus \Omega_j$.

It is easy to see that whenever $j_1 \neq j_2$ and both Ω_{j_1} and Ω_{j_2} are not discs, the partitions determined by Ω_{j_1} and Ω_{j_2} are different. Therefore, since the set of components of $\mathbf{S}^2 \setminus \Omega$, being finite, can be partitioned in only finitely many ways, it is clear that at most finitely many of the sets $\Omega_j (j = 1, 2, \ldots)$ can fail to be discs.[5]   **QED**

3.59. Theorem. *Let $\{\Omega_j \mid j = 1, 2, \ldots\}$ be a trim setting for a set F in a finitely connected region Ω, and let B be a component of $\partial(\Omega)$. Then B is wrapped by $\{\Omega_j \mid j = 1, 2, \ldots\}$ iff B is sealed by $\{\Omega_j \mid j = 1, 2, \ldots\}$.*

PROOF: The result follows at once from Theorem 3.58.   **QED**

3.60. Theorem. *Let φ be a flow in $\mathbf{S}^2$, and suppose $\mathscr{F}_s(\varphi)$ has only finitely many compoments.*
(a) *Every component of $\mathbf{S}^2 \setminus \mathscr{F}_s(\varphi)$ is finitely connected, and all but finitely many of these components are discs.*
(b) *Every component of $\mathbf{S}^2 \setminus \left(\mathscr{F}_s(\varphi) \cup \mathscr{G}_s(\varphi)\right)$ is finitely connected, and all but finitely many of these components are discs.*

PROOF: Because both $\mathscr{F}_s(\varphi)$ and $\mathscr{F}_s(\varphi) \cup \mathscr{G}_s(\varphi)$ have only finitely many components, both (a) and (b) follow from an argument similar to that used in the proof of the second part of Theorem 3.58.   **QED**

3.61. Theorem. *Let φ be a flow in $\mathbf{S}^2$ and suppose $\mathscr{S}(\varphi)$ is a finite set.*

(a) *$\mathscr{S}(\varphi)$ is an ATI subset of $\mathscr{F}(\varphi)$.*

(b) *Every component of $\mathbf{S}^2 \setminus \mathscr{F}_s(\varphi)$ is finitely connected, and all but finitely many of these components are discs.*

(c) *Every component of $\mathbf{S}^2 \setminus \left(\mathscr{F}_s(\varphi) \cup \mathscr{G}_s(\varphi)\right)$ is finitely connected, and all but finitely many of these components are discs.*

PROOF: Clear.   **QED**

3.62. Theorem. *Let φ be a flow in $\mathbf{S}^2$, and suppose $\mathscr{S}(\varphi) = \square$.*
Then $\mathscr{F}_s(\varphi) = \mathscr{G}_s(\varphi) = \square$, and $\mathscr{F}(\varphi)$ is a T-set of $\mathbf{S}^2$.

PROOF: Clear.   **QED**

---

[5] The reader may find some interest in proving that, if for each j, k_j is the connectivity of Ω_j, then

$$1 + \sum_j (k_j - 1) \leq n,$$

and that consequently, at most $n - 1$ of the sets Ω_j can fail to be discs.

3.63. Corollary [Beck [2], Theorem 5]. *Let φ be a flow in $\mathbf{S}^2$, and suppose every φ-orbit is compact.*

Then $\mathscr{F}_s(\varphi) = \mathscr{G}_s(\varphi) = \square$, and $\mathscr{F}(\varphi)$ is a T-set of $\mathbf{S}^2$.

PROOF: Clear.   **QED**

3.64. Theorem [Beck [2], Theorem 11]. *Let φ be a flow in $\mathbf{R}^2$, and suppose every φ-orbit is a closed[6] subset of $\mathbf{R}^2$. Suppose also that at least one φ-orbit is not compact.*

Then the union F of the family of unbounded components of $\mathscr{F}(\varphi)$, is a closed[6] subset of $\mathbf{R}^2$, and every component of $\mathbf{R}^2 \setminus F$ is a disc. Further, for each component Ω of $\mathbf{R}^2 \setminus F$, $\mathscr{F}(\varphi) \cap \Omega$ is a $\{\infty\}$-accessible T_σ-set of Ω.

PROOF: Let ψ be the usual extension of φ to a flow in $\mathbf{S}^2$, defined by making ∞ a fixed point. Then $\mathscr{F}(\psi) = \mathscr{F}(\varphi) \cup \{\infty\}$, and $\mathscr{S}(\psi) = \{\infty\}$, and the theorem will therefore follow at once when we have shown that

$$\mathscr{F}_s(\psi) = F \cup \{\infty\}.$$

But it is clear that $F \cup \{\infty\} \subseteq \mathscr{F}_s(\psi)$, and on the other hand, if $x \in \mathscr{F}_s(\psi)$, then either $x = \infty$, or we can choose a minimal continuum C from x to ∞ such that $C \subseteq \mathscr{F}_s(\psi)$ (see Theorem A. 7), and since $C \setminus \{\infty\}$ is a connected, unbounded subset of $\mathscr{F}(\varphi)$, we have $x \in F$.

This completes the proof of the theorem.   **QED**

Notes and Remarks to Chapter 3

The subject matter of Beck [2] which deals with flows in $\mathbf{R}^2$, all of whose orbits are closed as subsets of $\mathbf{R}^2$, is divided into two distinct parts. The first part deals exclusively with flows all of whose orbits are compact (*i.e.* periodic), while the second focuses attention on the altogether different behavior of non-compact orbits. In examining these two types of behavior we soon realized that the significant criterion that determines whether a given orbit should behave in one way or the other, is not so much the compactness or non-compactness of the orbit, but is rather the fact that if φ is a flow in the plane and all the φ-orbits are closed, then every non-compact φ-orbit has a *stagnation point* in $\mathbf{S}^2$ (namely, ∞), while every non-trivial compact orbit is contained in a neighborhood of points whose orbits (being compact) do not stagnate. This realization led us to the first generalization of this work. We showed (see Beck [3]) that if φ is a flow in the plane with no stagnation

---

[6] By "closed", we mean, of course, closed in the relative topology of $\mathbf{R}^2$ as a subspace of $\mathbf{S}^2$, *i.e.*, in the usual topology of $\mathbf{R}^2$.

points except possibly ∞ (equivalently, if φ is a flow in $\mathbf{S}^2$ with at most one stagnation point), then the behavior of φ is similar to the behavior of a flow in $\mathbb{R}^2$ with closed orbits. The sphere could be naturally divided into two parts: in the one part (the regular part) φ behaved like a flow with compact orbits, while in the other (the singular part), the behavior of φ was much more complicated. This led directly to the concept of regular point and singular point, which is described in Chapter 3.

In Beck [3], a detailed study is made of the behavior in its regular and singular parts of a flow φ in $\mathbf{S}^2$ which has only finitely many stagnation points. The theory parallels the development in this chapter, and climaxes with an analogue of Theorem 3.51. (The theory in Beck [3] then goes on to discuss the more difficult question of the sufficiency of the conditions in its analogue of Theorem 3.51. We shall discuss this in the notes and remarks to Chapters 6 and 7). In later work (which is mentioned in Beck [4] but has not previously been published), we went on to show that if a flow φ in $\mathbf{S}^2$ has the property that $\overline{\mathscr{S}(\varphi)}$ is countable, then it is still possible to divide $\mathbf{S}^2$ into φ-regular and φ-singular parts and to prove an analogue of Theorem 3.51. However, in this work, our definition of an accessible setting differed significantly from Definition 3.45, and was as follows:

Definition. Let F be a T_σ-set set of a connected open subset Ω of $\mathbf{S}^2$, and let Y be a subset of $\mathbf{S}^2 \setminus \Omega$.

A setting $\{\Omega_j \mid j = 1, 2, \ldots\}$ for F in Ω is said to be Y-accessible if

(i) The setting $\{\Omega_j \mid j = 1, 2, \ldots\}$ is trim,

and

(ii) given any component B of $\partial(\Omega)$ which is not sealed by $\{\Omega_j \mid j = 1, 2, \ldots\}$ and given any neighborhood U of B, there exists an arc in $\Omega \setminus \bigcup_j \overline{\Omega}_j$ which joins a point of $\Omega \setminus \bigcup_j \overline{\Omega}_j$ to a point of $Y \cap U$.

Since the set Y always had countable closure in that work, there was no need to define *weakly Y-accessible* setting. In addition, the concept of *wrapped* boundaries did not appear there. However, it can be shown that if the word "sealed" in condition (ii), shown above is replaced by the word "wrapped", the preceding definition is unaltered. Therefore it is clear that a setting which is accessible according to Definition 3.45 is certainly accessible according to the earlier definition. But unless Y is finite, the converse is not true, and Theorem 3.51 is a substantially stronger result than its analogue in Beck [4] for flows φ with $\overline{\mathscr{S}(\varphi)}$ countable. (More remarks about this can be found in the notes and remarks to Chapter 7.)

In formulating the theory of regular and singular points as it appears in Beck [3], our purpose was to describe those compact subsets of S^2 which are the fixed sets of flows which have, in some sense, only a few stagnation points. (For example, we showed that a necessary and sufficient condition that a compact subset F of S^2 be the fixed set of a flow φ in S^2 with no stagnation points at all, is that F be a T-set of S^2.) All of this work was done against the background of a general theorem to the effect that *every* non-empty compact subset of S^2 is the fixed set of a flow φ in S^2 which has only countably many stagnation points. (See Theorem 7.24.) Therefore, in order to obtain a family of flows with a non-trivial family of fixed sets, we confined our attention to flows φ in S^2 with $\overline{\mathscr{S}(\varphi)}$ countable. We postulated (see Beck [4]) that if $\varPhi$ is a family of flows determined by a restriction on the size of their sets of stagnation points, then the family $\{\mathscr{F}(\varphi) \mid \varphi \in \varPhi\}$ should either be all non-empty compact subsets of S^2, or should consist of all compact sets which satisfy a condition based on the notions of T-sets, T_σ-sets, *etc.*

In making the changes that bring the theory of regular and singular points into its present form, the Lewins took an altogether different attitude. Instead of describing the fixed sets of flows with only a few stagnation points, they saw this theory as a partial answer to the question

"Given a compact subset F of S^2 and a subset Y of F, when does there exist a flow φ in S^2 such that

$$\mathscr{F}(\varphi) = F \text{ and } \mathscr{S}(\varphi) \subseteq Y ?"$$

(see Theorem 3.54 and Remark 3.55). Because of this approach, the Lewins had no need to make a blanket assumption about the sets of stagnation points, and therefore the theory as it appears in Chapter 3 covers all flows in the sphere. For reasons which are explained in the notes and remarks to Chapter 7, the Lewins found it necessary to prove Theorems 3.41 and 3.44, to introduce the notion of wrapped boundaries, and to change our earlier definition of "accessible setting" into its present form (Definition 3.45). This led to Theorem 3.51 as it appears in this chapter. Then, after proving Theorem 3.53, the Lewins went on to present the main result of the chapter in Theorem 3.54.

We note finally from Theorem 3.51 that suitably interpreted, the postulate in Beck [4] mentioned above is true.

CHAPTER FOUR[1]

REPARAMETRIZATION I

4.1. Introduction. In this chapter, we study a process for changing a given quasi-flow by altering the time taken to go from one point in the underlying space to another. The orbits of the new quasi-flow sometimes coincide with those of the old quasi-flow. However, in general, if two points lie in an orbit of the old quasi-flow, the new time between them might be infinite, in which case they would lie in different orbits of the new quasi-flow.

Speaking more formally, if we have a quasi-flow φ in a Hausdorff space X, we can construct a new quasi-flow ψ which is related to φ in special ways, among which is always the property that every ψ-orbit is contained in a φ-orbit. This process, which we call *reparametrization*, is useful in altering flows to give flows with certain specified properties relating to speed, continuity, *etc.*

The control this gives over the new flows is very useful in constructing flows which will fulfill the promises made in Remark 3.55.

4.2. Theorem. *Let φ and ψ be quasi-flows in a Hausdorff space X, and suppose that each ψ-orbit is contained in a φ-orbit. For each $x \in X$ and $t \in \mathbb{R}$, let*

$$A_x(t) = \{\tau \in \mathbb{R} \mid \psi(t, x) = \varphi(\tau, x)\}.$$

Then we have

(a) *If $x \in X$ and $\mathrm{p}_\varphi(x) = 0$, then $A_x(t) = \mathbb{R}$ for every $t \in \mathbb{R}$,*

(b) *If $x \in X$ and $0 < \mathrm{p}_\varphi(x) < \infty$, then for each $t \in \mathbb{R}$, $A_x(t)$ has a least nonnegative member, and denoting this member by τ, we have*

$$A_x(t) = \{\tau + n\,\mathrm{p}_\varphi(x) \mid n \text{ is an integer}\},$$

[1] Chapters 4 and 5 constitute the major portion of the doctoral dissertation of Jonathan Walter Lewin (Ph. D., University of Wisconsin, 1970).

and

(c) *If* $x \in X$ *and* $p_\varphi(x) = \infty$, *then* $A_x(t)$ *consists of a single point.*

PROOF: Trivial.   **QED**

4.3. Theorem. *Let ψ be a quasi-flow and φ be a flow in a Hausdorff space X, and suppose each ψ-orbit is contained in a φ-orbit. For each $x \in X$ and $t \in \mathbb{R}$, define*

$$A_x(t) = \{\tau \in \mathbb{R} \mid \psi(t, x) = \varphi(\tau, x)\}.$$

Let the function $h : X \times \mathbb{R} \to \mathbb{R}$ be defined as follows:

For each $x \in X$ and $t \in \mathbb{R}$,

$h(x, t) = 0$ *if* $p_\varphi(x) = 0$,

$h(x, t)$ *is the least nonnegative member of $A_x(t)$, if $0 < p_\varphi(x) < \infty$,*

and

$h(x, t)$ *is the unique member of $A_x(t)$ if $p_\varphi(x) = \infty$.*

For each $x \in X$ let the function $h_x : \mathbb{R} \to \mathbb{R}$ be defined by

$$h_x(t) = h(x, t), \qquad \forall \, t \in \mathbb{R}.$$

Then we have

(a) $\psi(t, x) = \varphi\big(h_x(t), x\big), \qquad \forall \, x \in X \qquad \forall \, t \in \mathbb{R}.$

(b) $h_x(0) = 0, \qquad \forall \, x \in X.$

(c) *If $x \in X$ and $s, t \in \mathbb{R}$, then*

$$h_x(s + t) \equiv h_{\varphi(h_x(s), x)}(t) + h_x(s), \qquad \mathrm{mod} \; p_\varphi(x).$$

(d) *If $x \in X$ and $p_\varphi(x) = 0$, then $h_x = 0$ and $p_\psi(x) = 0$.*

(e) *If $x \in X$ and $p_\varphi(x) = \infty$, then h_x is continuous, and either $h_x = 0$, or h_x is strictly monotone, in which case $p_\psi(x) = \infty$.*

(f) *If $x \in X$ and $0 < p_\varphi(x) < \infty$, then h_x is continuous at each point t for which $h_x(t) \neq 0$.*

(g) *If $x \in X$ and $0 < p_\varphi(x) < \infty$, then for each point t for which $h_x(t) = 0$, and for each sequence $\{t_n\}$ of real numbers which converges to t, the sequence $\{h_x(t_n)\}$ has only two possible cluster-points, namely, 0 and $p_\varphi(x)$.*

PROOF: (a) and (b) are trivial, (c) is an immediate consequence of the identity:

$$\psi(s + t, x) = \psi\big(t, \psi(s, x)\big), \qquad \forall \, x \in X \qquad \forall \, s, t \in \mathbb{R},$$

and (d) is trivial.

Suppose that $x \in X$ and $p_\varphi(x) = \infty$. Let $t \in \mathbb{R}$, and suppose $t_n \to t$. To show that $h_x(t_n) \to h_x(t)$, we may assume without loss of generality that $t_n \in [t - 1, t + 1]$, $\quad \forall\, n = 1, 2, \ldots$. Let

$$C = \{\psi(t', x) \mid t - 1 \leq t' \leq t + 1\}.$$

Then C, being a compact, connected subset of $\mathcal{O}_\psi(x)$, is also a compact, connected subset of $\mathcal{O}_\varphi(x)$, and therefore, by Theorem 1.25,

$$\{\tau \in \mathbb{R} \mid \varphi(\tau, x) \in C\} \text{ is bounded.}$$

Therefore, since $\varphi\big(h_x(t_n), x\big) \in C$, $\quad \forall\, n = 1, 2, \ldots$, the sequence $\{h_x(t_n)\}$ must be bounded. We shall now show that $h_x(t_n) \to h_x(t)$ by showing that $h_x(t)$ is the only cluster-point of $\{h_x(t_n)\}$. Suppose τ is any cluster-point of $\{h_x(t_n)\}$. Then $\varphi(\tau, x)$ is a cluster-point of $\{\varphi(h_x(t_n), x)\}$. But since $\psi(t_n, x) \to \psi(t, x)$, we have

$$\varphi\big(h_x(t_n), x\big) \to \varphi\big(h_x(t), x\big),$$

and we therefore conclude that

$$\varphi(\tau, x) = \varphi\big(h_x(t), x\big), \textit{ i.e.,} \quad \tau = h_x(t).$$

This shows that h_x is continuous.

If $p_\psi(x) = \infty$, then h_x is one-one, and is therefore strictly monotone. On the other hand, if $p_\psi(x) < \infty$, then $\mathcal{O}_\psi(x)$, being a compact connected subset of $\mathcal{O}_\varphi(x)$, is homeomorphic to a compact subinterval of $\mathbb{R}$ (by Theorem 1.25). Therefore, since no compact subinterval of $\mathbb{R}$ can be homeomorphic to a Jordan curve, we must have $p_\psi(x) = 0$, and $h_x = 0$.

To prove (f) and (g) suppose that $x \in X$ and $0 < p_\varphi(x) < \infty$. Let $t \in \mathbb{R}$ and suppose $t_n \to t$. Then since

$$\psi(t_n, x) \to \psi(t, x),$$

we have

$$\varphi\big(h_x(t_n), x\big) \to \varphi\big(h_x(t), x\big),$$

and therefore, since $\{h_x(t_n)\}$ is bounded, we see that every cluster-point of $\{h_x(t_n)\}$ is congruent to $h_x(t)$ modulo $p_\varphi(x)$. (f) and (g) follow easily from this and the observation that $h_x(t_n) \in [0, p_\varphi(x)]$, $\quad \forall\, n = 1, 2, \ldots$.

$$\text{QED}$$

4.4. Theorem. *Let φ be a flow and ψ be a quasi-flow in a Hausdorff space X, and suppose each ψ-orbit is contained in a φ-orbit.*

Then there exists a unique function

$$f : X \times \mathbb{R} \to \mathbb{R}$$

such that, defining the function $f_x : \mathbb{R} \to \mathbb{R}$, *for each* $x \in X$, *by the identity:*

$$f_x(t) = f(x, t), \qquad \forall\, t \in \mathbb{R},$$

we have

(a) $\psi(t, x) = \varphi(f_x(t), x)$, $\qquad \forall\, t \in \mathbb{R}$, $\qquad \forall\, x \in X$.

(b) *For every* $x \in X$, f_x *is continuous and* $f_x(0) = 0$.

(c) *For every* $x \in X$ *and* $s, t \in \mathbb{R}$,

$$f_x(s + t) = f_{\varphi(f_x(s), x)}(t) + f_x(s).$$

(d) *For every* $x \in \mathscr{F}(\psi)$, $f_x = 0$, *i.e.,*

$$f_x(t) = 0, \qquad \forall\, t \in \mathbb{R}.$$

(e) *For every* $x \in \mathscr{G}(\psi)$, f_x *is strictly monotone, and either*

$$\mathcal{O}_\psi(x) = \mathcal{O}_\varphi(x) \text{ and } f_x(\mathbb{R}) = \mathbb{R},$$

or

$\mathcal{O}_\psi(x) \neq \mathcal{O}_\varphi(x)$ *and* $f_x(\mathbb{R})$ *is an open interval containing* 0, *whose length does not exceed* $p_\varphi(x)$, *and*

$$\mathcal{O}_\psi(x) = \{\varphi(\tau, x) \mid \tau \in f_x(\mathbb{R})\}.$$

PROOF: We shall first establish the uniqueness. Suppose f and $\bar{f}$ are any two functions from $X \times \mathbb{R}$ into $\mathbb{R}$, which satisfy the above conditions. If $x \in X$ and $p_\varphi(x) = 0$, then $x \in \mathscr{F}(\psi)$ and (d) implies that

$$f(x, t) = \bar{f}(x, t) = 0, \qquad \forall\, t \in \mathbb{R}.$$

If $x \in X$ and $p_\varphi(x) = \infty$, then, since

$$\varphi(f(x, t), x) = \varphi(\bar{f}(x, t), x), \qquad \forall\, t \in \mathbb{R},$$

we have

$$f(x, t) = \bar{f}(x, t), \qquad \forall\, t \in \mathbb{R}.$$

Finally, if $x \in X$ and $0 < p_\varphi(x) < \infty$, then, since

$$\varphi(f(x, t), x) = \varphi(\bar{f}(x, t), x), \qquad \forall\, t \in \mathbb{R},$$

we see that, for every $t \in \mathbb{R}$, $f_x(t) - \bar{f}_x(t)$ must be a multiple of $p_\varphi(x)$. But $f_x(0) = \bar{f}_x(0)$, and both f_x and $\bar{f}_x$ are continuous. Therefore

$$f_x(t) = \bar{f}_x(t), \qquad \forall\, t \in \mathbb{R}.$$

This completes the proof of the uniqueness. We now prove that the function f exists.

Let $h : X \times \mathbb{R} \to \mathbb{R}$ be defined for ψ and φ, as in Theorem 4.3. We shall define f by defining the functions f_x separately for each $x \in X$, and for the purpose of defining the functions f_x, we treat the following five cases separately.

CASE 1: If $p_\varphi(x) = 0$, let $f_x(t) = h_x(t)$, $\qquad \forall \, t \in \mathbb{R}$.

CASE 2: If $p_\varphi(x) = \infty$, let $f_x(t) = h_x(t)$, $\qquad \forall \, t \in \mathbb{R}$.

CASE 3: If $0 < p_\varphi(x) < \infty$ and $p_\psi(x) = 0$, let $f_x(t) = h_x(t)$, $\forall \, t \in \mathbb{R}$.

CASE 4: If $0 < p_\varphi(x) < \infty$ and $p_\psi(x) = \infty$, then noting that Theorem 1.13 implies that $\mathcal{O}_\psi(x)$ is not a Jordan curve, and consequently that $\mathcal{O}_\psi(x) \neq \mathcal{O}_\varphi(x)$, choose a number $\delta \in \big(0, p_\varphi(x)\big)$ such that $\varphi(\delta, x) \notin \mathcal{O}_\psi(x)$. Clearly, $h_x(t) \neq \delta$, $\qquad \forall \, t \in \mathbb{R}$.

Define

$$f_x(t) = \begin{cases} h_x(t), & \text{if } 0 \leq h_x(t) < \delta, \\ h_x(t) - p_\varphi(x), & \text{if } \delta < h_x(t). \end{cases}$$

CASE 5: Suppose $0 < p_\varphi(x) < \infty$ and $0 < p_\psi(x) < \infty$. We note first that whenever $t_1, t_2 \in \mathbb{R}$, we have

$$h_x(t_1) = h_x(t_2) \text{ iff } t_1 - t_2 \text{ is a multiple of } p_\psi(x).$$

From this it follows easily that the restriction of h_x to $\big(0, p_\psi(x)\big)$ is one-one and nonvanishing, and that

$$h_x(0) = h_x\big(p_\psi(x)\big) = 0.$$

It follows, using Theorem 4.3 (f), that the restriction of h_x to $\big(0, p_\psi(x)\big)$ is a continuous, strictly monotone function from $\big(0, p_\psi(x)\big)$ into $\big(0, p_\varphi(x)\big)$. It is now clear that h_x has a limit on the right at 0, and a limit on the left at $p_\psi(x)$, and it follows easily from Theorem 4.3 (g), that one of these limits must be 0 and the other must be $p_\varphi(x)$.

If $\lim\limits_{t \to 0+} h_x(t) = 0$, define f_x as follows:

$f_x\big(t + m p_\psi(x)\big) = h_x(t) + m p_\varphi(x)$, whenever $0 \leq t < p_\psi(x)$ and m is an integer.

If $\lim\limits_{t \to 0+} h_x(t) = p_\varphi(x)$, define f_x as follows:

$f_x\big(t + m p_\psi(x)\big) = h_x(t) - (m + 1) p_\varphi(x)$, whenever $0 < t \leq p_\psi(x)$ and m is an integer.

By checking each case separately, it is easy to see that f satisfies conditions (a), (b), (d) and (e).

To prove that f satisfies condition (c), we note firstly that, for every $x \in X$ and $s, t \in \mathbb{R}$,

$$f_x(s + t) \equiv f_{\varphi(f_x(s), x)}(t) + f_x(s), \qquad \operatorname{mod} p_\varphi(x),$$

and secondly, that for every $x \in X$ and $s \in \mathbb{R}$,

$$f_x(s + 0) = f_{\varphi(f_x(s), x)}(0) + f_x(s).$$

Condition (c) now follows at once from the continuity of the functions f_x.

This completes the proof of the theorem.   **QED**

4.5. Theorem. *Let φ be a flow and ψ be a quasi-flow in a Hausdorff space X, and suppose each ψ-orbit is contained in a φ-orbit. Let f be the unique function from $X \times \mathbb{R}$ into $\mathbb{R}$ which satisfies the conditions of Theorem 4.4, and let $\tilde{f}$ be any function from $X \times \mathbb{R}$ into $\mathbb{R}$, which satisfies conditions* (a) *and* (b) *of Theorem 4.4 $\left(\text{where, for each } x \in X \text{ and } t \in \mathbb{R}, \tilde{f}_x(t) = \tilde{f}(x, t)\right)$.*

Then for each $x \in \mathscr{G}(\varphi)$, $f_x = \tilde{f}_x$; in particular, $\tilde{f}$ satisfies condition (e) *of Theorem 4.4, as well as the identity*

$$\tilde{f}_x(s + t) = \tilde{f}_{\varphi(\tilde{f}_x(s), x)}(t) + \tilde{f}_x(s)$$

whenever $x \in \mathscr{G}(\varphi)$ and $s, t \in \mathbb{R}$.

PROOF: The proof is almost identical to the proof of the uniqueness of f in Theorem 4.4, and is omitted.   **QED**

4.6. Quasi-Reparametrizers. Let φ be a quasi-flow in a Hausdorff space X. Let f be a function from $X \times \mathbb{R}$ into $\mathbb{R}$, and for each $x \in X$, define a function f_x from $\mathbb{R}$ into $\mathbb{R}$ by

$$f_x(t) = f(x, t), \qquad \forall \, t \in \mathbb{R}.$$

Then f is said to be a *quasi φ-reparametrizer* if

(i) For every $x \in \mathscr{G}(\varphi)$, f_x is continuous and $f_x(0) = 0$,

and

(ii) For every $x \in \mathscr{G}(\varphi)$ and $s, t \in \mathbb{R}$,

$$f_x(s + t) = f_{\varphi(f_x(s), x)}(t) + f_x(s).$$

Given a quasi φ-reparametrizer f, the *quasi φ-reparametrization* corresponding to f is defined to be the function ψ from $\mathbb{R} \times X$ into X, given by

$$\psi(t, x) = \varphi\big(f_x(t), x\big), \qquad \forall \, t \in \mathbb{R} \qquad \forall \, x \in X.$$

If f is a quasi φ-reparametrizer, then its corresponding quasi φ-reparametrization is denoted by f rep φ.

4.7. Theorem. *Let φ be a flow and let ψ be a quasi-flow in a Hausdorff space X, and suppose each ψ-orbit is contained in a φ-orbit.*

Then there exists a quasi φ-reparametrizer f such that

$$\psi = f \operatorname{rep} \varphi.$$

PROOF: Immediate from Theorem 4.4.   **QED**

4.8. Theorem. *Let φ be a quasi-flow in a Hausdorff space X, let f be a quasi φ-reparametrizer, and let $\psi = f \operatorname{rep} \varphi$.*
Then

(a) *ψ is a quasi-flow in X, and each ψ-orbit is contained in a φ-orbit.*
(b) *For every $x \in \mathscr{F}(\psi) \cap \mathscr{G}(\varphi)$, we have $f_x = 0$.*
(c) *For every $x \in \mathscr{G}(\psi)$, f_x is strictly monotone, and either*

$$\mathcal{O}_\psi(x) = \mathcal{O}_\varphi(x) \text{ and } f_x(\mathbb{R}) = \mathbb{R},$$

or

$\mathcal{O}_\psi(x) \neq \mathcal{O}_\varphi(x)$, *$f_x(\mathbb{R})$ is an open interval containing 0 whose length does not exceed $\mathrm{p}_\varphi(x)$, and*

$$\mathcal{O}_\psi(x) = \{\varphi(\tau, x) \mid \tau \in f_x(\mathbb{R})\}.$$

PROOF: It follows from the identity

$$\psi(t, x) = \varphi\big(f_x(t), x\big),$$

that for every $x \in X$, the function $t \to \psi(t, x)$ is continuous. The identity

$$\psi(s + t, x) = \psi\big(t, \psi(s, x)\big)$$

is trivial for all $s, t \in \mathbb{R}$ if $x \in \mathscr{F}(\varphi)$, and on the other hand, if $x \in \mathscr{G}(\varphi)$ and $s, t \in \mathbb{R}$, we have

$$
\begin{aligned}
\psi(s + t, x) &= \varphi\big(f_x(s + t), x\big) \\
&= \varphi\big(f_{\varphi(f_x(s), x)}(t) + f_x(s), x\big) \\
&= \varphi\big(f_{\varphi(f_x(s), x)}(t), \varphi\big(f_x(s), x\big)\big) \\
&= \psi\big(t, \varphi\big(f_x(s), x\big)\big) \\
&= \psi\big(t, \psi(s, x)\big).
\end{aligned}
$$

This shows that ψ is a quasi-flow. It is obvious that each ψ-orbit is contained in a φ-orbit, and therefore (a) is proved.

To prove (b), suppose $x \in \mathscr{F}(\psi) \cap \mathscr{G}(\varphi)$. The fact that $f_x = 0$ follows at once from the fact that $f_x(0) = 0$, $f_x(t)$ is a multiple of $\mathrm{p}_\varphi(x)$, $\forall \, t \in \mathbb{R}$, and f_x is continuous.

To prove (c), suppose $x \in \mathscr{G}(\psi)$ and $\mathrm{p}_\varphi(x) < \infty$. Then by concerning ourselves only with the restriction of φ to $\mathcal{O}_\varphi(x)$, we can assume, by Theorem 1.19, that φ is a flow, and the fact that f_x satisfies (c) follows easily from Theorem 4.4 (e).

Now suppose $x \in \mathscr{G}(\psi)$ and $\mathrm{p}_\varphi(x) = \infty$. To show that f_x satisfies condition (c), it is clearly sufficient to show that f_x is strictly monotone, i.e., that f_x is one-one. But if f_x is not one-one, then $\mathrm{p}_\psi(x) \neq \infty$, and on the one hand, $\mathcal{O}_\psi(x)$ is a Jordan curve, while on the other hand, since

$$\mathcal{O}_\psi(x) = \big\{ \varphi(\tau, x) \mid \tau \in f_x([0, \mathrm{p}_\psi(x)]) \big\},$$

$\mathcal{O}_\psi(x)$ must be homeomorphic to a compact interval, a contradiction. Therefore f_x is strictly monotone, and the theorem is proved.   **QED**

4.9. Corollary. *Let φ be a quasi-flow in a Hausdorff space X, let f be a quasi φ-reparametrizer, and let $\psi = f \operatorname{rep} \varphi$. Let $x \in X$, and suppose $0 < \mathrm{p}_\psi(x) < \infty$.*
Then $0 < \mathrm{p}_\varphi(x) < \infty$.

PROOF: The result follows easily from Theorem 4.8 (c).   **QED**

4.10. Corollary. *Let φ be a quasi-flow in a Hausdorff space X, let f be a quasi φ-reparametrizer, and let $\psi = f \operatorname{rep} \varphi$.*
Then for every $x \in X$, the set

$$U_x = \{ \tau \in \mathbb{R} \mid \varphi(\tau, x) \in \mathscr{G}(\psi) \}$$

is open, and for each $\tau \in U_x$, the component of U_x which contains τ is $\tau + f_{\varphi(\tau, x)}(\mathbb{R})$.

PROOF: The result follows easily from Theorem 4.8 (c).   **QED**

4.11. Remark. Let φ be the quasi-flow defined in Example 1.20 (b), in the space X which is defined there. It is easy to define a quasi-flow ψ in X such that ψ is not a quasi φ-reparametrization, even though each ψ-orbit is contained in the solitary φ-orbit.

It is because of this type of pathology that the hypotheses of some of the above theorems demand that φ be a flow.

4.12. Definition. Let φ be a quasi-flow in a Hausdorff space X and let f be a quasi φ-reparametrizer. Then f is said to be an *isotonic quasi φ-reparametrizer* if, for every $x \in \mathscr{G}(f \operatorname{rep} \varphi)$, the function f_x is strictly *increasing*.

4.13. Definition. Let φ be a quasi-flow in a Hausdorff space X. Then $\text{Rep}_q(\varphi)$ denotes the family

$$\{f \text{ rep } \varphi \mid f \text{ is a quasi } \varphi\text{-reparametrizer}\},$$

and $\text{Rep}_q^+(\varphi)$ denotes the family

$$\{f \text{ rep } \varphi \mid f \text{ is an isotonic quasi } \varphi\text{-reparametrizer}\}.$$

4.14. Theorem. *Let φ be a quasi-flow in a Hausdorff space X, let f and $\tilde{f}$ be quasi φ-reparametrizers and let $\psi = f \text{ rep } \varphi$ and $\pi = \tilde{f} \text{ rep } \varphi$.*

Then the following four statements are equivalent:

(a) $\psi \in \text{Rep}_q(\pi)$.

(b) *Each ψ-orbit is contained in a π-orbit.*

(c) $\mathscr{F}(\pi) \subseteq \mathscr{F}(\psi)$.

(d) $f_x(\mathbb{R}) \subseteq \tilde{f}_x(\mathbb{R})$, $\quad \forall \, x \in \mathscr{G}(\varphi)$.

PROOF: It is obvious that (a) $\Rightarrow$ (b) and (b) $\Rightarrow$ (c).

To prove that (c) $\Rightarrow$ (d), assume that (c) holds, and suppose $x \in \mathscr{G}(\varphi)$. The inclusion $f_x(\mathbb{R}) \subseteq \tilde{f}_x(\mathbb{R})$ is trivial if $x \in \mathscr{F}(\psi)$, and we suppose therefore that $x \in \mathscr{G}(\psi)$. But now, $f_x(\mathbb{R})$ and $\tilde{f}_x(\mathbb{R})$ are, respectively, the components containing 0 of $\{\tau \in \mathbb{R} \mid \varphi(\tau, x) \in \mathscr{G}(\psi)\}$ and $\{\tau \in \mathbb{R} \mid \varphi(\tau, x) \in \mathscr{G}(\pi)\}$, and the fact that $\mathscr{F}(\pi) \subseteq \mathscr{F}(\psi)$ implies that $\{\tau \in \mathbb{R} \mid \varphi(\tau, x) \in \mathscr{G}(\psi)\} \subseteq \{\tau \in \mathbb{R} \mid \varphi(\tau, x) \in \mathscr{G}(\pi)\}$, and the inclusion $f_x(\mathbb{R}) \subseteq \tilde{f}_x(\mathbb{R})$ is clear. Therefore, (c) $\Rightarrow$ (d).

To prove that (d) $\Rightarrow$ (a), assume that (d) holds. We need to define a quasi π-reparametrizer h, such that $\psi = h \text{ rep } \pi$. We define h as follows:

$$\text{if } x \in \mathscr{F}(\psi), \text{ let } h(x, t) = 0, \quad \forall \, t \in \mathbb{R};$$

and

$$\text{if } x \in \mathscr{G}(\psi), \text{ let } h(x, t) = \tilde{f}_x^{-1}\big(f_x(t)\big), \quad \forall \, t \in \mathbb{R}.$$

It is clear that h is a well-defined quasi π-reparametrizer, and that $\psi = h \text{ rep } \pi$. **QED**

4.15. Theorem. *Let φ be a quasi-flow in a Hausdorff space X, and let f be a quasi φ-reparametrizer. Let $Y = \mathscr{G}(f \text{ rep } \varphi)$, and for each $x \in Y$, let (α_x, ω_x) be the range of f_x, and $g_x = f_x^{-1}$.*

(a) *For each $x \in Y$, g_x is a strictly monotone, continuous function from (α_x, ω_x) onto $\mathbb{R}$, and $g_x(0) = 0$.*

(b) *If $x \in Y$ and $\tau \in (\alpha_x, \omega_x)$, then $\alpha_{\varphi(\tau, x)} = \alpha_x - \tau$ and $\omega_{\varphi(\tau, x)} = \omega_x - \tau$.*

(c) *If* $x \in Y$ *and both* τ *and* $\tau + \sigma$ *lie in* (α_x, ω_x), *we have*

$$g_x(\sigma + \tau) = g_x(\tau) + g_{\varphi(\tau,x)}(\sigma).$$

PROOF: (a) is obvious, and (b) follows at once from Corollary 4.10.

To prove (c), suppose $x \in Y$ and $\tau, \tau + \sigma \in (\alpha_x, \omega_x)$. Then since $\sigma \in (\alpha_x - \tau, \omega_x - \tau) = (\alpha_{\varphi(\tau,x)}, \omega_{\varphi(\tau,x)})$, it is clear that $g_{\varphi(\tau,x)}(\sigma)$ is defined. Let $g_x(\tau) = t$ and $g_{\varphi(\tau,x)}(\sigma) = s$. Then $f_x(t) = \tau$ and $f_{\varphi(\tau,x)}(s) = \sigma$, i.e., $f_{\varphi(f_x(t),x)}(s) = \sigma$. Now $g_x(\tau) + g_{\varphi(\tau,x)}(\sigma) = s + t$, and therefore, to complete the proof, we need to show that

$$g_x(\sigma + \tau) = s + t, \quad i.e., \quad f_x(s + t) = \sigma + \tau.$$

But

$$f_x(s + t) = f_{\varphi(f_x(t),x)}(s) + f_x(t) = \sigma + \tau.$$

This completes the proof of the theorem. **QED**

4.16. Definition. Let φ be a quasi-flow in a Hausdorff space X.

(a) If f is a quasi φ-reparametrizer, then inv (f) denotes the family

$$\{f_x^{-1} \mid x \in \mathscr{G}(f \text{ rep } \varphi)\}.$$

(b) Let $Y \subseteq \mathscr{G}(\varphi)$, and suppose that, for every $x \in X$, the set $\{\tau \in \mathbb{R} \mid \varphi(\tau, x) \in Y\}$ is open. A family of functions $\{g_x \mid x \in Y\}$ is called an *inverse quasi φ-reparametrizer* if

(i) for each $x \in Y$, denoting the component of

$$\{\tau \in \mathbb{R} \mid \varphi(\tau, x) \in Y\}$$

which contains 0 by (α_x, ω_x), g_x is a strictly monotone, continuous function from (α_x, ω_x) onto $\mathbb{R}$, and $g_x(0) = 0$,

and

(ii) if $x \in Y$, and $\tau, \sigma + \tau \in (\alpha_x, \omega_x)$, then

$$g_x(\sigma + \tau) = g_x(\tau) + g_{\varphi(\tau,x)}(\sigma).$$

4.17. Theorem. *Let* φ *be a quasi-flow in a Hausdorff space* X.

(a) *If* f *is a quasi* φ-*reparametrizer, then* inv(f) *is an inverse quasi* φ-*reparametrizer.*

(b) *If* $\{g_x \mid x \in Y\}$ *is an inverse quasi* φ-*reparametrizer, then there exists a quasi* φ-*reparametrizer* f *such that* $Y = \mathscr{G}(f \text{ rep } \varphi)$ *and* $f_x^{-1} = g_x$, $\forall x \in Y$,

$$i.e., \quad \text{inv}(f) = \{g_x \mid x \in Y\}.$$

PROOF: (a) follows at once from Theorem 4.15.

To prove (b), let $\{g_x \mid x \in Y\}$ be an inverse quasi φ-reparametrizer. For each $x \in Y$, let (α_x, ω_x) be the domain of g_x, and let $f_x = g_x^{-1}$. For each $x \in X \setminus Y$, let $f_x(t) = 0$, $\forall t \in \mathbb{R}$. Now let $f : X \times \mathbb{R} \to \mathbb{R}$ be defined by:

$$f(x, t) = f_x(t), \qquad \forall x \in X \qquad \forall t \in \mathbb{R}.$$

We shall show that f is a quasi φ-reparametrizer. Clearly, for every $x \in \mathscr{G}(\varphi)$, f_x is continuous and $f_x(0) = 0$. Now suppose $x \in \mathscr{G}(\varphi)$ and $s, t \in \mathbb{R}$. Then if $x \in \mathscr{G}(\varphi) \setminus Y$, the identity

$$f_x(s + t) = f_{\varphi(f_x(s), x)}(t) + f_x(s)$$

is trivial. On the other hand, if $x \in Y$, let

$$\sigma = f_x(s) \quad \text{and} \quad \tau = f_{\varphi(f_x(s), x)}(t) = f_{\varphi(\sigma, x)}(t).$$

We need to show that $f_x(s + t) = \sigma + \tau$, and to do this, we must show that $g_x(\sigma + \tau) = s + t$. But it is easy to see that because $\sigma \in (\alpha_x, \omega_x)$, we have

$$\left(\alpha_{\varphi(\sigma, x)}, \omega_{\varphi(\sigma, x)}\right) = (\alpha_x - \sigma, \omega_x - \sigma)$$

and therefore $\sigma + \tau \in (\alpha_x, \omega_x)$. Therefore $g_x(\sigma + \tau)$ is defined, and

$$g_x(\sigma + \tau) = g_x(\sigma) + g_{\varphi(\sigma, x)}(\tau) = s + t,$$

as required. This shows that f is a quasi φ-reparametrizer.

Finally, to show that $\{g_x \mid x \in Y\} = \text{inv}(f)$, it remains to show only that $Y = \mathscr{G}(f \text{ rep } \varphi)$. But this is obvious, because $Y \subseteq \mathscr{G}(\varphi)$, and for each $x \in \mathscr{G}(\varphi)$, f_x is strictly monotone iff $x \in Y$.

This completes the proof of (b).   **QED**

4.18. Definition. Let φ be a flow in a Hausdorff space X, and let f be a quasi φ-reparametrizer.

If $f \text{ rep } \varphi$ is a flow, then f is said to be a *φ-reparametrizer*, and $f \text{ rep } \varphi$ is called the *φ-reparametrization* corresponding to f. Further, $\text{inv}(f)$ is said to be an *inverse φ-reparametrizer*.

If f is simultaneously an isotonic quasi φ-reparametrizer and a φ-reparametrizer, then we say f is an *isotonic φ-reparametrizer*.

4.19. Definition. Let φ be a flow in a Hausdorff space X.

Then $\text{Rep}(\varphi)$ denotes the family

$$\{f \text{ rep } \varphi \mid f \text{ is a } \varphi\text{-reparametrizer}\},$$

and Rep$^+$(φ) denotes the family

$$\{f \text{ rep } \varphi \mid f \text{ is an isotonic } \varphi\text{-reparametrizer}\}.$$

4.20. Continuity of Reparametrizers. Let φ be a flow in a Hausdorff space X. Theorems 4.4 and 4.5 made it mandatory for us to include in the definition of a quasi φ-reparametrizer f the condition that f_x be continuous for every $x \in \mathscr{G}(\varphi)$. In the next few results, we shall show that, to a certain extent, if f is a φ-reparametrizer then the joint continuity of f rep φ imparts a joint continuity to f, as a function from $X \times \mathbb{R}$ into $\mathbb{R}$. We shall also see that a similar result holds for inverse reparametrizers.

4.21. Lemma. *Let φ be a flow in a Hausdorff space X and f a φ-reparametrizer. Let $x \in X$, $t \in \mathbb{R}$, and suppose $|f_x(t)| < p_\varphi(x)$.*

Then f is continuous at the point $\langle x, t \rangle$ of $X \times \mathbb{R}$.

PROOF: To prove the lemma, we need to show that if $\{\langle x_\alpha, t_\alpha \rangle \mid \alpha \in A\}$ is a net in $X \times \mathbb{R}$ and $x_\alpha \to x$ and $t_\alpha \to t$, we have

$$f_{x_\alpha}(t_\alpha) \to f_x(t).$$

Let $\{\langle x_\alpha, t_\alpha \rangle \mid \alpha \in A\}$ be such a net. Choose τ_1 and τ_2 in $\mathbb{R}$ such that

$$\tau_1 < 0 < \tau_2, \quad \tau_1 < f_x(t) < \tau_2, \quad \text{and} \quad \tau_2 - \tau_1 < p_\varphi(x).$$

Let $\psi = f$ rep φ, and let

$$K = \left\{\varphi(\sigma, x) \mid \sigma \in [0, f_x(t)]\right\} = \left\{\psi(s, x) \mid s \in [0, t]\right\}.$$

Choose neighborhoods U, V and W, of $\varphi(\tau_1, x)$, $\varphi(\tau_2, x)$ and K, respectively, in such a way that no two of these neighborhoods intersect. Using Theorem 1.73 or Lemma 1.72 choose $\alpha_0 \in A$ such that, whenever $\alpha \geq \alpha_0$, we have

(i) $\varphi(\tau_1, x_\alpha) \in U$,

(ii) $\varphi(\tau_2, x_\alpha) \in V$,

and (iii) $\psi(s, x_\alpha) \in W$ for all $s \in [0, t_\alpha]$.

We now show that

$$\tau_1 < f_{x_\alpha}(t_\alpha) < \tau_2, \qquad \forall \, \alpha \geq \alpha_0.$$

To obtain a contradiction, suppose that $f_{x_\alpha}(t_\alpha) \geq \tau_2$ for some $\alpha \geq \alpha_0$. Then since $f_{x_\alpha}(0) = 0$ and f_{x_α} is continuous, we can choose $s_\alpha \in [0, t_\alpha]$ such that $f_{x_\alpha}(s_\alpha) = \tau_2$. Now on the one hand,

$$\varphi(f_{x_\alpha}(s_\alpha), x_\alpha) = \psi(s_\alpha, x_\alpha) \in W,$$

while on the other hand,

$$\varphi\big(f_{x_a}(s_a), x_a\big) = \varphi(\tau_2, x_a) \in V.$$

This contradicts the fact that $V \cap W = \square$, and we therefore have

$$f_{x_a}(t_a) < \tau_2, \qquad \forall\, \alpha \geq \alpha_0.$$

A similar argument shows that

$$f_{x_a}(t_a) > \tau_1, \qquad \forall\, \alpha \geq \alpha_0.$$

We shall now show that $f_{x_a}(t_a) \to f_x(t)$ by showing that $f_x(t)$ is the only cluster-point of the net $\{f_{x_a}(t_a) \mid \alpha \in A\}$. Suppose τ is a cluster-point of $\{f_{x_a}(t_a) \mid \alpha \in A\}$. Then $\varphi(\tau, x)$ is a cluster-point of $\{\varphi\big(f_{x_a}(t_a), x_a\big) \mid \alpha \in A\}$ and since

$$\varphi\big(f_{x_a}(t_a), x_a\big) = \psi(t_a, x_a) \to \psi(t, x) = \varphi\big(f_x(t), x\big),$$

it follows that $\varphi(\tau, x) = \varphi\big(f_x(t), x\big)$.

Therefore, since $\tau_1 \leq \tau \leq \tau_2$ and $\tau_1 < f_x(t) < \tau_2$ and $\tau_2 - \tau_1 < \mathrm{p}_\varphi(x)$, we must have $\tau = f_x(t)$, as required.

This completes the proof of the lemma.   **QED**

4.22. Theorem. *Let φ be a flow in a Hausdorff space X, and let f be a φ-reparametrizer.*
Then f is continuous on the subset $\mathscr{G}(\varphi) \times \mathbb{R}$ of $X \times \mathbb{R}$.

PROOF: Let $x \in \mathscr{G}(\varphi)$ and let t be any real number. Let $\{\langle x_a, t_a \rangle \mid \alpha \in A\}$ be a net in $X \times \mathbb{R}$ and suppose $x_a \to x$ and $t_a \to t$. We need to show that $f_{x_a}(t_a) \to f_x(t)$. Using the fact that $\mathrm{p}_\varphi(x) > 0$ and that f_x is uniformly continuous in $[0, t]$, choose a positive integer n so large that, whenever $s_1, s_2 \in [0, t]$ and $|s_1 - s_2| \leq \dfrac{|t|}{n}$, we have

$$|f_x(s_1) - f_x(s_2)| < \mathrm{p}_\varphi(x).$$

For each $\alpha \in A$, define $s_a = \dfrac{t_a}{n}$. Define $s = \dfrac{t}{n}$. We now complete the proof by showing, by induction, that for each $j = 1, \ldots, n$, we have

$$f_{x_a}(j \cdot s_a) \to f_x(j \cdot s).$$

But this is clear (by Lemma 4.21) if $j = 1$, because

$$|f_x(s)| = |f_x(s) - f_x(0)| < \mathrm{p}_\varphi(x).$$

Suppose, for some $j < n$, we have $f_{x_\alpha}(j \cdot s_\alpha) \to f_x(j \cdot s)$. Then we see that

$$f_{x_\alpha}((j+1) s_\alpha) = f_{x_\alpha}(j \cdot s_\alpha) + f_{\varphi(f_{x_\alpha}(j \cdot s_\alpha), x_\alpha)}(s_\alpha), \qquad \forall \, \alpha \in A,$$

and

$$f_x((j+1) s) = f_x(j \cdot s) + f_{\varphi(f_x(js), x)}(s).$$

Therefore, writing

$$y_\alpha = \varphi(f_{x_\alpha}(j \cdot s_\alpha), x_\alpha), \qquad \forall \, \alpha \in A, \quad \text{and} \quad y = \varphi(f_x(j \cdot s), x),$$

we will have

$$f_{x_\alpha}((j+1) s_\alpha) \to f_x((j+1) s)$$

as soon as we have shown that

$$f_{y_\alpha}(s_\alpha) \to f_y(s).$$

Since $y_\alpha \to y$, this will follow at once from Lemma 4.21 as soon as we have shown that $|f_y(s)| < p_\varphi(y)$. But $p_\varphi(y) = p_\varphi(x)$, and

$$|f_y(s)| = |f_x((j+1) s) - f_x(j \cdot s)| = |f_x\left(j \cdot s + \frac{t}{n}\right) - f_x(j \cdot s)| < p_\varphi(x),$$

and therefore the theorem is proved. **QED**

4.23. Theorem. *Let φ be a flow in a Hausdorff space X, let f be a φ-reparametrizer, let $\psi = f \operatorname{rep} \varphi$, and let*

$$\{g_x \mid x \in \mathscr{G}(\psi)\} = \operatorname{inv}(f).$$

Let $x \in \mathscr{G}(\psi)$ and $\tau \in f_x(\mathbb{R})$. Let $\{\langle x_\alpha, \tau_\alpha \rangle \mid \alpha \in A\}$ be a net in $X \times \mathbb{R}$, and suppose that $x_\alpha \to x$ and $\tau_\alpha \to \tau$.

Then $g_{x_\alpha}(\tau_\alpha) \to g_x(\tau)$.

PROOF: We note first that since $x \in \mathscr{G}(\psi)$ and $\mathscr{G}(\psi)$ is open, $x_\alpha \in \mathscr{G}(\psi)$ for all sufficiently large α, and we can therefore assume (for simplicity of notation) that $x_\alpha \in \mathscr{G}(\psi)$ for every α.

Write $t = g_x(\tau)$, and let $\delta > 0$. Then since $\tau \in (f_x(t - \delta), f_x(t + \delta))$, and since $\tau_x \to \tau, f_{x_\alpha}(t - \delta) \to f_x(t - \delta)$, and $f_{x_\alpha}(t + \delta) \to f_x(t + \delta)$, we must have

$$\tau_\alpha \in (f_{x_\alpha}(t - \delta), f_{x_\alpha}(t + \delta))$$

for all sufficiently large α. It follows that, for all sufficiently large α, $g_{x_\alpha}(\tau_\alpha)$ is defined and satisfies

$$t - \delta < g_{x_\alpha}(\tau_\alpha) < t + \delta.$$

This completes the proof of the theorem. **QED**

4.24. Theorem. *Let φ be a flow in a Hausdorff space X, let f be a φ-reparametrizer, and let $\bar{f}$ be the isotonic quasi φ-reparametrizer determined by f in the following natural way:*

(i) *If $x \in \mathscr{F}(f \operatorname{rep} \varphi)$, then $\bar{f}(x, t) = f(x, t)$, $\quad \forall\, t \in \mathbb{R}$;*

(ii) *If $x \in \mathscr{G}(f \operatorname{rep} \varphi)$, and f_x is strictly increasing, then*

$$\bar{f}(x, t) = f(x, t), \qquad \forall\, t \in \mathbb{R};$$

(iii) *If $x \in \mathscr{G}(f \operatorname{rep} \varphi)$, and f_x is strictly decreasing, then*

$$\bar{f}(x, t) = f(x, -t), \qquad \forall\, t \in \mathbb{R}.$$

Then $\bar{f}$ is an isotonic φ-reparametrizer.

PROOF: It is obvious that $\bar{f}$ is an isotonic quasi φ-reparametrizer. It is also clear that if $\psi = f \operatorname{rep} \varphi$ and $\tilde{\psi} = \bar{f} \operatorname{rep} \varphi$, then ψ and $\tilde{\psi}$ have the same orbits. We need to show that $\tilde{\psi}$ is continuous at each point $x \in X$.

Suppose $x \in \mathscr{F}(\tilde{\psi})$ and $t \in \mathbb{R}$, and suppose $\{\langle x_\alpha, t_\alpha \rangle \mid \alpha \in A\}$ is a net in $X \times \mathbb{R}$, such that $x_\alpha \to x$ and $t_\alpha \to t$. Since $x \in \mathscr{F}(\psi)$, and since, for each $\alpha \in A$, we have either $\tilde{\psi}(t_\alpha, x_\alpha) = \psi(t_\alpha, x_\alpha)$ or $\tilde{\psi}(t_\alpha, x_\alpha) = \psi(-t_\alpha, x_\alpha)$, it is clear that

$$\tilde{\psi}(t_\alpha, x_\alpha) \to x.$$

It follows that $\tilde{\psi}$ is continuous at each point $x \in \mathscr{F}(\tilde{\psi})$.

Now let

$$U = \{x \in \mathscr{G}(\tilde{\psi}) \mid f_x \text{ is strictly increasing}\}$$

and

$$V = \{x \in \mathscr{G}(\tilde{\psi}) \mid f_x \text{ is strictly decreasing}\}.$$

It is an easy consequence of Theorem 4.22 that both U and V are open, and from this it follows easily that $\tilde{\psi}$ is continuous at each point $x \in \mathscr{G}(\tilde{\psi})$.

This completes the proof of the theorem. **QED**

4.25. Example. Theorem 4.22 cannot be extended to assert that, if f is a φ-reparametrizer, where φ is a flow in X, then f is continuous in $X \times \mathbb{R}$, for let φ be the flow in $\mathbb{R}^2$, defined by

$$\varphi(t, \langle x, y \rangle) = \langle x + y^2 t, y \rangle, \qquad \forall\, t \in \mathbb{R} \qquad \forall\, \langle x, y \rangle \in \mathbb{R}^2,$$

and let $f : \mathbb{R}^2 \times \mathbb{R} \to \mathbb{R}$ be defined by

$$f(\langle x, y \rangle, t) = \frac{1}{y}\, t \text{ whenever } \langle x, y \rangle \in \mathbb{R}^2,\, y \neq 0,\, \text{and } t \in \mathbb{R}$$

and

$$f(\langle x, 0 \rangle, t) = 0, \qquad \forall\, x \in \mathbb{R} \qquad \forall\, t \in \mathbb{R}.$$

It is easy to see that f is a quasi φ-reparametrizer, and that if $\psi = f \operatorname{rep} \varphi$, then

$$\psi(t, \langle x, y \rangle) = \langle x + yt, y \rangle, \qquad \forall\, t \in \mathbb{R} \qquad \forall\, \langle x, y \rangle \in \mathbb{R}^2.$$

Since ψ is clearly a flow, it follows that f is a φ-reparametrizer. But f is not continuous in $\mathbb{R}^2 \times \mathbb{R}$, for if $x \in \mathbb{R}$ and $t \in \mathbb{R}$, and $y_n \to 0$ and $t_n \to t \neq 0$, then $|f(\langle x, y_n \rangle, t_n)| \to \infty$. Also, if $x \in \mathbb{R}$, and $y_n \to 0$, and we define t_n for every $n = 1, 2, \ldots$ by

$$t_n = y_n \text{ if } n \text{ is even,}$$

and

$$t_n = 2y_n \text{ if } n \text{ is odd,}$$

then although $t_n \to 0$, the sequence $\{f(\langle x, y_n \rangle, t_n)\}$ fails to converge, in spite of the fact that it is bounded.

4.26. Example. Given a flow φ in a Hausdorff space X, and a quasi φ-reparametrizer f which is continuous in $\mathscr{G}(\varphi) \times \mathbb{R}$, it is not necessarily true that f is a φ-reparametrizer, for let φ be the flow in $\mathbb{R}^2$ defined in Example 4.25, and let $f : \mathbb{R}^2 \times \mathbb{R} \to \mathbb{R}$ be defined by

$$f(\langle x, y \rangle, t) = \frac{1}{y^2} t \text{ whenever } \langle x, y \rangle \in \mathbb{R}^2, y \neq 0, \text{ and } t \in \mathbb{R},$$

and

$$f(\langle x, 0 \rangle, t) = 0, \qquad \forall\, x \in \mathbb{R} \qquad \forall\, t \in \mathbb{R}.$$

It is easy to see that f is a quasi φ-reparametrizer, and is continuous in $\mathscr{G}(\varphi) \times \mathbb{R}$. But if $\psi = f \operatorname{rep} \varphi$, then

$$\psi(t, \langle x, y \rangle) = \langle x + t, y \rangle \text{ whenever } t \in \mathbb{R}, \langle x, y \rangle \in \mathbb{R}^2, \text{ and } y \neq 0,$$

and

$$\psi(t, \langle x, 0 \rangle) = \langle x, 0 \rangle, \qquad \forall\, t \in \mathbb{R} \qquad \forall\, x \in \mathbb{R}.$$

Therefore ψ is not continuous at any point of $\mathscr{F}(\varphi)$, and therefore f is not a φ-reparametrizer.

4.27. Theorem. *Let φ be a flow in a Hausdorff space X, let f be a quasi φ-reparametrizer, and suppose*

(i) *f is continuous in $\mathscr{G}(\varphi) \times \mathbb{R}$,*

and

(ii) *f is bounded in a neighborhood of each point of $\mathscr{F}(\varphi) \times \mathbb{R}$.*

Then f is a φ-reparametrizer.

PROOF: The proof of this theorem is elementary, and will be omitted.

QED

4.28. Corollary. *Let φ be a flow in a Hausdorff space X, let f be a quasi φ-reparametrizer, and suppose f is continuous in $X \times \mathbb{R}$.*

Then f is a φ-reparametrizer.

PROOF: Obvious.   **QED**

4.29. Remark. Example 4.25 shows that the converse to Theorem 4.27 does not hold.

Flow Multipliers

4.30. Introduction. Given a quasi-flow φ in a Hausdorff space X, and a function u from X into $[-\infty, \infty]$, there is a natural way to define a "product" $u\varphi$, of φ by u, with the properties:

(i) $u\varphi$ and φ have the same sense whenever u is positive, and have opposite senses whenever u is negative.

(ii) For a large class of points $x \in X$, $u\varphi$ moves from x to a point near x, approximately $u(x)$ times as quickly as does φ.

We shall be interested in this process whenever the behavior of u and φ is such that $u\varphi$ is a quasi φ-reparametrization, and we will obtain from this an important category of quasi φ-reparametrizations.

The simplest situation occurs when u is a constant function. If u takes on the constant value c, then the obvious choice for the definition of $u\varphi$ is the quasi φ-reparametrization ψ which is given by:

$$\psi(t, x) = \varphi(ct, x), \qquad \forall\, t \in \mathbb{R} \qquad \forall\, x \in X.$$

We see at once that ψ satisfies conditions (i) and (ii), and that $\psi = f \operatorname{rep} \varphi$, where

$$f(x, t) = ct, \qquad \forall\, x \in X \qquad \forall\, t \in \mathbb{R}.$$

It is worth noting that for every $x \in \mathscr{G}(\psi)$ and $\tau \in \mathbb{R}$, we have

$$f_x^{-1}(\tau) = \frac{1}{c}\,\tau = \int_0^\tau \frac{1}{c}\, d\sigma,$$

and therefore

$$f_x^{-1}(\tau) = \int_0^\tau \frac{1}{u\big(\varphi(\sigma, x)\big)}\, d\sigma.$$

Let us now consider another very simple situation, and suppose that u is continuous, positive, bounded, and bounded away from 0. In order to determine what $u\varphi$ ought to be, let us suppose that $u\varphi$ has already been defined, that $\psi = u\varphi$, and that $\psi = f \operatorname{rep} \varphi$. Because of the assumptions we have made about u, it is reasonable to expect ψ and φ to have the same orbits, and consequently that, for every $x \in \mathscr{G}(\varphi)$, f_x is a strictly increasing, continuous function from $\mathbb{R}$ onto $\mathbb{R}$. Now if $x \in \mathscr{G}(\varphi)$ and $\tau \in \mathbb{R}$, then $f_x^{-1}(\tau)$ is the time taken by ψ to go from x to $\varphi(\tau, x)$, and therefore, by condition (ii) we see that if τ is small, $f_x^{-1}(\tau)$ is approximately $\tau/u(x)$. Therefore, if $\tau > 0$, and $\{\sigma_0, \sigma_1, \ldots, \sigma_n\}$ is a finite set of points in $[0, \tau]$ such that

$$0 = \sigma_0 < \sigma_1 < \ldots < \sigma_n = \tau,$$

then

$$\sum_{j=1}^{n} \frac{1}{u\big(\varphi(\sigma_j, x)\big)} (\sigma_j - \sigma_{j-1})$$

will be a reasonable approximation to $f_x^{-1}(\tau)$, provided that

$$\sup\{\sigma_j - \sigma_{j-1} \mid j = 1, \ldots, n\}$$

is sufficiently small. It is therefore reasonable to expect that, whenever $x \in \mathscr{G}(\varphi)$ and $\tau \in \mathbb{R}$, we have

$$f_x^{-1}(\tau) = \int_0^\tau \frac{1}{u\big(\varphi(\sigma, x)\big)} \, d\sigma.^2$$

The above argument suggests the following method of definition of $u\varphi$, in the case where u is continuous, positive, bounded, and bounded away from 0:

Step 1. For each $x \in \mathscr{G}(\varphi)$ define $g_x : \mathbb{R} \to \mathbb{R}$ by

$$g_x(\tau) = \int_0^\tau \frac{1}{u\big(\varphi(\sigma, x)\big)} \, d\sigma, \qquad \forall\, t \in \mathbb{R},$$

and note that $\{g_x \mid x \in \mathscr{G}(\varphi)\}$ is an inverse quasi φ-reparametrizer.

Step 2. Using Theorem 4.17(b), choose a quasi φ-reparametrizer f such that

$$\operatorname{inv}(f) = \{g_x \mid x \in \mathscr{G}(\varphi)\}.$$

Step 3. Define $u\varphi = f \operatorname{rep} \varphi$.

---

[2] We shall abide by the convention $\displaystyle\int_a^b = -\int_b^a$.

4.31. Definition of a Quasi φ-multiplier. Let φ be a quasi-flow in a Hausdorff space X, and let u be a function from X into $[-\infty, \infty]$. We say that u is a *quasi φ-multiplier* if

(i) The restriction of u to each φ-orbit is a Borel function.[3]

(ii) For every $x \in \mathscr{G}(\varphi)$, the set $\{\tau \in \mathbb{R} \mid |u(\varphi(\tau, x))| < \infty\}$ is a measure-dense subset of $\mathbb{R}$, *i.e.*, the intersection of $\{\tau \in \mathbb{R} \mid |u(\varphi(\tau, x))| < \infty\}$ with each non-empty open subset of $\mathbb{R}$ has positive Lebesgue measure.

(iii) For every $x \in \mathscr{G}(\varphi)$, we have

either

(a) $\displaystyle\int_0^\tau \frac{1}{u(\varphi(\sigma, x))} \, d\sigma$ diverges for every $\tau \neq 0$, and $u(x) = 0$,

or

(b) $\left\{ \tau \in [-\infty, \infty] \;\middle|\; \displaystyle\int_0^\tau \frac{1}{u(\varphi(\sigma, x))} \, d\sigma \quad \text{converges} \right\}$ is an open sub-

interval of $\mathbb{R}$, and on this interval, the function

$$\tau \to u(\varphi(\tau, x))$$

does not change sign.

4.32. Theorem. *Let φ be a quasi-flow in a Hausdorff space X, and let u be a quasi φ-multiplier. Let $x \in \mathscr{G}(\varphi)$, and suppose*

$$\left\{ \tau \in [-\infty, \infty] \;\middle|\; \int_0^\tau \frac{1}{u(\varphi(\sigma, x))} \, d\sigma \; \text{converges} \right\}$$

is an open subinterval (α_x, ω_x) of $\mathbb{R}$.

Then the function g_x defined on (α_x, ω_x) by

$$g_x(\tau) = \int_0^\tau \frac{1}{u(\varphi(\sigma, x))} \, d\sigma, \qquad \forall \, \tau \in (\alpha_x, \omega_x),$$

is a strictly monotone, continuous function from (α_x, ω_x) onto $\mathbb{R}$.

---

[3] If X is any topological space, then the family of *Borel* subsets of X is defined to be the smallest σ-algebra (σ-field) of subsets of X which contains all open subsets of X.

A function f from a topological space X to a topological space Y is said to be a *Borel function* if $f^{-1}(U)$ is a Borel subset of X whenever U is open in Y.

PROOF: It is obvious that g_x is a continuous, monotone function from (α_x, ω_x) into $\mathbb{R}$, and it follows easily from the fact that both

$$\int\limits_{\alpha_x}^{0} \frac{1}{u\big(\varphi(\sigma, x)\big)}\, d\sigma \quad \text{and} \quad \int\limits_{0}^{\omega_x} \frac{1}{u\big(\varphi(\sigma, x)\big)}\, d\sigma$$

diverge, that g_x carries (α_x, ω_x) *onto* $\mathbb{R}$. The *strict* monotonicity of g_x follows at once from the fact that in each open subinterval of (α_x, ω_x), the function $\sigma \to u\big(\varphi(\sigma, x)\big)$ is finite on a set of positive measure.

This completes the proof of the theorem.   **QED**

4.33. Remark. If φ is a quasi-flow in a Hausdorff space X, u is a quasi φ-multiplier, $x \in \mathscr{G}(\varphi)$, and

$$\left\{ \tau \in [-\infty, \infty] \,\middle|\, \int\limits_{0}^{\tau} \frac{1}{u\big(\varphi(\sigma, x)\big)}\, d\sigma \quad \text{converges} \right\} = (\alpha_x, \omega_x),$$

then at each $\tau \in (\alpha_x, \omega_x)$ at which $u\big(\varphi(\tau, x)\big) = 0$, the value of $\dfrac{1}{u\big(\varphi(\tau, x)\big)}$ will be understood to be whichever of the two numbers $-\infty$, ∞, is appropriate, in order that the function

$$\tau \to \frac{1}{u\big(\varphi(\tau, x)\big)}$$

does not change sign.

4.34. Theorem. *Let φ be a quasi-flow in a Hausdorff space X and let u be a quasi φ-multiplier. Let*

$$Y = \left\{ x \in \mathscr{G}(\varphi) \,\middle|\, \int\limits_{0}^{\tau} \frac{1}{u\big(\varphi(\sigma, x)\big)}\, d\sigma \quad \text{converges for some } \tau \neq 0 \right\}.$$

For each $x \in Y$, let

$$\left\{ \tau \in [-\infty, \infty] \,\middle|\, \int\limits_{0}^{\tau} \frac{1}{u\big(\varphi(\sigma, x)\big)}\, d\sigma \quad \text{converges} \right\} = (\alpha_x, \omega_x),$$

and let g_x be the function from (α_x, ω_x) onto $\mathbb{R}$ defined by

$$g_x(\tau) = \int\limits_{0}^{\tau} \frac{1}{u\big(\varphi(\sigma, x)\big)}\, d\sigma, \qquad \forall\, \tau \in (\alpha_x, \omega_x).$$

Then $\{g_x \mid x \in Y\}$ is an inverse quasi φ-reparametrizer.

PROOF: We note first that whenever $x \in X$, $\tau \in \mathbb{R}$, and $\varphi(\tau, x) \in Y$, we have $\varphi(\tau + \sigma, x) \in Y$ whenever $\sigma \in (\alpha_{\varphi(\tau,x)}, \omega_{\varphi(\tau,x)})$. It follows that for every $x \in X$, the set $\{\tau \in \mathbb{R} \mid \varphi(\tau, x) \in Y\}$ is open.

Now suppose $x \in Y$. Then either $\alpha_x = -\infty$ or $\varphi(\alpha_x, x) \in Y$, and either $\omega_x = \infty$ or $\varphi(\omega_x, x) \notin Y$. Therefore (α_x, ω_x) is the component of $\{\tau \in \mathbb{R} \mid \varphi(\tau, x) \in Y\}$ which contains 0. It is obvious that $g_x(0) = 0$, and Theorem 4.32 implies that g_x is a strictly monotone, continuous function from (α_x, ω_x) onto $\mathbb{R}$.

This shows that $\{g_x \mid x \in Y\}$ satisfies condition (i) of Definition 4.16(b).

To show that condition (ii) is also satisfied, suppose that $x \in Y$ and that τ and $\sigma + \tau$ belong to (α_x, ω_x). Then clearly $\sigma \in (\alpha_{\varphi(\tau,x)}, \omega_{\varphi(\tau,x)})$, and

$$g_x(\sigma + \tau) = \int\limits_0^{\sigma+\tau} \frac{1}{u\big(\varphi(\sigma', x)\big)}\, d\sigma'$$

$$= \int\limits_0^{\tau} \frac{1}{u\big(\varphi(\sigma', x)\big)}\, d\sigma' + \int\limits_\tau^{\sigma+\tau} \frac{1}{u\big(\varphi(\sigma', x)\big)}\, d\sigma'$$

$$= \int\limits_0^{\tau} \frac{1}{u\big(\varphi(\sigma', x)\big)}\, d\sigma' + \int\limits_0^{\sigma} \frac{1}{u\big(\varphi(\sigma' + \tau, x)\big)}\, d\sigma'$$

$$= g_x(\tau) + g_{\varphi(\tau,x)}(\sigma).$$

Thus condition (ii) is also satisfied, and the proof is complete.   **QED**

4.35. Definition. Let φ be a quasi-flow in a Hausdorff space X, and let u be a quasi φ-multiplier. Let Y and the functions $g_x (x \in Y)$ be defined as in Theorem 4.34.

Then $\{g_x \mid x \in Y\}$ is called *the inverse quasi φ-reparametrizer induced by u,* and every quasi φ-reparametrizer f for which $\{g_x \mid x \in Y\} = \mathrm{inv}(f)$ is said to be a *quasi φ-reparametrizer induced by u.* If f is a quasi φ-reparametrizer induced by u, then we call f rep φ the *product of φ by u,* and we denote f rep φ by $u\varphi$.

4.36. Remark. Let φ be a quasi-flow in a Hausdorff space X, and let u be a quasi φ-multiplier. Then $u\varphi$ is well-defined, for if $\{g_x \mid x \in Y\}$ is the inverse quasi φ-reparametrizer induced by u, then the quasi-flow f rep φ does not vary as f runs through the family of those quasi φ-reparametrizers satisfying the equation $\mathrm{inv}(f) = \{g_x \mid x \in Y\}$.

4.37. Definition. Let φ be a quasi-flow in a Hausdorff space X and let u be a quasi φ-multiplier. Then, noting that condition (iii) (a) of Definition 4.31 implies that

$$\mathscr{F}(u\varphi) \subseteq \mathscr{F}(\varphi) \cup \{x \in X \mid u(x) = 0\},$$

we say u is *proper* if

$$\mathscr{F}(u\varphi) = \mathscr{F}(\varphi) \cup \{x \in X \mid u(x) = 0\}.$$

4.38. Definition. Let φ be a flow in a Hausdorff space X, let u be a quasi φ-multiplier, and suppose $u\varphi$ is a flow in X.

Then u is said to be a φ-*multiplier*.

4.39. Definition. Let φ be a quasi-flow in a Hausdorff space X. We define

$$\mathrm{Prod}_q(\varphi) = \{u\varphi \mid u \text{ is a quasi } \varphi\text{-multiplier}\}$$

and

$$\mathrm{Prod}_q^+(\varphi) = \{u\varphi \mid u \text{ is a nonnegative quasi } \varphi\text{-multiplier}\}.$$

4.40. Definition. Let φ be a flow in a Hausdorff space X. We define

$$\mathrm{Prod}(\varphi) = \{u\varphi \mid u \text{ is a } \varphi\text{-multiplier}\},$$

and

$$\mathrm{Prod}^+(\varphi) = \{u\varphi \mid u \text{ is a nonnegative } \varphi\text{-multiplier}\}.$$

4.41. Remarks. Let φ be a quasi-flow in a Hausdorff space X. We remark that $\mathrm{Prod}_q(\varphi) \subseteq \mathrm{Rep}_q(\varphi)$, and $\mathrm{Prod}_q^+(\varphi) \subseteq \mathrm{Rep}_q^+(\varphi)$. Furthermore, if φ is a flow, then $\mathrm{Prod}(\varphi) \subseteq \mathrm{Rep}(\varphi)$, and $\mathrm{Prod}^+(\varphi) \subseteq \mathrm{Rep}^+(\varphi)$. We remark finally, that if φ is a flow and u is a φ-multiplier, then so is $|u|$.

4.42. Example. Let $u(x) = x$, $\quad \forall\, x \in \mathbb{R}$. Then u is a proper **1**-multiplier, (where **1** is, as usual, the flow in $\mathbb{R}$ defined by

$$\mathbf{1}(t, x) = t + x, \qquad \forall\, t \in \mathbb{R} \qquad \forall\, x \in \mathbb{R}).$$

Let $\psi = u\mathbf{1}$, and choose f such that $\psi = f \operatorname{rep} \mathbf{1}$. Then for every $x \in \mathbb{R}$ and $t \in \mathbb{R}$, we have

$$f_x(t) = x(e^t - 1) \text{ and } \psi(t, x) = xe^t.$$

4.43. Example. For each non-zero real number y, choose $\delta(y)$ to be that number in $(0, |y|)$, for which

$$\int_{\delta(y)}^{|y|} \frac{1}{t}\, dt = 3|y|^{-\frac{1}{3}}.$$

Let φ be the flow in $\mathbb{R}^2$, defined by

$$\varphi(t, \langle x, y \rangle) = \langle t + x, y \rangle, \qquad \forall\, t \in \mathbb{R} \qquad \forall\, \langle x, y \rangle \in \mathbb{R}^2,$$

and let u be the function from $\mathbb{R}^2$ into $\mathbb{R}$, which is defined by:

$$u(x, y) = 3\, x^{2/3} \qquad \text{whenever } |x| \geq |y|,$$

$$u(x, y) = 3|x|\, |y|^{-1/3} \quad \text{whenever } \delta(y) \leq |x| < |y|,$$

and

$$u(x, y) = 3\, \delta(y)\, |y|^{-1/3} \text{ whenever } |x| < \delta(y).$$

It is easy to see that u is a continuous nonnegative function from $\mathbb{R}^2$ into $\mathbb{R}$, and that $\langle 0, 0 \rangle$ is the only point at which u vanishes. It is easy to check that for each point $\langle x, y \rangle \in \mathbb{R}^2$, the integral

$$\int_0^\tau \frac{1}{u\big(\varphi(\sigma, \langle x, y \rangle)\big)}\, d\sigma$$

converges for all $\tau \in \mathbb{R}$, and that both of the integrals

$$\int_{-\infty}^0 \frac{1}{u\big(\varphi(\sigma, \langle x, y \rangle)\big)}\, d\sigma \text{ and } \int_0^\infty \frac{1}{u\big(\varphi(\sigma, \langle x, y \rangle)\big)}\, d\sigma$$

diverge. It is therefore easy to see that u is a continuous quasi φ-multiplier, but that u is not proper, because $\mathscr{F}(u\varphi) = \square$.

Let $\{g_{\langle x, y \rangle} \mid \langle x, y \rangle \in \mathbb{R}^2\}$ be the inverse quasi φ-reparametrizer induced by u, i.e., if $\langle x, y \rangle \in \mathbb{R}^2$ and $\tau \in \mathbb{R}$, then

$$g_{\langle x, y \rangle}(\tau) = \int_0^\tau \frac{1}{u\big(\varphi(\sigma, \langle x, y \rangle)\big)}\, d\sigma.$$

It would be a routine but tedious matter to evaluate $g_{\langle x, y \rangle}(\tau)$ explicitly for every point $\langle x, y \rangle \in \mathbb{R}^2$ and real number τ, but we are only interested in evaluating $g_{\langle x, 0 \rangle}(\tau)$ for every point $\langle x, 0 \rangle \in \mathbb{R}^2$ and $\tau \in \mathbb{R}$, and also $g_{\langle 0, y \rangle}(y)$ for every positive real number y. We leave it to the reader to show that

$$g_{\langle x, 0 \rangle}(\tau) = (\tau + x)^{1/3} - x^{1/3}, \qquad \forall\, \langle x, 0 \rangle \in \mathbb{R}^2, \qquad \forall\, \tau \in \mathbb{R},$$

and

$$g_{\langle 0, y \rangle}(y) = 1 + \frac{1}{3}\, y^{\frac{1}{3}} \text{ for all positive real numbers } y.$$

From this we deduce that if $\{y_n\}$ decreases to 0, we have

$$g_{\langle 0, y_n \rangle}(y_n) \to 1 \neq g_{\langle 0, 0 \rangle}(0) \quad \text{as} \quad n \to \infty.$$

It follows from Theorem 4.23 that $\{g_{\langle x, y \rangle} \mid \langle x, y \rangle \in \mathbb{R}^2\}$ is not an inverse φ-reparametrizer, and it follows that although u is a continuous quasi φ-multiplier, u is not a φ-multiplier. The next theorem shows that the fact that u is not proper is crucial in this example.

4.44. Theorem. *Let φ be a flow in a Hausdorff space X, let u be a function from X into $[-\infty, \infty]$, and suppose*

(i) *u is continuous at each point of $\mathscr{G}(\varphi)$,*

and

(ii) *For each $x \in \mathscr{F}(\varphi)$, there exists a neighborhood W_x of x such that u is bounded in $\mathscr{G}(\varphi) \cap W_x$.*

Then a necessary and sufficient condition that u be a proper φ-multiplier is that whenever $x \in \mathscr{G}(\varphi)$ and $u(x) \neq 0$, writing (α_x, ω_x) for that component of $\{\tau \in \mathbb{R} \mid u(\varphi(\tau, x)) \neq 0\}$ which contains 0, we have

(a) *$\{\tau \in (\alpha_x, \omega_x) \mid |u(\varphi(\tau, x))| < \infty\}$ is dense in (α_x, ω_x),*

and

(b) *both $\displaystyle\int_{\alpha_x}^{0} \frac{1}{u(\varphi(\sigma, x))}\, d\sigma$ and $\displaystyle\int_{0}^{\omega_x} \frac{1}{u(\varphi(\sigma, x))}\, d\sigma$ diverge.*

PROOF: It is easy to see that the condition is necessary, for if u is a proper φ-multiplier, $x \in \mathscr{G}(\varphi)$, and $u(x) \neq 0$, then it is clear that

$$(\alpha_x, \omega_x) = \left\{ \tau \in [-\infty, \infty] \,\middle|\, \int_{0}^{\tau} \frac{1}{u(\varphi(\sigma, x))}\, d\sigma \text{ converges} \right\}.$$

To show that the condition is sufficient, suppose that whenever $x \in \mathscr{G}(\varphi)$ and $u(x) \neq 0$, both (a) and (b) hold. It is clear that u satisfies conditions (i) and (ii) of Definition 4.31, (observe that an open, dense subset of an interval is measure-dense), and furthermore, u satisfies condition (iii) (b) of Definition 4.31 at each $x \in \mathscr{G}(\varphi)$ for which $u(x) \neq 0$. It will therefore follow at once that u is a proper quasi φ-multiplier, when we have shown that if $x \in \mathscr{G}(\varphi)$ and $u(x) = 0$, then

$$\int_{0}^{\tau} \frac{1}{u(\varphi(\sigma, x))}\, d\sigma$$

diverges for every $\tau \neq 0$.

Suppose $x \in \mathscr{G}(\varphi)$, $u(x) = 0$, $\tau \neq 0$, but

$$\int_0^\tau \frac{1}{u(\varphi(\sigma, x))} \, d\sigma$$

converges. Choose $\sigma_0 \in (0, \tau)$ such that $u(\varphi(\sigma_0, x)) \neq 0$. Clearly,

$$\int_0^{\sigma_0} \frac{1}{u(\varphi(\sigma, x))} \, d\sigma$$

converges. Let $y = \varphi(\sigma_0, x)$. Then $u(\varphi(-\sigma_0, y)) = u(x) = 0$, so that $-\sigma_0 \notin (\alpha_y, \omega_y)$, and therefore

$$\int_{-\sigma_0}^0 \frac{1}{u(\varphi(\sigma, y))} \, d\sigma$$

diverges. But

$$\int_0^{\sigma_0} \frac{1}{u(\varphi(\sigma, x))} \, d\sigma = \int_{-\sigma_0}^0 \frac{1}{u(\varphi(\sigma + \sigma_0, x))} \, d\sigma = \int_{-\sigma_0}^0 \frac{1}{u(\varphi(\sigma, y))} \, d\sigma,$$

and therefore

$$\int_{-\sigma_0}^0 \frac{1}{u(\varphi(\sigma, y))} \, d\sigma$$

converges, giving a contradiction.

This shows that u is a proper quasi φ-multiplier. Let f be that quasi φ-reparametrizer induced by u which has the property that, whenever $x \in \mathscr{F}(\varphi)$, we have $f_x(t) = 0$, $\forall t \in \mathbb{R}$. We shall show that u is a proper φ-multiplier by showing that f satisfies the hypotheses of Theorem 4.27.

To show that f satisfies hypothesis (ii) of Theorem 4.27, let $x \in \mathscr{F}(\varphi)$ and $t \in \mathbb{R}$. Choose $\lambda > 0$ such that $|u(y)| < \lambda$, $\forall y \in W_x \cap \mathscr{G}(\varphi)$. Since $x \in \mathscr{F}(\varphi)$, we can choose a neighborhood V of x such that

$$\varphi(\sigma, y) \in W_x \text{ whenever } |\sigma| \leq \lambda(1 + |t|) \text{ and } y \in V.$$

We can now show that whenever $y \in V$ and $|s - t| < 1$, we have $|f_y(s)| \leq \lambda(1 + |t|)$, for suppose not, i.e., suppose $y \in V$ and $|s - t| < 1$, but $|f_y(s)| > \lambda(1 + |t|)$. Then clearly $y \in \mathscr{G}(\varphi)$, and

$$
|s| = \left| \int_0^{f_y(s)} \frac{1}{u(\varphi(\sigma, y))} \, d\sigma \right| = \left| \int_0^{f_y(s)} \frac{1}{|u(\varphi(\sigma, y))|} \, d\sigma \right|
$$

$$
\geq \left| \int_0^{\lambda(1+|t|) \cdot \operatorname{sgn}(f_y(s))} \frac{1}{|u(\varphi(\sigma, y))|} \, d\sigma \right|
$$

$$
\geq \lambda(1 + |t|) \cdot \frac{1}{\lambda},
$$

contradicting the fact that $|s - t| < 1$.

We have therefore shown that f satisfies hypothesis (ii) of Theorem 4.27.

We shall now complete the proof by showing that f is continuous in $\mathscr{G}(\varphi) \times \mathbb{R}$. To do this, let $x \in \mathscr{G}(\varphi)$, $t \in \mathbb{R}$, and $\varepsilon > 0$. We shall find a neighborhood V of x and a number $\delta > 0$ such that

$$
|f_x(t) - f_y(s)| < \varepsilon, \text{ whenever } y \in V \text{ and } |s - t| < \delta.
$$

CASE 1: $u(x) > 0$.

Since $f_x(t) \in (\alpha_x, \omega_x)$, we can choose η satisfying $0 < \eta < \varepsilon$, such that

$$
[f_x(t) - \eta, \, f_x(t) + \eta] \subseteq (\alpha_x, \omega_x).
$$

Let I be the smallest closed interval containing the three numbers $f_x(t) - \eta$, 0, $f_x(t) + \eta$. Then clearly $I \subseteq (\alpha_x, \omega_x)$, and therefore $u(\varphi(\tau, x)) > 0$, $\quad \forall \tau \in I$. Choose $\delta > 0$ such that

$$
3\delta < \int_{f_x(t)-\eta}^{f_x(t)} \frac{1}{u(\varphi(\sigma, x))} \, d\sigma \quad \text{and} \quad 3\delta < \int_{f_x(t)}^{f_x(t)+\eta} \frac{1}{u(\varphi(\sigma, x))} \, d\sigma.
$$

Choose a neighborhood V of x such that $V \subseteq \mathscr{G}(\varphi)$, and such that for all $y \in V$, we have both

$$
u(\varphi(\sigma, y)) > 0, \quad \forall \sigma \in I,
$$

and

$$
\left| \frac{1}{u(\varphi(\sigma, y))} - \frac{1}{u(\varphi(\sigma, x))} \right| < \frac{\delta}{m(I)}, \quad \forall \sigma \in I.
$$

Now suppose $y \in V$ and $|s - t| < \delta$. Then we must have $f_y(s) > f_x(t) - \eta$, for otherwise, since $u(y) > 0$, and consequently

$u\big(\varphi(\sigma, y)\big) > 0,$ $\forall\, \sigma \in (\alpha_y, \omega_y),$ we would have

$$s = \int_0^{f_y(s)} \frac{1}{u\big(\varphi(\sigma, y)\big)}\, d\sigma$$

$$\leq \int_0^{f_x(t)-\eta} \frac{1}{u\big(\varphi(\sigma, y)\big)}\, d\sigma$$

$$\leq \int_0^{f_x(t)-\eta} \frac{1}{u\big(\varphi(\sigma, x)\big)}\, d\sigma + \left| \int_0^{f_x(t)-\eta} \left| \frac{1}{u\big(\varphi(\sigma, x)\big)} - \frac{1}{u\big(\varphi(\sigma, y)\big)} \right|\, d\sigma \right|$$

$$\leq \int_0^{f_x(t)} \frac{1}{u\big(\varphi(\sigma, x)\big)}\, d\sigma - \int_{f_x(t)-\eta}^{f_x(t)} \frac{1}{u\big(\varphi(\sigma, x)\big)}\, d\sigma + \delta$$

$$\leq t - 3\delta + \delta < t - \delta,$$

contradicting the fact that $|s - t| < \delta$.

A similar argument shows that we must have $f_y(s) < f_x(t) + \eta$, and therefore we have

$$|f_x(t) - f_y(s)| < \eta < \varepsilon.$$

CASE 2: $u(x) < 0$.

This case is similar to Case 1, and will be omitted.

CASE 3: $u(x) = 0$.

Let $\delta = 1$. We shall find V by first finding a neighborhood V^+ of x such that whenever $y \in V^+$, we have,

either $\displaystyle\int_0^\varepsilon \frac{1}{u\big(\varphi(\sigma, y)\big)}\, d\sigma$ diverges,   or $\displaystyle\left| \int_0^\varepsilon \frac{1}{u\big(\varphi(\sigma, y)\big)}\, d\sigma \right| > 1 + |t|.$

After this, we shall find a neighborhood V^- of x, such that, whenever $y \in V^-$, we have

either $\displaystyle\int_{-\varepsilon}^0 \frac{1}{u\big(\varphi(\sigma, y)\big)}\, d\sigma$ diverges,   or $\displaystyle\left| \int_{-\varepsilon}^0 \frac{1}{u\big(\varphi(\sigma, y)\big)}\, d\sigma \right| > 1 + |t|.$

It is easy to see that if we then let $V = V^+ \cap V^-$, we will have $|f_y(s)| < \varepsilon$ whenever $y \in V$ and $|s - t| < 1$; i.e., $|f_y(s) - f_x(t)| < \varepsilon$ whenever $y \in V$ and $|s - t| < \delta$.

To find V^+, let

$$A = \{\sigma \in [0, \varepsilon] \mid u(\varphi(\sigma, x)) = 0\} \text{ and } B = [0, \varepsilon] \setminus A.$$

We note that A is compact. We shall consider separately the possibilities $m(A) > 0$, and $m(A) = 0$.

If $m(A) > 0$, choose V^+ such that, whenever $y \in V^+$, we have

$$|u(\varphi(\sigma, y)) - u(\varphi(\sigma, x))| < \frac{m(A)}{1 + |t|}, \qquad \forall \sigma \in A,$$

$$i.e., \qquad |u(\varphi(\sigma, y))| < \frac{m(A)}{1 + |t|}, \qquad \forall \sigma \in A.$$

Clearly, if $y \in V^+$, we have

$$\int_0^\varepsilon \frac{1}{|u(\varphi(\sigma, y))|} \, d\sigma \geq \int_A \frac{1}{|u(\varphi(\sigma, y))|} \, d\sigma > 1 + |t|,$$

and therefore

$$\text{either} \int_0^\varepsilon \frac{1}{u(\varphi(\sigma, y))} \, d\sigma \text{ diverges, or} \left| \int_0^\varepsilon \frac{1}{u(\varphi(\sigma, y))} \, d\sigma \right| > 1 + |t|.$$

This shows how to define V^+ when $m(A) > 0$.

Now suppose $m(A) = 0$. For each $n = 1, 2, \ldots,$ let

$$B_n = \left\{ \sigma \in [0, \varepsilon] \,\middle|\, |u(\varphi(\sigma, x))| \geq \frac{1}{n} \right\}.$$

Then $B_1 \subseteq B_2 \subseteq \ldots,$ $\bigcup_{n=1}^\infty B_n = B$, and for each $n = 1, 2, \ldots,$ B_n is compact. Since $u(x) = 0$ and $m(A) = 0$, we see that

$$\int_B \frac{1}{|u(\varphi(\sigma, x))|} \, d\sigma = \int_0^\varepsilon \frac{1}{|u(\varphi(\sigma, x))|} \, d\sigma = \infty.$$

Choose n such that

$$\int_{B_n} \frac{1}{|u(\varphi(\sigma, x))|} \, d\sigma > 2 + |t|.$$

Now using the facts that B_n is compact, and that

$$\left| \frac{1}{u\big(\varphi(\sigma, x)\big)} \right| \leq n, \qquad \forall \, \sigma \, \epsilon \, B_n,$$

let V^+ be a neighborhood of x such that, whenever $y \, \epsilon \, V^+$, we have

$$\left| \frac{1}{u\big(\varphi(\sigma, x)\big)} - \frac{1}{u\big(\varphi(\sigma, y)\big)} \right| < \frac{1}{m(B_n)}, \qquad \forall \, \sigma \, \epsilon \, B_n.$$

Clearly, if $y \, \epsilon \, V^+$, we have

$$\int_0^\varepsilon \frac{1}{|u\big(\varphi(\sigma, y)\big)|} \, d\sigma \geq \int_{B_n} \frac{1}{|u\big(\varphi(\sigma, y)\big)|} \, d\sigma$$

$$\geq \int_{B_n} \frac{1}{|u\big(\varphi(\sigma, x)\big)|} \, d\sigma - 1 > 1 + |t|$$

and therefore

either $\displaystyle\int_0^\varepsilon \frac{1}{u\big(\varphi(\sigma, y)\big)} \, d\sigma$ diverges, or $\displaystyle\left| \int_0^\varepsilon \frac{1}{u\big(\varphi(\sigma, y)\big)} \, d\sigma \right| > 1 + |t|.$

This shows that V^+ can also be defined when $m(A) = 0$. In a similar manner, we can define V^-, and therefore the proof is complete.   **QED**

4.45. Corollary. *Let φ be a flow in a Hausdorff space X.*

Then every continuous, proper quasi φ-multiplier is a proper φ-multiplier.

4.46. Corollary. *Let φ be a flow in a Hausdorff space X, let u be a continuous, bounded function from X into $\mathbb{R}$, and suppose $u(x) \neq 0$, $\forall \, x \, \epsilon \, \mathscr{G}(\varphi)$.*

Then u is a proper φ-multiplier.

PROOF: This result follows easily from Theorem 4.44.   **QED**

Reparametrizers which are Induced by Multipliers

4.47. Introduction. Let φ be a quasi-flow in a Hausdorff space X. We have already seen that, given a quasi φ-multiplier u, the equations

$$g_x(\tau) = \int_0^\tau \frac{1}{u\big(\varphi(\sigma, x)\big)} \, d\sigma$$

determine an inverse quasi φ-reparametrizer, and from this a quasi φ-reparametrizer. It is of interest to note that the above equations determine u in terms of the functions g_x $\big($for $x \in \mathcal{G}(u\varphi)\big)$, since for each $x \in \mathcal{G}(u\varphi)$, we have

$$u\big(\varphi(\tau, x)\big) = \frac{1}{g_x'(\tau)} = \frac{1}{g_{\varphi(\tau, x)}(0)}$$

for almost all τ in the domain of g_x. It is therefore reasonable to expect the equation

$$u(x) = \frac{1}{g_x'(0)}$$

to hold for a large family of points in $\mathcal{G}(u\varphi)$.

In view of this, we can see that given a quasi φ-reparametrizer f, writing $\psi = f \operatorname{rep} \varphi$ and $\operatorname{inv}(f) = \{g_x \mid x \in \mathcal{G}(\psi)\}$, the question of whether f is induced by a quasi φ-multiplier reduces to the question of whether f is induced by a quasi φ-multiplier u which satisfies the equation

$$u(x) = \frac{1}{g_x'(0)}$$

for a large family of points in $\mathcal{G}(\psi)$.

4.48. Theorem. *Let φ be a quasi-flow in a Hausdorff space X, let f be a quasi φ-reparametrizer, let $\psi = f \operatorname{rep} \varphi$, and let $\operatorname{inv}(f) = \{g_x \mid x \in \mathcal{G}(\psi)\}$.*

Then a necessary and sufficient condition that f be a quasi φ-reparametrizer induced by a quasi φ-multiplier is that for each $x \in \mathcal{G}(\psi)$, writing the domain of g_x as (α_x, ω_x) (as usual), g_x is an absolutely continuous function from (α_x, ω_x) onto $\mathbb{R}$.

PROOF: We remind the reader that a monotone, continuous, real function defined on an interval is differentiable almost everywhere in that interval, and is absolutely continuous iff it is the integral of its derivative [see, *e.g.*, Rudin [1] Chapter 8].

The condition is clearly necessary, for if f is induced by a quasi φ-multiplier u, then for each $x \in (\mathcal{G}\psi)$, the fact that

$$g_x(\tau) = \int_0^\tau \frac{1}{u\big(\varphi(\sigma, x)\big)}\, d\sigma, \qquad \forall\, \tau \in (\alpha_x, \omega_x),$$

implies at once that g_x is absolutely continuous.

Suppose now that g_x is absolutely continuous, for every $x \in \mathscr{G}(\psi)$. Define $u : X \to \mathbb{R}$ as follows:

$u(x) = \dfrac{1}{g_x'(0)}$ whenever $g_x'(0)$ exists and is non-zero, where, of course,

if $g_x'(0) = \pm \infty$, we agree to write $\dfrac{1}{g_x'(0)} = 0$,

$u(x) = \infty$ whenever $g_x'(0) = 0$ and g_x is increasing,

$u(x) = -\infty$ whenever $g_x'(0) = 0$ and g_x is decreasing,

and

$u(x) = 0$ whenever $g_x'(0)$ does not exist.

(The last case includes all points $x \in \mathscr{F}(\psi)$.)

We remark that if $x \in \mathscr{G}(\varphi)$ and $f_x'(0)$ exists, then $u(x) = f_x'(0)$.

Now given any $x \in \mathscr{G}(\psi)$, we know that $g_x'(\tau)$ exists for almost all $\tau \in (\alpha_x, \omega_x)$, and therefore, observing the correct sign convention for

$$\frac{1}{g_x'(\tau)}$$

in case $g_x'(\tau) = 0$, we have

$$u\big(\varphi(\tau, x)\big) = \frac{1}{g_x'(\tau)}, \text{ for almost all } \tau \in (\alpha_x, \omega_x).$$

It is easy to see that each φ-orbit contains only countably many non-trivial ψ-orbits. Therefore, since orbits are always σ-compact and u takes the value 0 everywhere in $\mathscr{F}(\psi)$, we see that in order to prove that the restriction of u to each φ-orbit is a Borel function, it is sufficient to show that the restriction of u to every non-trivial ψ-orbit is a Borel function.

Suppose $x \in \mathscr{G}(\psi)$. Since g_x' is a Borel function on (α_x, ω_x), and the set

$$B = \left\{ \tau \in (\alpha_x, \omega_x) \,\bigg|\, u\big(\varphi(\tau, x)\big) \neq \frac{1}{g_x'(\tau)} \right\}$$

is a Borel set, and $u\big(\varphi(\tau, x)\big) = 0$, $\forall \tau \in B$, it is clear that the function $\tau \to u\big(\varphi(\tau, x)\big)$ is a Borel function on (α_x, ω_x). Now (α_x, ω_x) is a countable union of compact subintervals, each of length less than $p_\varphi(x)$. Therefore, to prove that the restriction of u to $\mathscr{O}_\psi(x)$ is a Borel function, it is sufficient to show that, for each compact subinterval I of (α_x, ω_x) which has length less than $p_\varphi(x)$, the restriction of u to $\{\varphi(\tau, x) \mid \tau \in I\}$ is a Borel function. But this is clear for each such I, because the map

$\tau \to u\big(\varphi(\tau, x)\big)$ is a Borel function, and the map $\tau \to \varphi(\tau, x)$ $(\tau \in I)$, is a homeomorphism.

This shows that the restriction of u to each φ-orbit is a Borel function, *i.e.*, u satisfies condition (i) of Definition 4.31. It is easy to see that u also satisfies condition (ii) of Definition 4.31. Furthermore, it is clear that if $x \in \mathscr{G}(\psi)$, then since g_x is absolutely continuous, we have

$$(\alpha_x, \omega_x) = \left\{ \tau \in [-\infty, \infty] \middle| \int_0^\tau \frac{1}{u\big(\varphi(\sigma, x)\big)}\, d\sigma \text{ converges} \right\},$$

and that consequently u satisfies condition (iii)(b) of Definition 4.31 at x.

To show that u is a quasi φ-multiplier, we need only show that whenever $x \in \mathscr{F}(\psi) \cap \mathscr{G}(\varphi)$, the integral

$$\int_0^\tau \frac{1}{u\big(\varphi(\sigma, x)\big)}\, d\sigma$$

diverges for every $\tau \neq 0$. Suppose, to obtain a contradiction, that $x \in \mathscr{F}(\psi) \cap \mathscr{G}(\varphi)$, $\tau \neq 0$, but

$$\int_0^\tau \frac{1}{u\big(\varphi(\sigma, x)\big)}\, d\sigma$$

converges. Proceeding as in the proof of Theorem 4.44, choose $\sigma_0 \in (0, \tau)$ such that $u\big(\varphi(\sigma_0, x)\big) \neq 0$, and let $y = \varphi(\sigma_0, x)$. Then $y \in \mathscr{G}(\psi)$, and since $\varphi(-\sigma_0, y) = x \in \mathscr{F}(\psi)$, we have $-\sigma_0 \notin (\alpha_y, \omega_y)$,

$$i.e. \int_{-\sigma_0}^0 \frac{1}{u\big(\varphi(\sigma, y)\big)}\, d\sigma \text{ diverges.}$$

But, as we saw in the proof of Theorem 4.44, this contradicts the fact that

$$\int_0^{\sigma_0} \frac{1}{u\big(\varphi(\sigma, x)\big)}\, d\sigma \text{ converges.}$$

Thus u is a quasi φ-multiplier, and the absolute continuity of the functions g_x, $\big(x \in \mathscr{G}(\psi)\big)$ now implies at once that f is induced by u.

This completes the proof of the theorem.   **QED**

4.49. Lemma. *Let* $-\infty \leq \alpha < \beta \leq \infty$, *and suppose g is a continuous, strictly increasing function from* (α, β) *into* $\mathbb{R}$. *Let* c_1 *and* c_2 *be real numbers, let A be a Borel subset of* (α, β), *and suppose*

$$c_1 \leq g'(\tau) \leq c_2, \qquad \forall\ \tau \in A.$$

Then

$$c_1 \cdot m(A) \leq m\big(g(A)\big) \leq c_2 \cdot m(A).$$

PROOF: Let μ be the nonnegative Borel measure defined on (α, β) by:

$$\mu(E) = m\big(g(E)\big), \text{ for all Borel subsets } E \text{ of } (\alpha, \beta).$$

It is well-known [see *e.g.* Rudin [1], Chapter 8] that there exists a nonnegative, singular Borel measure μ_s on (α, β) such that

$$\mu(E) = \mu_s(E) + \int_E g'\, dm$$

for all Borel subsets E of (α, β). From this it is clear that

$$c_1 \cdot m(A) \leq \int_A g'\, dm \leq \mu(A) = m\big(g(A)\big).$$

To obtain the second half of the desired inequality, let $g(A) = B$, and $f = g^{-1}$. Then since

$$\frac{1}{c_2} \leq f'(t) \leq \frac{1}{c_1}, \qquad \forall\ t \in B,$$

the argument we have used above, when applied to f, shows that

$$\frac{1}{c_2} \cdot m(B) \leq m\big(f(B)\big), \qquad i.e., \qquad \frac{1}{c_2} \cdot m\big(g(A)\big) \leq m(A),$$

as required. This completes the proof of the lemma.   **QED**

4.50. Lemma. *Let* $-\infty \leq \alpha < \beta \leq \infty$, *and let g be a continuous, strictly monotone function from* (α, β) *into* $\mathbb{R}$. *Let*

$$A = \{\tau \in (\alpha, \beta) \mid |g'(\tau)| = \infty\}.$$

Then g is absolutely continuous iff $m\big(g(A)\big) = 0$.

PROOF: There is no loss of generality in assuming that g is increasing. Let μ be the nonnegative Borel measure defined on (α, β) by

$$\mu(E) = m\big(g(E)\big), \text{ for every Borel subset } E \text{ of } (\alpha, \beta).$$

Then the condition that g be absolutely continuous is the condition that

$$\mu(E) = \int_E g' \, dm \quad \text{for all Borel subsets } E \text{ of } (\alpha, \beta),$$

and this is equivalent to the assertion that $\mu \ll m$. Clearly, if g is absolutely continuous, then since $m(A) = 0$,

$$m\big(g(A)\big) = \mu(A) = \int_A g' \, dm = 0.$$

Suppose now that $m\big(g(A)\big) = 0$. To show that $\mu \ll m$, let E be a Borel subset of (α, β), and suppose $m(E) = 0$.

Let $E_0 = \{\tau \in E \mid g'(\tau) \text{ does not exist or } g'(\tau) = 0\}$,

and $E_\infty = \{\tau \in E \mid g'(\tau) = \infty\}$,

and for each positive integer n,

let $E_n = \left\{\tau \in E \,\middle|\, \dfrac{1}{n} \leq g'(\tau) \leq n\right\}$.

Since g^{-1} has a finite derivative almost everywhere, it is clear that $\mu(E_0) = 0$. Also, since $E_\infty \subseteq A$ and $\mu(A) = 0$, we have $\mu(E_\infty) = 0$. Therefore, to prove that $\mu(E) = 0$, it is sufficient to show that $\mu(E_n) = 0$ for every positive integer n. But for each positive integer n, we have

$$\mu(E_n) \leq n \cdot m(E_n) \quad \text{by Lemma 4.49,}$$
$$\leq n \cdot m(E) = 0.$$

This completes the proof of the lemma.   **QED**

4.51. Theorem. *Let φ be a quasi-flow in a Hausdorff space X, let f be a quasi φ-reparametrizer, and let $\psi = f \operatorname{rep} \varphi$.*

Then a necessary and sufficient condition that f be a quasi φ-reparametrizer induced by a quasi φ-multiplier is that for each $x \in \mathscr{G}(\psi)$ the set

$$\{t \in \mathbb{R} \mid f'_x(t) = 0\}$$

has measure zero.

PROOF: The theorem is immediate from Theorem 4.48 and Lemma 4.50.
                                                          QED

4.52. Theorem. *Let φ be a flow in a Hausdorff space X, let f be a φ-reparametrizer, let $\psi = f \operatorname{rep} \varphi$, let $\operatorname{inv}(f) = \{g_x \mid x \in \mathscr{G}(\psi)\}$, and for each $x \in \mathscr{G}(\psi)$, let the domain of g_x be denoted by (α_x, ω_x).*

(a) *A necessary and sufficient condition that f be a φ-reparametrizer, induced by a φ-multiplier whose restriction to $\mathscr{G}(\varphi)$ is continuous, is that*

  (i) *Whenever $x \in \mathscr{G}(\varphi)$ and $t \in \mathbb{R}$, f_x has a (possibly infinite) derivative at t.*

  (ii) *Whenever $x \in \mathscr{G}(\psi)$, the set $\{t \in \mathbb{R} \mid f_x{}'(t) = 0\}$ has measure zero.*

  (iii) *The map $\langle x, t \rangle \to f_x'(t)$ is a continuous function from $\mathscr{G}(\varphi) \times \mathbb{R}$ into $[-\infty, \infty]$.*

Furthermore, if f is induced by a φ-multiplier u whose restriction to $\mathscr{G}(\varphi)$ is continuous, we have

(b) $g_x'(\tau) = \dfrac{1}{u\big(\varphi(\tau, x)\big)}, \qquad \forall\, x \in \mathscr{G}(\psi) \qquad \forall\, \tau \in (\alpha_x, \omega_x),$

and

(c) $f_x'(t) = u\big(\psi(t, x)\big), \qquad \forall\, x \in \mathscr{G}(\varphi) \qquad \forall\, t \in \mathbb{R}.$

PROOF: Suppose f is induced by a φ-multiplier u, whose restriction to $\mathscr{G}(\varphi)$ is continuous. Then it is easy to see that (b), (c), and (a)(i) hold. It follows from Theorem 4.51 that (a)(ii) also holds, and the fact that (a)(iii) holds is an elementary consequence of (c).

Suppose now that f satisfies conditions (a)(i), (a)(ii) and (a)(iii). Let $u : X \to \mathbb{R}$ be defined as in the proof of Theorem 4.48. Then

$$u(x) = \frac{1}{g_x'(0)}, \qquad \forall\, x \in \mathscr{G}(\psi),$$

and

$$u(x) = 0, \qquad \forall\, x \in \mathscr{F}(\psi).$$

It follows that $u(x) = f_x'(0), \qquad \forall\, x \in \mathscr{G}(\varphi)$, and therefore (iii) implies at once that the restriction of u to $\mathscr{G}(\varphi)$ is continuous.

This completes the proof of the theorem.   **QED**

We conclude this chapter with an example.

4.53. Example. Let $0 < \delta \leq \tfrac{1}{3}$, and let C be the δ-Cantor subset of $[0, 1]$ (see Appendix A). Let f_0 be the continuous, strictly increasing function from $\mathbb{R}$ onto $\mathbb{R}$, defined by

$$f_0(t) = \int_0^t \big(1 \wedge d(s, C)\big)\, ds, \qquad \forall\, t \in \mathbb{R}.$$

Let $g_0 = f_0^{-1}$, and for each $x \in \mathbb{R}$, define the function g_x by

$$g_x(\tau) = g_0(x + \tau) - g_0(x), \qquad \forall\, \tau \in \mathbb{R}.$$

For each $x \in \mathbb{R}$, let $f_x = g_x^{-1}$. Now define $f : \mathbb{R} \times \mathbb{R} \to \mathbb{R}$ by

$$f(x, t) = f_x(t), \qquad \forall\, x \in \mathbb{R} \qquad \forall\, t \in \mathbb{R}.$$

It is easy to see that f is a 1-reparametrizer, and that $\operatorname{inv}(f) = \{g_x \mid x \in \mathbb{R}\}$. Let $u : \mathbb{R} \to \mathbb{R}$ be defined by

$$u(x) = 1 \wedge d(g_0(x), C).$$

It is easy to see that, for every $x \in \mathbb{R}$, we have

$$u(x) = \frac{1}{g_x'(0)},$$

and it is not hard to show that u is a 1-multiplier. But since $f_0'(t) = 0$ iff $t \in C$, it is clear from Theorem 4.51 that f is induced by u only if $\delta = \frac{1}{3}$.   •

Thus, if $0 < \delta < \frac{1}{3}$, f is not induced by a multiplier, in spite of the fact that each function f_x is continuously differentiable, and the function

$$x \to \frac{1}{g_x'(0)}$$

is a 1-multiplier.

More examples exhibiting this kind of behavior will be given in a later chapter.

Notes and Remarks to Chapter 4

In Beck [1] it is shown that if a flow φ in a metric space X has no fixed points, and if F is any closed subset of X, then it is possible to slow φ down near F to obtain a new flow whose fixed set is F. The theory of reparametrization as it appears in Chapters 4 and 5 was formulated by the Lewins in order to investigate the techniques of this theorem more closely, and also to provide the technical machinery which is required to clarify the proofs of some interpolation theorems in Beck [3]. (Modifications of these interpolation theorems may be found in Chapter 6).

In this theory of reparametrization, the slowing down procedure used in Beck [1] is manifested as the multiplication of the flow by an appropriate multiplier. Theorem 4.44, which is one of the central theorems on multiplication in this chapter, is motivated by the proof of the main result in Beck [1].

Chapters 4 and 5 constitute the major portion of the doctoral dissertation "Reparametrization of Continuous Flows" of Jonathan Walter Lewin, (University of Wisconsin, 1970).

CHAPTER FIVE

REPARAMETRIZATION II

For the purpose of our first theorem, we need the concept of a *zero set*.
We refer the reader to Gillman and Jerison [1] for details.

5.1. Definition. A subset H of a topological space X is said to be a
zero subset of X if there exists a continuous function h from X into $\mathbb{R}$,
such that
$$H = \{x \in X \mid h(x) = 0\}.$$

When this happens, we say that H is the *zero set of* the function h.

5.2. Lemma
(a) *A subset H of a topological space X is a zero subset of X iff there exists
a nonnegative, bounded, continuous function h from X into $\mathbb{R}$ such that*
$$H = \{x \in X \mid h(x) = 0\}.$$

(b) *A zero subset of a topological space is closed.*

(c) *Every closed subset of a metric space X is a zero subset of X.*

(d) *The family of zero subsets of a topological space is closed under the
formation of finite unions and countable intersections.*

PROOF: These results are elementary; their proofs appear in Gillman and Jerison [1]. **QED**

5.3. Theorem. *Let φ be a flow in a Hausdorff space X.*
(a) *Given any continuous, proper φ-multiplier u, there exists a zero subset
H of X, such that*
$$\mathscr{F}(u\varphi) = H \cup \mathscr{F}(\varphi).$$

(b) *Given any zero subset H of X, there exists a continuous, bounded,
nonnegative, proper φ-multiplier u, such that*
$$\mathscr{F}(u\varphi) = H \cup \mathscr{F}(\varphi).$$

PROOF: To prove (a), let u be a continuous, proper φ-multiplier, and let H be the zero set of u. Since u is proper, we have $\mathscr{F}(u\varphi) = H \cup \mathscr{F}(\varphi)$, and therefore (a) is proved.

To prove (b), we note that if $H = \square$, then the function u defined by $u(x) = 1$, $\quad \forall\, x \in X$, satisfies the requirements of the theorem.

Suppose now that $H \neq \square$. Choose a nonnegative, bounded continuous function h from X into $\mathbb{R}$, such that H is the zero set of h. Let g be the function from X into $[0, \infty]$, defined by

$$g(x) = \inf\{|t| \mid \varphi(t, x) \in H\}, \qquad \forall\, x \in X.$$

Since H is closed, it is clear that $H = \{x \in X \mid g(x) = 0\}$. For each $n = 1, 2, \ldots$, let f_n be the function from X into $[0, \infty)$, defined by

$$f_n(x) = \inf\left\{h\big(\varphi(t, x)\big) \,\middle|\, |t| \le \frac{1}{n}\right\}, \qquad \forall\, x \in X.$$

We note that, for every $n = 1, 2, \ldots$, and every $x \in X$, $f_n(x) \le h(x)$. Further, for each $n = 1, 2, \ldots$, f_n is continuous; for if $x \in X$ and $\varepsilon > 0$, we can choose a neighborhood V of x such that

$$|h\big(\varphi(t, y)\big) - h\big(\varphi(t, x)\big)| < \varepsilon, \quad \text{whenever } y \in V \text{ and } |t| \le \frac{1}{n};$$

and clearly

$$|f_n(y) - f_n(x)| \le \varepsilon, \qquad \forall\, y \in V.$$

Next, we note that for every $n = 1, 2, \ldots$, we have

$$\{x \in X \mid f_n(x) = 0\} = \left\{x \in X \,\middle|\, g(x) \le \frac{1}{n}\right\}.$$

Now define the function f from X into $[0, \infty)$ by

$$f = \sum_{n=1}^{\infty} \frac{f_n}{2^n}.$$

Since $f_n \le h$ for every $n = 1, 2, \ldots$, and h is bounded, it is clear that f is continuous, and that $f \le h$. Furthermore,

$$\{x \in X \mid f(x) = 0\} = \bigcap_{n=1}^{\infty} \{x \in X \mid f_n(x) = 0\} = \{x \in X \mid g(x) = 0\} = H.$$

Thus f, like h, is a nonnegative, bounded, continuous function from X into $\mathbb{R}$, whose zero set is H. But as we shall now show, f enjoys the additional property, that given $\varepsilon > 0$, there exists $\delta > 0$ such that $f(y) < \varepsilon$ whenever $g(y) < \delta$. To show this, let $\varepsilon > 0$. Since h is bounded, we can choose a positive integer N such that

$$\sum_{n=N}^{\infty} \frac{f_n(x)}{2^n} < \varepsilon, \qquad \forall \; x \in X.$$

It is clear that whenever $g(y) < \dfrac{1}{N}$, we have $f(y) < \varepsilon$.

We shall now define a function u from X into $\mathbb{R}$, which, we shall show, has the properties stipulated in (b). Define

$$u(x) = \inf \left\{ |f(x) - f(y)| + g(y) \mid y \in X \right\}, \qquad \forall \; x \in X.$$

We note firstly that by taking $y \in H$, we can see that $u(x) \leq f(x)$, $\forall \; x \in X$, and consequently that u is bounded.

Secondly, by taking $y = x$, we can see that $u(x) \leq g(x)$, $\qquad \forall \, x \in X$. Thirdly, if $x \in H$, it is clear that $u(x) = 0$, and conversely, whenever $u(x) = 0$, we must have $x \in H$, because if we choose a sequence $\{y_n\}$ such that $|f(y_n) - f(x)| + g(y_n) \to 0$, then since $f(y_n) \to f(x)$ and $g(y_n) \to 0$, the above-mentioned property of f implies that $f(y_n) \to 0$, i.e., $f(x) = 0$, i.e., $x \in H$.

Fourthly, u is continuous, for given any points $x_1, x_2 \in X$, we have, for every $y \in X$,

$$u(x_1) \leq |f(x_1) - f(y)| + g(y) \leq |f(x_1) - f(x_2)| + |f(x_2) - f(y)| + g(y)$$

and therefore

$$u(x_1) \leq |f(x_1) - f(x_2)| + u(x_2).$$

From this, we see that given any points $x_1, x_2 \in X$, we have

$$|u(x_1) - u(x_2)| \leq |f(x_1) - f(x_2)|,$$

and therefore the continuity of f implies that u is continuous.

We shall complete the proof by showing that u is a proper φ-multiplier. This will follow at once from Theorem 4.44 when we have shown that whenever $x \in \mathcal{G}(\varphi)$ and $u(x) \neq 0$, writing (α_x, ω_x) for the component of $\{\tau \in \mathbb{R} \mid u(\varphi(\tau, x)) \neq 0\}$ which contains 0, we have that both

$$\int_{\alpha_x}^{0} \frac{1}{u(\varphi(\sigma, x))} \, d\sigma \quad \text{and} \quad \int_{0}^{\omega_x} \frac{1}{u(\varphi(\sigma, x))} \, d\sigma$$

diverge. Suppose $x \in \mathscr{G}(\varphi)$ and $u(x) \neq 0$. Since u is bounded, it is clear that in case $\alpha_x = -\infty$, we have

$$\int\limits_{\alpha_x}^{0} \frac{1}{u(\varphi(\sigma, x))} \, d\sigma = \infty.$$

If $-\infty < \alpha_x < 0$, then since $\varphi(\alpha_x, x) \in H$, we must have

$$u(\varphi(\sigma, x)) \leq g(\varphi(\sigma, x)) \leq \sigma - \alpha_x, \qquad \forall \, \sigma \in (\alpha_x, 0],$$

and therefore

$$\int\limits_{\alpha_x}^{0} \frac{1}{u(\varphi(\sigma, x))} \, d\sigma \geq \int\limits_{\alpha_x}^{0} \frac{1}{\sigma - \alpha_x} \, d\sigma = \infty.$$

Similarly, we can show that

$$\int\limits_{0}^{\omega_x} \frac{1}{u(\varphi(\sigma, x))} \, d\sigma = \infty,$$

and this completes the proof of the theorem.   **QED**

5.4. Corollary. *Let φ be a flow in a metric space X, let F be a closed subset of X, and suppose $\mathscr{F}(\varphi) \subseteq F$.*

Then there exists a continuous proper φ-multiplier u, such that $F = \mathscr{F}(u\varphi)$.

PROOF: This is immediate from Theorem 5.3, because $F = F \cup \mathscr{F}(\varphi)$, and F is a zero subset of X.   **QED**

5.5. Corollary [Beck [1]]. *Let X be a metric space, and suppose there exists a flow in X, which has no fixed points.*

Then given a closed subset F of X, there exists a flow φ in X, such that $F = \mathscr{F}(\varphi)$.

5.6. Corollary. *Let X be a normed linear space, and let F be a closed subset of X.*

Then there exists a simple flow φ in X, such that $F = \mathscr{F}(\varphi)$.

PROOF: Choose a non-zero point $y \in X$, and let ψ be the flow in X defined by

$$\psi(t, x) = x + ty, \qquad \forall \, t \in \mathbb{R} \qquad \forall \, x \in X.$$

Then ψ is a simple flow, and $\mathscr{F}(\psi) = \square$. The result now follows at once from Corollary 5.5, and the trivial observation that every member of $\text{Rep}_q(\psi)$ is simple.   **QED**

The following theorem is an analogue of Theorem 5.3 for quasi-flows.

5.7. Theorem. *Let φ be a quasi-flow in a Hausdorff space X.*
(a) *Let u be a proper quasi φ-multiplier, and suppose that for each $x \in X$, the function*

$$t \to u\big(\varphi(t, x)\big) \qquad (t \in \mathbb{R})$$

is continuous.
Then there exists a subset H of X and a function h from X into $\mathbb{R}$ such that

(i) $\mathscr{F}(u\varphi) = H \cup \mathscr{F}(\varphi)$,

(ii) *for each $x \in X$, the function $t \to h\big(\varphi(t, x)\big)$ is continuous,*

(iii) $H = \{x \in X \mid h(x) = 0\}$.

(b) *Let H be a subset of X, and h a function from X into $\mathbb{R}$. Suppose that $H = \{x \in X \mid h(x) = 0\}$, and that for each $x \in X$, the function $t \to h\big(\varphi(t, x)\big)$ is continuous. Then there exists a bounded, nonnegative, proper quasi φ-multiplier u such that*

(i) $\mathscr{F}(u\varphi) = H \cup \mathscr{F}(\varphi)$,

(ii) *for each $x \in X$, the function $t \to u\big(\varphi(t, x)\big)$ is continuous.*

PROOF: The proof of (a) is elementary, and will be left to the reader.

To prove (b), we note, as in the proof of Theorem 5.3, that it is sufficient to confine our attention to the case $H \neq \square$.

Suppose then, that H is a non-empty subset of X, h is a function from X into $\mathbb{R}$, $H = \{x \in X \mid h(x) = 0\}$, and for each $x \in X$, the function $t \to h\big(\varphi(t, x)\big)$ is continuous. There is clearly no loss of generality in assuming that h is nonnegative and bounded.

As in the proof of Theorem 5.3, let g be the function from X into $[0, \infty]$ defined by

$$g(x) = \inf\big\{|t| \,\big|\, \varphi(t, x) \in H\big\}, \qquad \forall\, x \in X.$$

Since

$$g(x) = \inf\big\{|t| \,\big|\, h\big(\varphi(t, x)\big) = 0\big\}, \qquad \forall\, x \in X,$$

it is clear that

$$H = \{x \in X \mid g(x) = 0\}.$$

For each $n = 1, 2, \ldots$, let f_n be the function from X into $[0, \infty)$, defined by

$$f_n(x) = \inf\left\{h\big(\varphi(t, x)\big) \,\middle|\, |t| \le \frac{1}{n}\right\}, \qquad \forall\, x \in X.$$

Then for each n, $f_n \leq h$. Further, for each $n = 1, 2, \ldots$, and $x \in X$, the function $t \to f_n\big(\varphi(t, x)\big)$ is continuous, for given $\varepsilon > 0$ and $t_0 \in \mathbb{R}$, we can choose $\delta > 0$ such that

$$|h\big(\varphi(t_0 + t + s, x)\big) - h\big(\varphi(t_0 + t, x)\big)| < \varepsilon,$$

whenever $|s| < \delta$ and $|t| \leq \dfrac{1}{n}$; and clearly

$$|f_n\big(\varphi(t_0 + s, x)\big) - f_n\big(\varphi(t_0, x)\big)| \leq \varepsilon, \quad \text{whenever } |s| < \delta.$$

The rest of the proof is similar to the proof of Theorem 5.3, and will be left to the reader. **QED**

5.8. Theorem.

(a) *Let φ be a quasi-flow in a normed linear space X, let u be a quasi φ-multiplier, let $\psi = u\varphi$, and suppose that u is continuous and finite at x, and that $\dot{\varphi}(x)$ exists.*

Then $\dot{\psi}(x) = u(x)\dot{\varphi}(x)$.

(b) *Let φ be a quasi-flow in a metric space X, let u be a quasi φ-multiplier, let $\psi = u\varphi$, and suppose that u is continuous at x, and that $|\dot{\varphi}|(x)$ exists.*

Then unless the expression $|u(x)| \, |\dot{\varphi}|(x)$ is of the form $0 \times \infty$ or $\infty \times 0$, we have

$$|\dot{\psi}|(x) = |u(x)| \, |\dot{\varphi}|(x).$$

PROOF: Since the theorem is trivial if $x \in \mathscr{F}(\psi)$, we shall suppose that $x \in \mathscr{G}(\psi)$.

(a) Let f be a quasi φ-reparametrizer induced by u, and let $g_x = f_x^{-1}$.

It is easy to see that

$$g_x'(0) = \frac{1}{u(x)}$$

(subject to the usual sign convention for $\dfrac{1}{u(x)}$ in case $u(x) = 0$). Therefore

$$f_x'(0) = u(x).$$

Now for any $t \neq 0$, we have

$$\frac{\psi(t, x) - x}{t} = \frac{\varphi\big(f_x(t), x\big) - x}{f_x(t) - 0} \cdot \frac{f_x(t) - f_x(0)}{t - 0}$$

$$\to \dot{\varphi}(x) \cdot f_x'(0) \qquad \text{as } t \to 0.$$

This completes the proof of (a).

The proof of (b) is similar, and will be omitted. **QED**

Canonical Reparametrizations

5.9. Introduction. Suppose φ is a quasi-flow in a subspace Y of a metric space X, suppose K is a non-empty closed subset of X, and suppose $|\dot{\varphi}|$ exists and is continuous, finite and non-zero everywhere in Y. Suppose further that the function u from Y into $[0, \infty]$, defined by

$$u(x) = \frac{d(x, K)}{|\dot{\varphi}|(x)}, \qquad \forall\, x \in Y$$

is a quasi φ-multiplier. Then by Theorem 5.8, it is clear that $u\varphi$ is canonical with respect to K.

This observation suggests the following question:

Let φ be a quasi-flow in a subspace Y of a metric space X, let K be a non-empty closed subset of X, and suppose $|\dot{\varphi}|$ exists and is finite everywhere in $Y \setminus K$. Does there exist an isotonic quasi φ-reparametrizer f such that f rep φ is canonical with respect to K?

If we do not have $\mathscr{F}(\varphi) \subseteq K$, it is clear that the answer to this question is always NO. We shall therefore concern ourselves only with the case $\mathscr{F}(\varphi) \subseteq K$, and we begin by giving examples which illustrate three important ways in which the answer to the above question can still be NO.

5.10. Example and Remark. Let $X = \mathbb{R}^2$, and let

$$Y = \left\{ \langle x, y \rangle \in \mathbb{R}^2 \,\middle|\, y = x^2 \sin\left(\frac{1}{x^2}\right) \right\},$$

where we agree to write $0^2 \sin\left(\dfrac{1}{0^2}\right) = 0$. Let K be any non-empty, closed subset of $\mathbb{R}^2$ which does not meet Y. Let φ be the flow in Y defined by

$$\varphi\left(t, \left\langle x, x^2 \sin\left(\frac{1}{x^2}\right) \right\rangle\right) = \left\langle x + t, (x + t)^2 \sin\left(\frac{1}{(x + t)^2}\right) \right\rangle,$$

for every $t \in \mathbb{R}$ and $x \in \mathbb{R}$.

It is easy to see that $|\dot{\varphi}|$ exists and is finite everywhere in Y, and that $|\dot{\varphi}|$ is continuous at each point of Y, except the point $\langle 0, 0 \rangle$. However, since the arc length of Y, between any two points of Y which lie on opposite sides of $\langle 0, 0 \rangle$, is clearly infinite, there cannot exist a quasi-reparametrization of φ which is canonical with respect to K. (See Corollary 1.60.)

It is clear that to avoid the pathology that occurs here, we need to assume that for each $x \in Y$ and $t \in \mathbb{R}$, either the segment $\{\varphi(s, x) \mid s \in [0, t]\}$ meets K, or its arc-length is finite, *i.e.*, either

$$\{\varphi(s, x) \mid s \in [0, t]\} \cap K \neq \square,$$

or

$$\int_0^t |\dot{\varphi}| (s, x)\, ds \qquad \text{converges. (See Corollary 1.60.)}$$

Unfortunately, this assumption is not good enough, as we shall now show:

Let $X = \mathbb{R}^2$, and let

$$Y = \left\{ \langle x, y \rangle \in \mathbb{R}^2 \,\middle|\, y = x^2 \sin \frac{1}{x} \right\}.$$

Let K be any non-empty, closed subset of $\mathbb{R}^2$ which does not meet Y, and let φ be the flow in Y defined by

$$\varphi(t, \langle x, y \rangle) = \left\langle x + t, (x + t)^2 \sin \frac{1}{(x + t)} \right\rangle.$$

It is clear that $|\dot{\varphi}|$ exists everywhere, but is not continuous at $\langle 0, 0 \rangle$. But if, for each point $z \in Y$, we let $\mathscr{L}(z)$ denote the arc length of Y between z and $\langle 0, 0 \rangle$, it can be shown that

$$\lim_{z \to \langle 0,0 \rangle} \frac{\mathscr{L}(z)}{d(z, \langle 0, 0 \rangle)} = \sqrt{2},$$

and it follows easily that no flow in Y can have continuous finite non-vanishing speed.

It is not hard to show that if this kind of pathology does not occur, then some quasi-reparametrization of φ has speed which varies continuously with t. We shall therefore confine our attention to quasi-flows of this form.

5.11. Example. Let $X = \mathbb{R}^2$, and let

$$Y = \{ \langle x, y \rangle \in \mathbb{R}^2 \mid -1 < x < 1 \ \text{ and } \ y = 0 \}.$$

Let $K = \{ \langle x, y \rangle \in \mathbb{R}^2 \mid y = 1 \}$. Let h be the homeomorphism of $\mathbb{R}$ onto Y defined by

$$h(x) = \left\langle \frac{2}{\pi} \arctan x, 0 \right\rangle, \qquad \forall\, x \in \mathbb{R},$$

and let $\varphi = h[\mathbf{1}]$.

It is obvious that no quasi-reparametrization of φ can be canonical with respect to K.

We shall now state a positive result.

5.12. Theorem. *Let φ be a quasi-flow in a subspace Y of a metric space X, and let K be a non-empty, closed subset of X. Suppose that for every $x \in Y \setminus K$, $|\dot{\varphi}|(x)$ exists and*

$$0 < |\dot{\varphi}|(x) < \infty.$$

Suppose also that for each $x \in Y$, the function

$$t \to |\dot{\varphi}|(t, x)$$

is continuous at each number t for which $\varphi(t, x) \in Y \setminus K$.

Then the following are equivalent:

(a) There exists an isotonic quasi φ-reparametrizer f such that $f \operatorname{rep} \varphi$ is canonical with respect to K.

(b) There exists a nonnegative proper quasi φ-multiplier w such that $w\varphi$ is canonical with respect to K.

(c) There exists a quasi φ-reparametrizer f such that $f \operatorname{rep} \varphi$ is canonical with respect to K.

(d) Whenever $x \in Y \setminus K$ and $\mathrm{p}_\varphi(x) = \infty$, writing $\mathscr{L}(\tau)$ for the arc length of $\{\varphi(\sigma, x) \mid \sigma \in [0, \tau]\}$ for every $\tau \in \mathbb{R}$, we have:

(i) either $\inf\{d(\varphi(\tau, x), K) \mid \tau \geq 0\} = 0$, or $\mathscr{L}(\tau) \to \infty$ as $\tau \to \infty$ and

(ii) Either $\inf\{d(\varphi(\tau, x), K) \mid \tau \leq 0\} = 0$, or $\mathscr{L}(\tau) \to \infty$ as $\tau \to -\infty$.

(e) If, using Theorem 5.7, we choose a bounded, nonnegative proper quasi φ-multiplier v which satisfies

(i) for each $x \in Y$, the function $t \to v(\varphi(t, x))$ is continuous, and

(ii) $\mathscr{F}(v\varphi) = Y \cap K$,

and if, writing $\pi = v\varphi$, we define $u : Y \to [0, \infty)$, by

$$u(x) = \begin{cases} \dfrac{d(x, K)}{|\dot{\pi}|(x)} & \text{whenever } x \in Y \setminus K, \\[2mm] 0 & \text{whenever } x \in Y \cap K, \end{cases}$$

then u is a nonnegative proper quasi π-multiplier, and $u\pi$ is canonical with respect to K.

PROOF: We shall prove the theorem by showing that (e)$\Rightarrow$(a)$\Rightarrow$(b) $\Rightarrow$(c)$\Rightarrow$(d)$\Rightarrow$(e).

The fact that (e)$\Rightarrow$(a) is easy to see, for if v, π and u are defined as in the statement of (e), then the fact that π is an isotonic quasi φ-reparametrization, and $u\pi$ is an isotonic quasi π-reparametrization, implies at once that $u\pi$ is an isotonic quasi φ-reparametrization.

To see that (a)$\Rightarrow$(b), suppose f is an isotonic quasi φ-reparametrizer, and f rep φ is canonical with respect to K. Let $\psi = f$ rep φ. Then since ψ is canonical with respect to K, and $|\dot\psi|$ is non-zero and finite everywhere in $Y \setminus K$, the formula

$$\psi(t, x) = \varphi\big(f_x(t), x\big), \qquad \forall\, t \in \mathbb{R} \qquad \forall\, x \in Y,$$

can easily be used to show that

$$f'_x(0) = \frac{d(x, K)}{|\dot\varphi|(x)}, \qquad \forall\, x \in Y \setminus K.$$

It now follows from Theorem 4.51 that f is induced by some quasi φ-multiplier. Therefore, by Theorem 4.48 and its method of proof, the function w from Y into $[0, \infty)$, defined by

$$w(x) = \begin{cases} f'_x(0) & \text{whenever } x \in Y \setminus K \\ 0 & \text{whenever } x \in Y \cap K, \end{cases}$$

is a quasi φ-multiplier, and f is induced by w. It is obvious that w is proper. Thus (a)$\Rightarrow$(b).

It is obvious that (b)$\Rightarrow$(c).

To see that (c)$\Rightarrow$(d), suppose f is a quasi φ-reparametrizer, and f rep φ is canonical with respect to K. Let $\psi = f$ rep φ. Now suppose $x \in Y$, $p_\varphi(x) = \infty$, and

$$\inf \big\{d\big(\varphi(\tau, x), K\big) \mid \tau \geq 0\big\} = \delta > 0.$$

For simplicity of notation, assume without loss of generality that f_x is strictly *increasing*. Now clearly, to prove that $\mathscr{L}(\tau) \to \infty$ as $\tau \to \infty$, it is sufficient to show that $\mathscr{L}\big(f_x(t)\big) \to \infty$ as $t \to \infty$. But for each $t > 0$, $\mathscr{L}\big(f_x(t)\big)$ is the arc-length of $\{\psi(s, x) \mid 0 \leq s \leq t\}$, and therefore

$$\mathscr{L}\big(f_x(t)\big) = \int_0^t |\dot\psi|(s, x)\, ds = \int_0^t d\big(\psi(s, x), K\big)\, ds \geq \delta t,$$

and therefore $\mathscr{L}\big(f_x(t)\big) \to \infty$ as $t \to \infty$. A similar argument shows that if $x \in Y$, $p_\varphi(x) = \infty$, and

$$\inf \big\{d\big(\varphi(\tau, x), K\big) \mid \tau \leq 0\big\} > 0,$$

then $\mathscr{L}(\tau) \to \infty$ as $\tau \to -\infty$. Thus (c)$\Rightarrow$(d).

We shall now show that (d)$\Rightarrow$(e). Assume that (d) holds, and let v, π and u be chosen as in the statement of (e). It is clear that to prove (e), we need only show that u is a nonnegative quasi π-multiplier, for the rest of (e) will then follow at once.

Now u clearly satisfies conditions (i) and (ii) of Definition 4.31. Furthermore, it is clear that whenever $x \in \mathcal{G}(\pi)$ and $\tau \in \mathbb{R}$, the integral

$$\int\limits_0^\tau \frac{1}{u(\pi(\sigma, x))} \, d\sigma$$

converges. Therefore, to show that u is a quasi π-multiplier, it is sufficient to show that for every $x \in \mathcal{G}(\pi)$, we have

$$\int\limits_{-\infty}^0 \frac{1}{u(\pi(\sigma, x))} \, d\sigma = \int\limits_0^\infty \frac{1}{u(\pi(\sigma, x))} \, d\sigma = \infty.$$

Suppose $x \in \mathcal{G}(\pi)$ and $p_\pi(x) < \infty$. Then since

$$u(y) = \frac{d(y, K)}{v(y) \, |\dot{\varphi}|(y)}, \qquad \forall \, y \in \mathcal{O}_\pi(x),$$

it is easy to see that the restriction of u to $\mathcal{O}_\pi(x)$ is bounded. From this, it follows at once that

$$\int\limits_{-\infty}^0 \frac{1}{u(\pi(\sigma, x))} \, d\sigma = \int\limits_0^\infty \frac{1}{u(\pi(\sigma, x))} \, d\sigma = \infty,$$

as required.

Suppose now that $x \in \mathcal{G}(\pi)$ and $p_\pi(x) = \infty$. Let $g_x : \mathbb{R} \to \mathbb{R}$ be defined by

$$g_x(\tau) = \int\limits_0^\tau \frac{1}{u(\pi(\sigma, x))} \, d\sigma, \qquad \forall \, \tau \in \mathbb{R}.$$

Then g_x is strictly increasing and continuous, and for every $\tau \in \mathbb{R}$ we have

$$g_x'(\tau) = \frac{1}{u(\pi(\tau, x))}.$$

We wish to show that the range of g_x is the whole of $\mathbb{R}$. Let f_x be the inverse function of g_x, and let ψ_x be the function from $g_x(\mathbb{R})$ into $\mathcal{O}_\pi(x)$ defined by

$$\psi_x(t) = \pi(f_x(t), x), \qquad \forall \, t \in g_x(\mathbb{R}).$$

It is clear that for every $t \in g_x(\mathbb{R})$, we have

$$(D\psi_x)(t) = \lim_{s \to t} \frac{d(\psi_x(s), \psi_x(t))}{|s - t|} \qquad \text{(see Definition 1.52)}$$

$$= |\dot{\pi}| (f_x(t), x) \, | f'_x(t) |$$

$$= |\dot{\pi}| (f_x(t), x) \, u(\pi(f_x(t), x))$$

$$= d(\pi(f_x(t), x), K)$$

$$= d(\psi_x(t), K).$$

We shall now show that

$$\int_0^\infty \frac{1}{u(\pi(\sigma, x))} \, d\sigma = \infty$$

i.e., $g_x(\tau) \to \infty$ as $\tau \to \infty$. To do this, we consider two cases:

CASE 1: Suppose $\inf \{d(\pi(\tau, x), K) \mid \tau \geq 0\} = 0$.
Choose an increasing sequence $\{\tau_n\}$ of real numbers, such that

$$d(\pi(\tau_{n+1}, x), K) < \frac{1}{2} d(\pi(\tau_n, x), K), \qquad \forall \, n = 1, 2, \ldots.$$

For each $n = 1, 2, \ldots$, let $t_n = g_x(\tau_n)$. Then we have

$$d(\psi_x(t_n), K) > 2 \cdot d(\psi_x(t_{n+1}), K), \qquad \forall \, n = 1, 2, \ldots.$$

Therefore, by Corollary 1.68, we have

$$|t_{n+1} - t_n| > 1/2, \qquad \forall \, n = 1, 2, \ldots,$$

and from this it follows that $t_n \to \infty$ as $n \to \infty$, and therefore $g_x(\tau) \to \infty$ as $\tau \to \infty$, as required.

CASE 2: Suppose $\inf \{d(\pi(\tau, x), K) \mid \tau \geq 0\} > 0$. Then it is easy to see that

$$\{\pi(\tau, x) \mid \tau \geq 0\} = \{\varphi(\tau, x) \mid \tau \geq 0\}.$$

Therefore $p_\varphi(x) = \infty$, $\inf \{d(\varphi(\tau, x), K) \mid \tau \geq 0\} > 0$, and consequently $\mathcal{L}(\tau) \to \infty$ as $\tau \to \infty$. It follows that the arc-length of $\{\pi(\sigma, x) \mid 0 \leq \sigma \leq \tau\}$ tends to ∞ as $\tau \to \infty$. Therefore, defining $L(t)$ to be the arc-length of $\{\psi_x(s) \mid 0 \leq s \leq t\}$ for every nonnegative $t \in g_x(\mathbb{R})$, and writing $\omega = \sup g_x(\mathbb{R})$, we have

$$\lim_{t \to \omega} L(t) = \infty.$$

But since

$$L(t) = \int\limits_0^t (D\psi_x)(s)\,ds, \qquad \forall\, t \in [0, \omega),$$

(see Theorem 1.59), we have that for all $t \in (0, \omega)$,

$$L'(t) = (D\psi_x)(t) = d\big(\psi_x(t), K\big)$$

$$\leq d\big(\psi_x(t), x\big) + d(x, K) \leq L(t) + d(x, K).$$

Therefore, since $L(t) \to \infty$ as $t \to \omega$, we can choose $t_0 \in (0, \omega)$ such that

$$L'(t) < 2 \cdot L(t), \quad \text{whenever } t_0 < t < \omega.$$

It is clear that whenever $t_0 < t < \omega$, we have

$$\int\limits_{t_0}^t \frac{L'(s)}{L(s)}\,ds \;\leq\; \int\limits_{t_0}^t 2\,ds,$$

and therefore

$$L(t) \;\leq\; L(t_0) \cdot e^{2(t-t_0)} \;\leq\; L(t_0)\, e^{2(\omega - t_0)}.$$

But this would contradict the fact that $L(t) \to \infty$ as $t \to \omega$, unless we have $\omega = \infty$. Thus $\omega = \infty$, and it follows that $g_x(\tau) \to \infty$ as $\tau \to \infty$, as required.

We have therefore shown that, in either case, $g_x(\tau) \to \infty$ as $\tau \to \infty$. A similar argument can be used to show that $g_x(\tau) \to -\infty$ as $\tau \to -\infty$, and this completes the proof that (d)$\Rightarrow$(e).   **QED**

We now state an analogue of Theorem 5.12 for flows.

5.13. Theorem. *Let φ be a flow in a subspace Y of a metric space X. Let K be a non-empty closed subset of X. Suppose that, for every $x \in Y \setminus K$, $|\dot\varphi|(x)$ exists and*

$$0 < |\dot\varphi|(x) < \infty.$$

Suppose also, that $|\dot\varphi|$ is continuous at each point of $Y \setminus K$.

Then the following are equivalent:

(a) There exists an isotonic φ-reparametrizer f such that f rep φ is canonical with respect to K.

(b) There exists a nonnegative proper φ-multiplier w such that $w\varphi$ is canonical with respect to K.

(c) There exists a φ-reparametrizer f such that f rep φ is canonical with respect to K.

(d) *Whenever* $x \in Y \setminus K$ *and* $\mathrm{p}_\varphi(x) = \infty$, *writing* $\mathscr{L}(\tau)$ *for the arc-length of* $\{\varphi(\sigma, x) \mid \sigma \in [0, \tau]\}$ *for every* $\tau \in \mathbb{R}$, *we have*

(i) *either* $\inf \{d(\varphi(\tau, x), K) \mid \tau \geq 0\} = 0$, *or* $\mathscr{L}(\tau) \to \infty$ *as* $\tau \to \infty$,

and

(ii) *either* $\inf \{d(\varphi(\tau, x), K) \mid \tau \leq 0\} = 0$, *or* $\mathscr{L}(\tau) \to \infty$ *as* $\tau \to -\infty$.

(e) *If, using Theorem 5.3, we choose a continuous bounded, nonnegative, proper φ-multiplier v such that*

$$\mathscr{F}(v\varphi) = Y \cap K;$$

and if, writing $\pi = v\varphi$, *we define* $u: Y \to [0, \infty)$, *by*

$$u(x) = \begin{cases} \dfrac{d(x, K)}{|\dot{\pi}|(x)} & \text{whenever } x \in Y \setminus K, \\[2mm] 0 & \text{whenever } x \in Y \cap K, \end{cases}$$

then u is a nonnegative, proper π-multiplier, and $u\pi$ is canonical with respect to K.

PROOF: The proofs that (e)$\Rightarrow$(a), (a)$\Rightarrow$(b), (b)$\Rightarrow$(c), and (c)$\Rightarrow$(d) are similar to the corresponding proofs in Theorem 5.12, and will therefore be omitted.

Suppose now that (d) holds, and let v, π, and u be chosen as in the statement of (e). It follows from Theorem 5.12, that u is a nonnegative, proper quasi π-multiplier, and $u\pi$ is canonical with respect to K. The proof will therefore be complete when we have shown that $u\pi$ is a flow.

Now by Corollary 1.69 and Theorem 1.65, we see that $u\pi$ is continuous at each point of $Y \cap K$. Therefore, since $Y \setminus K$ is open in Y, and invariant of $u\pi$, the proof will be complete when we have shown that the restriction of $u\pi$ to $Y \setminus K$, is a flow in $Y \setminus K$.

Let $\tilde{u}$ and $\tilde{\pi}$ respectively be the restrictions of u and π to $Y \setminus K$. It is easy to see that $\tilde{u}$ is a nonnegative proper quasi $\tilde{\pi}$-multiplier, and that $\tilde{u}\tilde{\pi}$ is the restriction of $u\pi$ to $Y \setminus K$. But since v is continuous, it follows from Theorem 5.8(b) that $|\dot{\pi}|$ is continuous in $Y \setminus K$, and it follows from this that $\tilde{u}$ is continuous. Therefore by Corollary 4.45, $\tilde{u}$ is a $\tilde{\pi}$-multiplier. This completes the proof of the theorem. **QED**

5.14. Theorem. *Let φ be a flow in a subspace Y of a metric space X, and suppose that the closure in Y of each φ-orbit is a complete metric space. Let K be a non-empty, closed subset of X. Suppose that, for every $x \in Y \setminus K$, $|\dot{\varphi}|(x)$ exists and*

$$0 < |\dot{\varphi}|(x) < \infty,$$

and suppose also that $|\dot{\varphi}|$ is continuous at each point of $Y \setminus K$.

Then there exists a nonnegative proper φ-multiplier w such that $w\varphi$ is canonical with respect to K.

PROOF: Suppose $x \in Y \setminus K$ and $p_\varphi(x) = \infty$. For each $\tau \in \mathbb{R}$, let $\mathscr{L}(\tau)$ be the arc-length of $\{\varphi(\sigma, x) \mid \sigma \in [0, \tau]\}$. Suppose $\{\mathscr{L}(\tau) \mid \tau \geq 0\}$ is bounded. Then it is easy to see that as $\sigma, \tau \to \infty$, we have

$$d\big(\varphi(\sigma, x), \varphi(\tau, x)\big) \to 0.$$

Therefore, since the closure in Y of $\mathcal{O}_\varphi(x)$ is complete, there must exist a point $y \in Y$ such that

$$\varphi(\tau, x) \to y \qquad \text{as} \qquad \tau \to \infty.$$

Since $\omega_\varphi(x) = \{y\}$, and $\omega_\varphi(x)$ is invariant of φ, it is clear that $y \in \mathscr{F}(\varphi)$, and it follows that $y \in K$. Therefore

$$\inf \big\{ d\big(\varphi(\tau, x), K\big) \mid \tau \geq 0 \big\} = 0.$$

We can show similarly that if $\{\mathscr{L}(\tau) \mid \tau \leq 0\}$ is bounded, we must have

$$\inf \big\{ d\big(\varphi(\tau, x), K\big) \mid \tau \leq 0 \big\} = 0.$$

The theorem now follows immediately from Theorem 5.12.   **QED**

5.15. Corollary. *Let φ be a flow in a complete subspace Y of a metric space X, let K be a non-empty, closed subset of X, suppose that for every $x \in Y \setminus K$, $|\dot\varphi|(x)$ exists and*

$$0 < |\dot\varphi|(x) < \infty,$$

and suppose that $|\dot\varphi|$ is continuous at each point of $Y \setminus K$.
Then there exists a nonnegative proper φ-multiplier w, such that $w\varphi$ is canonical with respect to K.

PROOF: This follows at once from Theorem 5.14.   **QED**

5.16. Remark. Theorem 5.13 is no longer true if the hypothesis that $|\dot\varphi|$ is continuous at each point of $Y \setminus K$, is replaced by the weaker hypothesis used in Theorem 5.12, *viz.*, that for each $x \in Y$, the function

$$t \to |\dot\varphi|(t, x)$$

is continuous at each number t for which $\varphi(t, x) \in Y \setminus K$.

In the following example, we exhibit a flow φ which satisfies this weaker hypothesis, and which has the property that although there exists a quasi φ-reparametrizer f such that $f \operatorname{rep} \varphi$ is canonical with respect to K, we cannot also demand that $f \operatorname{rep} \varphi$ be a flow.

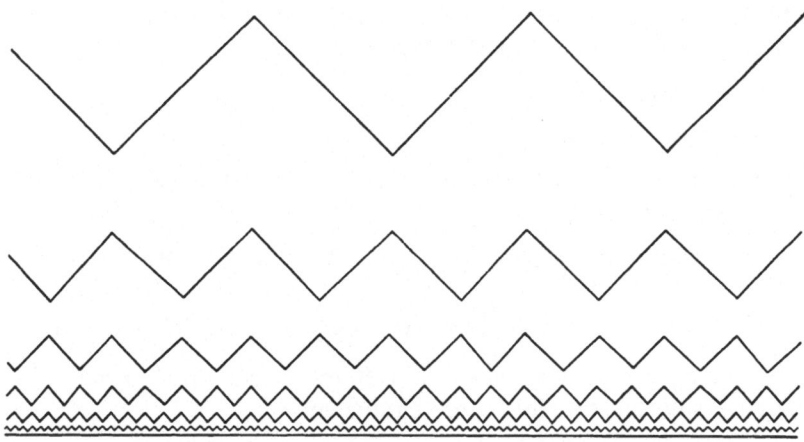

Fig. 5.1

5.17. Example. Let Y be a subset of $\mathbb{R}^2$ of the type illustrated in Fig. 5.1 where each segment of each of the countably many broken lines makes an angle of 45° with the horizontal.

Let φ be the flow in Y defined as follows: Whenever $\langle x, y \rangle \in Y$ and $t \in \mathbb{R}$, $\varphi(t, \langle x, y \rangle)$ is defined to be $\langle x + t, y' \rangle$ where y' is chosen in such a way that $\langle x, y \rangle$ and $\langle x + t, y' \rangle$ belong to the same component of Y.

Choose K to be any horizontal line which does not meet Y.

We leave it to the reader to show that φ is a well-defined flow, and has the properties described in Remark 5.16.

The Time Measure of a Quasi-Flow

5.18. Definition. Let φ be a quasi-flow in a Hausdorff space X, and let $x \in \mathscr{G}(\varphi)$.

Then given a subset A of $\mathcal{O}_\varphi(x)$, *the total φ-time preimage of A, based at x* is defined to be the set $\{t \in \mathbb{R} \mid \varphi(t, x) \in A\}$.

$\{t \in \mathbb{R} \mid \varphi(t, x) \in A\}$ is also said to be *a total φ-time preimage of A*.

5.19. Theorem. *Let φ be a quasi-flow in a Hausdorff space X, let $x \in \mathscr{G}(\varphi)$, let $y \in \mathcal{O}_\varphi(x)$, and suppose A is any subset of $\mathcal{O}_\varphi(x)$.*

Then the total φ-time preimage of A, based at x, is a translate of the total φ-time preimage of A, based at y.

PROOF: Choose $s \in \mathbb{R}$ such that $x = \varphi(s, y)$. Then it is easy to see that

$$\{t \in \mathbb{R} \mid \varphi(t, x) \in A\} \ = \ s + \{t \in \mathbb{R} \mid \varphi(t, y) \in A\}. \qquad \textbf{QED}$$

5.20. Theorem. *Let φ be a quasi-flow in a Hausdorff space X, let $x \in \mathcal{G}(\varphi)$, and suppose A is any subset of $\mathcal{O}_\varphi(x)$.*

Then the following are equivalent:

(a) *A is a Borel subset of X.*

(b) *A is a Borel subset of $\mathcal{O}_\varphi(x)$ (where, of course, $\mathcal{O}_\varphi(x)$ is topologized as a subspace of X).*

(c) *Every total φ-time preimage of A is a Borel subset of $\mathbb{R}$.*

PROOF: Since $\mathcal{O}_\varphi(x) = \bigcup\limits_{n=1}^{\infty} \{\varphi(t, x) \mid -n \leq t \leq n\}$, it is clear that $\mathcal{O}_\varphi(x)$, being σ-compact, must be a Borel subset of X. From this it follows easily that (a) and (b) are equivalent.

The implication (b)$\Rightarrow$(c) is an elementary consequence of the continuity of the function $t \rightarrow \varphi(t, x)$, and the previous theorem.

Now suppose that (c) holds, and let E be the total φ-time preimage of A, based at x. Choose a number δ such that $0 < \delta < p_\varphi(x)$, and for each integer n, let

$$E_n = E \cap [n\delta, (n + 1)\delta].$$

Then it is easy to see that for each n, the restriction to E_n of the function $t \rightarrow \varphi(t, x)$ is a homeomorphism, and that consequently, the image $A_n = \{\varphi(t, x) \mid t \in E_n\}$ is a Borel subset of $\mathcal{O}_\varphi(x)$. But

$$A = \bigcup_{n=1}^{\infty} A_n,$$

and therefore A is a Borel subset of $\mathcal{O}_\varphi(x)$.   **QED**

5.21. Definition. *Let φ be a quasi-flow in a Hausdorff space X, let $x \in \mathcal{G}(\varphi)$, and suppose A is a Borel subset of $\mathcal{O}_\varphi(x)$.*

A simple φ-time preimage of A, based at x is a Borel subset E of $\mathbb{R}$ which has the property that the restriction to E of the function $t \rightarrow \varphi(t, x)$, is a one-one map of E onto A. Such a set E is also said to be a simple φ-time preimage of A.

5.22. Theorem. *Let φ be a quasi-flow in a Hausdorff space X, let $x \in \mathcal{G}(\varphi)$, and suppose A is a Borel subset of $\mathcal{O}_\varphi(x)$.*

(a) *There exists a simple φ-time preimage of A, based at x.*

(b) *All simple φ-time preimages of A have the same Lebesgue measure.*

PROOF: To prove (a), we note that if $p_\varphi(x) = \infty$, then the total φ-time preimage of A, based at x, is a simple φ-time preimage of A, based at x. (In fact, there is only one.) On the other hand, if $p_\varphi(x) < \infty$, then

$$\{t \in [0, p_\varphi(x)) \mid \varphi(t, x) \in A\}$$

can easily be seen to be a simple φ-time preimage of A, based at x.

To prove (b), we note first that if $y \in \mathcal{O}_\varphi(x)$, and E is a simple φ-time preimage of A, based at y, then some translate of E is a simple φ-time preimage of A, based at x. Therefore, to prove (b), it is sufficient to show that all simple φ-time preimages of A, based at x, have the same Lebesgue measure. This is trivial if $p_\varphi(x) = \infty$, and we therefore assume that $p_\varphi(x) < \infty$.

Let $E' = \{t \in [0, p_\varphi(x)) \mid \varphi(t, x) \in A\}$, and suppose E is any simple φ-time preimage of A, based at x. For each integer n, let

$$E_n = E \cap [n p_\varphi(x), (n + 1) p_\varphi(x)).$$

Then

$$m(E) = \sum_{n=-\infty}^{\infty} m(E_n) = \sum_{n=-\infty}^{\infty} m(E_n - n p_\varphi(x)) = m\left(\bigcup_{n=-\infty}^{\infty} (E_n - n p_\varphi(x)) \right)$$
$$= m(E').$$

This completes the proof of the theorem.   **QED**

5.23. Definition. Let φ be a quasi-flow in a Hausdorff space X, let $x \in \mathscr{G}(\varphi)$, and suppose A is a Borel subset of $\mathcal{O}_\varphi(x)$.

Then the *φ-time taken by x to traverse A* is defined to be the Lebesgue measure of any simple φ-time preimage of A. In view of Theorem 5.22, we are clearly entitled to refer to the φ-time taken by x to traverse A, as the *φ-time taken to traverse A*.

5.24. Definition. Let φ be a quasi-flow in a Hausdorff space X. Then $\Sigma(\varphi)$ denotes the σ-algebra (σ-field) of all subsets A of X with the property that $A \cap \mathcal{O}_\varphi(x)$ is a Borel subset of $\mathcal{O}_\varphi(x)$, for every $x \in X$.

5.25. Definition. Let φ be a quasi-flow in a Hausdorff space X. Then the *time measure of φ*, which we denote by μ_φ, is defined on $\Sigma(\varphi)$ as follows:

If A is a Borel subset of a non-trivial φ-orbit, then $\mu_\varphi(A)$ is defined to be the φ-time taken to traverse A.

If $A = \{x\}$, where $x \in \mathscr{F}(\varphi)$, then $\mu_\varphi(A) = \infty$.

In general, if $A \in \Sigma(\varphi)$, we define

$$\mu_\varphi(A) = \sum_{\mathcal{O} \in O(\varphi)} \mu_\varphi(A \cap \mathcal{O}).$$

(It is clear that μ_φ is uniquely defined.)

5.26. Remark. Let φ be a quasi-flow in a Hausdorff space X. Then since μ_φ is clearly determined by its values on the Borel subsets of X, we can identify μ_φ with its restriction to the family of Borel sets, and we shall do this whenever such an identification is convenient.

We remark that, if f is a function from X into $\mathbb{R}$, then f is $\boldsymbol{\Sigma}(\varphi)$ measurable iff its restriction to every φ-orbit is a Borel function.

5.27. Theorem. *Let φ be a quasi-flow in a Hausdorff space X.*

Then μ_φ is a nonnegative, countably additive measure.

PROOF: To prove the theorem, it is clearly sufficient to show that for each $x \in X$, the restriction of μ_φ to $\mathcal{O}_\varphi(x)$ is countably additive. This is trivial if $\mathrm{p}_\varphi(x) = 0$. Suppose that $\mathrm{p}_\varphi(x) > 0$, let $\{A_n \mid n = 1, 2, \ldots\}$ be a pairwise disjoint family of Borel subsets of $\mathcal{O}_\varphi(x)$, and let

$$A = \bigcup_{n=1}^\infty A_n.$$

For each $n = 1, 2, \ldots$, let E_n be a simple φ-time peimage of A_n, based at x. Then $\{E_n \mid n = 1, 2, \ldots\}$ is pairwise disjoint, and $\bigcup_{n=1}^\infty E_n$ is a simple φ-time preimage of A, based at x. Clearly,

$$\mu_\varphi(A) = m\left(\bigcup_{n=1}^\infty E_n\right) = \sum_{n=1}^\infty m(E_n) = \sum_{n=1}^\infty \mu_\varphi(A_n).$$

This completes the proof of the theorem.   **QED**

5.28. Theorem. *Let φ be a quasi-flow in a Hausdorff space X, and let $x \in X$.*

(a) *If $\mathrm{p}_\varphi(x) = 0$, then $\mu_\varphi(\{x\}) = \infty$.*

(b) *If $0 < \mathrm{p}_\varphi(x) < \infty$, then the restriction of μ_φ to $\mathcal{O}_\varphi(x)$ is a nonnegative, continuous, finite Borel measure, which is non-zero on all those non-empty Borel subsets of $\mathcal{O}_\varphi(x)$ whose total φ-time preimages are open.*[1] *Furthermore,*

$$\mu_\varphi\big(\mathcal{O}_\varphi(x)\big) = \mathrm{p}_\varphi(x).$$

(c) *If $\mathrm{p}_\varphi(x) = \infty$, then the restriction of μ_φ to $\mathcal{O}_\varphi(x)$ is a nonnegative, continuous, σ-finite Borel measure, which is non-zero on all those non-empty Borel subsets of $\mathcal{O}_\varphi(x)$ whose total φ-time preimages are open, and is finite on all those Borel subsets of $\mathcal{O}_\varphi(x)$ whose total φ-time preimages are compact. Furthermore,*

$$\mu_\varphi\big(\{\varphi(t, x) \mid t \geq 0\}\big) \;=\; \mu_\varphi\big(\{\varphi(t, x) \mid t \leq 0\}\big) \;=\; \infty.$$

---

[1] A measure is said to be continuous if its value on every countable set is zero.

PROOF: The proof of this theorem is elementary and will be left to the reader.   **QED**

5.29. Theorem. *Let φ be a quasi-flow in a Hausdorff space X, let $x \in \mathcal{G}(\varphi)$, let g be a nonnegative Borel function defined on $\mathcal{O}_\varphi(x)$, and let A be a Borel subset of $\mathcal{O}_\varphi(x)$.*

Then given any simple φ-time preimage E of A, based at x, we have

$$\int_A g \, d\mu_\varphi \;=\; \int_E g\big(\varphi(t, x)\big) \, dt.$$

PROOF: Since a nonnegative Borel function g on $\mathcal{O}_\varphi(x)$ is the pointwise limit of an increasing sequence of nonnegative, simple Borel functions defined on $\mathcal{O}_\varphi(x)$, it is sufficient, by the Lebesgue monotone convergence theorem, to prove the theorem while assuming that g is the characteristic function of a Borel subset B of $\mathcal{O}_\varphi(x)$.

Let E be a simple φ-time preimage of A based at x. Then $\{t \in E \mid \varphi(t, x) \in B\}$ is a simple φ-time preimage of $A \cap B$ based at x, and therefore

$$\int_A g \, d\mu_\varphi = \int_{A \cap B} 1 \, d\mu_\varphi = \mu_\varphi(A \cap B)$$

$$= m\big(\{t \in E \mid \varphi(t, x) \in B\}\big)$$

$$= \int_{\{t \in E \mid \varphi(t,x) \in B\}} 1 \, dt = \int_E g\big(\varphi(t, x)\big) \, dt.$$

This completes the proof of the theorem.   **QED**

The Time Measure of a Quasi-Reparametrization

5.30. Theorem. *Let φ be a quasi-flow in a Hausdorff space X and let ψ be a quasi φ-reparametrization.*

Then

(a) $\mathbf{\Sigma}(\varphi) \subseteq \mathbf{\Sigma}(\psi)$.

(b) *For any set A which belongs to $\mathbf{\Sigma}(\psi)$, we have*

$$A \in \mathbf{\Sigma}(\varphi) \quad \text{iff} \quad A \cap \mathscr{F}(\psi) \in \mathbf{\Sigma}(\varphi).$$

PROOF: (a) is an immediate consequence of the fact that every ψ-orbit is a Borel set. (b) follows from the observation that the intersection of $\mathcal{G}(\psi)$ with each φ-orbit is a Borel set, and is a union of countably many ψ-orbits (see Corollary 4.10), and that both $\mathcal{G}(\psi)$ and $\mathscr{F}(\psi)$ belong to $\mathbf{\Sigma}(\varphi)$.   **QED**

5.31. Theorem. *Let φ be a quasi-flow in a Hausdorff space X, let f be a quasi φ-reparametrizer, and let $\psi = f \operatorname{rep} \varphi$. Let $x \in \mathcal{G}(\varphi)$, and let A be a Borel subset of $\mathcal{O}_\psi(x)$.*

Then

(a) *Given any simple ψ-time preimage E of A, based at x, we have $f_x(E)$ is a simple φ-time preimage of A, based at x.*

(b) *There exists a simple φ-time preimage of A, based at x, which is included in the range of f_x.*

(c) *If E is any simple φ-time preimage of A, based at x, which is contained in the range of f_x, then unless $x \in \mathcal{F}(\psi)$, we have $f_x^{-1}(E)$ is a simple ψ-time preimage of A, based at x.*

(d) *If E is any simple φ-time preimage of A, based at x, which is contained in the range of f_x, we have*

$$\mu_\psi(A) = m\big(f_x^{-1}(E)\big).$$

PROOF: (a) is clear. (b) is clear if $x \in \mathcal{F}(\psi)$, and is an immediate consequence of (a) if $x \in \mathcal{G}(\psi)$. (c) is clear. (d) is clear if $x \in \mathcal{F}(\psi)$, and is an immediate consequence of (c) if $x \in \mathcal{G}(\psi)$.   **QED**

5.32. Corollary. *Let φ be a quasi-flow in a Hausdorff space X, let f be a quasi φ-reparametrizer, let $\psi = f \operatorname{rep} \varphi$, and suppose $x \in \mathcal{G}(\psi)$.*

Then whenever $t \in \mathbb{R}$ and $|t| \leq \mathrm{p}_\psi(x)$, we have

$$\mu_\psi\big(\{\varphi(\tau, x) \mid \tau \in [0, f_x(t)]\}\big) = |t|.$$

PROOF: Immediate from the above theorem.   **QED**

5.33. Theorem. *Let φ be a quasi-flow in a Hausdorff space X, and let μ be a nonnegative Borel measure on X.*

Then a necessary and sufficient condition that μ be the time measure of a quasi φ-reparametrization is that μ has the properties (a) *and* (b) *described below:*

(a) *For every Borel subset A of X,*

$$\mu(A) = \sum_{\mathcal{O} \in O(\varphi)} \mu(A \cap \mathcal{O}).$$

(b) *For each $x \in X$, exactly one of the following,* (i), (ii), (iii), *is satisfied:*

(i) *$\mu(\{x\}) = \infty$.*

(ii) *$0 < \mathrm{p}_\varphi(x) < \infty$, and the restriction of μ to $\mathcal{O}_\varphi(x)$ is a continuous finite Borel measure, which is non-zero on all those non-empty Borel subsets of $\mathcal{O}_\varphi(x)$ whose total φ-time preimages are open.*

(iii) *There exist α_x and ω_x in $[-\infty, \infty]$ such that*

$$-\infty \leq \alpha_x < 0 < \omega_x \leq \infty \quad and \quad \omega_x - \alpha_x \leq p_\varphi(x),$$

and such that the restriction of μ to $\{\varphi(\tau, x) \mid \alpha_x < \tau < \omega_x\}$ is a continuous Borel measure, which is non-zero on all those non-empty Borel subsets of $\{\varphi(\tau, x) \mid \alpha_x < \tau < \omega_x\}$ whose total φ-time preimages are open, and is finite on all those Borel subsets of $\{\varphi(\tau, x) \mid \alpha_x < \tau < \omega_x\}$ whose total φ-time preimages, based at x, have compact intersections with (α_x, ω_x), and has the property that

$$\mu\big(\{\varphi(\tau, x) \mid 0 \leq \tau < \omega_x\}\big) = \mu\big(\{\varphi(\tau, x) \mid \alpha_x < \tau \leq 0\}\big) = \infty.$$

PROOF: To prove the necessity, suppose ψ is a quasi φ-reparametrization, and $\mu = \mu_\psi$.

It is clear that (a) holds, and it is not hard to show that the conditions (b) (i),(b) (ii), and (b) (iii) correspond exactly to the conditions $p_\psi(x) = 0$, $0 < p_\psi(x) < \infty$, and $p_\psi(x) = \infty$. Thus (b) also holds, and this completes the proof of the necessity.

To prove the sufficiency, suppose that (a) and (b) are satisfied. Let

$$Y_1 = \{x \in X \mid x \text{ satisfies (b) (i)}\},$$

$$Y_2 = \{x \in X \mid x \text{ satisfies (b) (ii)}\},$$

$$Y_3 = \{x \in X \mid x \text{ satisfies (b) (iii)}\}.$$

It is clear that $Y_2 \cup Y_3 \subseteq \mathscr{G}(\varphi)$. Now it is easy to see that if $x \in Y_2$, then $\varphi(\tau, x) \in Y_2$ for all $\tau \in \mathbb{R}$. On the other hand, we see that if $x \in Y_3$, then (α_x, ω_x) is that component of $\{\tau \in \mathbb{R} \mid \varphi(\tau, x) \in Y_3\}$, which contains 0, and furthermore, for each $\tau \in (\alpha_x, \omega_x)$, we have

$$\alpha_{\varphi(\tau, x)} = \alpha_x - \tau, \quad and \quad \omega_{\varphi(\tau, x)} = \omega_x - \tau.$$

We shall now define a family of functions

$$\{g_x \mid x \in Y_2 \cup Y_3\}$$

which we shall prove is an inverse quasi φ-reparametrizer. Before doing this, we note that the above observations imply that $Y_2 \cup Y_3$ meets the requirement of Definition 4.16(b) that, for every $x \in X$, the set $\{\tau \in \mathbb{R} \mid \varphi(\tau, x) \in Y_2 \cup Y_3\}$ is open.

If $x \in Y_2$, we define g_x from $\mathbb{R}$ onto $\mathbb{R}$ as follows: Given $\tau \in \mathbb{R}$, write $\tau = \tau_0 + n \cdot p_\varphi(x)$, where n is an integer, and $0 \leq \tau_0 < p_\varphi(x)$, and let

$$g_x(\tau) = \mu\big(\{\varphi(\sigma, x) \mid 0 \leq \sigma \leq \tau_0\}\big) + n \cdot \mu\big(\mathcal{O}_\varphi(x)\big).$$

If $x \in Y_3$, define g_x from (α_x, ω_x) onto $\mathbb{R}$, by

$$g_x(\tau) = \begin{cases} \mu\big(\{\varphi(\sigma, x) \mid 0 \leq \sigma \leq \tau\}\big) & \text{whenever } 0 \leq \tau < \omega_x, \\ -\mu\big(\{\varphi(\sigma, x) \mid \tau \leq \sigma \leq 0\}\big) & \text{whenever } \alpha_x < \tau \leq 0. \end{cases}$$

It is easy to see that, in either case, g_x is a strictly increasing, continuous function, and its range is $\mathbb{R}$. Furthermore, it is clear that, for every $x \in Y_2 \cup Y_3$, $g_x(0) = 0$. This shows that $\{g_x \mid x \in Y_2 \cup Y_3\}$ satisfies condition (i) of Definition 4.16(b).

Now suppose $x \in Y_2$, and σ and τ belong to $\mathbb{R}$. We wish to show that

$$g_x(\sigma + \tau) = g_x(\tau) + g_{\varphi(\tau, x)}(\sigma).$$

For this purpose, write $\sigma = \sigma_0 + m\,\mathrm{p}_\varphi(x)$ and $\tau = \tau_0 + n\,\mathrm{p}_\varphi(x)$, where m and n are integers, $0 \leq \sigma_0 < \mathrm{p}_\varphi(x)$, and $0 \leq \tau_0 < \mathrm{p}_\varphi(x)$. Since

$$g_x(\sigma + \tau) = g_x\big(\sigma_0 + \tau_0 + (m + n)\,\mathrm{p}_\varphi(x)\big)$$
$$= g_x(\sigma_0 + \tau_0) + (m + n) \cdot \mu\big(\mathcal{O}_\varphi(x)\big),$$

and since

$$g_x(\tau) + g_{\varphi(\tau, x)}(\sigma) = g_x\big(\tau_0 + n\,\mathrm{p}_\varphi(x)\big) + g_{\varphi(\tau_0, x)}\big(\sigma_0 + m \cdot \mathrm{p}_\varphi\big(\varphi(\tau_0, x)\big)\big)$$
$$= g_x(\tau_0) + n \cdot \mu\big(\mathcal{O}_\varphi(x)\big) + g_{\varphi(\tau_0, x)}(\sigma_0) + m \cdot \mu\big(\mathcal{O}_\varphi(x)\big),$$

we will have the required equation as soon as we have shown that

$$g_x(\sigma_0 + \tau_0) = g_x(\tau_0) + g_{\varphi(\tau_0, x)}(\sigma_0).$$

Now

$$g_x(\tau_0) + g_{\varphi(\tau_0, x)}(\sigma_0) = \mu\big(\{\varphi(\lambda, x) \mid 0 \leq \lambda \leq \tau_0\}\big)$$
$$+ \mu\big(\{\varphi(\lambda, \varphi(\tau_0, x)) \mid 0 \leq \lambda \leq \sigma_0\}\big)$$
$$= \mu\big(\{\varphi(\lambda, x) \mid 0 \leq \lambda \leq \tau_0\}\big)$$
$$+ \mu\big(\{\varphi(\lambda, x) \mid \tau_0 \leq \lambda \leq \tau_0 + \sigma_0\}\big).$$

In the case $\tau_0 + \sigma_0 < \mathrm{p}_\varphi(x)$, the latter expression reduces to

$$\mu\big(\{\varphi(\lambda, x) \mid 0 \leq \lambda \leq \tau_0 + \sigma_0\}\big) = g_x(\tau_0 + \sigma_0).$$

Otherwise, we have $\mathrm{p}_\varphi(x) \leq \tau_0 + \sigma_0 < 2\mathrm{p}_\varphi(x)$, and the abovementioned expression reduces to

$$\mu\big(\mathcal{O}_\varphi(x)\big) + \mu\big(\{\varphi(\lambda, x) \mid 0 \leq \lambda \leq \tau_0 + \sigma_0 - \mathrm{p}_\varphi(x)\}\big)$$
$$= \mu\big(\mathcal{O}_\varphi(x)\big) + g_x\big(\tau_0 + \sigma_0 - \mathrm{p}_\varphi(x)\big)$$
$$= g_x(\tau_0 + \sigma_0).$$

We have therefore shown that whenever $x \in Y_2$, and σ and τ belong to $\mathbb{R}$, we have

$$g_x(\sigma + \tau) = g_x(\tau) + g_{\varphi(\tau, x)}(\sigma).$$

Suppose now that $x \in Y_3$ and that τ and $\sigma + \tau$ belong to (α_x, ω_x). The reader can easily verify that the formula

$$g_x(\sigma + \tau) = g_x(\tau) + g_{\varphi(\tau, x)}(\sigma)$$

holds, by considering separately, each of the following six cases:

(i) $\tau \geq 0, \sigma \geq 0$,

(ii) $\tau \leq 0, \sigma \leq 0$,

(iii) $\tau > 0, \sigma < 0, \sigma + \tau \geq 0$,

(iv) $\tau > 0, \sigma < 0, \sigma + \tau < 0$,

(v) $\tau < 0, \sigma > 0, \sigma + \tau \geq 0$,

(vi) $\tau < 0, \sigma > 0, \sigma + \tau < 0$.

We have therefore shown that $\{g_x \mid x \in Y_2 \cup Y_3\}$ is an inverse quasi φ-reparametrizer. Choose a quasi φ-reparametrizer f such that

$$\text{inv}(f) = \{g_x \mid x \in Y_2 \cup Y_3\},$$

and let $\psi = f \text{ rep } \varphi$. It is clear that $\mathscr{F}(\psi) = Y_1$.

The proof will be complete when we have shown that $\mu_\psi(A) = \mu(A)$ for every Borel subset A of X. Since μ satisfies (a), it is clear that this will follow when we have shown that $\mu_\psi(A) = \mu(A)$ for every Borel subset A of any one φ-orbit. But

$$\mu_\psi(\{x\}) = \mu(\{x\}) = \infty, \qquad \forall\, x \in \mathscr{F}(\psi),$$

and in any given φ-orbit, there are only countably many non-trivial ψ-orbits. We therefore need only show that $\mu_\psi(A) = \mu(A)$ for every Borel subset A of any one non-trivial ψ-orbit.

Now suppose x is an arbitrary member of $\mathscr{G}(\psi)$. We see firstly that whenever $-\infty < t_1 < t_2 < \infty$ and $t_2 - t_1 < p_\psi(x)$, we have on the one hand

$$\mu_\psi\big(\{\psi(t, x) \mid t_1 < t < t_2\}\big) = t_2 - t_1,$$

and on the other hand,

$$\mu\big(\{\psi(t,x) \mid t_1 < t < t_2\}\big) \;=\; \mu\big(\{\psi(t,y) \mid 0 < t < t_2 - t_1\}\big),$$

$$\text{where } y = \psi(t_1, x),$$

$$= \mu\big(\{\varphi(\tau, y) \mid \tau \in (0, f_y(t_2 - t_1))\}\big)$$

$$= |g_y(f_y(t_2 - t_1))|, \quad \text{since } |f_y(t_2 - t_1)| < \mathrm{p}_\varphi(y),$$

$$= t_2 - t_1.$$

It follows easily that whenever

$$-\infty < t_1 < t_2 < \infty \quad \text{and} \quad t_2 - t_1 < \mathrm{p}_\psi(x),$$

and U is an open subset of (t_1, t_2), we have

$$\mu_\psi\big(\{\psi(t,x) \mid t \in U\}\big) = \mu\big(\{\psi(t,x) \mid t \in U\}\big),$$

and from this we deduce at once that for each compact subset C of (t_1, t_2), we have

$$\mu_\psi\big(\{\psi(t,x) \mid t \in C\}\big) = \mu\big(\{\psi(t,x) \mid t \in C\}\big).$$

Now suppose A is any Borel subset of $\{\psi(t,x) \mid t_1 < t < t_2\}$, and let E be the simple ψ-time preimage of A, based at x, which is contained in (t_1, t_2), i.e.,

$$E = \{t \in (t_1, t_2) \mid \psi(t, x) \in A\}.$$

Using the regularity of Lebesgue measure, choose an expanding sequence $\{C_n\}$ of compact subsets of E, and a contracting sequence $\{U_n\}$ of open supersets of E, all contained in (t_1, t_2), such that

$$m(C_n) \to m(E) \quad \text{and} \quad m(U_n) \to m(E), \quad \text{as} \quad n \to \infty.$$

For each $n = 1, 2, \ldots$, let

$$\tilde{C}_n = \{\psi(t,x) \mid t \in C_n\} \quad \text{and} \quad \tilde{U}_n = \{\psi(t,x) \mid t \in U_n\}.$$

Then for each $n = 1, 2, \ldots$,

$$m(C_n) = \mu_\psi(\tilde{C}_n) = \mu(\tilde{C}_n) \le \mu(A),$$

and

$$\mu(A) \le \mu(\tilde{U}_n) = \mu_\psi(\tilde{U}_n) = m(U_n),$$

and we deduce at once that

$$\mu(A) = m(E) = \mu_\psi(A).$$

Thus we have shown that whenever $-\infty < t_1 < t_2 < \infty$ and $t_2 - t_1 < \mathrm{p}_\psi(x)$, we have $\mu_\psi(A) = \mu(A)$, whenever A is a Borel subset of $\{\psi(t, x) \mid t_1 < t < t_2\}$.

It follows easily that the formula $\mu_\psi(A) = \mu(A)$ holds for all Borel subsets A of $\mathcal{O}_\psi(x)$, and this completes the proof of the theorem. **QED**

5.34. Theorem. *Let φ be a quasi-flow in a Hausdorff space X, and let ψ be a quasi φ-reparametrization.*

(a) *Given any subset A of X, which is invariant of ψ, if we define the algebraic flow π by*

$$\pi(t, x) = \begin{cases} \psi(t, x), & \forall\, t \in \mathbb{R} & \forall\, x \in A, \\ \psi(-t, x), & \forall\, t \in \mathbb{R} & \forall\, x \in X \setminus A; \end{cases}$$

then π is a quasi φ-reparametrization, and

$$\mu_\psi = \mu_\pi.$$

(b) *Given any quasi φ-reparametrization π with the property that $\mu_\psi = \mu_\pi$, there exists a subset A of X, which is invariant of φ, such that*

$$\pi(t, x) = \begin{cases} \psi(t, x), & \forall\, t \in \mathbb{R} & \forall\, x \in A, \\ \psi(-t, x), & \forall\, t \in \mathbb{R} & \forall\, x \in X \setminus A. \end{cases}$$

PROOF: (a) is obvious, and (b) is an elementary consequence of Theorem 5.32. **QED**

5.35. Corollary. *Let φ be a quasi-flow in a Hausdorff space X, and let μ be the time measure of a quasi φ-reparametrization.*

Then μ is the time measure of exactly one isotonic quasi φ-reparametrization.

5.36. Remark. Let φ be a flow in a Hausdorff space X, let ψ be a quasi φ-reparametrization, and let π be the unique isotonic quasi φ-reparametrization whose time measure is μ_ψ.

Then in view of Theorem 4.24, it is clear that if ψ is a flow, π must also be a flow.

5.37. Example. In this example, we exhibit two flows φ and ψ in $\mathbb{R}^2$, whose orbits are different, but whose time measures agree on the Borel subsets of $\mathbb{R}^2$.

Let φ be the flow in $\mathbb{R}^2$ defined by

$$\varphi(t, \langle x, y \rangle) = \langle x + t, y \rangle, \qquad \forall\, t \in \mathbb{R}, \qquad \forall\, \langle x, y \rangle \in \mathbb{R}^2.$$

Let f be the continuous, increasing function from $[0, 1]$ onto $[0, 1]$ which is constant on each component of $[0, 1] \smallsetminus C$, where C is the $\frac{1}{3}$-Cantor subset of $[0, 1]$, and which satisfies:

$$f(x) = \frac{1}{2} \quad \text{whenever} \quad \frac{1}{3} \leq x \leq \frac{2}{3},$$

$$f(x) = \frac{1}{4} \quad \text{whenever} \quad \frac{1}{9} \leq x \leq \frac{2}{9}$$

and

$$f(x) = \frac{3}{4} \quad \text{whenever} \quad \frac{7}{9} \leq x \leq \frac{8}{9}$$

and generally, for each positive integer n, in the 2^{n-1} components of $[0, 1] \smallsetminus C$ which have length 3^{-n}, f takes the values

$$\frac{1}{2^n}, \frac{3}{2^n}, \frac{5}{2^n}, \ldots, \frac{2^n - 1}{2^n}.$$

Extend f to an increasing, continuous function from $\mathbb{R}$ onto $\mathbb{R}$ by the identity

$$f(x + n) = f(x) + n, \qquad \forall \, x \in [0, 1] \text{ and all integers } n.$$

We now define ψ as follows: Given any point $\langle x, y \rangle \in \mathbb{R}^2$ and given any real number t, choose α such that $y = f(x) + \alpha$, and let

$$\psi(t, \langle x, y \rangle) = \langle x + t, f(x + t) + \alpha \rangle.$$

It is not hard to see that ψ is a flow in $\mathbb{R}^2$, that every ψ-orbit meets $2^{\aleph_0}$ φ-orbits, that every φ-orbit meets $2^{\aleph_0}$ ψ-orbits, and that μ_φ and μ_ψ agree on the Borel subsets of $\mathbb{R}^2$.

Note, however, that neither of the families $\boldsymbol{\Sigma}(\varphi)$, $\boldsymbol{\Sigma}(\psi)$, contains the other: for if A is a non-Borel subset of C, then

$$\{\langle x, 0 \rangle \mid x \in A\} \in \boldsymbol{\Sigma}(\psi) \smallsetminus \boldsymbol{\Sigma}(\varphi),$$

and

$$\{\langle x, f(x) \rangle \mid x \in A\} \in \boldsymbol{\Sigma}(\varphi) \smallsetminus \boldsymbol{\Sigma}(\psi).$$

In view of this example, the following theorem is interesting, although it relies on the continuum hypothesis:

5.38. Theorem. *Let φ and ψ be quasi-flows in a Hausdorff space X. Suppose that*

$$\boldsymbol{\Sigma}(\varphi) \subseteq \boldsymbol{\Sigma}(\psi),$$

and suppose that μ_ψ and μ_φ agree on the Borel subsets of X.

Then $\boldsymbol{\Sigma}(\varphi) = \boldsymbol{\Sigma}(\psi)$, $\mu_\varphi = \mu_\psi$, and ψ is a quasi φ-reparametrization.

PROOF: Our first main task is to show that every ψ-orbit is contained in a φ-orbit.

Suppose $x \in X$ and, to obtain a contradiction, assume that $\mathcal{O}_\psi(x)$ meets more than one φ-orbit. Then we can choose a positive integer m such that $\{\psi(t, x) \mid -m \leq t \leq m\}$ meets more than one φ-orbit. Now since

$$\mu_\psi\big(\{\psi(t, x) \mid -m \leq t \leq m\}\big) < \infty,$$

it is clear that for every $y \in X$ satisfying $p_\varphi(y) = \infty$, there exists a sequence $\{t_n\}$ $(n = 0, \pm 1, \pm 2, \ldots)$ such that

$$t_n < t_{n+1} \qquad \text{for all integers } n,$$

$$t_n \to -\infty \text{ as } n \to -\infty \text{ and } t_n \to \infty \text{ as } n \to \infty,$$

and

$$\varphi(t_n, y) \notin \{\psi(t, x) \mid -m \leq t \leq m\} \text{ for all integers } n.$$

It is clear that for each $y \in X$, if we choose $\{t_n\}$ satisfying these conditions then $\{\psi(t, x) \mid -m \leq t \leq m\} \cap \mathcal{O}_\varphi(y)$ is the countable disjoint union of its compact subsets

$$\{\psi(t, x) \mid -m \leq t \leq m\} \cap \{\varphi(t, y) \mid t_n \leq t \leq t_{n+1}\}, \quad (n = 0, \pm 1, \pm 2, \ldots).$$

Therefore, since the intersection of $\{\psi(t, x) \mid -m \leq t \leq m\}$ with every periodic φ-orbit is compact, it follows from Theorem A. 11 that $\{\psi(t, x) \mid -m \leq t \leq m\}$ meets uncountably many φ-orbits, and this shows that $\mathcal{O}_\psi(x)$ meets uncountably many φ-orbits.

Since the restriction of μ_ψ to $\mathcal{O}_\psi(x)$ is σ-finite, it is clear that $\mathcal{O}_\psi(x)$ cannot meet more than countably many φ-orbits in a set of positive μ_ψ-measure. It follows that, if we denote by $\{E_\alpha \mid \alpha \in A\}$ the family of those intersections of $\mathcal{O}_\psi(x)$ with a φ-orbit which have μ_ψ-measure zero, then A is uncountable, and if

$$E = \bigcup_{\alpha \in A} E_\alpha,$$

then E is a Borel set. It is clear that the intersection of E with every φ-orbit has μ_φ-measure zero, and it follows from this that

$$\mu_\psi(E) = \mu_\varphi(E) = 0.$$

For each $\alpha \in A$, let A_α be the total ψ-time preimage of E_α, based at x, and let A be the total ψ-time preimage of E, based at x. Then it is clear that

$$A = \bigcup_{\alpha \in A} A_\alpha,$$

that $\{A_\alpha \mid \alpha \in A\}$ is a pairwise disjoint family of Borel sets, and that A is a Borel set of Lebesgue measure zero.

Using the continuum hypothesis, we can assert that the cardinality of A is at least $2^{\aleph_0}$ (and therefore exactly $2^{\aleph_0}$), and we therefore deduce from Cantor's inequality[2] that A has more than $2^{\aleph_0}$ subsets. But it is well-known that the set of Borel subsets of $\mathbb{R}$ has cardinality $2^{\aleph_0}$, and we can therefore choose a subset A' of A such that if

$$A' = \bigcup_{\alpha \in A'} A_\alpha,$$

then A' is not a Borel set. Let

$$E' = \{\psi(t, x) \mid t \in A'\}.$$

Then since E' is a subset of $\mathcal{O}_\psi(x)$ and E' is not a Borel set, we have $E' \notin \Sigma(\psi)$. But on the other hand,

$$E' = \bigcup_{\alpha \in A'} E_\alpha,$$

and it is therefore clear that $E' \in \Sigma(\varphi)$.

This contradicts the fact that $\Sigma(\varphi) \subseteq \Sigma(\psi)$, and we have therefore shown that every ψ-orbit is contained in a φ-orbit.

Now to prove that ψ is a quasi φ-reparametrization, we shall restrict our attention to a single φ-orbit. It is easy to see that if this orbit is periodic, then since φ is a flow, the result follows at once from Theorem 4.7. Suppose then that this φ-orbit is aperiodic, and let x be any point in it. For each $t \in \mathbb{R}$, let $f_x(t)$ be the unique real number for which

$$\psi(t, x) = \varphi\big(f_x(t), x\big).$$

The proof will be complete when we have shown that f_x is continuous. But whenever $t_n \to t$, it is clear that $f_x(t)$ is the only possible finite cluster-point of the sequence $\{f_x(t_n)\}$. Therefore the continuity of f_x will have been established when we have shown that, for any real number t, the restriction of f_x to $[t - 1, t + 1]$ is bounded. But if $t \in \mathbb{R}$, and $A = \{\psi(s, x) \mid t - 1 \leq s \leq t + 1\}$, then A is a compact, connected subset of $\mathcal{O}_\varphi(x)$, and since

$$\mu_\varphi(A) = \mu_\psi(A) < \infty,$$

_____

[2] Cantor's inequality states that for any set A, A has lower cardinality than the set of subsets of A. To see this, we need only note that if B is the set of subsets of A, and f is a function from A into B, then f cannot be *onto* B, because the range of f does not contain in the set

$$\{x \in A \mid x \notin f(x)\}.$$

there must exist a sequence $\{\tau_n\}$ $(n = 0, \pm 1, \pm 2, \ldots)$, such that

$$\tau_n < \tau_{n+1} \text{ for all integers } n,$$

$$\tau_n \to -\infty \text{ as } n \to -\infty \text{ and } \tau_n \to \infty \text{ as } n \to \infty,$$

and

$$\varphi(\tau_n, x) \notin A \text{ for all integers } n.$$

It therefore follows from Theorem A.11 that since

$$A = \bigcup_{n=-\infty}^{\infty} \left(A \cap \{\varphi(\sigma, x) \mid \tau_n \leq \sigma \leq \tau_{n+1}\} \right),$$

we must have $A \subseteq \{\varphi(\sigma, x) \mid \tau_n \leq \sigma \leq \tau_{n+1}\}$ for some integer n. Thus $f_x(s) \in [\tau_n, \tau_{n+1}]$ for all $s \in [t-1, t+1]$, and the proof is complete.

QED

5.39. Corollary. *Let φ and ψ be quasi-flows in a Hausdorff space X, suppose $\Sigma(\varphi) \subseteq \Sigma(\psi)$, and suppose that μ_φ and μ_ψ agree on the Borel subsets of X.*

Then there exists a quasi φ-multiplier u such that

(a) *u is a quasi ψ-multiplier,*

(b) *$|u(x)| = 1$ for all $x \in \mathcal{G}(\varphi)$,*

and

(c) *$\psi = u\varphi$ and $\varphi = u\psi$.*

Furthermore, if φ is a flow, then ψ is a flow if and only if u is constant on the components of $\mathcal{G}(\varphi)$.

PROOF: This result is an immediate consequence of Theorems 5.34 and 5.38, and Theorem 4.22.   **QED**

5.40. Theorem. *Let φ be a quasi-flow in a Hausdorff space X and let μ be the time measure of a quasi φ-reparametrization. Let λ be a Borel measure on X and suppose there exist positive real numbers a and b such that*

$$a\mu \leq \lambda \leq b\mu.$$

Then λ is the time measure of a quasi φ-reparametrization.

PROOF: It is easy to see that since μ satisfies condition (b) of Theorem 5.33, this condition is also satisfied by λ.

It therefore remains to show that λ satisfies condition (a) of Theorem 5.33. Suppose A is any Borel subset of X. Then clearly,

$$\lambda(A) \geq \sum_{\mathcal{O} \in \mathbf{O}(\varphi)} \lambda(A \cap \mathcal{O}).$$

To obtain a contradiction, suppose that

$$\lambda(A) > \sum_{\mathcal{O} \in \boldsymbol{O}(\varphi)} \lambda(A \cap \mathcal{O}).$$

Then since

$$\sum_{\mathcal{O} \in \boldsymbol{O}(\varphi)} \lambda(A \cap \mathcal{O}) < \infty,$$

there are at most countably many $\mathcal{O} \in \boldsymbol{O}(\varphi)$ for which $\lambda(A \cap \mathcal{O}) > 0$. Let

$$B = \cup \left\{ A \cap \mathcal{O} \mid \mathcal{O} \in \boldsymbol{O}(\varphi) \text{ and } \lambda(A \cap \mathcal{O}) = 0 \right\}.$$

Then since A and $A \setminus B$ are Borel sets, B must be a Borel set. Now on the one hand, whenever $\lambda(A \cap \mathcal{O}) = 0$, we have

$$a\mu(A \cap \mathcal{O}) = 0, \quad i.e., \quad \mu(A \cap \mathcal{O}) = 0.$$

It follows that $\mu(B) = 0$, and therefore that

$$\lambda(B) \leq b\mu(B) = 0, \quad i.e., \quad \lambda(B) = 0.$$

But on the other hand,

$$\lambda(A \setminus B) = \sum_{\mathcal{O} \in \boldsymbol{O}(\varphi)} \lambda(A \cap \mathcal{O}) < \lambda(A),$$

and therefore

$$\lambda(B) = \lambda(A) - \lambda(A \setminus B) > 0,$$

a contradiction.

We have therefore shown that λ satisfies condition (a) of Theorem 5.33, and this completes the proof.   **QED**

The Time Measure of a Product

5.41. Theorem. *Let φ be a quasi-flow in a Hausdorff space X, let u be a quasi φ-multiplier, and let $\psi = u\varphi$.*

Then for every set A in $\boldsymbol{\Sigma}(\varphi)$ for which $A \subseteq \mathscr{G}(\psi)$, we have

$$u_\psi(A) = \int_A \frac{1}{|u|} \, d\mu_\varphi.$$

PROOF: Let f be a quasi φ-reparametrizer induced by u, and let $\mathrm{inv}(f) = \{g_x \mid x \in \mathscr{G}(\psi)\}$. Let μ be the measure defined on the $\boldsymbol{\Sigma}(\varphi)$-measurable subsets of $\mathscr{G}(\psi)$ by

$$\mu(A) = \int_A \frac{1}{|u|} \, d\mu_\varphi,$$

for all $\Sigma(\varphi)$-measurable subsets A of $\mathscr{G}(\psi)$. (The reader should recall that inside $\mathscr{G}(\psi)$, $\Sigma(\varphi)$-measurability is the same as $\Sigma(\psi)$-measurability.)

It is easy to see that, for every $\Sigma(\varphi)$-measurable subset A of $\mathscr{G}(\psi)$, we have

$$\mu(A) = \sum_{\mathcal{O} \in O(\psi)} \mu(A \cap \mathcal{O});$$

and from this we see that, to prove the theorem, it is sufficient to show that $\mu_\psi(A) = \mu(A)$ whenever A is a Borel subset of any one non-trivial ψ-orbit.

Let x be an arbitrary member of $\mathscr{G}(\psi)$. It is clear that given any subset A of $\mathcal{O}_\psi(x)$ of the form

$$A = \{\psi(t, x) \mid t_1 < t < t_2\},$$

where $-\infty < t_1 < t_2 < \infty$ and $t_2 - t_1 < p_\psi(x)$, then on the one hand we have $\mu_\psi(A) = t_2 - t_1$, and on the other hand we have

$$\mu(A) = \int_A \frac{1}{|u|} \, d\mu_\varphi$$

$$= \left| \int_{f_x(t_1)}^{f_x(t_2)} \frac{1}{|u(\varphi(\sigma, x))|} \, d\sigma \right|, \qquad \text{by Theorem 5.29,}$$

$$= \left| \int_{f_x(t_1)}^{f_x(t_2)} \frac{1}{u(\varphi(\sigma, x))} \, d\sigma \right|$$

$$= |g_x(f_x(t_2)) - g_x(f_x(t_1))|$$

$$= t_2 - t_1.$$

Now, arguing as we did in the proof of Theorem 5.33, we can show that $\mu_\psi(A) = \mu(A)$ whenever A is a Borel subset of $\mathcal{O}_\psi(x)$.

This completes the proof of the theorem.   **QED**

5.42. Theorem. *Let* φ *be a quasi-flow in a Hausdorff space* X*, let* f *be a quasi* φ*-reparametrizer, and let* $\psi = f$ *rep* φ*. Let* u *be a* $\Sigma(\varphi)$*-measurable function from* X *into* $[-\infty, \infty]$*, and suppose*

(i) $u(x) = 0, \qquad \forall\ x \in \mathscr{G}(\varphi) \cap \mathscr{F}(\psi).$

(ii) *For each* $x \in \mathscr{G}(\psi)$*, the restriction of* u *to* $\mathcal{O}_\psi(x)$ *does not change sign.*

(iii) *For every $\Sigma(\varphi)$-measurable subset A of $\mathscr{G}(\psi)$, we have*

$$\mu_\psi(A) = \int_A \frac{1}{|u|} \, d\mu_\varphi.$$

Then u is a quasi φ-multiplier.

Furthermore, if we assume in addition to (ii), *that whenever $x \in \mathscr{G}(\psi)$ and f_x is increasing we have $u(x) \geq 0$, and whenever $x \in \mathscr{G}(\psi)$ and f_x is decreasing, we have $u(x) \leq 0$, then we have*

$$\psi = u\varphi.$$

PROOF: It is clear that, if $x \in \mathscr{G}(\psi)$ and (α_x, ω_x) is the range of f_x, then for every $\tau \in (\alpha_x, \omega_x)$ for which $|\tau| < p_\varphi(x)$, we have

$$|g_x(\tau)| = \mu_\psi\big(\{\varphi(\sigma, x) \mid \sigma \in [0, \tau]\}\big)$$

$$= \int\limits_{\{\varphi(\sigma, x) \mid \sigma \in [0, \tau]\}} \frac{1}{|u|} \, d\mu_\varphi$$

$$= \left| \int_0^\tau \frac{1}{u\big(\varphi(\sigma, x)\big)} \, d\sigma \right|,$$

and it is therefore easy to see that the formula

$$|g_x(\tau)| = \left| \int_0^\tau \frac{1}{u\big(\varphi(\sigma, x)\big)} \, d\sigma \right|$$

holds for every $\tau \in (\alpha_x, \omega_x)$.

The theorem follows easily from this observation, and we leave the details to the reader. **QED**

The following theorem complements Theorems 4.48 and 4.51.

5.43. Theorem. *Let φ be a quasi-flow in a Hausdorff space X, let f be a quasi φ-reparametrizer, and let $\psi = f$ rep φ.*

Then the following conditions are equivalent:

(a) *There exists a quasi φ-multiplier u such that $\psi = u\varphi$.*

(b) *Inside the set $\mathscr{G}(\psi)$, μ_ψ is absolutely continuous with respect to μ_φ.*

(c) $\mu_\psi\big(\{x \in \mathscr{G}(\psi) \mid f_x'(0) = 0\}\big) = 0.$

PROOF: The assertion (a) $\Rightarrow$ (b) follows at once from Theorem 5.41. Now for each $x \in \mathcal{G}(\psi)$, we have $|(f_x^{-1})'(\tau)| < \infty$ for almost all $\tau \in f_x(\mathbb{R})$, i.e., $f_x'(f_x^{-1}(\tau)) \neq 0$ for almost all $\tau \in f_x(\mathbb{R})$, i.e., $f_{\varphi(\tau,x)}'(0) \neq 0$ for almost all $\tau \in f_x(\mathbb{R})$.

It follows easily that

$$\mu_\varphi\big(\{x \in \mathcal{G}(\psi) \mid f_x'(0) = 0\}\big) = 0,$$

and from this we deduce at once that (b) $\Rightarrow$ (c).

(c) clearly implies that, for each $x \in \mathcal{G}(\psi)$, $f_{\psi(t,x)}'(0) \neq 0$ for almost all $t \in \mathbb{R}$, i.e., $f_x'(t) \neq 0$ for almost all $t \in \mathbb{R}$.

The assertion (c) $\Rightarrow$ (a) therefore follows at once from Theorem 4.51.

QED

5.44. Theorem. *Let φ be a quasi-flow in a Hausdorff space X, let f be a quasi φ-reparametrizer induced by a quasi φ-multiplier u, and let $\psi = u\varphi$.*

Then

(a) $u(x) = f_x'(0)$ *for μ_φ-almost all $x \in \mathcal{G}(\varphi)$.*

(b) $u(x) = f_x'(0)$ *for μ_ψ-almost all $x \in \mathcal{G}(\varphi)$.*

(c) $u(x) \neq 0$ *for μ_φ-almost all $x \in \mathcal{G}(\psi)$.*

PROOF: Since $u(x) = 0 = f_x'(0)$ for all $x \in \mathcal{G}(\varphi) \cap \mathcal{F}(\psi)$, and since condition (b) of Theorem 5.43 holds, (a) and (b) will be proved as soon as we have shown that

$$u(x) = f_x'(0) \text{ for } \mu_\varphi\text{-almost all } x \in \mathcal{G}(\psi).$$

But this is equivalent to saying that for each $x \in \mathcal{G}(\psi)$,

$$u\big(\varphi(\tau, x)\big) = f_{\varphi(\tau,x)}'(0) \text{ for almost all } \tau \in f_x(\mathbb{R}),$$

i.e., $(f_x^{-1})'(\tau) = \dfrac{1}{u\big(\varphi(\tau, x)\big)}$ for almost all $\tau \in f_x(\mathbb{R})$.

But this follows at once from the equation

$$(f_x^{-1})(\tau) = \int\limits_0^\tau \frac{1}{u\big(\varphi(\sigma, x)\big)} \, d\sigma, \qquad \forall \, \tau \in f_x(\mathbb{R}).$$

Thus (a) and (b) are proved.

Now as in the proof of Theorem 5.43, we can show that $f'_x(0) \neq 0$ for μ_φ-almost all $x \in \mathscr{G}(\psi)$, and in view of this, (c) is a consequence of (a).   **QED**

5.45. Theorem. *Let φ be a quasi-flow in a Hausdorff space X, let v be a quasi φ-multiplier, and let u be a quasi $v\varphi$-multiplier.*

Then uv is a quasi φ-multiplier, and

$$(uv)\, \varphi = u\,(v\varphi).$$

PROOF: First, we need to show that the restriction of uv to each φ-orbit is a Borel function. Suppose $x \in \mathscr{G}(\varphi)$. Then since $u(v\varphi)$ is a quasi φ-reparametrization, $\mathcal{O}_\varphi(x) \cap \mathscr{F}\big(u(v\varphi)\big)$ is a Borel set, and $\mathcal{O}_\varphi(x)$ meets only countably many non-trivial $u(v\varphi)$-orbits. Therefore, since

$$u(y)\, v(y) = 0, \qquad \forall\, y \in \mathcal{O}_\varphi(x) \cap \mathscr{F}\big(u(v\varphi)\big),$$

and since the restriction of uv to each $u(v\varphi)$-orbit is a Borel function, the restriction of uv to $\mathcal{O}_\varphi(x)$ must be a Borel function.

Thus we see that uv is $\Sigma(\varphi)$-measurable. Now define $\psi = u(v\varphi)$, and choose a quasi φ-reparametrizer f such that $\psi = f \operatorname{rep} \varphi$.

It is easy to see that, whenever $x \in \mathscr{G}(\psi)$ and f_x is increasing, we have $u(x)\, v(x) \geq 0$; and whenever $x \in \mathscr{G}(\psi)$ and f_x is decreasing, we have $u(x)\, v(x) \leq 0$.

It is also easy to see that

$$u(x)\, v(x) = 0, \qquad \forall\, x \in \mathscr{G}(\varphi) \cap \mathscr{F}(\psi).$$

Finally, if A is a $\Sigma(\varphi)$-measurable subset of $\mathscr{G}(\psi)$, then since A is also a subset of $\mathscr{G}(v\varphi)$, we have

$$\mu_\psi(A) = \int_A \frac{1}{|u|}\, d\mu_{v\varphi} = \int_A \frac{1}{|u|}\, \frac{1}{|v|}\, d\mu_\varphi.$$

It therefore follows from Theorem 5.42 that uv is a quasi φ-multiplier, and that

$$(uv)\, \varphi = \psi = u\,(v\varphi).$$ **QED**

5.46. Theorem. *Let φ be a quasi-flow in a Hausdorff space X, let v and w be quasi φ-multipliers, and suppose that $|v(x)| < \infty, \qquad \forall\, x \in X$.*

Then the following two conditions are equivalent:

(a) $w\varphi \in \mathrm{Rep}_q(v\varphi)$.

(b) *There exists a quasi $v\varphi$-multiplier u such that*

$$w\varphi = (uv)\,\varphi = u(v\varphi).$$

PROOF: The assertion (b) $\Rightarrow$ (a) is trivial.

To prove that (a) $\Rightarrow$ (b), assume that $w\varphi \in \mathrm{Rep}_q(v\varphi)$, and define the function u from X into $[-\infty, \infty]$ as follows:

$$u(x) = \begin{cases} 0 & \text{whenever } v(x) = 0 \\[2mm] \dfrac{w(x)}{v(x)} & \text{whenever } v(x) \neq 0. \end{cases}$$

It is easy to see that u is $\Sigma(v\varphi)$-measurable, (in fact, u is $\Sigma(\varphi)$-measurable). Furthermore, we can see easily that $u(x) = 0$ whenever $x \in \mathcal{G}(v\varphi) \cap \mathcal{F}(w\varphi)$, and that u does not change sign in any $w\varphi$-orbit.

The theorem will therefore follow from Theorems 5.45 and 5.42 when we have shown that

$$\mu_{w\varphi}(A) = \int_A \frac{1}{|u|}\, d\mu_{v\varphi}$$

for every $\Sigma(v\varphi)$-measurable subset A of $\mathcal{G}(w\varphi)$. Suppose A is a $\Sigma(v\varphi)$-measurable subset of $\mathcal{G}(w\varphi)$. Let $B = \{x \in A \mid v(x) = 0\}$, and $C = A \setminus B$. Then since $\mu_\varphi(B) = 0$, we have $\mu_{v\varphi}(B) = 0$, and therefore

$$\mu_{w\varphi}(B) = \int_B \frac{1}{|w|}\, d\mu_\varphi = 0 = \int_B \frac{1}{|u|}\, d\mu_{v\varphi}.$$

Further, since $w(x) = u(x)\,v(x)$, $\quad \forall\, x \in C$, we have

$$\mu_{w\varphi}(C) = \int_C \frac{1}{|w|}\, d\mu_\varphi = \int_C \frac{1}{|uv|}\, d\mu_\varphi$$

$$= \int_C \frac{1}{|u|}\frac{1}{|v|}\, d\mu_\varphi = \int_C \frac{1}{|u|}\, d\mu_{v\varphi}.$$

Thus,

$$\mu_{w\varphi}(A) = \int_A \frac{1}{|u|}\, d\mu_{v\tau},$$

and the proof is complete. **QED**

The Speed of a Reparametrization

5.47. Lemma. *Let φ be a quasi-flow in a metric space X, let f be a quasi φ-reparametrizer, and let $\psi = f$ rep φ. Let $x \in X$, and suppose $f'_x(0)$ and $|\dot\varphi|(x)$ exist.*

Then unless the expression $|\dot\varphi|(x) \cdot |f'_x(0)|$ is of the form $0 \times \infty$ or $\infty \times 0$, $|\dot\psi|(x)$ exists, and

$$|\dot\psi|(x) = |\dot\varphi|(x) \, |f'_x(0)|.$$

PROOF: For each non-zero real number t, we have

$$\frac{d\big(\psi(t, x), x\big)}{|t|} = \frac{d\big(\varphi\big(f_x(t), x\big), x\big)}{|f_x(t) - f_x(0)|} \cdot \left| \frac{f_x(t) - f_x(0)}{t - 0} \right|.$$

From this, the result follows at once.   **QED**

5.48. Theorem. *Let φ be a quasi-flow in a metric space X, let f be a quasi φ-reparametrizer, and let $\psi = f$ rep φ.*

(a) *Suppose that $|\dot\varphi|$ exists and is non-zero in a $\Sigma(\varphi)$-measurable subset A of $\mathscr{G}(\varphi)$.*

Then for μ_φ-almost every $x \in A$, $|\dot\psi|(x)$ exits and $|\dot\psi|(x) = |\dot\varphi|(x) \cdot |f'_x(0)|$.

(b) *Suppose that $|\dot\varphi|$ exists and is finite in a $\Sigma(\varphi)$-measurable subset A of $\mathscr{G}(\varphi)$.*

Then for μ_ψ-almost every $x \in A$, $|\dot\psi|(x)$ exists, and

$$|\dot\psi|(x) = |\dot\varphi|(x) \cdot |f'_x(0)|.$$

PROOF:

(a) We note first that the equation

$$|\dot\psi|(x) = |\dot\varphi|(x) \cdot |f'_x(0)|$$

clearly holds at each $x \in A \cap \mathscr{F}(\psi)$.

Now for each $x \in \mathscr{G}(\psi)$, the set $\{\tau \in f_x(\mathbb{R}) \mid f'_x\big(f_x^{-1}(\tau)\big) = 0\}$ has Lebesgue measure 0, and consequently,

$$\mu_\varphi\big(\{x \in A \cap \mathscr{G}(\psi) \mid f'_x(0) = 0\}\big) = 0.$$

(a) follows easily from these observations and Lemma 5.47.

(b) For each $x \in \mathscr{G}(\psi)$, the set $\{t \in \mathbb{R} \mid |f'_x(t)| = \infty\}$ has Lebesgue measure 0, and consequently,

$$\mu_\psi\big(\{x \in A \cap \mathscr{G}(\psi) \mid |f'_x(0)| = \infty\}\big) = 0.$$

From this and Lemma 5.47, (b) follows easily.   **QED**

5.49. Theorem. *Let φ be a quasi-flow in a metric space X, let f be a quasi φ-reparametrizer, and let $\psi = f \operatorname{rep} \varphi$. Suppose that*

$$0 < |\dot{\varphi}|(x) < \infty \quad \text{for } \mu_\psi\text{-almost all } x \in \mathscr{G}(\psi).$$

Then the following are equivalent:

(a) *There exists a quasi φ-multiplier u such that $\psi = u\varphi$.*

(b) $|\dot{\psi}|(x) \ne 0$ *for μ_ψ-almost all $x \in \mathscr{G}(\psi)$.*

PROOF: The theorem follows at once from Theorem 5.48(b) and Theorem 5.43. **QED**

5.50. Theorem. *Let φ be a quasi-flow in a metric space X, let u be a non-negative quasi φ-multiplier, and let $\psi = u\varphi$. Suppose that*

$$0 < |\dot{\varphi}|(x) < \infty \quad \text{for } \mu_\varphi\text{-almost all } x \in \mathscr{G}(\psi).$$

Then defining the expression $\dfrac{|\dot{\psi}|(x)}{|\dot{\varphi}|(x)}$ to be zero at each $x \in X$ at which it would not otherwise be defined, we have

$$\psi = \frac{|\dot{\psi}|}{|\dot{\varphi}|}\, \varphi.$$

PROOF: It is clear that $\dfrac{|\dot{\psi}|}{|\dot{\varphi}|}$ never changes sign, and that

$$\frac{|\dot{\psi}|(x)}{|\dot{\varphi}|(x)} = 0 \quad \text{for all } x \in \mathscr{G}(\varphi) \cap \mathscr{F}(\psi).$$

Therefore, by Theorems 5.41 and 5.42, it suffices to prove that

$$\frac{|\dot{\psi}|(x)}{|\dot{\varphi}|(x)} = u(x) \quad \text{for } \mu_\varphi\text{-almost all } x \in \mathscr{G}(\psi).$$

But by Theorem 5.48(a),

$$\frac{|\dot{\psi}|(x)}{|\dot{\varphi}|(x)} = f'_x(0) \quad \text{for } \mu_\varphi\text{-almost all } x \in \mathscr{G}(\psi)$$

and by Theorem 5.44(a),

$$u(x) = f'_x(0) \quad \text{for } \mu_\varphi\text{-almost all } x \in \mathscr{G}(\psi).$$

This completes the proof of the theorem. **QED**

5.51. Remark. Example 4.53 shows that there exists a flow φ with continuous, bounded, non-vanishing speed, and an isotonic φ-reparametrization ψ which has continuous, bounded speed such that $\dfrac{|\dot\psi|}{|\dot\varphi|}$ is a φ-multiplier, but $\psi \neq \dfrac{|\dot\psi|}{|\dot\varphi|}\,\varphi$.

Notes and Remarks to Chapter 5

Chapter 5 may be divided into two distinct parts. The first of these comprises sections 5.1 to 5.17, and consists of applications of the theory that was developed in Chapter 4, and the second part of the chapter introduces the time measure, which is the basic tool of Chapters 8 and 9.

Sections 5.3 to 5.7. Theorem 5.3 is a sharper form of the main result of Beck [1], and the proof of Theorem 5.3 closely models the techniques which are used in Beck [1]. The precise result stated in Beck [1] is Corollary 5.5, but the proof given in Beck [1] also implies Corollary 5.4.

Sections 5.9 to 5.17. Theorem 5.13 is precisely the result that is needed in the proofs of the interpolation theorems of Chapter 6. It was while examining the proofs of these interpolation theorems as they originally appeared in Beck [3], that the Lewins first felt the need for a reparametrization theory that would yield a theorem like Theorem 5.13, or its counterparts Theorems 5.14 and Corollary 5.15, in complete metric spaces.

Section 5.18. For the circumstances surrounding the definition of "time measure", see the notes and remarks to Chapters 8 and 9.

Section 5.38. Theorem 5.38 is no longer dependent on the Continuum Hypothesis if φ is a flow. In Beck [5] it is shown that if φ is a flow, and a subset I of $\mathscr{G}(\varphi)$ is homeomorphic to $[0, 1]$, then either I is contained in a single φ-orbit, or I meets exactly $2^{\aleph_0}$ orbits. From this it follows at once that in Theorem 5.38, if φ is a flow, then the cardinality of A is $2^{\aleph_0}$.

The results contained in Beck [5] were obtained for the specific purpose of showing that the use by the Lewins of the Continuum Hypothesis is unnecessary in Theorem 5.38, if φ is a flow.

CHAPTER SIX

EXISTENCE THEOREMS I

6.1. Introduction. In Chapters 6 and 7, we shall attempt to answer the question which we raised in Section 3.55, and we shall show that, under certain circumstances, if F is a compact subset of $\mathbf{S}^2$ and Y is a WATI subset [respectively ATI subset] of F, then there exists a flow φ in $\mathbf{S}^2$, such that

$$\mathscr{F}(\varphi) = F \text{ and } \mathscr{S}(\varphi) \subseteq Y.$$

Unfortunately, given a WATI subset Y of a compact set F, we do not know whether such a flow φ can always be found.[1]

Now suppose we are given a compact subset F of $\mathbf{S}^2$, and a WATI subset Y of F. The enormous task of defining a flow φ in $\mathbf{S}^2$ satisfying

$$\mathscr{F}(\varphi) = F \text{ and } \mathscr{S}(\varphi) \subseteq Y$$

will be undertaken in three main steps:

(a) Partition $\mathbf{S}^2$ into a family $\{X_\alpha \mid \alpha \in A\}$ of subsets, in such a way that the required properties of φ take on an especially simple form in each of the sets X_α.

(b) For each $\alpha \in A$, define a flow φ_α in $\overline{X_\alpha}$, with the required properties, and do this in such a way that for any $\alpha_1, \alpha_2 \in A$, φ_{α_1} and φ_{α_2} agree on $\partial(X_{\alpha_1}) \cap \partial(X_{\alpha_2})$.

(c) Combine all the flows φ_α $(\alpha \in A)$.

6.2. Definition. Let X be a Hausdorff space, let $\{X_\alpha \mid \alpha \in A\}$ be a family of subsets of X, and suppose that $X = \bigcup_{\alpha \in A} X_\alpha$. For each $\alpha \in A$, let φ_α be an algebraic flow in X_α, and suppose that, whenever $\alpha_1, \alpha_2, \in A$,

---

[1] (Footnote added in proof). The answer to this question is YES. The proof will be published later.

we have

$$\varphi_{\alpha_1}(t, x) = \varphi_{\alpha_2}(t, x), \qquad \forall\, t \in \mathbb{R}, \qquad \forall\, x \in X_{\alpha_1} \cap X_{\alpha_2}.$$

Then the algebraic flow φ in X, defined by:

$$\varphi(t, x) = \varphi_{\alpha}(t, x) \text{ whenever } t \in \mathbb{R} \text{ and } x \in X_{\alpha}, \ (\alpha \in A)$$

is called the *conjunction* of $\{\varphi_{\alpha} \mid \alpha \in A\}$, and is denoted by $\bigcup_{\alpha \in A} \varphi_{\alpha}$.

6.3. Theorem. *The conjunction of a family of quasi-flows is a quasi-flow.*

PROOF: Clear.   **QED**

6.4. Theorem. *Let X be a Hausdorff space, and let $X = \bigcup_{\alpha \in A} X_{\alpha}$. For each $\alpha \in A$, let φ_{α} be a flow in X_{α}, and suppose that the conjunction $\varphi = \bigcup_{\alpha \in A} \varphi_{\alpha}$ exists.*
Then φ is continuous at each point x which belongs to the interior of some X_{α} $(\alpha \in A)$.

PROOF: Clear.   **QED**

6.5. Theorem. *Let $X = \bigcup_{\alpha \in A} X_{\alpha}$ be a Hausdorff space and suppose that, for each $\alpha \in A$, X_{α} is closed and φ_{α} is a flow in X_{α}. Suppose that the conjunction $\varphi = \bigcup_{\alpha \in A} \varphi_{\alpha}$ exists. Let $x \in X$, and suppose that x has a neighborhood which is contained in the union of finitely many of the sets X_{α}.*
Then φ is continuous at x.

PROOF: Clear.   **QED**

6.6. Example. Let $X = \mathbb{R}^2$, let $X_0 = \{\langle x, y \rangle \in \mathbb{R}^2 \mid y \leq 0\}$, let $X_1 = \{\langle x, y \rangle \in \mathbb{R}^2 \mid y \geq 1\}$, and for each integer $n \geq 2$, let

$$X_n = \left\{ \langle x, y \rangle \in \mathbb{R}^2 \,\middle|\, \frac{1}{n} \leq y \leq \frac{1}{n-1} \right\}.$$

Let φ_0 be the flow in X_0 which leaves every point of X_0 fixed, and for each $n = 1, 2, \ldots$, let φ_n be the flow in X_n defined by

$$\varphi_n(t, \langle x, y \rangle) = \langle x + t, y \rangle, \qquad \forall\, t \in \mathbb{R}, \qquad \forall\, \langle x, y \rangle \in X_n.$$

Then although each X_n is closed, and $\bigcup_n \varphi_n$ exists, $\bigcup_n \varphi_n$ is not continuous at any point $\langle x, 0 \rangle$.

6.7. Theorem. *Let* $X = \bigcup_{\alpha \in A} X_\alpha$ *be a metric space, and suppose that, for each* $\alpha \in A$, X_α *is closed and* φ_α *is a flow in* X_α. *Suppose* $\varphi = \bigcup_{\alpha \in A} \varphi_\alpha$ *exists. Let F be a closed subset of X, and suppose that* φ *is bounded with respect to F (see Section 1.62). Suppose further that for every* $x \in X$, *we have*

either

(i) $x \in F$

or

(ii) *x has a neighborhood which is contained in the union of finitely many of the sets* X_α.

Then φ *is a flow.*

PROOF: Combine Theorems 1.65 and 6.5.   **QED**

6.8. Corollary. *Let X be a metric space, let* $\{X_\alpha \mid \alpha \in A\}$ *be a family of closed subsets of X, and suppose* $X = \bigcup_{\alpha \in A} X_\alpha$. *Let F be a closed subset of X, and for each* $\alpha \in A$, *let* φ_α *be a flow in* X_α, *which is canonical with respect to F. Suppose that* $\varphi = \bigcup_{\alpha \in A} \varphi_\alpha$ *exists, and suppose further that, for every* $x \in X$, *we have*

either

(i) $x \in F$

or

(ii) *x has a neighborbood which is contained in the union of finitely many of the sets* X_α.

Then φ *is a flow, which is canonical with respect to F.*

PROOF: Combine Corollary 1.69 and Theorem 6.7.   **QED**

6.9. Note. When we refer to the speed $|\dot\varphi|$ of a quasi-flow φ defined in a subspace of $\mathbf{S}^2$, we shall always mean the speed of φ defined in terms of the chordal metric ϱ, rather than the Euclidean metric d, (see Section C.1). On those occasions when we need the Euclidean speed, we shall denote it by $|\dot\varphi|_d$.

It follows from Section C.1 that if φ is a quasi-flow in a subspace X of $\mathbb{R}^2$, and $x = re^{i\theta} \in X$, then we have

$$|\dot\varphi|(x) = \frac{4}{4 + r^2} |\dot\varphi|_d(x)$$

whenever either $|\dot\varphi|(x)$ or $|\dot\varphi|_d(x)$ exists.

The word "canonical" will always refer to the chordal metric, instead of the Euclidean metric.

6.10. Interpolations. We shall now concern ourselves with the problem of constructing a flow φ_1 in a region Ω, whose conjunction φ with a given flow φ_0 in $\partial(\Omega)$, is a flow in $\bar{\Omega}$.

Such a flow φ is called an *interpolation of φ_0 into Ω*.

6.11. Definition. Let φ be a flow in a subspace X of $\mathbf{S}^2$.

φ is said to be *orbit-analytic* if for every $x \in X$, $\mathcal{O}_\varphi(x)$ is either a single point, an analytic Jordan curve, or an analytic arc.

6.12. Definition. Let φ_0 be a flow in the boundary of a region Ω.

φ_0 is said to be *weakly Ω-accessible* if:

(a) For every $x \in \mathcal{G}(\varphi_0)$, $\mathcal{O}_{\varphi_0}(x)$ is a relatively open subset of $\partial(\Omega)$, and

(b) For every $x \in \mathcal{G}(\varphi_0)$, we have

$$\alpha_{\varphi_0}(x) \cup \omega_{\varphi_0}(x) \subseteq \mathcal{F}(\varphi_0).$$

φ_0 is said to be *Ω-accessible* if φ_0 is weakly Ω-accessible and simple. (See Section 1.44.)

6.13. Definition. Let φ_0 be a flow in the boundary of a region Ω, and let φ be an interpolation of φ_0 into Ω.

(a) φ is said to be a *conservative* interpolation of φ_0 into Ω if $\mathcal{S}(\varphi) = \mathcal{S}(\varphi_0)$, and every simple stagnation point of φ_0 is a simple stagnation point of φ.

(b) φ is said to be an *ultra-conservative* interpolation of φ_0 into Ω if whenever $x \in \Omega$ and $\mathrm{p}_\varphi(x) = \infty$, there exists $x' \in \partial(\Omega)$ such that

(i) either $\alpha_\varphi(x) \subseteq \alpha_{\varphi_0}(x')$ or $\alpha_\varphi(x) \subseteq \omega_{\varphi_0}(x')$ and

(ii) either $\omega_\varphi(x) \subseteq \alpha_{\varphi_0}(x')$ or $\omega_\varphi(x) \subseteq \omega_{\varphi_0}(x')$.

It is obvious that every ultra-conservative interpolation is conservative.

6.14. Definition. Let φ_0 be a flow in the boundary of a region Ω, let φ be an interpolation of φ_0 into Ω, and suppose that φ_0 is canonical with respect to a set K.

If φ also is canonical with respect to K, then we say that φ is obtained by *interpolating φ_0 canonically* with respect to K.

Interpolation Theorems

6.15. Remark on Notation. Let h be a conformal equivalence from a finitely connected region Ω onto an elementary region Δ. (See Appendix C.) Let Λ be an analytic arc, which is a relatively open subset of $\partial(\Omega)$, and let γ be one of the (at most 2) homeomorphic images of Λ, which comprise $C_\Omega(h, \Lambda)$ (see Theorem C.58). Theorem C.58 gives us a homeomorphism h_γ from Λ onto γ, whose inverse h_γ^{-1} can be extended to a conformal equivalence from a neighborhood of γ onto a neighborhood of Λ, with the property that the restriction of this conformal equivalence to Δ coincides with h^{-1}.

This function h_γ will always be referred to as "the homeomorphism corresponding to h and γ in the sense of Theorem C.58".

6.16. Theorem. *Let φ_0 be an orbit-analytic, weakly accessible flow in the boundary of a finitely connected region Ω, and let h be a conformal equivalence from Ω onto an elementary region Δ. Let $\mathscr{C}$ be the family of all images under h of non-trivial φ_0-orbits, in the sense of Theorem C.58, and for each $\gamma \in \mathscr{C}$, let h_γ be the homeomorphism corresponding to h and γ in the sense of Theorem C.58.*

For each $\gamma \in \mathscr{C}$, let φ_γ be the restriction of φ_0 to $h_\gamma^{-1}(\gamma)$, and let $\psi_\gamma = h_\gamma(\varphi_\gamma)$. Let ψ_F be the flow in $\partial(\Delta) \setminus \bigcup \mathscr{C}$ which leaves every point fixed, and let

$$\psi_0 = \bigcup_{\gamma \in \mathscr{C}} \psi_\gamma \cup \psi_F.$$

Then ψ_0 is a flow in $\partial(\Delta)$.

PROOF: The proof of this theorem is elementary, and will be left to the reader. **QED**

6.17. Definition. Let φ_0 be a weakly accessible flow in the boundary of a finitely connected region Ω, let h be a conformal mapping from Ω onto an elementary region Δ, and let ψ_0 be the flow in $\partial(\Delta)$ defined as in Theorem 6.16.

Then we say that ψ_0 is the *image of φ_0 under h*, and we write $\psi_0 = h[\varphi_0]$. We also call φ_0 the *image under h^{-1} of ψ_0*, and write $\varphi_0 = h^{-1}[\psi_0]$.

We remark that although the notation $\varphi_0 = h^{-1}[\psi_0]$ is ambiguous whenever Ω is itself an elementary region, no inconsistency occurs.

6.18. Definition. Let φ_0 be a weakly accessible flow in the boundary of a disc (or annulus) Ω.

We say that φ_0 is *unidirectional* if there exists a conformal equivalence h from Ω onto the open unit disc $U = \mathrm{ins}\,(\Gamma_1)$, (or an annulus $A = \{z \in \mathbb{R}^2 \mid r_1 < |z| < r_2\}$ where $0 \le r_1 < r_2 \le \infty$), such that every non-trivial orbit of $h[\varphi_0]$ is oriented clockwise, or every nontrivial orbit of $h[\varphi_0]$ is oriented counter-clockwise.

6.19. Theorem. *Let φ_0 be a unidirectional, orbit-analytic flow in the boundary of an annulus Ω. Let K be a non-empty compact subset of $\mathbf{S}^2 \setminus \Omega$, and suppose that φ_0 is canonical with respect to K.*

Then there exists an ultra-conservative interpolation φ of φ_0 into Ω, which is orbit-analytic, canonical with respect to K, and satisfies $\mathrm{p}_\varphi(x) < \infty$, $\forall\, x \in \Omega$.

PROOF: Choose a conformal equivalence h from Ω onto $A = \{z \in \mathbb{R}^2 \mid r_1 < |z| < r_2\}$ where $0 \le r_1 < r_2 \le \infty$, and let h be chosen in such a way that, if $\mathscr{G}(\varphi_0) \ne \square$, then $r_2 = 1$ and $h[\varphi_0]$ is unidirectional. Assume, for simplicity of notation, that every non-trivial orbit of $h[\varphi_0]$ is oriented counter-clockwise; the other case is dual. Choose a point of K to play the role of ∞.

Let $\bar{\chi}$ be the flow in $\bar{A}$ defined by

$$\bar{\chi}(t, re^{i\theta}) = re^{i(\theta + t)}, \qquad \forall\, t \in R, \qquad \forall\, re^{i\theta} \in \bar{A},$$

where we agree to write $\infty\, e^{i\theta} = \infty$.

Let $\psi_0 = h[\varphi_0]$. Now using Theorem 5.3, choose a continuous, proper $\bar{\chi}$-multiplier $v \ge 0$ such that $\mathscr{F}(v\bar{\chi}) = \mathscr{F}(\psi_0)$. Let χ be the restriction of $v\bar{\chi}$ to $\bar{A} \setminus \mathscr{F}(\psi_0)$. It is clear that $|\dot\chi|$ is continuous and non-vanishing on $\bar{A} \setminus \mathscr{F}(\psi_0)$, and that every χ-orbit is oriented counter-clockwise.

Theorem C. 58 (b) implies that since φ_0 is unidirectional, every non-trivial φ_0-orbit is an arc in $\partial(\Omega)$ of type 1, and we can therefore extend h to a conformal equivalence from a neighborhood of $\Omega \cup \mathscr{G}(\varphi_0)$ onto a neighborhood of $A \cup \mathscr{G}(\psi_0)$. Let $\tilde{\varphi} = h^{-1}(\chi)$. Then $\tilde{\varphi}$ is a flow in $\Omega \cup \mathscr{G}(\varphi_0)$, and $\tilde{\varphi}$ has continuous, non-vanishing speed. Furthermore every aperiodic $\tilde{\varphi}$ orbit, being a φ_0-orbit, has all its endpoints in K. We can therefore apply Theorem 5.13 to find an isotonic $\tilde{\varphi}$-reparametrization φ which is canonical with respect to K. It is clear that the restriction of φ to $\mathscr{G}(\varphi_0)$ is an isotonic reparametrization of φ_0, and that in $\mathscr{G}(\varphi_0)$, φ and φ_0 have the same continuous non-vanishing speed. We therefore deduce from Theorems 5.49 and 5.50 that φ and φ_0 agree in $\mathscr{G}(\varphi_0)$. Finally we note, using Corollary 1.69 and Theorem 1.65, that we can extend φ continuously to a flow in $\bar{\Omega}$ which leaves every point of $\mathscr{F}(\varphi_0)$ fixed. This completes the proof.   **QED**

6.20. Corollary. *Let Ω be an open subset of $\mathbf{S}^2$, and let F be a T-set of Ω.*

Then there exists an orbit-analytic flow φ in Ω satisfying

(a) $\mathscr{F}(\varphi) = F$,

(b) $p_\varphi(x) < \infty, \qquad \forall\, x \in \Omega$,

and

(c) *φ is canonical with respect to $F \cup \partial(\Omega)$.*

PROOF: For each component A of $\Omega \setminus F$, apply Theorem 6.19 to obtain an interpolation φ_A into A of the everywhere fixed flow in $\partial(A)$. Let φ_F be the flow in F that leaves every point fixed.

It follows from Corollary 6.8 that if

$$\varphi = \varphi_F \cup \bigcup_A \varphi_A$$

then the restriction of φ to Ω has all the desired properties.   **QED**

We are now in a position to prove a very special case of our main result.

6.21. Theorem. *Let F be a compact subset of $\mathbf{S}^2$.*

Then the following conditions are equivalent:

(a) *There exists a flow φ in $\mathbf{S}^2$ such that $\mathscr{F}(\varphi) = F$ and $\mathscr{S}(\varphi) = \square$.*

(b) *$\square$ is a WATI-subset of F.*

(c) *F is a T-set of $\mathbf{S}^2$.*

(d) *There exists an orbit-analytic flow φ in $\mathbf{S}^2$ which is canonical with respect to F, and satisfies $p_\varphi(x) < \infty, \qquad \forall\, x \in \mathbf{S}^2$.*

PROOF: The assertion (a) $\Rightarrow$ (b) follows from Theorem 3.54, (b) and (c) are obviously equivalent, the assertion (c) $\Rightarrow$ (d) follows from Theorem 6.20, and the assertion (d) $\Rightarrow$ (a) is trivial.   **QED**

6.22. Examples.

(a) Let A be the annulus between Γ_1 and Γ_2, (where, as usual, Γ_δ is the circle $\{z \in \mathbb{R}^2 \mid |z| = \delta\}$, for every $\delta > 0$). Let φ_0 be the flow in $\partial(A)$ defined by:

$$\varphi_0(t, e^{i\vartheta}) = e^{i(\vartheta+t)}, \qquad \forall\, t \in \mathbb{R} \qquad \forall\, \vartheta \in \mathbb{R},$$

and

$$\varphi_0(t, 2e^{i\vartheta}) = 2e^{i(\vartheta-t)}, \qquad \forall\, t \in \mathbb{R} \qquad \forall\, \vartheta \in \mathbb{R}.$$

Then φ_0 is not unidirectional, and it is easy to see that there does not exist an ultra-conservative interpolation φ of φ_0 into A satisfying

$\mathscr{F}(\varphi) = \square$. Note, however, that the standard order-reversing spiral between Γ_1 and Γ_2 (see Section 3.14) is a conservative interpolation of φ_0 into A.

(b) Let φ_0 be the flow in Γ_1 defined by

$$\varphi_0(t, e^{i\vartheta}) = e^{i(\vartheta+t)}, \qquad \forall\, t \in \mathbb{R} \qquad \forall\, \vartheta \in \mathbb{R}.$$

Then it follows easily from Corollary 1.33 (b) that there does not exist an interpolation φ of φ_0 into ins (Γ_1), satisfying $\mathscr{F}(\varphi) = \square$.

(c) Let φ_0 be the flow in Γ_1 which leaves every point fixed. Then there does not exist a conservative interpolation φ of φ_0 into ins (Γ_1) satisfying $\mathscr{F}(\varphi) = \mathscr{F}(\varphi_0)$, for if φ were such an interpolation, we could extend φ to a flow in $\mathbf{S}^2$ by defining φ to be everywhere fixed outside Γ_1, and this extension would violate Theorem 6.21.

6.23. Theorem. *Let φ_0 be a weakly accessible, orbit-analytic flow in the boundary of a disc Ω, suppose that both $\mathscr{F}(\varphi_0)$ and $\mathscr{G}(\varphi_0)$ are non-empty, let K be a non-empty compact subset of $\mathbf{S}^2 \setminus \Omega$, and suppose that φ_0 is canonical with respect to K.*

Then there exists an ultra-conservative interpolation φ of φ_0 into Ω, which is orbit-analytic and canonical with respect to K.

PROOF: Choose a point of K to play the role of ∞.

We begin by defining a useful flow $\bar{\chi}$ in the closure of the open unit disc $U = \text{ins}(\Gamma_1)$, and this we do as follows:

For each $\alpha \in (0, 1]$, let

$$C_\alpha = \left\{ \alpha(1 + e^{i\vartheta})^\alpha - 1 \mid -\pi < \vartheta < \pi \right\}.^2$$

(See Fig. 6.1.)

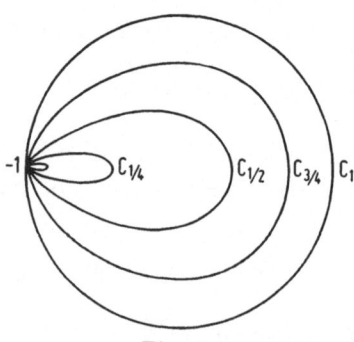

Fig. 6.1

---

[2] We agree that $(re^{i\vartheta})^\alpha = r^\alpha e^{i\alpha\vartheta}$ whenever $\alpha > 0, r > 0$, and $-\dfrac{\pi}{2} < \vartheta < \dfrac{\pi}{2}$.

We assert that the family $\{C_\alpha \mid 0 < \alpha \leq 1\}$ is a pairwise disjoint family of analytic arcs, whose union is $\bar{U} \setminus \{-1\}$:

Certainly, each C_α is an analytic arc, being the image of $C_1 = \{e^{i\vartheta} \mid -\pi < \vartheta < \pi\}$ under the conformal equivalence $z \to \alpha(1+z)^\alpha - 1$, which maps the half-plane $\{z \in \mathbb{C} \mid \mathrm{Re}\,(z) > -1\}$ into itself.

Secondly the family $\{C_\alpha \mid 0 < \alpha \leq 1\}$ is pairwise disjoint, for if $\alpha_1 < \alpha_2$ and C_{α_1} meets C_{α_2}, we can obtain a contradiction as follows: Choose ϑ_1 and ϑ_2 in $(-\pi, \pi)$ such that

$$\alpha_1(1 + e^{i\vartheta_1})^{\alpha_1} - 1 = \alpha_2(1 + e^{i\vartheta_2})^{\alpha_2} - 1.$$

Then we have

$$\alpha_1 \cdot 2^{\alpha_1} \cdot \cos^{\alpha_1}\left(\frac{\vartheta_1}{2}\right) e^{\frac{i\alpha_1\vartheta_1}{2}} = \alpha_2 \cdot 2^{\alpha_2} \cdot \cos^{\alpha_2}\left(\frac{\vartheta_2}{2}\right) e^{\frac{i\alpha_2\vartheta_2}{2}}$$

and therefore

$$\alpha_1\vartheta_1 = \alpha_2\vartheta_2 \quad \text{and} \quad \alpha_1 2^{\alpha_1} \cos^{\alpha_1}\left(\frac{\vartheta_1}{2}\right) = \alpha_2 2^{\alpha_2} \cos^{\alpha_2}\left(\frac{\vartheta_2}{2}\right).$$

It is clear that both ϑ_1 and ϑ_2 are non-zero, and since ϑ_1 and ϑ_2 clearly have the same sign, we can (by replacing them by $-\vartheta_1$ and $-\vartheta_2$ if necessary) assume that ϑ_1 and ϑ_2 are positive. We now have $0 < \vartheta_2 < \vartheta_1 < \pi$. Define

$$f(t) = \left(\frac{\alpha_2\vartheta_2}{t}\right) \cdot 2^{\frac{\alpha_2\vartheta_2}{t}} \cos^{\frac{\alpha_2\vartheta_2}{t}}\left(\frac{t}{2}\right), \qquad \forall\, \vartheta_2 \leq t < \pi.$$

Then we have $f(\vartheta_1) = f(\vartheta_2)$. We shall contradict this last statement by showing that $f'(t) < 0$ whenever $\vartheta_2 \leq t < \pi$. Now, for $\vartheta_2 < t < \pi$, we have

$$\frac{1}{f(t)}\,f'(t) = -\frac{1}{t} - \frac{\alpha_2\vartheta_2}{t^2}\log 2 - \frac{\alpha_2\vartheta_2}{t^2}\log\cos\left(\frac{t}{2}\right) - \frac{\alpha_2\vartheta_2}{2t}\tan\left(\frac{t}{2}\right),$$

and since $f(t) > 0$, the requirement that $f'(t) < 0$ for $\vartheta_2 \leq t < \pi$ becomes the requirement that $h(t) > 0$ for $\vartheta_2 \leq t < \pi$, where we define

$$h(t) = t + \alpha_2\vartheta_2 \log 2 + \alpha_2\vartheta_2 \log\cos\left(\frac{t}{2}\right) + \frac{1}{2}\alpha_2\vartheta_2 t \tan\left(\frac{t}{2}\right),$$

$$\forall\, 0 \leq t < \pi.$$

But this is clear, since $h(0) = \alpha_2 \vartheta_2 \log 2 > 0$, and

$$h'(t) = 1 + \frac{1}{4}\, \alpha_2 \vartheta_2 t \sec^2\left(\frac{t}{2}\right) \geq 0, \qquad \forall\, 0 \leq t < \pi.$$

We have therefore shown that the family $\{C_\alpha \mid 0 < \alpha \leq 1\}$ is pairwise disjoint, and we deduce from this that since each C_α $(0 < \alpha < 1)$ contains some points in U and does not meet $\partial(U)$, we must have $C_\alpha \subseteq U$ for every $\alpha \in (0, 1)$.

Thirdly, every point z of $\overline{U} \setminus \{-1\}$ lies in one of the curves C_α $(0 < \alpha \leq 1)$. This is clear if $|z| = 1$, for then $z \in C_1$. Suppose $z \in U$. The condition that $z \in C_\alpha$ is the condition that $z = \alpha(1 + w)^\alpha - 1$ for some $w \in C_1$, i.e., that

$$\left| \left(\frac{1}{\alpha}\right)^{\frac{1}{\alpha}} (1 + z)^{\frac{1}{\alpha}} - 1 \right| = 1.$$

But the left hand expression is less than 1 when $\alpha = 1$, and greater than 1 when α is small, and therefore must take the value 1 for some $\alpha \in (0, 1)$.

Finally, we note that for each $\alpha \in (0, 1)$ all points on the curve C_α close to -1 lie inside the Stolz angle of U at -1, of opening $\dfrac{\alpha\pi}{2}$, (see Definition C.53).

Let $\bar{\chi}_1$ be a flow in C_1 which is obtained by an application of Corollary 5.4 to the flow in Γ_1 which sends the point $e^{i\theta}$, in time t, to the point $e^{i(\theta + t)}$, with F replaced by $\{-1\}$. $\bar{\chi}_1$ is a flow in C_1 which is oriented counterclockwise and has continuous, non-vanishing speed.

For each $\alpha \in (0, 1)$, define a flow $\bar{\chi}_\alpha$ in C_α by:

$$\bar{\chi}_\alpha(t, z) = \alpha\left[1 + \bar{\chi}_1\left(t, \left(\frac{1}{\alpha}\right)^{\frac{1}{\alpha}} (1 + z)^{\frac{1}{\alpha}} - 1\right) \right]^\alpha - 1,$$

$$\forall\, t \in \mathbb{R} \qquad \forall\, z \in C_\alpha,$$

i.e., let $\bar{\chi}_\alpha$ be the image of $\bar{\chi}_1$ under the function $z \to \alpha(1 + z)^\alpha - 1$.

Let $\bar{\chi}_0$ be the trivial flow in $\{-1\}$. Now let

$$\bar{\chi} = \bigcup_{\alpha \in [0,1]} \bar{\chi}_\alpha.$$

It is easy to show that $\bar{\chi}$ is a flow in $\overline{U}$ whose only fixed point is -1, and whose other orbits are the curves C_α $(0 < \alpha \leq 1)$.

CASE 1: Suppose φ_0 is unidirectional.

Let h be a conformal equivalence from Ω onto U, so chosen that $h[\varphi_0]$ is unidirectional, and let $\psi_0 = h[\varphi_0]$. We may clearly assume that every non-trivial ψ_0-orbit is oriented counter-clockwise, as the other case is dual, and since ψ_0 has fixed and moving points, we may also assume that -1 is an endpoint of some ψ_0-orbit. Using Theorem 5.3, choose a continuous, proper $\bar{\chi}$-multiplier $v \geq 0$ such that $\mathscr{F}(v\bar{\chi}) = \mathscr{F}(\psi_0)$, and let χ be the restriction of $v\bar{\chi}$ to $\overline{U} \setminus \mathscr{F}(\psi_0)$. It is clear that $|\dot{\chi}|$ is continuous and non-vanishing on $\overline{U} \setminus \mathscr{F}(\psi_0)$.

Theorem C.58, implies that since φ_0 is unidirectional, every non-trivial φ_0-orbit is an arc in $\partial(\Omega)$ of type 1, and we can therefore extend h to a conformal equivalence from a neighborhood of $\Omega \cup \mathscr{G}(\varphi_0)$ onto a neighborhood of $U \cup \mathscr{G}(\psi_0)$. Let $\tilde{\varphi} = h^{-1}[\chi]$. Then $\tilde{\varphi}$ is a flow in $\Omega \cup \mathscr{G}(\varphi_0)$, and $\tilde{\varphi}$ has continuous, non-vanishing speed. Furthermore, Theorem C.59, together with the fact that each curve C_α begins and ends in a Stolz angle of U at -1, imply that each $\tilde{\varphi}$-orbit has all its endpoints in K. We can therefore apply Theorem 5.13 to find an isotonic $\tilde{\varphi}$-reparametrization φ which is canonical with respect to K. Arguing as we did in the proof of Theorem 6.19, we see that if we extend φ to a flow in $\bar{\Omega}$ by making every point of $\mathscr{F}(\varphi_0)$ fixed, then φ is an interpolation of φ_0 into Ω. From Theorem C.59, we see that this interpolation is ultra-conservative, and the proof of Case 1 is complete.

CASE 2: General Case.

Let h be a conformal equivalence from Ω onto U, and let $\psi_0 = h[\varphi_0]$. Denote by $\{\gamma_n \mid n = 1, 2, \ldots\}$, the set of ψ_0-orbits which are oriented clockwise, and for each $n = 1, 2, \ldots$, let $\alpha_{\psi_0}(\gamma_n) = a_n$ and $\omega_{\psi_0}(\gamma_n) = b_n$. For each $n = 1, 2, \ldots$, let Λ_n be the straight line arc (b_n, a_n). Let J be the Jordan curve obtained from $\partial(U)$ by replacing γ_n by Λ_n for each $n = 1, 2, \ldots$, and for each $n = 1, 2, \ldots$, let J_n be the Jordan curve $\gamma_n \cup \Lambda_n \cup \{a_n, b_n\}$. Let $\tilde{\Omega} = h^{-1}(\text{ins}(J))$, and for each $n = 1, 2, \ldots$, let $\Omega_n = h^{-1}(\text{ins}(J_n))$. For each $n = 1, 2, \ldots$, let ψ_n be the flow in Λ_n satisfying

(i) $h^{-1}[\psi_n]$ is a flow in $h^{-1}(\Lambda_n)$ which is canonical with respect to K;

and

(ii) ψ_n is so oriented that $\alpha_{\psi_n}(\Lambda_n) = b_n$ and $\omega_{\psi_n}(\Lambda_n) = a_n$.

(It is not hard to establish the existence of ψ_n, using Corollary 5.4 and Theorem 5.13).

It is easy to see that $\varphi_0 \cup \bigcup_n h^{-1}[\psi_n]$ is well-defined and continuous, and that the restrictions of this flow to $\partial(\tilde{\Omega})$, $\partial(\Omega_1)$, $\partial(\Omega_2)$, $\ldots$, are all unidirectional. Let $\tilde{\varphi}$ be an ultra-conservative interpolation into $\tilde{\Omega}$ of the restriction of $\varphi_0 \cup \bigcup_n h^{-1}[\psi_n]$ to $\partial(\tilde{\Omega})$, which is orbit-analytic, and canonical with respect to K. For each $n = 1, 2, \ldots,$ let φ_n be an ultra-conservative interpolation into Ω_n of the restriction of $\varphi_0 \cup \bigcup_n h^{-1}[\psi_n]$ to $\partial(\Omega_n)$ which is orbit-analytic and canonical with respect to K. (The existence of $\tilde{\varphi}$ and φ_n $(n = 1, 2, \ldots)$ follow from Case 1.)

It is easy to see that the conjunction

$$\varphi = \varphi_0 \cup \tilde{\varphi} \cup \bigcup_n \varphi_n$$

is well-defined and continuous, and satisfies all the requirements of the theorem.   **QED**

6.24. Corollary. *Let φ_0 be a simple, accessible, orbit-analytic flow in the boundary of a disc Ω, suppose $\alpha_{\varphi_0}(x) = \omega_{\varphi_0}(x)$ for every $x \in \mathscr{G}(\varphi_0)$, and suppose that both $\mathscr{F}(\varphi_0)$ and $\mathscr{G}(\varphi_0)$ are non-empty. Let K be a non-empty compact subset of $\mathbf{S}^2 \setminus \Omega$, and suppose that φ_0 is canonical with respect to K.*

Then there exists an ultra-conservative interpolation φ of φ_0 into Ω which is simple, orbit-analytic, and canonical with respect to K, and satisfies $\alpha_\varphi(x) = \omega_\varphi(x)$, for every $x \in \mathscr{G}(\varphi)$.

PROOF: This follows at once from Theorem 6.23.   **QED**

6.25. Theorem. *Let φ_0 be a weakly accessible, orbit-analytic flow in the boundary of an annulus Ω, and suppose that $\mathscr{F}(\varphi_0)$ is not empty. Let K be a non-empty compact subset of $\mathbf{S}^2 \setminus \Omega$, and suppose that φ_0 is canonical with respect to K.*

Then there exists an ultra-conservative interpolation φ of φ_0 into Ω, which is orbit-analytic, and canonical with respect to K.

PROOF:

CASE 1: Suppose that one component of $\partial(\Omega)$ is a periodic φ_0-orbit, and the other component meets both $\mathscr{F}(\varphi_0)$ and $\mathscr{G}(\varphi_0)$.

Let h be a conformal equivalence from Ω onto an annulus $A = \{z \in \mathbf{R}^2 \mid r < |z| < 1\}$, where $0 < r < 1$, and h is chosen in such a way that the restriction of $h[\varphi_0]$ to Γ_1 has both fixed and moving points. We may also suppose that -1 is an endpoint of some orbit of $h[\varphi_0]$. Choose $\alpha \in (0, 1)$ such that the Jordan curve

$$J = \{-1\} \cup \{\alpha(1 + e^{i\vartheta})^\alpha - 1 \mid -\pi < \vartheta < \pi\}$$

contains Γ_r in its inside. Let φ_J be the flow in $h^{-1}(J \smallsetminus \{-1\})$ which is canonical with respect to K, and which is so oriented that $h[\varphi_J]$ has the same orientation as the restriction of $h[\varphi_0]$ to Γ_r.

Let $\Omega_1 = h^{-1}(\text{ins}(J) \cap A)$ and $\Omega_2 = h^{-1}(\text{outs}(J) \cap A)$. Then Ω_1 is an annulus and Ω_2 is a disc, and the restrictions of $\varphi_0 \cup \varphi_J$ to $\partial(\Omega_1)$ and $\partial(\Omega_2)$ satisfy the hypotheses of Theorems 6.19 and 6.23 respectively. We can therefore complete the proof of Case 1 by interpolating separately into Ω_1 and Ω_2.

CASE 2: General Case:

If either component of $\partial(\Omega)$ is a periodic φ_0-orbit, the result follows from Theorem 6.19, or Case 1 of this theorem. We shall therefore suppose that both components of $\partial(\Omega)$ meet $\mathscr{F}(\varphi_0)$. Choose an analytic Jordan curve $J \subseteq \Omega$ such that J separates $\partial(\Omega)$, and let φ_J be a flow in J which is canonical with respect to K. Let Ω_1 and Ω_2 be the intersections of Ω with the two sides of J. We now complete the proof by interpolating separately into Ω_1 and Ω_2 (using either Theorem 6.19 or Case 1 of this theorem).   **QED**

6.26. Theorem. *Let φ_0 be an orbit-analytic, weakly accessible flow in the boundary of a finitely connected region Ω, and suppose that some component B_1 of $\partial(\Omega)$ meets both $\mathscr{F}(\varphi_0)$ and $\mathscr{G}(\varphi_0)$. Let K be a non-empty compact subset of $\mathbf{S}^2 \smallsetminus \Omega$, and suppose φ_0 is canonical with respect to K.*

Then there exists an ultra-conservative interpolation φ of φ_0 into Ω which is orbit-analytic and canonical with respect to K.

PROOF: The theorem is already known if the connectivity of Ω is 1 or 2. Let n be an integer, $n \geq 2$, suppose Ω has connectivity n, and that the theorem has been proved for all regions of connectivity less than n.

Choose $x \in \mathscr{G}(\varphi_0) \cap B_1$, and suppose for the moment that we have constructed an analytic arc $\Lambda \subseteq \Omega$ which joins a subset of $\omega_{\varphi_0}(x)$ to a subset of $\omega_{\varphi_0}(x)$ in such a way that $\Omega \smallsetminus \Lambda$ has two components, both of which have connectivity less than n.

Choose a flow φ_Λ in Λ which is canonical with respect to K, and complete the proof by interpolating separately into the two components of $\Omega \smallsetminus \Lambda$.

The theorem will therefore be proved when we have constructed the arc Λ, and to obtain Λ we proceed as follows: Choose an arc $L \subseteq \Omega$ which joins x to a point of some component B_2 of $\partial(\Omega)$, where $B_1 \neq B_2$. Let

$$H = L \cup [B_1 \text{ side} : \Omega] \cup [B_2 \text{ side} : \Omega].$$

Then H is compact, connected and simply connected, and we can choose a conformal equivalence h from $\mathbf{S}^2 \setminus H$ onto $U = \text{ins}(\Gamma_1)$. Let γ be one of the at most two possible homeomorphic images of $\{\varphi_0(t, x) \mid t > 0\}$ under h, in the sense of Theorem C. 58, and let z be the endpoint of γ which corresponds to $\omega_{\varphi_0}(x)$. By adjusting h if necessary, we may clearly assume that $z = -1$. Choose $\alpha \in (0, 1)$ such that the Jordan curve

$$\{-1\} \cup \left\{\alpha(1 + e^{i\vartheta})^\alpha - 1 \mid -\pi < \vartheta < \pi\right\}$$

contains $h\left(\mathbf{S}^2 \setminus (H \cup \Omega)\right)$ in its inside. Let

$$\Lambda = h^{-1}(J \setminus \{-1\}).$$

Using Theorem C. 59, we see that Λ joins a subset of $\omega_{\varphi_0}(x)$ to a subset of $\omega_{\varphi_0}(x)$, and it is easy to see that Λ satisfies our other requirements.

QED

6.27. Theorem. *Let φ_0 be an orbit-analytic, weakly accessible flow in the boundary of a finitely connected region Ω. Let K be a compact subset of $\mathbf{S}^2$, suppose $\Omega \cap K$ is a non-empty T_σ-set of Ω, and suppose φ_0 is canonical with respect to K. Suppose there exists a finite, weakly $\mathscr{S}(\varphi_0)$-accessible setting $\{\Omega_j \mid j = 1, 2, \ldots, n\}$ for $K \cap \Omega$ in Ω.*

Then there exists an ultra-conservative interpolation φ of φ_0 into Ω, which is orbit-analytic and canonical with respect to K.

PROOF: For each component γ of the boundary of each of the sets Ω_j, define a flow φ_γ in γ as follows: If γ is a component of $\partial(\Omega)$, then φ_γ is the restriction of φ_0 to γ, and if γ is an analytic Jordan curve contained in Ω, then φ_γ is a flow in γ which is canonical with respect to K. We make the one further requirement that if Ω_j is an annulus and $\Omega_j \cap K = \square$, and γ is that component of $\partial(\Omega_j)$ that lies in Ω, and the other component of $\partial(\Omega_j)$ (which must of course be a component of $\partial(\Omega)$) consists of a periodic φ_0-orbit, then the orientation of φ_γ is chosen in such a way as to make the flow in $\partial(\Omega_j)$ unidirectional.

For each $j = 1, 2, \ldots, n$, we can interpolate the flow in $\partial(\Omega_j)$ into Ω_j by applying either Theorem 6.19 or Theorem 6.26 to the components of $\Omega_j \setminus K$.

Now given a component A of $\Omega \setminus \bigcup_{j=1}^{n} \overline{\Omega_j}$, it is clear that A is a finitely connected region, and that each component of $\partial(A)$ is either a component of $\partial(\Omega_j)$ for some j, or is a component of $\partial(\Omega)$ which is not sealed by $\{\Omega_j \mid j = 1, \ldots, n\}$. Furthermore, since the setting $\{\Omega_j \mid j = 1, \ldots, n\}$ is weakly $\mathscr{S}(\varphi_0)$-accessible, at least one component of $\partial(A)$ must be of the latter type, and since this component meets $\mathscr{S}(\varphi_0)$ and consequently contains both fixed and moving points of φ_0, we may apply Theorem 6.26 to interpolate the flow in $\partial(A)$ into A.

We complete the proof of the theorem by taking the conjunction of these interpolations.   **QED**

6.28. Remark. In the notation of the above theorem, it is easy to see that if φ_0 is unidirectional on each component of $\partial(\Omega)$ which is sealed by $\{\Omega_j \mid j = 1, \ldots, n\}$ then the interpolation φ can be chosen in such a way that

$$\bigcup_{j=1}^{n} \Omega_j \subseteq \mathscr{F}_r(\varphi) \cup \mathscr{G}_r(\varphi).$$

6.29. Theorem. *Let φ_0 be an orbit-analytic, weakly accessible flow in the boundary of a disc Ω, suppose that $\mathscr{G}(\varphi_0)$ consists of a single φ_0-orbit, and that $\mathscr{F}(\varphi_0)$ is not empty. Let K be a compact subset of $\mathbf{S}^2$, suppose that $K \cap \Omega$ is a weakly $\mathscr{S}(\varphi_0)$-accessible T_σ-set of Ω, and suppose that φ_0 is canonical with respect to K.*

Then there exists an ultra-conservative interpolation φ of φ_0 into Ω, which is orbit-analytic and canonical with respect to K.

PROOF: Since the theorem is known whenever $K \cap \Omega = \square$ or some weakly $\mathscr{S}(\varphi_0)$-accessible setting for $K \cap \Omega$ in Ω is finite, we shall assume that $K \cap \Omega \neq \square$, and that $\{\Omega_j \mid j = 1, 2, \ldots\}$ is an infinite weakly $\mathscr{S}(\varphi_0)$-accessible setting for $K \cap \Omega$ in Ω.

Theorem 3.58 implies that each Ω_j is a disc. By trimming each set Ω_j around $\Omega_j \cap K$, if necessary, (see Definition 3.35) we can assume that for each j, and $x \in \partial(\Omega_j)$, we have $\varrho(x, \Omega_j \cap K) < \dfrac{1}{j}$.

This latter assumption assures that no compact subset of Ω can meet infinitely many of the sets $\overline{\Omega_j}$ $(j = 1, 2, \ldots)$, for if H is a compact subset of Ω and H meets infinitely many of the sets $\overline{\Omega_j}$, we can deduce a contradiction as follows: Suppose, for convenience of notation, that H meets each set $\overline{\Omega_j}$ in a point x_j, and that the sequence $\{x_j\}$ converges to a point $x \in H$. For each j, the straight line segment $[x_j, x_{j+1}]$ meets $\partial(\overline{\Omega_j})$ in a point y_j (say) and it is clear that $y_j \to x$. Therefore, since $\varrho(y_j, K) < \dfrac{1}{j}$ for all j, we must have $x \in K$, and there must exist j_0 such that $x \in \Omega_{j_0}$. But this implies that $x_j \in \Omega_{j_0}$ for all sufficiently large j, giving us a contradiction.

It is now easy to see that $\bigcup_j \overline{\Omega_j}$ is relatively closed in Ω, and we deduce from Corollary C.12 that $\Omega \setminus \bigcup_j \overline{\Omega_j}$ is a region.

Let $\varLambda_0$ be the sole moving orbit of φ_0.

CASE 1: Λ_0 is an arc of type 1 in $\partial(\Omega)$:

We need to construct a family $\{\Lambda_n \mid n = 1, 2, \ldots\}$ of analytic arcs, which has the following properties:

(i)   For each $n = 1, 2, \ldots$, $\Lambda_n \subseteq \Omega$, and joins a subset of $\alpha_{\varphi_0}(\Lambda_0)$ to a subset of $\omega_{\varphi_0}(\Lambda_0)$.

(ii) For each $n = 1, 2, \ldots$, $\Omega \smallsetminus \Lambda_n$ is the union of two mutually disjoint discs P_n^- and P_n^+ with

$$\bigcup_{m=1}^{n-1} \Lambda_m \subseteq P_n^-, \quad \bigcup_{m=n+1}^{\infty} \Lambda_m \subseteq P_n^+, \text{ and } \Lambda_0 \subseteq \overline{P_n^-}.$$

(iii) For each $n = 1, 2, \ldots$, Λ_n does not meet $\bigcup_{j=1}^{\infty} \overline{\Omega_j}$.

(iv) For each $n = 1, 2, \ldots$, P_n^- contains only finitely many of the sets $\overline{\Omega_j}$.

and

(v) $\Omega = \bigcup_{n=1}^{\infty} P_n^-$.

When this family of arcs has been constructed, we can complete the proof of Case 1 by first defining a flow φ_n in each Λ_n which is canonical with respect to K, then using either Theorem 6.26, or Theorem 6.27 to interpolate into P_1^- and into each of the discs $P_{n+1}^- \cap P_n^+$, and finally taking the conjunction of φ_0 and all these interpolations.

We shall transfer the problem of finding this family of arcs to the open unit disc $U = \text{ins}(\Gamma_1)$ as follows:

Choose a conformal equivalence h from Ω onto U. Then the image of Λ_0 under h in the sense of Theorem C.58, is an arc contained in Γ_1, and we shall denote this arc by L_0. L_0 joins a point α of Γ_1 to a point ω of Γ_1. For each $j = 1, 2, \ldots$, let $\Delta_j = h(\Omega_j)$; then it is clear that no compact subset of U can meet infinitely many of the sets $\overline{\Delta_j}$ $(j = 1, 2, \ldots)$. We shall obtain the required arcs $\Lambda_1, \Lambda_2, \ldots$, by constructing a family $\{L_n \mid n = 1, 2, \ldots\}$ of analytic arcs, which has the following properties:

(i)* For each $n = 1, 2, \ldots$, $L_n \subseteq U$, and joins α to ω.

(ii)* For each $n = 1, 2, \ldots$, $U \smallsetminus L_n$ is the union of two mutually disjoint discs Q_n^- and Q_n^+ with

$$\bigcup_{m=1}^{n-1} L_m \subseteq Q_n^-, \quad \bigcup_{m=n+1}^{\infty} L_m \subseteq Q_n^+, \quad \text{and} \quad L_0 \subseteq \overline{Q_n^-}.$$

(iii)* For each $n = 1, 2, \ldots$, L_n does not meet $\bigcup_{j=1}^{\infty} \overline{\Delta_j}$.

(iv)* For each $n = 1, 2, \ldots$, Q_n^- contains only finitely many of the sets $\overline{\Delta_j}$.

(v)* $U = \bigcup\limits_{n=1}^{\infty} Q_n^-$,

and

(vi)* For each $n = 1, 2, \ldots$, all points of L_n which are sufficiently close to either α or ω, must lie in the segment W of U, which is bounded by L_0, and the straight line $[\alpha, \omega]$.

When the family $\{L_n \mid n = 1, 2, \ldots\}$ has been defined, the family $\{\Delta_n \mid n = 1, 2, \ldots\}$ can be obtained at once by defining $\Delta_n = h^{-1}(L_n)$ for every $n = 1, 2, \ldots$. Theorem C.59 together with condition (vi)* ensure that condition (i) is satisfied.

Now to construct the family $\{L_n \mid n = 1, 2, \ldots\}$, we proceed as follows: It is easy to see that a compact subset of L_0 is a positive distance from $\bigcup\limits_{j=1}^{\infty} \overline{\Delta_j}$, and it is therefore possible to construct an arc $C_\infty \subseteq W$ which joins α to ω, and lies so close to L_0 that neither C_∞, nor the disc D bounded by the Jordan curve $L_0 \cup C_\infty \cup \{\alpha, \omega\}$ can meet $\bigcup\limits_{j=1}^{\infty} \overline{\Delta_j}$.[3] Let $L_\infty = \partial(U) \setminus L_0$. (See Fig. 6.2.)

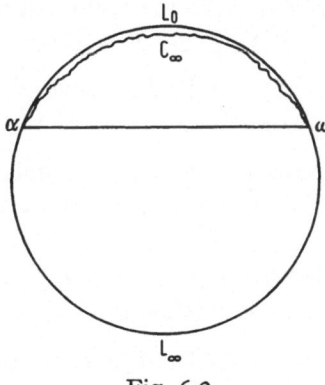

Fig. 6.2

Now since $\partial(D)$ is a Jordan curve, it is easy to find a pairwise disjoint family $\{C_n \mid n = 1, 2, \ldots\}$ of arcs joining α to ω, such that, for each $n \geq 2$, C_n separates C_{n-1} from C_{n+1}, and C_1 separates L_0 from C_2.

---

[3] The construction of such an arc C_∞ is elementary but tedious. It would have been an easy matter to define C_∞ if we had required C_∞ to join one point of L_0 to another and the arc that we want can be made by suitable small corrections to a family of arcs of this form.

We now construct a family $\{ \check{L}_n \mid n = 1, 2, \ldots \}$ of arcs joining α to ω, as follows:

Using the fact that $\partial(U)$ is one component of the boundary of the region $U \smallsetminus \cup_j \overline{\varDelta}_j$, we may apply Theorem C.30 to obtain a Jordan curve $J_1 \subseteq U \smallsetminus \cup_j \overline{\varDelta}_j$ which lies so close to $\partial(U)$ that some subarc of J_1 which lies close to L_∞ does not meet C_1 and joins a point of C_1 near α to a point of C_1 near ω. Let C_1^α and C_1^ω respectively be the two short subarcs of C_1 which join those two points to α and ω, and let $\check{L}_1$ be the union of C_1^α, C_1^ω, these two points, and the subarc of J_1 mentioned above. (See Figs. 6.3 (a) and (b).)

$\check{L}_1$ is an arc which joins α to ω, and the disc bounded by the Jordan curve $L_0 \cup \check{L}_1 \cup \{\alpha, \omega\}$ contains only finitely many of the sets $\overline{\varDelta}_j$.

Now in the disc bounded by the Jordan curve $\check{L}_1 \cup L_\infty \cup \{\alpha, \omega\}$, choose a Jordan curve J_2 which does not meet any of the sets $\overline{\varDelta}_j$, and which is close to $\check{L}_1 \cup L_\infty$. Let $\check{L}_2$ be an arc which is the union of a subarc of J_2 which is close to $\check{L}_1$ and joins a point of C_2 near α to a point of C_2 near ω, and two small subarcs C_2^α, C_2^ω of C_2 which are respectively close to α and ω. By starting with J_2 sufficiently close to $\check{L}_1 \cup L_\infty$, we can ensure that $\check{L}_2$ is an arc which joins α to ω and is disjoint from $\check{L}_1$; and since the disc bounded by $\check{L}_1 \cup \check{L}_2 \cup \{\alpha, \omega\}$ contains only finitely many of the sets $\overline{\varDelta}_j$, we can (and do) make the further stipulation that this disc does not meet any of the sets $\overline{\varDelta}_j$. (See Figs. 6.4 (a) and (b).)

Now in the disc bounded by the Jordan curve $\check{L}_2 \cup L_\infty \cup \{\alpha, \omega\}$, apply a similar process to construct an arc $\check{L}_3$ in $U \smallsetminus \cup_j \overline{\varDelta}_j$ which begins and ends along C_3, joins α to ω, lies close to L_∞, and has the property that the disc bounded by $\check{L}_2 \cup \check{L}_3 \cup \{\alpha, \omega\}$ contains only finitely many of the sets $\overline{\varDelta}_j$.

Now in the disc bounded by $\check{L}_3 \cup L_\infty \cup \{\alpha, \omega\}$, construct an arc $\check{L}_4$ in $U \smallsetminus \cup_j \overline{\varDelta}_j$ which begins and ends along C_4, joins α to ω, and lies so close to $\check{L}_3$ that the disc bounded by $\check{L}_3 \cup \check{L}_4 \cup \{\alpha, \omega\}$ meets none of the sets $\overline{\varDelta}_j$.

We continue constructing the sets $\check{L}_1, \check{L}_2, \ldots$, in pairs in such a way that each $\check{L}_n$ begins and ends along C_n, for each n, the disc bounded by $\check{L}_{2n-1} \cup \check{L}_{2n} \cup \{\alpha, \omega\}$ contains none of the sets $\overline{\varDelta}_j$, and for each n, the disc bounded by $\check{L}_n \cup L_0 \cup \{\alpha, \omega\}$ contains only finitely many of the sets $\overline{\varDelta}_j$; and we do this in such a way that the discs bounded by $\check{L}_n \cup L_0 \cup \{\alpha, \omega\}$ $(n = 1, 2, \ldots)$ exhaust U.[4] (See Fig. 6.5.)

---

[4] It is our intention that U should be exhausted before the reader is.

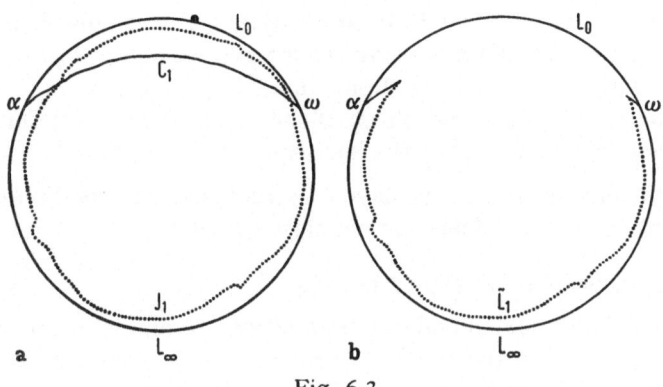

Fig. 6.3

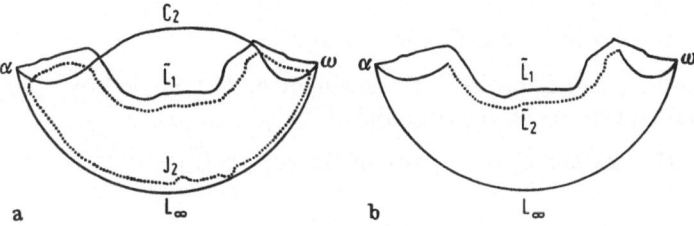

Fig. 6.4

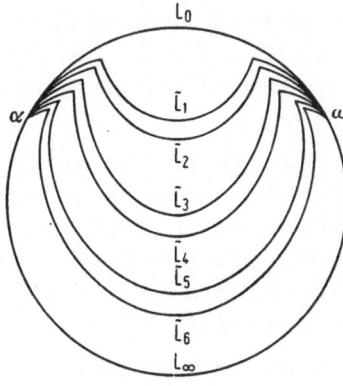

Fig. 6.5

For each $n = 1, 2, \ldots$, let L_n be an analytic arc in the disc bounded by $\hat{L}_{2n-1} \cup \hat{L}_{2n} \cup \{\alpha, \omega\}$. (If n is a positive integer, we obtain L_n by choosing a conformal equivalence g_n from $\operatorname{ins}(\hat{L}_{2n-1} \cup \hat{L}'_{2n} \cup \{\alpha, \omega\})$ onto U, extending g_n to a homeomorphism of the closures, and then defining L_n to be the image under g_n^{-1} of the line segment $(g_n(\alpha), g_n(\omega))$.

It is clear that the arcs L_n we have constructed satisfy conditions (i)* to (vi)*, and the proof of Case 1 is therefore complete.

CASE 2: Λ_0 is an arc of type 2 in $\partial(\Omega)$:

Let h be a conformal equivalence from Ω onto U, and let γ and γ' be the two images of Λ_0, in the sense of Theorem C. 58. γ joins a point α of Γ_1 to a point ω of Γ_1, and γ' joins a point α' of Γ_1 to a point ω' of Γ_1. From Theorem C. 58(b), we see that of the four points $\alpha, \alpha', \omega, \omega'$, of Γ_1, α and α' are adjacent, and ω and ω' are adjacent.

Now using techniques similar to those used in the proof of Case 1, construct two mutually disjoint analytic arcs L_1 and L_2 in U, such that

(i) L_1 joins α to α' and L_2 joins ω to ω',

(ii) both L_1 and L_2 begin in the segment of U bounded by $\gamma \cup [\alpha, \omega]$, and end in the segment bounded by $\gamma' \cup [\alpha', \omega']$,

(iii) neither L_1 nor L_2 meets any of the sets $h(\overline{\Omega_j})$, and

(iv) the disc bounded by $L_1 \cup L_2 \cup \gamma \cup \gamma'$ contains only finitely many of the sets $h(\overline{\Omega_j})$.

Let $\Lambda_1 = h^{-1}(L_1)$ and $\Lambda_2 = h^{-1}(L_2)$, and on Λ_1 and Λ_2, define flows φ_1 and φ_2 which are canonical with respect to K. (This is possible by Theorem 5.13.)

It is easy to see that $\Omega \setminus (\Lambda_1 \cup \Lambda_2)$ is the disjoint union of three discs, and that we can interpolate into each of these discs by using either Theorem 6.26, Theorem 6.27, or Case 1 of this theorem. The conjunction of these three interpolations satisfies all the requirements of the theorem.

QED

6.30. Theorem. *Let φ_0 be an orbit-analytic, weakly accessible flow in the boundary of a disc Ω, suppose that φ_0 has at least one, but only finitely many, moving orbits, and suppose that $\mathscr{F}(\varphi_0) \neq \square$. Let K be a non-empty, compact subset of $\mathbf{S}^2$, and suppose φ_0 is canonical with respect to K. Let Y be a non-empty subset of $\mathscr{F}(\varphi_0)$ which contains $\mathscr{S}(\varphi_0)$, and suppose $K \cap \Omega$ is a weakly Y-accessible T_σ-set of Ω.*

Then there exists a conservative interpolation φ of φ_0 into Ω, which is orbit-analytic and canonical with respect to K.

PROOF: Choose a weakly Y-accessible setting $\{\Omega_j \mid j = 1, 2, \ldots\}$ for $K \cap \Omega$ in Ω, and trim each disc Ω_j around $K \cap \Omega_j$ in such a way that (with the new setting again called $\{\Omega_j \mid j = 1, 2, \ldots\}$), no compact subset of Ω can meet infinitely many of the sets $\overline{\Omega_j}$.

Let h be a conformal equivalence from Ω onto the open unit disc $U = \mathrm{ins}\,(\Gamma_1)$, and let $\{\gamma_1, \ldots, \gamma_n\}$ be the set of images under h of the moving φ_0-orbits, in the sense of Theorem C.58. We may assume that $\gamma_1, \ldots, \gamma_n$ are listed in the order in which they appear, as we move around Γ_1 counter-clockwise. For each $m = 1, \ldots, n$, write $\gamma_m = \{e^{i\vartheta} \mid \alpha_m < \vartheta < \omega_m\}$, where to avoid difficulties in notation, we assume that $0 = \alpha_1 < \omega_1 \leq \alpha_2 < \ldots < \omega_n \leq 2\pi$. For convenience, define $\alpha_{n+1} = 2\pi$, $\omega_{n+1} = \omega_1 + 2\pi$, and $\gamma_{n+1} = \gamma_1$. Now, for each $m = 1, \ldots n$, construct an analytic arc $L_m \subseteq U \setminus \bigcup_j h(\overline{\Omega_j})$ which joins $e^{i\omega_m}$ to $e^{i\alpha_{m+1}}$, and construct these arcs in such a way that:

(i)   the family $\{L_1, \ldots, L_n\}$ is pairwise disjoint,

(ii) for each m, L_m begins in the segment of U bounded by $\gamma_m \cup [e^{i\alpha_m}, e^{i\omega_m}]$, and ends in the segment bounded by $\gamma_{m+1} \cup [e^{i\alpha_{m+1}}, e^{i\omega_{m+1}}]$,

and

(iii) the disc bounded by $\bigcup_{m=1}^{n} L_m \cup \bigcup_{m=1}^{n} \gamma_m$ contains only finitely many of the sets $h(\overline{\Omega_j})$.

Let $\Lambda_1, \ldots, \Lambda_n$ be the images under h^{-1} of $L_1, \ldots, L_n$ respectively, and for each $m = 1, \ldots, n$, define a flow φ_m in Λ_m which is canonical with respect to K. Now it is clear that $\{\Omega_j \mid j = 1, 2, \ldots\}$ is a weakly $\mathcal{S}(\varphi_0)$-accessible setting for $K \cap \Omega$ in Ω, and we can therefore interpolate separately into the $n + 1$ subdiscs that comprise $\Omega \setminus \bigcup_{m=1}^{n} \Lambda_m$, using Theorem 6.26, Theorem 6.27 or Theorem 6.29. The conjunction of these interpolations satisfies the requirements of the theorem. **QED**

6.31. Note. The interpolation obtained in the proof of Theorem 6.30 is not in general ultra-conservative.

6.32. Theorem. *Let φ_0 be an orbit-analytic, weakly accessible flow in the boundary of a finitely connected region Ω, suppose that φ_0 has only finitely many moving orbits, and suppose that $\mathcal{F}(\varphi_0) \neq \square$. Let K be a non-empty, compact subset of $\mathbf{S}^2$, and suppose φ_0 is canonical with respect to K. Let Y be a subset of $\mathcal{F}(\varphi_0)$ which contains $\mathcal{S}(\varphi_0)$, and suppose that $K \cap \Omega$ is a weakly Y-accessible T_σ-set of Ω.*

Then there exists an interpolation φ of φ_0 into Ω which has the following properties:

(a) *φ is orbit-analytic, and canonical with respect to K.*

(b) $\mathscr{S}(\varphi) \subseteq Y$.

and

(c) *If $x \in \Omega$ and $\mathrm{p}_\varphi(x) = \infty$, then all the φ-endpoints of x lie in the same component of $\partial(\Omega)$.*

Furthermore, given any weakly Y-accessible setting $\{\Omega_j \mid j = 1, 2, \ldots\}$ for $K \cap \Omega$ in Ω, if we define for each component B of $\partial(\Omega)$, a subset Y_B of $B \cap Y$ by

(i)   *if B is sealed by $\{\Omega_j \mid j = 1, 2, \ldots\}$, then $Y_B = \square$,*

(ii) *if B is not sealed by $\{\Omega_j \mid j = 1, 2, \ldots\}$, and $B \subseteq \mathscr{F}(\varphi_0)$, then Y_B is a compact, connected subset of B which can be joined to a point of $\Omega \setminus \bigcup_j \overline{\Omega_j}$ by an arc in $\Omega \setminus \bigcup_j \overline{\Omega_j}$,*

(iii) *if B is not sealed by $\{\Omega_j \mid j = 1, 2, \ldots\}$ and φ_0 has moving orbits in B, then $Y_B = B \cap \mathscr{S}(\varphi_0)$,*

then the interpolation φ can be so constructed that, for every $x \in \overline{\Omega}$ with $\mathrm{p}_\varphi(x) = \infty$, there exists a component B of $\partial(\Omega)$ such that all φ-endpoints of x lie in Y_B.

PROOF: Let $\{\Omega_j \mid j = 1, 2, \ldots\}$ be a weakly Y-accessible setting for $K \cap \Omega$ in Ω. By trimming each set Ω_j around $\Omega_j \cap K$, if necessary, we may suppose that for each j we have $\varrho\bigl(x, K \cup \partial(\Omega)\bigr) < \dfrac{1}{j}$ for every $x \in \partial(\Omega_j)$. This assumption ensures that no compact subset of Ω can meet infinitely many of the sets $\overline{\Omega_j}$.

For each component B of $\partial(\Omega)$, choose a subset Y_B of $B \cap Y$ by the procedure outlined in the statement of the theorem. Let $\mathscr{B}$ be the family of components of $\partial(\Omega)$ which are not sealed by $\{\Omega_j \mid j = 1, 2, \ldots\}$. We shall construct a family $\{\Lambda_B \mid B \in \mathscr{B}\}$ of analytic arcs, which has the following properties (see Fig. 6.6):

(1) For each $B \in \mathscr{B}$, Λ_B joins a subset of Y_B to a subset of Y_B.

(2) For each $B \in \mathscr{B}$, $\Lambda_B \subseteq \Omega \setminus \bigcup_j \overline{\Omega_j}$.

(3) For each $B \in \mathscr{B}$, $\Omega \setminus \Lambda_B$ is the disjoint union of a disc D_B whose boundary is $B \cup \Lambda_B$, and a finitely connected region Δ_B whose boundary is $\overline{\Lambda_B} \cup \bigl(\partial(\Omega) \setminus B\bigr)$.

(4) For each $B \in \mathscr{B}$ and each j, if $\Omega_j \subseteq D_B$ then Ω_j is a disc.

$$\square \ \underset{j}{\bigcup} \Omega_j \qquad \blacksquare \ S^2 \setminus \Omega$$

Fig. 6.6

(5) For each $B \in \mathscr{B}$, there exists $\delta_B > 0$ such that for each j, if $\Omega_j \subseteq \Lambda_B$ then $\varrho(\Omega_j, B) \geq \delta_B$.

(6) The family $\{\Lambda_B \mid B \in \mathscr{B}\}$ is pairwise disjoint.

To see how these arcs may be constructed, suppose $B \in \mathscr{B}$. Choose an arc $L_B \subseteq \Omega \setminus \bigcup_j \overline{\Omega_j}$ which joins a point of Ω to a subset of Y_B. (If $B \subseteq \mathscr{F}(\varphi_0)$, the existence of L_B is automatic, and otherwise, L_B may be found by techniques similar to those used to find C_∞ in the proof of Theorem 6.29.) Choose a continuous function f from $\mathbf{S}^2$ onto itself whose restriction to Ω is a homeomorphism, and which carries L_B onto the interval $(0, 1)$ with 1 corresponding to the set of endpoints of L_B in Y_B. Since every compact subinterval of L_B is a positive distance from $B \cup \bigcup_j \overline{\Omega_j}$, we can construct an arc $A_B \subseteq f(\Omega \setminus \bigcup_j \overline{\Omega_j})$ which joins 1 to 1, such that the Jordan curve $A_B \cup \{1\}$ contains $(0, 1)$ in its inside, and lies so close to $(0, 1)$ that,

firstly,

for each $x \in \mathrm{ins}\,(A_B \cup \{1\})$, there exists $y \in (0, 1)$ such that

$$\varrho\big(f^{-1}(x), f^{-1}(y)\big) < \frac{1}{2}\,\varrho\big(f^{-1}(y), B\big),$$

and secondly,

for each $y \in (0, 1)$, there exists $z \in A_B$ such that

$$\varrho\big(f^{-1}(y), f^{-1}(z)\big) < \frac{1}{2}\,\varrho\big(f^{-1}(y), B\big).$$

Let J be a Jordan curve in $f(\Omega \setminus \bigcup_j \overline{\Omega_j})$ which lies close to $f(B)$, and let l be an arc joining 1 to 1, which is everywhere close to $f(B)$, begins and ends along A_B, and is otherwise contained in J, using all of J except a small arc near 1.

Let $\Lambda'_B = f^{-1}(l)$. Λ'_B is not analytic, but apart from this, it has all the properties we require. Λ_B can be found by choosing an analytic arc which parallels Λ'_B sufficiently closely. (See the proof of Theorem 6.29.)

Having constructed the arc Λ_B for each $B \in \mathscr{B}$, choose a flow φ_B in each arc Λ_B, which is canonical with respect to K. Into each disc D_B we can interpolate using Theorem 6.30. The rest of Ω consists of $\bigcap_{B \in \mathscr{B}} \Lambda_B$, and we can interpolate into this set using either Theorem 6.26 or Theorem 6.27.

The conjunction of these interpolations satisfies the requirements of the theorem.   **QED**

6.33. Corollary. *Let φ_0 be an orbit-analytic, accessible flow in the boundary of a finitely-connected region Ω, suppose that φ_0 has only finitely many moving orbits, and that $\mathscr{F}(\varphi_0) \neq \square$. Let K be a non-empty, compact subset of $\mathbf{S}^2$ and suppose φ_0 is canonical with respect to K. Let Y be a subset of $\mathscr{F}(\varphi_0)$ which contains $\mathscr{S}(\varphi_0)$, and suppose $K \cap \Omega$ is a Y-accessible T_σ-set of Ω.*

Then there exists an interpolation φ of φ_0 into Ω which satisfies the conditions (a), (b), *and* (c) *of Theorem 6.32, and in addition, satisfies* (d) *φ is simple.*

PROOF: Since the hypotheses of Theorem 6.32 are satisfied, we know that an interpolation φ exists satisfying (a), (b) and (c) of that theorem. Condition (d) may be seen to be satisfied by choosing a Y-accessible setting $\{\Omega_j \,|\, j = 1, 2, \ldots\}$ for $K \cap \Omega$ in Ω, and then noting that each of the sets Y_B (defined as in Theorem 6.32) is finite.   **QED**

6.34. Corollary. *Let* φ_0, Ω, K, *and* Y *be as in Corollary* 6.33, *and suppose in addition that each component of* $\partial(\Omega)$ *contains at most one point of* $\mathscr{S}(\varphi_0)$.

Then there exists an interpolation φ *of* φ_0 *into* Ω *which in addition to satisfying the conditions of Theorem* 6.32 *and Corollary* 6.33, *satisfies*

(e) *Each component of* $\partial(\Omega)$ *contains at most one point of* $\mathscr{S}(\varphi)$,

and

(f) *Whenever* $x \in \bar{\Omega}$ *and* $\mathrm{p}_\varphi(x) = \infty$, *we have* $\alpha_\varphi(x) = \omega_\varphi(x)$.

PROOF: Clear.   **QED**

6.35. Note. Using the above interpolation theorems, we can now give a partial answer to the question raised in Section 3.55. This we do in the following theorem, which is the principal theorem of this chapter.

6.36. Theorem. *Let* F *be a compact subset of* $\mathbf{S}^2$, *let* Y *be a subset of* F, *and suppose that each component of* $\mathbf{S}^2 \setminus \mathscr{C}(F, \bar{Y})$ *is finitely connected.*

Then the following conditions are equivalent:

(a) *There exists a flow* φ *in* $\mathbf{S}^2$ *such that* $\mathscr{F}(\varphi) = F$ *and* $\mathscr{S}(\varphi) \subseteq Y$.

(b) *There exists a flow* φ *in* $\mathbf{S}^2$ *which is orbit-analytic and canonical with respect to* F, *and has the property that whenever* $x \in \mathbf{S}^2$ *and* $\mathrm{p}_\varphi(x) = \infty$, $\alpha_\varphi(x) \cup \omega_\varphi(x)$ *is contained in a compact connected subset of* Y.

(c) Y *is a WATI subset of* F.

PROOF: The implication (b) $\Rightarrow$ (a) is trivial, and the implication (a) $\Rightarrow$ (c) follows from Theorem 3.54.

To prove that (c) $\Rightarrow$ (b), suppose that Y is a WATI subset of F. Let φ_0 be the flow in $\mathscr{C}(F, \bar{Y})$ which leaves every point fixed, and for each component Ω of $\mathbf{S}^2 \setminus \mathscr{C}(F, \bar{Y})$, choose an interpolation of φ_0 into Ω in accordance with the requirements of Theorem 6.32. The conjunction of φ_0 and these interpolations is a flow with all the required properties.
          QED

6.37. Theorem. *Let* F *be a compact subset of* $\mathbf{S}^2$, *let* Y *be a subset of* F, *and suppose that each component of* $\mathbf{S}^2 \setminus \mathscr{C}(F, \bar{Y})$ *is finitely connected.*

Then the following conditions are equivalent:

(a) *There exists a semi-simple flow φ in $\mathbf{S}^2$ such that $\mathscr{F}(\varphi) = F$ and $\mathscr{S}(\varphi) \subseteq Y$.*

(b) *There exists a simple flow φ in $\mathbf{S}^2$ which is orbit-analytic and canonical with respect to F, satisfies $\mathscr{S}(\varphi) \subseteq Y$, and has the property that whenever $x \in \mathbf{S}^2$ and $\mathrm{p}_\varphi(x) = \infty$, we have $\alpha_\varphi(x) = \omega_\varphi(x)$.*

(c) *Y is an ATI subset of F.*

PROOF: The proof of this theorem is similar to that of Theorem 6.36, and will be omitted.   **QED**

6.38. Note. If F is a compact subset of $\mathbf{S}^2$, Y is a subset of F, and $\mathscr{C}(F, \overline{Y})$ has finitely many components, then each component of $\mathbf{S}^2 \setminus \mathscr{C}(F, \overline{Y})$ is finitely connected, and Theorems 6.36 and 6.37 hold. In view of this, we have the following theorem.

6.39. Theorem. *Let F be a compact subset of $\mathbf{S}^2$ and let Y be a finite subset of F.*

Then the following conditions are equivalent:

(a) *There exists a flow φ in $\mathbf{S}^2$ such that $\mathscr{F}(\varphi) = F$ and $\mathscr{S}(\varphi) \subseteq Y$.*

(b) *There exists a simple flow φ in $\mathbf{S}^2$ which is orbit-analytic and canonical with respect to F, satisfies $\mathscr{S}(\varphi) \subseteq Y$, and has the property that whenever $x \in \mathbf{S}^2$ and $\mathrm{p}_\varphi(x) = \infty$, we have $\alpha_\varphi(x) = \omega_\varphi(x)$.*

(c) *Y is a WATI subset of F.*

(d) *Y is an ATI subset of F.*

(e) *In every component Ω of $\mathbf{S}^2 \setminus \mathscr{C}(F, \overline{Y})$, $F \cap \Omega$ is a Y-accessible T_σ-set of Ω.*

Notes and Remarks to Chapter 6

Chapter 6 parallels the theory which is developed in Beck [3] in order to prove that if Y is a finite ATI subset of a compact set $F \subseteq \mathbf{S}^2$, then there exists a flow φ in $\mathbf{S}^2$ such that $\mathscr{F}(\varphi) = F$ and $\mathscr{S}(\varphi) \subseteq Y$. (See Theorem 6.39.) The main difference between the theory presented in Beck [3] and the theory given in the chapter is that in the chapter we do not assume that Y is finite, but we assume instead that each component of $\mathbf{S}^2 \setminus \mathscr{C}(F, \overline{Y})$ is finitely connected. This change was made by the Lewins as part of an attempt to answer the question raised in Remark 3.55, but the basic approach to the proofs of the theorems is unaltered.

The technique of the chapter hinges on the interpolation theorems. We originally proved these interpolation theorems for bounded flows, but later rewrote them in a considerably simpler form, confining our attention to canonical flows. The more general theorems for bounded flows are not required for the main results of Chapters 6 and 7, and have therefore been omitted here. However, many of these theorems can be easily reconstructed using the techniques developed in Chapters 4 and 5, (which were not available at that time). The tools developed in Chapters 4 and 5 have also been used in Chapter 6 to make the proofs of the interpolation theorems precise, and in some cases, to simplify them.

Section 6.12. Part (a) of Definition 6.12, implies that no φ_0-orbit can have moving endpoints in a different φ_0-orbit. We do not know, however, whether an orbit of a plane flow can have endpoints in itself.[5] We therefore do not know whether or not condition (b) of Theorem 6.12 is redundant.

Section 6.13. The principal theorems of Chapters 6 and 7 differ, in a way not already discussed, from the theorems in Beck [3], this difference being the added information given as the final conditions in the parts (b) of Theorems 6.36. and 6.37. These extra conditions were added by the Lewins, and the notion of ultra-conservative interpolation was introduced for this purpose.

Sections 6.19. and 6.23. Theorems 6.19 and 6.23 are the two basic interpolation theorems, and are the building blocks for all the others. Theorem 6.19 appears much in the same form as it appeared in Beck [3] (except for the precision gained in the proof by applying results from Chapter 5). The proof of Theorem 6.23 is due to the Lewins, and replaces the proof in Beck [3] which applies only when φ_0 has only finitely many moving orbits. (While the proof in Beck [3] is fully adequate for all the results in Beck [3] and all the results in Chapter 6, we require the more general statement in Chapter 7.)

---

[5] Theorem 2.13 implies that if φ is a flow in an *open* subset of S^2, then no φ-orbit can have endpoints in itself. At the time of going to press, we have seen a proof for the non-open case by Lawrence D. Crowson. Thus, we see that condition (b) of Theorem 6.12 *is* redundant.

CHAPTER SEVEN

EXISTENCE THEOREMS II

7.1. Introduction. Our main purpose in this chapter is to show that Theorems 6.36 and 6.37 remain true even if we replace the assumption of finite connectivity of the components of $\mathbf{S}^2 \setminus \mathscr{C}(F, \overline{Y})$ by the weaker assumption that each of these components be countably connected. We shall do this by an inductive method, starting with finite connectivity of a component Ω of $\mathbf{S}^2 \setminus \mathscr{C}(F, \overline{Y})$, and then allowing the family of components of $\mathbf{S}^2 \setminus \Omega$ to become more and more complicated.

In order to carry out this inductive process, we need some topological machinery, which we shall develop before proving our main result.

7.2. Motivation. Suppose Y is a WATI subset of a compact subset F of $\mathbf{S}^2$, and let Ω be a component of $\mathbf{S}^2 \setminus \mathscr{C}(F, \overline{Y})$. If Ω is not finitely connected, then the next simplest state of affairs one might consider is the situation in which $\mathbf{S}^2 \setminus \Omega$ consists of a convergent sequence and its limit. Having considered a set Ω of this type, one might go a step further and consider the situation in which $\mathbf{S}^2 \setminus \Omega$ consists of a convergent $\{x_n\}$ and its limit x, together with countably many convergent sequences $\{y_{n,j}\}$ where for each $n = 1, 2, \ldots,$ $y_{n,j} \to x_n$ as $j \to \infty$.

Continuing transfinitely in this way, we can treat every region Ω whose complement is countable.

Disconnection Kernels

7.3. Definition. If X is a compact Hausdorff space, then for each ordinal α, we define the α-*disconnection kernel* $X^{(\alpha)}$ of X by the following recurrence formula:[1]

(i) $X^{(0)} = X$,

---

[1] The process of forming disconnection kernels of totally disconnected spaces is sometimes called differentiation.

(ii) For every ordinal α, $X^{(\alpha+1)}$ is obtained from $X^{(\alpha)}$ by removing those components of $X^{(\alpha)}$ which are relatively open subsets of $X^{(\alpha)}$;

and

(iii) For every limit ordinal α,

$$X^{(\alpha)} = \bigcap_{\beta < \alpha} X^{(\beta)}.$$

7.4. Theorem. *Let X be a compact Hausdorff space, and for each ordinal α, let $X^{(\alpha)}$ be the α-disconnection kernel of X.*

Then

(a) *For each ordinal α, $X^{(\alpha)}$ is compact, and is a union of components of X.*

(b) *Given any ordinals α and β with $\alpha < \beta$, we have $X^{(\beta)} \subseteq X^{(\alpha)}$.*

(c) *For any ordinal α, the following conditions are equivalent:*

(i) $X^{(\alpha)} = X^{(\alpha+1)}$,

(ii) $X^{(\alpha)} = X^{(\beta)}$ *for some* $\beta > \alpha$,

(iii) $X^{(\alpha)} = X^{(\beta)}$ *for every* $\beta > \alpha$,

and

(iv) *No component of $X^{(\alpha)}$ is relatively open in $X^{(\alpha)}$.*

PROOF: If the first part of (a) is false, let α be the least ordinal for which $X^{(\alpha)}$ is not compact. Then we cannot have $X^{(\alpha)} = \bigcap_{\beta < \alpha} X^{(\beta)}$ and therefore α is not a limit ordinal. Since X is compact, $\alpha \neq 0$, and therefore $\alpha = \beta + 1$ for some β. But $X^{(\beta)}$ is compact, and $X^{(\alpha)} = X^{(\beta)} \setminus U$ where U is a relatively open subset of $X^{(\beta)}$, contradicting the fact that $X^{(\alpha)}$ is not compact.

The proof of the second part of (a) is similar, and is omitted.

To prove (b), let α be any ordinal, and to obtain a contradiction, suppose that $X^{(\beta)} \not\subseteq X^{(\alpha)}$ for some ordinal $\beta > \alpha$. Let β be the least ordinal greater than α for which $X^{(\beta)} \not\subseteq X^{(\alpha)}$. Since $\beta > \alpha$, we have $\beta \neq 0$. We cannot have $X^{(\beta)} = \bigcap_{\gamma < \beta} X^{(\gamma)}$, and therefore β is not a limit ordinal. But if $\beta = \gamma + 1$, then since $\alpha \leq \gamma < \beta$, we have $X^{(\beta)} \subseteq X^{(\gamma)} \subseteq X^{(\alpha)}$; and therefore β is not a successor ordinal either, giving a contradiction. This proves (b).

To prove (c), we note first that the assertion (iii) $\Rightarrow$ (ii) is obvious, the assertion (ii) $\Rightarrow$ (i) follows easily from (b), and the assertion (i) $\Leftrightarrow$ (iv) is obvious. To show that (iv) $\Rightarrow$ (iii), suppose that (iv) holds, and to obtain a contradiction, suppose that $X^{(\beta)} \neq X^{(\alpha)}$ for some $\beta > \alpha$. Let β be the least ordinal greater than α for which $X^{(\beta)} \neq X^{(\alpha)}$. As before, we obtain a contradiction by showing that $\beta \neq 0$, β is not a limit ordinal, and β is not a successor ordinal.   **QED**

7.5. Theorem. *Let X be a compact Hausdorff space, and for each ordinal α, let $X^{(\alpha)}$ be the α-disconnection kernel of X.*

Then

(a) *There exists an ordinal α such that $X^{(\alpha)} = X^{(\alpha+1)}$.*

(b) *If α is the least ordinal for which $X^{(\alpha)} = X^{(\alpha+1)}$, and if $\beta < \gamma \leq \alpha$, then $X^{(\gamma)}$ is a proper subset of $X^{(\beta)}$.*

PROOF: In view of Theorem 7.4, it suffices to prove (a). Choose an ordinal α_0 such that $\{\alpha \mid \alpha < \alpha_0\}$ has larger cardinality than the set of all subsets of X. If $X^{(\alpha)} \neq X^{(\alpha+1)}$ for every $\alpha < \alpha_0$, then by Theorem 7.4, $\{X^{(\alpha)} \mid \alpha < \alpha_0\}$ is a set of subsets of X with the same cardinality as $\{\alpha \mid \alpha < \alpha_0\}$. But this is impossible, and therefore $X^{(\alpha)} = X^{(\alpha+1)}$ for some $\alpha < \alpha_0$.   **QED**

7.6. Definition. Let X be a compact Hausdorff space, and for each ordinal α, let $X^{(\alpha)}$ be the α-disconnection kernel of X. Let α_0 be the least ordinal for which $X^{(\alpha_0)} = X^{(\alpha_0+1)}$.

Then α_0 is called the *disconnection order of X*, and $X^{(\alpha_0)}$ is called the *disconnection kernel of X*. If X has no open component, then X is said to be *disconnection perfect*.

7.7. Some Examples.

(a) Let $X = \square$. Then the disconnection order of X is 0, and the disconnection kernel of X is $\square$.

(b) Let X be a compact Hausdorff space with finitely many components. Then the disconnection order of X is 1, and the disconnection kernel of X is empty.

(c) Let C be the Cantor set, and let

$$X = C \cup \left\{ 1 + \frac{1}{2^m} + \frac{1}{2^n} \,\middle|\, m, n = 1, 2, \ldots \right\}.$$

It is easy to see that

$$X^{(1)} = C \cup \left\{ 1 + \frac{1}{2^n} \,\middle|\, n = 1, 2, \ldots \right\}, \quad \text{and} \quad X^{(\alpha)} = C, \quad \forall \, \alpha \geq 2.$$

The disconnection order of X is 2, and the disconnection kernel of X is C.

(d) For every countable successor ordinal α, there is a countable compact subset of $\mathbb{R}$ whose disconnection order is α. (It will be shown in the next few theorems that the disconnection order of a non-empty countable compact Hausdorff space is always a countable successor ordinal, and that the disconnection kernel of such a space is always empty.)

We shall prove this by induction on α. Suppose α is a countable successor ordinal and that for every countable successor ordinal $\beta < \alpha$, there is a countable compact subset of $\mathbb{R}$ whose disconnection order is β. Denote the immediate predecessor of α by $\alpha - 1$.

CASE 1: $\alpha - 1 = 0$.

As we saw above, a finite set has disconnection order 1.

CASE 2: $\alpha - 1$ is a successor ordinal.

Denote the immediate predecessor of $\alpha - 1$ by $\alpha - 2$, and choose a strictly increasing sequence $\{x_n\}$ which converges to a real number x. For each $n = 1, 2, \ldots$, choose a countable compact subset C_n of (x_n, x_{n+1}) whose disconnection order is $\alpha - 1$. Let

$$X = \{x, x_1, x_2, x_3, \ldots\} \cup \bigcup_{n=1}^{\infty} C_n.$$

It is easy to see that for each $n = 1, 2, \ldots$, $X^{(\alpha-2)}$ contains at least one, but only finitely many, points between x_n and x_{n+1}. Therefore $X^{(\alpha-1)} = \{x\}$, and we see that X has disconnection order α.

CASE 3: $\alpha - 1$ is a limit ordinal.

Since α is countable, we may write $\{\beta < \alpha \mid \beta$ is a successor ordinal$\}$ as $\{\beta_n \mid n = 1, 2, \ldots\}$.

Choose a strictly increasing sequence $\{x_n\}$ which converges to a real number x, and for each $n = 1, 2, \ldots$, choose a countable compact subset C_n of (x_n, x_{n+1}) whose disconnection order is β_n. Let

$$X = \{x, x_1, x_2, x_3, \ldots\} \cup \bigcup_{n=1}^{\infty} C_n.$$

It is easy to see that for each $\beta < \alpha - 1$, we have $x \in X^{(\beta)}$, and it follows that $X^{(\alpha-1)} \neq \square$. But since $X^{(\beta_n)}$ does not meet C_n for any $n = 1, 2, \ldots$, we have $X^{(\alpha)} = \square$.

(e) By replacing $\{x\}$ by a Cantor set in the above example, it is possible to find compact subsets of $\mathbb{R}$ whose disconnection order is any given countable ordinal.

7.8. Theorem. *If a compact Hausdorff space X has an empty disconnection kernel, then its disconnection order is not a limit ordinal.*

PROOF: This follows at once from the fact that the intersection of a chain of non-empty compact sets is non-empty.   **QED**

7.9. Theorem. *Let X be a non-empty compact Hausdorff space, and suppose that the disconnection order of X is 0.*

Then X has at least $\aleph = 2^{\aleph_0}$ components.

PROOF: Since X is non-empty, and no component of X is open, X must have infinitely many components. Therefore, since X contains no continuum from any one component to any other, Theorem A.8 implies that X can be partitioned into two non-empty, compact open subsets A_0 and A_1. It is clear that both A_0 and A_1 are unions of infinitely many components of X, and that neither A_0 nor A_1 has any open components. Therefore each of the sets A_0, A_1 can be partitioned into the union of two non-empty, compact open subsets.

Continuing this process, we obtain $\aleph$ distinct, strictly decreasing sequences of compact open subsets of X. The intersection of each of these sequences is a non-empty, compact union of components of X, and any two distinct such sequences have disjoint intersections. From this we deduce at once that X has at least $\aleph$ components.   **QED**

7.10. Theorem. *Let X be a non-empty compact Hausdorff space with only countably many components.*

Then the disconnection order of X is a countable successor ordinal, and the disconnection kernel of X is empty. Furthermore, if $\alpha + 1$ is the disconnection order of X, then the α-disconnection kernel $X^{(\alpha)}$ of X has only finitely many components.

PROOF: Let Y be the disconnection kernel of X. If $Y \neq \square$, then since the disconnection order of Y is 0, Theorem 7.9 contradicts the assumption that X has only countably many components. Therefore the disconnection kernel of X is empty, and we deduce from Theorem 7.8 that the disconnection order of X is a successor ordinal. Denote the disconnection order of X by $\alpha + 1$, and let $X^{(\beta)}$ denote the β-disconnection kernel of X, for each ordinal β.

Since $\{X^{(\beta)} \setminus X^{(\beta+1)} \mid \beta < \alpha + 1\}$ is a pairwise disjoint family of non-empty unions of components of X, and X has only countably many components, it is clear that $\alpha + 1$ is a countable ordinal.

Finally, $X^{(\alpha)}$ is a compact Hausdorff space, all of whose components are open. The family of components of $X^{(\alpha)}$, being an open cover of

$X^{(\alpha)}$, must have a finite subcover, and we deduce that $X^{(\alpha)}$ has only finitely many components.   **QED**

7.11. Lemma. *Let X be a compact metric space.*

Then there exists a countable family $\{W_1, W_2, \ldots\}$ of open compact subsets of X such that, given any component C of X, and any neighborhood U of C, we have $C \subseteq W_n \subseteq U$ for some n.

PROOF: Let $\{B_1, B_2, \ldots\}$ be a countable base for the topology on X. Since the family of all finite unions of these sets is also a countable base for the topology on X, there is no loss of generality in assuming that $\{B_1, B_2, \ldots\}$ is closed under the formation of finite unions.

For each pair $\langle m, n \rangle$ of positive integers, choose, if possible, an open compact set U_{mn} such that $B_m \subseteq U_{mn} \subseteq B_n$. The family of these sets U_{mn} is countable, and we can therefore rename it $\{W_1, W_2, \ldots\}$.

Now suppose C is a component of X and U is a neighborhood of C. For each $x \in C$, choose a member B_x of the base such that $x \in B_x \subseteq U$. Since C is compact, finitely many of the sets B_x cover C, and their union B is a member of the base satisfying $C \subseteq B \subseteq U$.

Since $C \subseteq B$ and C is a component of X, it is clear that X contains no continuum from C to $X \setminus B$, and we deduce from Theorem A. 8 that there exists a compact open subset V of X such that $C \subseteq V \subseteq B$. Since V is a neighborhood of C, we can repeat the argument given above to find a member $\tilde{B}$ of the base such that $C \subseteq \tilde{B} \subseteq V$. Between $\tilde{B}$ and B we have the compact open set V, and therefore, for some n, we must have $\tilde{B} \subseteq W_n \subseteq B$. Clearly, $C \subseteq W_n \subseteq U$, and this completes the proof.   **QED**

7.12. Lemma. *The set of open components of a compact metric space is countable.*

PROOF: Let X be a compact metric space, and choose a family $\{W_1, W_2, \ldots\}$ of compact open subsets of X which satisfies the conditions of Lemma 7.11. Then since each open component of X must coincide with one of the sets W_n, the result follows at once.   **QED**

7.13. Theorem. *Let X be a compact metric space.*

Then

(a) *The disconnection order of X is countable.*

(b) *The disconnection kernel Y of X is either empty, or contains at least $\aleph$ components of X.*

(c) *$X \setminus Y$ is a countable union of components of X.*

PROOF: For each ordinal α, let $X^{(\alpha)}$ be the α-disconnection kernel of X. Let $\{W_1, W_2, \ldots\}$ satisfy the conditions of Lemma 7.11, and for each ordinal α, and integer $n \geq 1$, let $W_n^{(\alpha)}$ be the α-disconnection kernel of W_n.

It is easy to see that, for each α and each n,

$$W_n^{(\alpha)} = X^{(\alpha)} \cap W_n.$$

Let Π_0 be the set of those positive integers n for which W_n has only countably many components, and let Π_1 be the set of those positive integers n for which W_n has uncountably many components. For each $n \in \Pi_0$, let α_n be the disconnection order of W_n. We see from Theorem 7.10, that α_n is countable and that $W_n^{(\alpha_n)} = \square$ for every $n \in \Pi_0$.

Let $\alpha = \bigcup_{n \in \Pi_0} \alpha_n$. Then α is a countable ordinal, and for each $n \in \Pi_0$, we have $W_n^{(\alpha)} = \square$. Also, using Lemma 7.12, it is easy to see that for every $n \in \Pi_1$, $W_n^{(\alpha)}$ has uncountably many components.

We shall now show that α is the disconnection order of X. Once this has been done, the theorem will follow easily. It is clear that the disconnection order of X is at least α, since it must be at least α_n for every $n \in \Pi_0$. On the other hand, $X^{(\alpha)}$ has no open components, for if C is an open component of $X^{(\alpha)}$, we obtain a contradiction as follows: Choose an open subset U of X such that $U \cap X^{(\alpha)} = C$, and having done this, choose n such that $C \subseteq W_n \subseteq U$. Then

$$W_n^{(\alpha)} = W_n \cap X^{(\alpha)} = U \cap X^{(\alpha)} = C,$$

which is impossible, since $W_n^{(\alpha)}$ must either be empty or have uncountably many components. This completes the proof of the theorem.

QED

7.14. Corollary. *A compact metric space has an empty disconnection kernel iff it has only countably many components.*

7.15. Corollary. *An uncountable closed subset of* $\mathbb{R}$ *contains a homeomorphic image of the Cantor set.*

PROOF: It is clearly sufficient to prove this result for compact subsets of $\mathbb{R}$. Suppose X is an uncountable compact subset of $\mathbb{R}$. If the interior of X is non-empty, the result follows at once from A.14. Suppose X has no interior. Then X is totally disconnected, and therefore has uncountably many components. Let Y be the disconnection kernel of X. It is easy to apply an analogue of the proof of Theorem 7.9 to show that Y contains a homeomorphic image of the Cantor set. (In fact, a slightly harder argument can be used to show that Y is itself homeomorphic to the Cantor set.)

QED

The Main Result

7.16. Lemma. *Let Ω be either the whole sphere $\mathbf{S}^2$, or a finitely connected region, and let φ_0 be an orbit-analytic, weakly accessible flow in $\partial(\Omega)$. Suppose that each component of $\partial(\Omega)$ meets $\mathscr{G}(\varphi_0)$, and suppose also that for each component B of $\partial(\Omega)$, $B \cap \mathscr{G}(\varphi_0)$ contains only finitely many φ_0-orbits. Let K be a compact subset of $\mathbf{S}^2$ such that $K \cap \bar{\Omega} \neq \square$, suppose that no component of K meets more than one component of $\partial(\Omega)$, and suppose that φ_0 is canonical with respect to K. Let Y be a subset of K which contains $\mathscr{F}(\varphi_0)$, suppose that for each component B of $\partial(\Omega)$, $B \cap \mathscr{F}(\varphi_0)$ is contained in a compact, connected subset of Y, and suppose that $\mathscr{C}(K, \bar{Y} \cap \bar{\Omega})$ has only countably many components. Let $\Delta = \Omega \setminus \mathscr{C}(K, \bar{Y} \cap \bar{\Omega})$ and $F = K \cap \Delta$, suppose that Δ is connected, and suppose that F is a weakly Y-accessible T_σ-set of Δ.*

Then there exists an interpolation φ of φ_0 into Ω which is orbit-analytic, canonical with respect to K, and has the property that whenever $x \in \bar{\Omega}$ and $p_\varphi(x) = \infty$, $\alpha_\varphi(x) \cup \omega_\varphi(x)$ is contained in a compact connected subset of Y.

PROOF: Suppose that the lemma is false. Then there is a least ordinal α such that Ω, φ_0, K and Y can be found satisfying the hypotheses of the lemma but not its conclusions, and having the further property that the disconnection order of $\mathscr{C}(K, \bar{Y} \cap \bar{\Omega})$ is α.

We see at once from Theorem 6.27 that $\alpha \neq 0$, and we see from Theorem 6.32 that $\alpha \neq 1$. We therefore have $\alpha \geq 2$ and it follows from Theorem 7.10 that α is a successor ordinal. Denote by $\alpha - 1$ the immediate predecessor of α. Then Theorem 7.10 implies that whenever Ω, φ_0, K and Y satisfy the hypotheses of the lemma and the disconnection order of $\mathscr{C}(K, \bar{Y} \cap \bar{\Omega})$ is α, the $(\alpha - 1)$-disconnection kernel of $\mathscr{C}(K, \bar{Y} \cap \bar{\Omega})$ has only finitely many components (but at least one).

Let n be the least positive integer for which Ω, φ_0, K, and Y can be found satisfying the hypotheses of the lemma but not its conclusions, and with the further property that the disconnection order of $\mathscr{C}(K, \bar{Y} \cap \bar{\Omega})$ is α, and the $(\alpha - 1)$-disconnection kernel of $\mathscr{C}(K, \bar{Y} \cap \bar{\Omega})$ has exactly n components.

Now choose Ω, φ_0, K and Y satisfying the hypotheses of the lemma but not its conclusions, and having the property that the disconnection order of $\mathscr{C}(K, \bar{Y} \cap \bar{\Omega})$ is α, and the $(\alpha - 1)$-disconnection kernel of $\mathscr{C}(K, \bar{Y} \cap \bar{\Omega})$ has exactly n components.

Let us examine the components of $\mathbf{S}^2 \setminus \Delta$. If a component H of $\mathbf{S}^2 \setminus \Delta$ does not meet $\mathbf{S}^2 \setminus \Omega$, then H is clearly a component of $\mathscr{C}(K, \bar{Y} \cap \bar{\Omega})$.

Now suppose a component H of $\mathbf{S}^2 \setminus \Delta$ meets $\mathbf{S}^2 \setminus \Omega$. Then H is a union of some components of $\mathscr{C}(K, \bar{Y} \cap \bar{\Omega})$ and some components of $\mathbf{S}^2 \setminus \Omega$. H cannot contain more than one component of $\mathbf{S}^2 \setminus \Omega$, for if H contained two distinct components B_1, B_2 of $\mathbf{S}^2 \setminus \Omega$, we could deduce a contradiction as follows:

Choose two mutually disjoint compact subsets C_1 and C_2 of $\mathbf{S}^2 \setminus \Omega$ such that $C_1 \cup C_2 = \mathbf{S}^2 \setminus \Omega$, $B_1 \subseteq C_1$ and $B_2 \subseteq C_2$. H, being a continuum from C_1 to C_2, contains a minimal continuum H_0 from C_1 to C_2 (by Theorem A.7). It follows from Theorem A.9 that $H_0 \setminus (C_1 \cup C_2)$ is a connected subset of K which has limit points in two different components of $\mathbf{S}^2 \setminus \Omega$, and this contradicts the assumption that no component of K can meet more than one component of $\partial(\Omega)$. Now using Theorem A.9, it is easy to see that each component of $\mathscr{C}(K, \overline{Y} \cap \overline{\Omega})$ which is contained in H must meet $\mathbf{S}^2 \setminus \Omega$.

We have therefore shown that a component of $\mathbf{S}^2 \setminus \Delta$ is either a component of $\mathscr{C}(K, \overline{Y} \cap \overline{\Omega})$ contained in Ω, or consists of a component of $\mathbf{S}^2 \setminus \Omega$ together with at most one component of $\mathscr{C}(K, \overline{Y} \cap \overline{\Omega})$.

Choose a weakly Y-accessible setting $\{\Delta_j \mid j = 1, 2, \ldots\}$ for F in Δ. Let Q be a component of the $(\alpha - 1)$-disconnection kernel of $\mathscr{C}(K, \overline{Y} \cap \overline{\Omega})$ and let $D = \partial([Q \text{ side} : \Delta])$. Then D is a component of $\partial(\Delta)$.

We shall obtain a contradiction by dividing Ω into countably many subregions, and defining an appropriate flow in the boundary of each of them, in such a way that each subregion together with the flow in its boundary and K and Y satisfy the hypotheses of the lemma. Our construction will be made in such a way that if $\tilde{\Omega}$ is any one of the sub-regions, the $(\alpha - 1)$-disconnection kernel of $\mathscr{C}(K, \overline{Y} \cap \overline{\tilde{\Omega}})$ will not contain Q, and from this it will follow at once that either the disconnection order of $\mathscr{C}(K, \overline{Y} \cap \overline{\tilde{\Omega}})$ will be less than α, or the disconnection order of $\mathscr{C}(K, \overline{Y} \cap \overline{\tilde{\Omega}})$ is α and the $(\alpha - 1)$-disconnection kernel of $\mathscr{C}(K, \overline{Y} \cap \overline{\tilde{\Omega}})$ will have fewer than n components. By the minimality of α and n, the conclusions of the lemma will necessarily hold in each subregion, and we can therefore interpolate into each subregion in accordance with the lemma. The conjunction of these interpolations will be an interpolation of φ_0 into Ω in accordance with the conclusions of the lemma for Ω, φ_0, K and Y. This will contradict our assumption that the lemma is false.

In order to construct these subregions, we need to consider separately five cases:

CASE 1: $Q \cap \partial(\Omega) \neq \square$.

Let B be that component of $\partial(\Omega)$ which is contained in $[Q \text{ side} : \Delta]$. It is easy to see that

$$B \cap \mathscr{G}(\varphi_0) \subseteq D \subseteq B \cup Q.$$

D is not sealed by $\{\Delta_j \mid j = 1, 2, \ldots\}$, for it it were, we could deduce a contradiction as follows:

Choose j such that D is a component of $\partial(\Delta_j)$. Since no point of $D \cap \mathscr{G}(\varphi_0)$ is a limit point of F, it follows from Theorem 3.32 that D is a component of the boundary of some component annulus of $\Delta_j \setminus F$, and that $\Delta_j \cup [Q \text{ side}: \Delta_j]$ is a neighborhood of Q. This neighborhood of Q does not meet any other component of $\mathscr{C}(K, \overline{Y} \cap \overline{\Omega})$, and therefore Q is an open component of $\mathscr{C}(K, \overline{Y} \cap \overline{\Omega})$, contradicting the fact that $\alpha \geq 2$.

Using Theorem 3.41 (d), we can now see that D is not wrapped by $\{\Delta_j \mid j = 1, 2, \ldots\}$.

Trim each set Δ_j around F at $\partial([D \text{ side}: \Delta_j])$ in such a way that, with the new sets (again called Δ_j for simplicity of notation), we have

$$\varrho\left(x, (\mathbf{S}^2 \setminus \Delta) \cup F\right) < \frac{1}{j},$$

for every $x \in \partial([D \text{ side}: \Delta_j])$ and every $j = 1, 2, \ldots$.

For each $j = 1, 2, \ldots$, let $\Delta_j^D = \mathbf{S}^2 \setminus [D \text{ side}: \Delta_j]$. Let h be a conformal equivalence from $\mathbf{S}^2 \setminus [Q \text{ side}: \Delta]$ onto the open unit disc U. Denote by $\gamma_1, \ldots, \gamma_M$ the set of images under h, in the sense of Theorem C.58, of the non-trivial φ_0-orbits which lie in B. Suppose that $\gamma_1, \gamma_2, \ldots, \gamma_M$ are listed in the order in which they appear as we move round Γ_1 counter-clockwise, suppose that for each $m = 1, \ldots, M$, $\gamma_m = \{e^{i\vartheta} \mid a_m < \vartheta < b_m\}$, and suppose without loss of generality that $0 = a_1 < b_1 \leq a_2 \leq \ldots \leq a_M < b_M \leq 2\pi$. For convenience of notation, define $a_{M+1} = 2\pi$, $b_{M+1} = b_1 + 2\pi$, and $\gamma_{M+1} = \gamma_1$.

We need to construct a family $\{\Lambda_{mk} \mid m = 1, 2, \ldots, M, \ k = 1, 2, \ldots\}$ of analytic arcs in U, which has the following properties:

(i)   For each $m = 1, \ldots, M$, and $k = 1, 2, \ldots$, Λ_{mk} joins e^{ib_m} to $e^{ia_{m+1}}$.

(ii) The family $\{\Lambda_{mk} \mid m = 1, \ldots, M, \ k = 1, 2, \ldots\}$ is pairwise disjoint.

(iii) For each $m = 1, \ldots, M$ and each $k \geq 2$, the Jordan curve $\Lambda_{mk} \cup \{e^{i\vartheta} \mid b_m \leq \vartheta \leq a_{m+1}\}$ has $\Lambda_{m,k+1}$ in its inside and $\Lambda_{m,k-1}$ in its outside.

(iv) For each $m = 1, \ldots, M$, no point of U can lie inside all of the Jordan curves $\Lambda_{mk} \cup \{e^{i\vartheta} \mid b_m \leq \vartheta \leq a_{m+1}\}$, $k = 1, 2, \ldots$.

(v)   For each $m = 1, \ldots, M$, and $k = 1, 2, \ldots$, Λ_{mk} begins in the segment of U bounded by $\gamma_m \cup [e^{ia_m}, e^{ib_m}]$ and ends in the segment of U bounded by $\gamma_{m+1} \cup [e^{ia_{m+1}}, e^{ib_{m+1}}]$.

(vi) For each $m = 1, \ldots, M$, and $k = 1, 2, \ldots$,

$$h^{-1}(\Lambda_{mk}) \subseteq \Delta \setminus \bigcup_j \overline{\Delta_j^D}.$$

(vii) For each $m = 1, \ldots, M$ and $k = 1, 2, \ldots,$ all points x outside $\Lambda_{mk} \cup \{e^{i\vartheta} \mid b_m \leq \vartheta \leq a_{m+1}\}$ and sufficiently close to either e^{ib_m} or $e^{ia_{m+1}}$ satisfy

$$h^{-1}(x) \in \Delta \setminus \bigcup_j \overline{\Delta_j^p}.$$

Since no compact subset of Δ can meet infinitely many maximal sets $\overline{\Delta_j^p}$, we deduce from Corollary C.12 that $\Delta \setminus \bigcup_j \overline{\Delta_j^p}$ is a region. The arcs Λ_{mk} can therefore be obtained by the same techniques that were used in the proofs of Theorems 6.29 and 6.30.

For each $m = 1, \ldots, M$ and $k = 1, 2, \ldots,$ $h^{-1}(\Lambda_{mk})$ is an analytic arc, and all its endpoints lie in K, (in fact, they lie in $B \cap Y$). We can therefore choose, in each of the arcs $h^{-1}(\Lambda_{mk})$, a flow which is canonical with respect to K.

If $\tilde{\Omega}$ is any component of the open set

$$\Omega \setminus \left(\bigcup_{m=1}^{M} \bigcup_{k=1}^{\infty} h^{-1}(\Lambda_{mk}) \right)$$

then $\tilde{\Omega}$ is clearly a finitely connected region and Q is a relatively open subset of $\mathscr{C}(K, \overline{Y} \cap \overline{\tilde{\Omega}})$. Further, $\tilde{\Omega}$ and the flow in its boundary, together with K and Y satisfy the hypotheses of the lemma, and must therefore by the minimality of α and n, also satisfy its conclusions. Thus we may interpolate separately into each of these components in accordance with the lemma, and the conjunction of these interpolations clearly satisfies the conclusions of the lemma for Ω, φ_0, K and Y. But this contradicts our choice of Ω, φ_0, K and Y, and we conclude that Case 1 cannot occur.

CASE 2: $Q \cap \partial(\Omega) = \square$.

It is clear that in this case, Q is a component of $\mathbf{S}^2 \setminus \Delta$ and $D = \partial(Q)$. We divide Case 2 into three parts, (a), (b), and (c), according as D is respectively sealed by $\{\Delta_j \mid j = 1, 2, \ldots\}$, wrapped but not sealed by $\{\Delta_j \mid j = 1, 2, \ldots\}$, and not wrapped by $\{\Delta_j \mid j = 1, 2, \ldots\}$.

CASE 2(a): D is sealed by $\{\Delta_j \mid j = 1, 2, \ldots\}$.

Choose j such that D is a component of $\partial(\Delta_j)$. Since $\alpha \geq 2$, it is easy to see that $F \cap \Delta_j \neq \square$, i.e., $F \cap \Delta_j$ is a non-empty T-set of Δ_j. It follows that each component annulus of $\Delta_j \setminus F$ has at least one component of its boundary contained in F. Denote by $\{\gamma_m \mid m = 1, 2, \ldots\}$ the family of those components of $\partial(\Delta_j)$ which are contained in Δ.

For each m, γ_m is an analytic Jordan curve, and we choose in γ_m a flow which is canonical with respect to K. Using Theorem 6.25 to interpolate into each component annulus of $\Delta_j \setminus F$, we obtain an interpolation into Δ_j of the flow in $\partial(\Delta_j)$.

For each $m = 1, 2, \ldots,$ the side of γ_m away from Δ_j meets Ω in a finitely connected region Ω_m, and it is easy to see that Ω_m and the flow in its boundary, together with K and Y, satisfy the hypotheses of the lemma. Since Q does not meet $\overline{\Omega_m}$, the minimality of α and n allows us to interpolate into Ω_m in accordance with the lemma.

It is clear that the conjunction of all these interpolations satisfies the conclusions of the lemma for Ω, φ_0, K and Y. But this contradicts our choice of Ω, φ_0, K and Y, and we conclude that Case 2(a) cannot occur.

CASE 2(b): D is wrapped but not sealed by $\{\Delta_j \mid j = 1, 2, \ldots\}$.
For each $j = 1, 2, \ldots,$ let $\Delta_j^D = \mathbf{S}^2 \setminus [D \text{ side}: \Delta_j]$. Using Theorem 3.41, choose a sequence $j_1, j_2, \ldots,$ such that $j_1 < j_2 < j_3 < \ldots,$

$$\Delta_{j_1}^D \subseteq \Delta_{j_2}^D \subseteq \Delta_{j_3}^D \subseteq \ldots$$

and

$$\mathbf{S}^2 \setminus Q = \bigcup_{m=1}^{\infty} \Delta_{j_m}^D.$$

To simplify the picture and the notation, choose a point of Q to play the role of ∞. For each $m = 1, 2, \ldots,$ let $\gamma_m = \partial(\Delta_{j_m}^D)$. Then for each m, γ_m is an analytic Jordan curve contained in Δ, and for each $m = 2, 3, \ldots,$ we have

$$\gamma_{m-1} \subseteq \text{ins} (\gamma_m) \quad \text{and} \quad Q \cup \gamma_{m+1} \subseteq \text{outs} (\gamma_m).$$

In each γ_m, define a flow, canonical with respect to K.

Let $\Omega_1 = \Omega \cap \text{ins} (\gamma_1)$ and for each $m = 2, 3, \ldots,$ let $\Omega_m = \Omega \cap \text{ins} (\gamma_m) \cap \text{outs} (\gamma_{m-1})$. Then for each $m = 1, 2, \ldots,$ Ω_m and the flow in its boundary, together with K and Y, satisfy the hypotheses of the lemma, and therefore the minimality of α and n allows us to interpolate into each set Ω_m in accordance with the lemma.

It is clear that the conjunction of all these interpolations satisfies the conclusions of the lemma for Ω, φ_0, K and Y. But this contradicts our choice of Ω, φ_0, K and Y, and we conclude that Case 2(b) cannot occur.

CASE 2(c): D is not wrapped by $\{\Delta_j \mid j = 1, 2, \ldots\}$.
For each $j = 1, 2, \ldots,$ let $\Delta_j^D = \mathbf{S}^2 \setminus [D \text{ side}: \Delta_j]$. As in Case 1, we shall assume that

$$\varrho\left(x, (\mathbf{S}^2 \setminus \Delta) \cup F\right) < \frac{1}{j} \quad \text{whenever } x \in \partial(\Delta_j^D), j = 1, 2, \ldots.$$

We now divide Case 2(c) into two subcases (this is the last division into subcases):

CASE 2(c)(i): There exists an arc L in $\varDelta \setminus \cup\, \overline{\varDelta_j^p}$ which joins a point of $\varDelta \setminus \underset{j}{\cup}\, \overline{\varDelta_j^p}$ to a compact connected subset Y_D of $Y \cap D$.

Using the techniques that were used in the proof of Theorem 6.32. to construct the arcs $\varLambda_B$ there, construct an analytic arc $\varLambda$ in $\varDelta \setminus \underset{j}{\cup}\, \overline{\varDelta_j^p}$ which has the following properties (see Fig. 7.1):

(i) $\varLambda$ joins a subset of Y_D to a subset of Y_D,

(ii) $\varLambda \cup Q$ is a compact subset of $\varOmega$ whose complement in $\varOmega$ has two components $\varOmega_1$ and $\varOmega_2$, each of which is a finitely connected region,

(iii) $\varLambda \cup D$ is a component of $\partial(\varOmega_1)$ which lies in the $(\alpha - 1)$-disconnection kernel of $\mathscr{C}(K, \overline{Y} \cap \overline{\varOmega_1})$,

and

(iv) Q is a relatively open subset of $\mathscr{C}(K, \overline{Y} \cap \overline{\varOmega_2})$.

Since all endpoints of $\varLambda$ lie in K, we can define in $\varLambda$ a flow canonical with respect to K. It is clear by the minimality of α and n, that the flow in $\partial(\varOmega_2)$ can be interpolated into $\varOmega_2$ in accordance with the lemma. It is also clear that because of the minimality of α and n, the argument we used for $\varOmega$ in Case 1 can be applied to $\varOmega_1$, to show that the flow in $\partial(\varOmega_1)$ can be interpolated into $\varOmega_1$ in accordance with the lemma.

Once more, the conjunction of these interpolations satisfies the conclusions of the lemma for $\varOmega$, φ_0, K and Y, and we conclude that Case 2(c)(i) cannot occur.

CASE 2(c)(ii): This is the final case, and consists of the remaining possibility in Case 2(c), namely that Case 2(c)(i) does not occur.

In this case, the definition of a weakly accessible setting implies that given any neighborhood V of Q, there is an arc in $\varDelta \setminus \cup\, \overline{\varDelta_j^p}$ which joins a point of $\varDelta \setminus \underset{j}{\cup}\, \overline{\varDelta_j^p}$ to a subset of $(Y \cap V) \setminus Q$.

We need to construct a family $\{ \varLambda_m \mid m = 1, 2, \ldots \}$ of analytic arcs which has the following properties (see Fig. 7.2):

(i)   For each $m = 1, 2, \ldots$, there exists a compact connected subset Y_m of Y and a component Q_m of $\mathscr{C}(K, \overline{Y} \cap \overline{\varOmega})$ such that $Y_m \subseteq Q_m \subseteq \varOmega$ and $\varLambda_m$ joins a subset of Y_m to a subset of Y_m.

(ii) For each $m = 1, 2, \ldots$, $\varLambda_m \subseteq \varDelta \setminus \underset{j}{\cup}\, \overline{\varDelta_j^p}$.

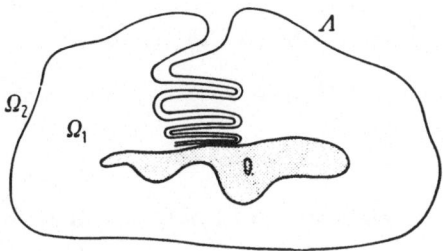

Fig. 7.1

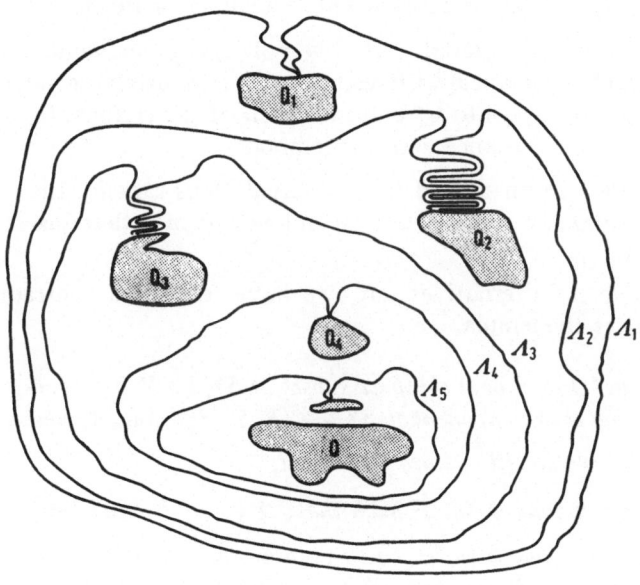

Fig. 7.2

(iii) For each $m = 1, 2, \ldots,$ $\Omega \setminus (\Lambda_m \cup Q_m)$ has two components, Ω_m^- and Ω_m^+. Both Ω_m^- and Ω_m^+ are finitely connected regions, and we have

$$Q \subseteq \Omega_m^+, \ \Lambda_{m+1} \cup Q_{m+1} \subseteq \Omega_m^+,$$

and for $m \geq 2$, $\Lambda_{m-1} \cup Q_{m-1} \subseteq \Omega_m^-$.

(iv) $\bigcap\limits_{m=1}^{\infty} \Omega_m^+ = Q$.

To obtain Λ_1, choose a neighborhood V_1 of Q such that $V_1 \subseteq \Omega$. Using Theorem C.30, choose a Jordan curve $J_1 \subseteq \Delta \setminus \bigcup\limits_{j} \overline{\Delta_j^D}$ such that $[Q$ side $: J_1] \subseteq V_1$. Choose an arc $L_1 \subseteq [Q$ side $: J_1] \cap (\Delta \setminus \bigcup\limits_{j} \overline{\Delta_j^D})$ which joins a point of $\Delta \setminus \bigcup\limits_{j} \overline{\Delta_j^D}$ to a subset Y_1 of Y. Y_1 is contained in a component Q_1 of $\mathscr{C}(K, \overline{Y} \cap \bar{\Omega})$, and it is clear that $Q_1 \neq Q$. In the region $\Delta \setminus \bigcup\limits_{j} \overline{\Delta_j^D}$, extend L_1 if necessary to make it join a point of J_1 to Y_1. Now, using the techniques we have used in the past, replace $J_1 \cup L_1$ by an analytic arc Λ_1 which has the properties we require of it. The arcs $\Lambda_2, \Lambda_3, \ldots$, can be constructed in a similar manner.

Having constructed the family $\{\Lambda_m \mid m = 1, 2, \ldots\}$, define in each arc Λ_m a flow which is canonical with respect to K. Now using the minimality of α and n, interpolate into Ω_1^- and into each of the regions $\Omega_m^+ \cap \Omega_{m+1}^-$ ($m = 1, 2, \ldots$) in accordance with the lemma.

Once more, the conjunction of these interpolations satisfies the conclusions of the lemma for Ω, φ_0, K and Y, and we conclude that Case 2(c)(ii) cannot occur.

Since we have no alternatives left, we have reached a contradiction, and this proves the lemma.   **QED**

7.17. Theorem.[2] *Let F be a compact subset of $\mathbf{S}^2$, let Y be a subset of F, and suppose that each component of $\mathbf{S}^2 \setminus \mathscr{C}(F, \overline{Y})$ is countably connected.*

Then the following conditions are equivalent.

(a) *There exists a flow φ in $\mathbf{S}^2$ such that $\mathscr{F}(\varphi) = F$ and $\mathscr{S}(\varphi) \subseteq Y$.*

(b) *There exists a flow φ in $\mathbf{S}^2$ which is orbit-analytic and canonical with respect to F, and has the property that whenever $x \in \mathbf{S}^2$ and $p_\varphi(x) = \infty$, $\alpha_\varphi(x) \cup \omega_\varphi(x)$ is contained in a compact connected subset of Y.*

(c) *Y is a WATI subset of F.*

---

[2] cf. Theorem 6.36.

PROOF: The implication (b) $\Rightarrow$ (a) is trivial, and the implication (a) $\Rightarrow$ (c) follows from Theorem 3.54.

To prove that (c) $\Rightarrow$ (b), suppose that Y is a WATI subset of F. For each component Δ of $\mathbf{S}^2 \setminus \mathscr{C}(F, \overline{Y})$, let φ_Δ be the flow in Δ defined as follows:

Let $\Omega = \mathbf{S}^2$ and $K = F \cup (\mathbf{S}^2 \setminus \Delta)$. Applying Lemma 7.16 to Ω and the empty flow in its boundary, together with K and Y, choose a flow ψ_Δ in Ω. Now let φ_Δ be the restriction of ψ_Δ to Δ.

The conjunction of all the flows φ_Δ together with the fixed flow in F, clearly has all the required properties.   **QED**

7.18. Note. We note that if F is a compact subset of $\mathbf{S}^2$, Y is a subset of F, and $\mathscr{C}(F, \overline{Y})$ has only countably many components, (in particular, if either F or $\overline{Y}$ has only countably many components) then Theorem C.13 implies that each component of $\mathbf{S}^2 \setminus \mathscr{C}(F, \overline{Y})$ is countably connected.

7.19. Lemma. *Let Ω be either the whole sphere $\mathbf{S}^2$, or a finitely connected region, and let φ_0 be an orbit-analytic, accessible flow in $\partial(\Omega)$. Suppose that each component of $\partial(\Omega)$ meets $\mathscr{G}(\varphi_0)$, and suppose also that for each component B of $\partial(\Omega)$, $B \cap \mathscr{F}(\varphi_0)$ consists of at most one point, and $B \cap \mathscr{G}(\varphi_0)$ contains only finitely many orbits.*

Let K be a compact subset of $\mathbf{S}^2$ such that $K \cap \overline{\Omega} \neq \square$, suppose that no component of K meets more than one component of $\partial(\Omega)$, and suppose that φ_0 is canonical with respect to K. Let Y be a subset of K which contains $\mathscr{F}(\varphi_0)$, and suppose that $\mathscr{C}(K, \overline{Y} \cap \overline{\Omega})$ has only countably many components. Let $\Delta = \Omega \setminus \mathscr{C}(K, \overline{Y} \cap \overline{\Omega})$ and $F = K \cap \Delta$. Suppose that Δ is connected, and suppose that F is a Y-accessible T_σ-set of Δ.

Then there exists an interpolation φ of φ_0 into Ω which is simple, orbit-analytic, canonical with respect to K, and has the property that $\mathscr{S}(\varphi) \subseteq Y$ and whenever $x \in \mathbf{S}^2$ and $\mathrm{p}_\varphi(x) = \infty$, we have $\alpha_\varphi(x) = \omega_\varphi(x)$.

PROOF: The proof of this lemma is identical to that of Lemma 7.16, and is omitted.   **QED**

7.20. Theorem.[3] *Let F be a compact subset of $\mathbf{S}^2$, let Y be a subset of F, and suppose that each component of $\mathbf{S}^2 \setminus \mathscr{C}(F, \overline{Y})$ is countably connected.*

Then the following conditions are equivalent:

(a) There exists a semi-simple flow φ in $\mathbf{S}^2$ such that $\mathscr{F}(\varphi) = F$ and $\mathscr{S}(\varphi) \subseteq Y$.

---

[3] cf. Theorem 6.37.

(b) *There exists a simple flow φ in $\mathbf{S}^2$ which is orbit-analytic and canonical with respect to F, and has the properties that $\mathscr{S}(\varphi) \subsetneq Y$ and whenever $x \in \mathbf{S}^2$ and $p_\varphi(x) = \infty$, we have $\alpha_\varphi(x) = \omega_\varphi(x)$.*

(c) *Y is an ATI subset of F.*

PROOF: The proof is identical, *mutatis mutandis*, with that of Theorem 7.17, and is omitted.   **QED**

7.21. Theorem. *Let F be a compact subset of $\mathbf{S}^2$, let Y be a subset of F, and suppose that Y has countable closure.*

Then the following conditions are equivalent:

(a) *There exists a flow φ in $\mathbf{S}^2$ such that $\mathscr{F}(\varphi) = F$ and $\mathscr{S}(\varphi) \subseteq Y$.*

(b) *There exists a simple flow φ in $\mathbf{S}^2$ which is orbit-analytic and canonical with respect to F, satisfies $\mathscr{S}(\varphi) \subseteq Y$, and has the property that whenever $x \in \mathbf{S}^2$ and $p_\varphi(x) = \infty$, we have $\alpha_\varphi(x) = \omega_\varphi(x)$.*

(c) *Y is a WATI subset of F.*

(d) *Y is an ATI subset of F.*

(e) *In every component Ω of $\mathbf{S}^2 \setminus \mathscr{C}(F, \overline{Y})$, $F \cap \Omega$ is a Y-accessible T_σ-set of Ω.*

PROOF: The result follows directly from Theorem 7.20 and Note 7.18.
                                                                     QED

7.22. Remark. It is an elementary consequence of Corollary 5.6 that if F is any non-empty compact subset of $\mathbf{S}^2$ then *some* flow φ can be found in $\mathbf{S}^2$ satisfying $\mathscr{F}(\varphi) = F$. However, if we construct the flow φ by suitably reparametrizing flows of the type illustrated in the proof of Corollary 5.6 (choosing a point of F to play the role of ∞), then we obtain a flow in which far too many orbits stagnate blindly into F instead of steering a nice smooth path around it.

This picture is very unlike a fluid flowing around a stationary obstacle in its path, since for such a flow, only the very few orbits that are unable to turn will stagnate into the obstacle. (See Fig. 7.3.)

There is therefore good reason to ask, given some non-empty compact subset F of $\mathbf{S}^2$, what constraints we can impose on the set of stagnation points of a flow φ in $\mathbf{S}^2$ satisfying $\mathscr{F}(\varphi) = F$. Under a wide variety of circumstances, Theorem 7.17 provides the answer, *viz*, $\mathscr{S}(\varphi)$ can be required to be contained in any given WATI subset of F.

7.23. Example. Let Q_1 be the boundary of the unit square. Let Q_2 be the union of Q_1 with the two straight line segments required to quarter the unit square. For each $n \geq 2$, if Q_n has been defined, let Q_{n+1} be the union of Q_n with all straight line segments required to quarter those squares determined by Q_n which lie adjacent to Q_1.

Let $Q = \bigcup\limits_{n=1}^{\infty} Q_n$ and let $F = Q \cup \text{outs}(Q_1)$. (See Fig. 7.4.)

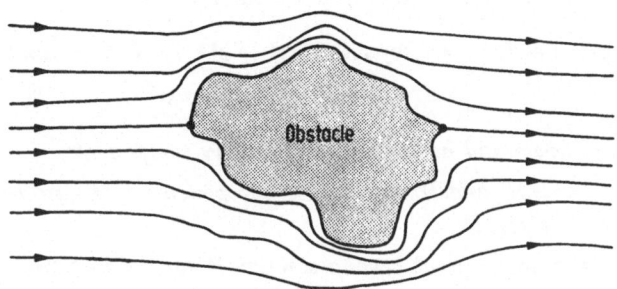

Fig. 7.3

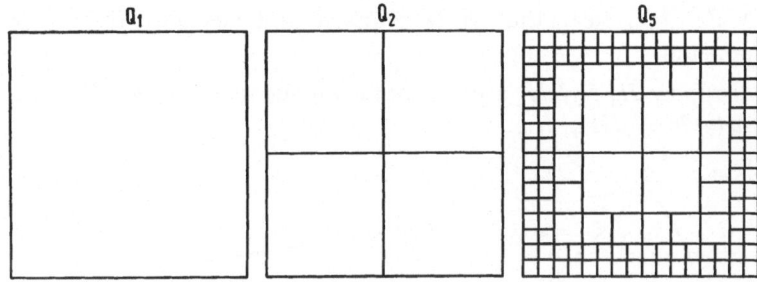

Fig. 7.4

It is easy to see that every WATI subset of F must contain points in the boundary of every component of $\mathbf{S}^2 \setminus F$, and must therefore have uncountable closure. We deduce that if φ is a flow in $\mathbf{S}^2$ and $\mathscr{F}(\varphi) = F$, then $\overline{\mathscr{S}(\varphi)}$ is uncountable.

However, it is easy to find a countable ATI subset of F, and therefore there does exist a simple flow φ in $\mathbf{S}^2$ such that $\mathscr{F}(\varphi) = F$ and $\mathscr{S}(\varphi)$ is countable.

7.24. Theorem. *Let F be a non-empty compact subset of $\mathbf{S}^2$. Then there exists a simple, orbit-analytic flow φ in $\mathbf{S}^2$ which is canonical with respect to F, and which satisfies*

(a) *$\mathscr{S}(\varphi)$ is countable*

and

(b) *$\mathscr{G}_r(\varphi) = \mathscr{F}_r(\varphi) = \square$*

Before proving this theorem, we require some lemmas:

7.25. Lemma. *Let a, b, p, and q be points in $\mathbb{R}^2$, no three of which are collinear, and suppose that (a, b) meets (p, q).*

Then either $d(p, b) < d(a, b)$ or $d(a, q) < d(p, q)$.

PROOF: We omit the proof of this elementary result.   **QED**

7.26. Lemma. *Let K be a compact subset of $\mathbb{R}^2$, let $\delta > 0$, and let*

$$H = \{x \in \mathbb{R}^2 \mid d(x, K) \leq \delta\}.$$

Then H is compact and has only finitely many components.

Suppose now that H_1 and H_2 are distinct components of H, $a \in H_1$, $b \in H_2$, and $d(a, b) = d(H_1, H_2)$.

Then the projection of $[b, a]$ beyond a meets K for the first time in a point x whose distance from a is δ, and for any other point y of K, we have $d(y, a) > \delta$.

PROOF: It is clear that H is compact and has only finitely many components.

Now suppose H_1 and H_2 are distinct components of H, $a \in H_1$, $b \in H_2$, and $d(a, b) = d(H_1, H_2)$.

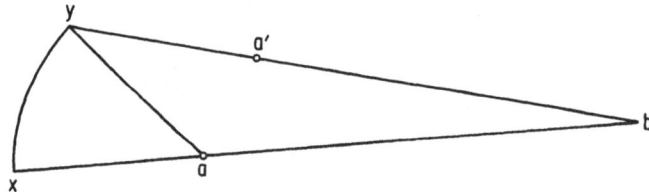

Fig. 7.5

Let x be that point on the projection of $[b, a]$ beyond a, whose distance from a is δ. (See Fig. 7.5.) It is clear that $d(a, K) = \delta$, and consequently there exist points $y \in K$ such that $d(a, y) = \delta$, and we shall prove the lemma by showing that each such point y coincides with x.

Suppose that $y \in K$ and $d(a, y) = \delta$. Since

$$d(y, b) \leq d(y, a) + d(a, b) = d(x, a) + d(a, b) = d(x, b),$$

we have $d(y, b) \leq d(x, b)$, with equality holding iff $y = x$. But since $y \in K \cap H_1$ and $b \notin H_1$, it is clear that $d(y, b) > \delta$. Let a' be the point in (y, b) satisfying $d(y, a') = \delta$. Then since $a' \in H_1$, we have $d(a', b) \geq d(a, b)$, and consequently $d(y, b) \geq d(x, b)$.

Thus $y = x$, and the lemma is proved.   **QED**

7.27. Lemma. *Let F be a non-empty compact subset of $\mathbf{S}^2$.*

Then there exists a countable pairwise disjoint family $\{ \Lambda_n \mid n = 1, 2, \ldots \}$ of analytic arcs, which has the following properties:

(a) *For each n, $\Lambda_n \subseteq \mathbf{S}^2 \setminus F$, and joins a point of F to a point of F.*

(b) *For each n, Λ_n is a relatively open subset of $F \cup \bigcup_j \Lambda_j$.*

(c) *$F \cup \bigcup_j \Lambda_j$ is compact and connected.*

PROOF:[4] If F is connected, the result is trivial.

Suppose that F is not connected. Then $F \neq \mathbf{S}^2$, and by choosing a point of $\mathbf{S}^2 \setminus F$ to play the role of ∞, we can regard F as a compact subset of $\mathbb{R}^2$.

For each $n = 1, 2, \ldots$, define $H_n = \{ x \in \mathbb{R}^2 \mid d(x, F) \leq 2^{-n} \}$. For each n, and each pair of distinct components of H_n, choose a shortest line segment joining a point of the one to a point of the other. List the set of these shortest line segments in order of increasing length, and call this ordered finite set W_n.

Let $B_{1,1}$ be the first member of W_1, and having defined $B_{k,1}$ $(k = 1, 2, \ldots)$ let $B_{k+1,1}$ be the first member of W_1 not already selected (if there is one) whose union with $H_1 \cup \bigcup_{j=1}^{k} B_{j,1}$ has fewer components than $H_1 \cup \bigcup_{j=1}^{k} B_{j,1}$ itself. In this way, we obtain a finite sequence $B_{1,1}, \ldots, B_{k,,1}$ of line segments whose union with H_1 is connected.

Now using Lemmas 7.25 and 7.26, it is easy to construct an argument to show that if $[a, b]$ and $[p, q]$ are any two of the segments chosen above, then

(i) $[a, b]$ and $[p, q]$ do not intersect

and

(ii) If a', b', p' and q' are the points of F whose distances from a, b, p, and q, respectively, are 2^{-1}, then (a', b') and (p', q') do not intersect.

We now define finitely many straight line segments $B_{k,2}$ as follows: For $k = 1, \ldots, k_1$, $B_{k,2}$ is obtained from $B_{k,1}$ by extending $B_{k,1}$ by 2^{-2} in each direction. (Note that, for $k = 1, \ldots, k_1$, $B_{k,2} \subseteq \mathbb{R}^2 \setminus H_2$, and $B_{k,2}$ joins two components of H_2 which lie in distinct components of H_1.)

Let $B_{k_1+1,2}$ be the first member of W_2 whose union with $H_2 \cup \bigcup_{j=1}^{k_1} B_{j,2}$ has fewer components than $H_2 \cup \bigcup_{j=1}^{k_1} B_{j,2}$ itself. Having defined $B_{k,2}$ $(k > k_1)$,

---

[4] This proof stems from an idea of Jerome Dancis.

define $B_{k+1,2}$ (if we can) to be the first member of W_2 whose union with $H_2 \cup \bigcup_{j=1}^{k} B_{j,2}$ has fewer components than $H_2 \cup \bigcup_{j=1}^{k} B_{j,2}$ itself. In this way, we obtain a finite sequence $B_{1,2}, \dots B_{k_2,2}$ whose union with H_2 is connected.

Continuing in this way, we obtain for each $n = 1, 2, \dots$, finitely many straight line segments $B_{k,n}$ ($k = 1, \dots, k_n$) such that

(1) For each $n \geq 2$ and $k = 1, \dots, k_{n-1}$, $B_{k,n}$ is obtained from $B_{k,n-1}$ by extending $B_{k,n-1}$ by 2^{-n} in each direction.

(2) For $k = k_{n-1} + 1, \dots, k_n$, the segments $B_{k,n}$ are chosen successively from W_n by selecting at each stage, the first member of W_n whose union with H_n and the sets $B_{j,n}$ ($j < k$) will have fewer components than $H_n \cup \bigcup_{j=1}^{k-1} B_{j,n}$ itself.

Using Lemmas 7.25 and 7.26, it is easy to verify that

(i) For each $n \geq 2$, each segment $B_{k,n}$ ($k \geq k_{n-1}$) lies inside some component of H_{n-1}, and its length does not exceed 2^{-n+1}.

(ii) For each n, the open segments obtained from the segments $B_{k,n}$ ($k = 1, \dots, k_n$), by extending them by 2^{-n} in each direction, form a pairwise disjoint family.

Now for each $k = 1, 2, \dots$, define $\Lambda_k = \bigcup_n B_{k,n}$, the union being taken over those n for which $B_{k,n}$ are defined. Note that, if $B_{k,n}$ is defined, then Λ_k is the open segment obtained by extending $B_{k,n}$ by 2^{-n} in each direction. Discard those sets Λ_k which are empty. The family $\{\Lambda_k \mid k = 1, 2, \dots\}$ is a pairwise disjoint family of open straight line segments, each of which joins a point of F to a point of F. Further, it is clear that for each k, Λ_k is open in $F \cup \bigcup_j \Lambda_j$.

To see that $F \cup \bigcup_j \Lambda_j$ is compact and connected, we need only note that

$$F \cup \bigcup_j \Lambda_j = \bigcap_{n=1}^{\infty} \left(H_n \cup \bigcup_{j=1}^{k_n} B_{j,n} \right),$$

and that the family $\left\{ H_n \cup \bigcup_{j=1}^{k_n} B_{j,n} \;\middle|\; n = 1, 2, \dots \right\}$ is a chain of compact, connected sets.

This completes the proof of the lemma.   **QED**

PROOF OF THEOREM 7.24: Choose a pairwise disjoint family $\{\Lambda_n \mid n = 1, 2, \dots\}$ of analytic arcs in $\mathbf{S}^2 \setminus F$, each of which joins a point of F to a point of F and is relatively open in $F \cup \bigcup_n \Lambda_n$, with the property that $F \cup \bigcup_n \Lambda_n$ is connected.

Let φ_F be the flow in F which leaves every point fixed, and in each arc Λ_n define a flow canonical with respect to F.

If Ω is a component of $\mathbf{S}^2 \setminus (F \cup \bigcup_n \Lambda_n)$, then Ω is a disc, and $\partial(\Omega)$ is a union of some (possibly none) of the arcs Λ_n, and a non-empty subset of F. If $\partial(\Omega)$ does not meet $\bigcup_n \Lambda_n$, then by drawing an analytic cross-cut in Ω, and extending the family $\{\Lambda_n \mid n = 1, 2, \ldots\}$ to include this cross-cut, we may break Ω into two smaller regions, each of whose boundaries meets $\bigcup_n \Lambda_n$.

We may therefore assume that for each component Ω of $\mathbf{S}^2 \setminus (F \cup \bigcup_n \Lambda_n)$, $\partial(\Omega)$ contains at least one of the arcs Λ_n. For each component Ω of $\mathbf{S}^2 \setminus (F \cup \bigcup_n \Lambda_n)$, let φ_Ω be the interpolation into Ω of the flow in $\partial(\Omega)$, obtained by an application of Theorem 6.23. It is clear that the conjunction of φ_F and all of the flows φ_Ω satisfies the requirements of the theorem.

QED

7.28. Corollary. *Every non-empty compact subset of* $\mathbf{S}^2$ *contains a countable* ATI *subset.*

Notes and Remarks to Chapter 7

In work which is mentioned in Beck [4] but which has not previously been published, we devised the transfinite induction proof of Lemma 7.19, and used an analogue of this lemma to prove that if an ATI subset Y of a compact set $F \subseteq \mathbf{S}^2$ has countable closure, then there exists a simple flow φ in $\mathbf{S}^2$ which is orbit-analytic and canonical with respect to F, and satisfies $\mathscr{S}(\varphi) \subseteq Y$.

Strictly speaking, the original form (Beck [3]) of Lemma 7.19 was false; a flaw occured in the treatment of case 2 (c) (ii). Fortunately, however, this flaw could be easily eliminated by changing the original definition of accessible setting (see the notes and remarks to Chapter 3), and replacing it by Definition 3.45. Although this new definition resulted in some substantial changes in Chapter 3, it did not necessitate any essential alteration in the proof of Lemma 7.19.

As in Chapter 6, the earlier condition that $\overline{Y}$ be countable has been replaced by the condition that every component of $\mathbf{S}^2 \setminus \mathscr{C}(F, \overline{Y})$ be countably connected.

Theorem 7.17 is the principal result of Chapter 7, and is the closest statement we know to a converse of Theorem 3.54. We conjecture that "If Y is a subset of a compact set $F \subseteq \mathbf{S}^2$, then a necessary and sufficient condition for the existence of a flow φ in $\mathbf{S}^2$ satisfying $\mathscr{F}(\varphi) = F$ and $\mathscr{S}(\varphi) \subseteq Y$ is that Y be a WATI subset of F." Theorem 7.17 shows that this conjecture is true if every component of $\mathbf{S}^2 \setminus \mathscr{C}(F, \overline{Y})$ is countably connected, but apart from some progress that has been made by the Lewins in a few isolated special cases, the uncountable case is an open question.[5]

---

[5] At the time of going to the press, we have found a proof that this conjecture is true in general.

CHAPTER EIGHT[1]

ALGEBRAIC COMBINATIONS OF FLOWS I

A Partial Order in $\mathrm{Rep}_q^+(\varphi)$

8.1. Introduction and Definition. Suppose φ is a quasi-flow in a Hausdorff space X and that u and v are nonnegative quasi φ-multipliers. If it happens that $u + v$ is also a quasi φ-multiplier, then it is tempting to think of $(u + v)\varphi$ as the "sum" of $u\varphi$ and $v\varphi$. With this in mind, it is natural to ask what the "sum" $\psi + \pi$ of two members ψ and π of a family $\mathrm{Rep}_q^+(\varphi)$ ought to be, and when such a sum can be defined.

If, instead of looking at $u + v$, we consider the maximum $u \vee v$ and minimum $u \wedge v$ of u and v, a similar situation arises. $(u \vee v)\varphi$ and $(u \wedge v)\varphi$ should in some sense be the "maximum" and "minimum" of $u\varphi$ and $v\varphi$. Now in order for the notion of the maximum or minimum of two members of $\mathrm{Rep}_q^+(\varphi)$, to be meaningful, there must be an appropriate partial order in $\mathrm{Rep}_q^+(\varphi)$, and we begin this chapter by defining this partial order.

If φ is a quasi-flow in a Hausdorff space X and $\psi, \pi \in \mathrm{Rep}_q^+(\varphi)$, then we say that ψ *is slower than* π, and write $\psi \leq \pi$, if $\mu_\pi \leq \mu_\psi$.[2]

It is clear that $\leq$ is a partial order in $\mathrm{Rep}_q^+(\varphi)$.

8.2. Theorem. *Let φ be a quasi-flow in a Hausdorff space X, let f and $\tilde{f}$ be isotonic quasi φ-reparametrizers, and let ψ and π be f rep φ and $\tilde{f}$ rep φ respectively. Let*

$$\mathrm{inv}\,(f) = \{g_x \mid x \in \mathscr{G}(\psi)\} \quad and \quad \mathrm{inv}\,(\tilde{f}) = \{\tilde{g}_x \mid x \in \mathscr{G}(\pi)\}.$$

---

[1] Chapters 8 and 9 constitute the major portion of the doctoral dissertation of Mirit Hope Lewin (Ph. D., University of Wisconsin, 1970). The thesis is a development and extension of some joint work of the author and Aryeh Dvoretsky, presented by the author at the International Congress of Mathematicians (Moscow, 1966) under the title, "Addition of generalized velocities".
[2] The statement that $\mu_\pi \leq \mu_\psi$ means that $\mu_\pi(A) \leq \mu_\psi(A)$, for every Borel set A, or equivalently, for every $\Sigma(\varphi)$-measurable set A.

Then the following statements are equivalent:

(a) $\psi \leq \pi$.

(b) *Whenever* $x \in \mathcal{G}(\psi)$ *and the real number* τ *lies in the domain of* g_x, *we have*

$$|\tilde{g}_x(\tau)| \leq |g_x(\tau)|.$$

(c) *For every* $x \in \mathcal{G}(\varphi)$ *and* $t \in \mathbb{R}$ *we have*

$$|f_x(t)| \leq |\tilde{f}_x(t)|.$$

(d) *There exists an isotonic quasi* π-*reparametrizer* h *such that*

$$|h_x(t)| \leq |t| \text{ whenever } x \in X \text{ and } t \in \mathbb{R},$$

and $\psi = h \operatorname{rep} \pi$.

PROOF: We first show that (a) $\Rightarrow$ (b). Suppose (a) holds. We see at once that $\mathcal{F}(\pi) \subseteq \mathcal{F}(\psi)$, and therefore Theorem 4.14 implies that $\psi \in \operatorname{Rep}_q^+(\pi)$ and also that whenever $x \in \mathcal{G}(\psi)$, the domain of $\tilde{g}_x$ contains the domain of g_x. Now suppose τ is any positive real number in the domain of g_x, and write $\tau = \tau_0 + n \cdot \mathrm{p}_\varphi(x)$, where $0 \leq \tau_0 < \mathrm{p}_\varphi(x)$, and $n \geq 0$ is an integer. (We remark here that n is necessarily zero if $\mathrm{p}_\psi(x) = \infty$.) We now have

$$\begin{aligned}
g_x(\tau) &= g_x(\tau_0) + n \cdot \mathrm{p}_\psi(x) \\
&= \mu_\psi\big(\{\varphi(\sigma, x) \mid 0 \leq \sigma \leq \tau_0\}\big) + n \cdot \mu_\psi\big(\mathcal{O}_\varphi(x)\big) \\
&\geq \mu_\pi\big(\{\varphi(\sigma, x) \mid 0 \leq \sigma \leq \tau_0\}\big) + n \cdot \mu_\pi\big(\mathcal{O}_\varphi(x)\big) \\
&= \tilde{g}_x(\tau) \quad \geq 0.
\end{aligned}$$

A similar argument shows that, if $\tau < 0$, and τ lies in the domain of g_x, then $g_x(\tau) \leq \tilde{g}_x(\tau) \leq 0$. This completes the proof that (a) $\Rightarrow$ (b).

In view of Theorem 4.14, it is not hard to see that (b), (c) and (d) are equivalent.

We now complete the proof by showing that (d) $\Rightarrow$ (a). Choose an isotonic quasi π-reparametrizer h such that $\psi = h \operatorname{rep} \pi$ and

$$|h_x(t)| \leq |t|, \qquad \forall\, x \in X \qquad \forall\, t \in \mathbb{R}.$$

To show that $\mu_\pi(A) \leq \mu_\psi(A)$ for every $\Sigma(\varphi)$-measurable set A, it is clearly sufficient to show that this inequality holds for every Borel subset of a non-trivial ψ-orbit (for we may clearly reduce the problem to a single φ-orbit, and such an orbit is the union of a subset of $\mathcal{F}(\psi)$ and countably many non-trivial ψ-orbits).

If A is a set of the form $\{\psi(s, x) \mid 0 < s < t\}$, where $x \in \mathscr{G}(\psi)$ and $0 < t < p_\psi(x)$, then we have

$$\mu_\pi(A) = h_x(t) \leq t = \mu_\psi(A).$$

Therefore, the required inequality holds for all sets A of the form $\{\psi(s, x) \mid 0 < s < t\}$ where $x \in \mathscr{G}(\psi)$ and $0 < t < p_\psi(x)$. Suppose now that A is of the form $\{\psi(s, x) \mid t_1 < s < t_2\}$ where $x \in \mathscr{G}(\psi)$ and $t_1 < t_2 < t_1 + p_\psi(x)$. Then since

$$A = \{\psi(s, y) \mid 0 < s < t_2 - t_1\},$$

where $y = \psi(t_1, x)$, we again have $\mu_\pi(A) \leq \mu_\psi(A)$.

Now let A be a Borel subset of $\{\psi(s, x) \mid t_1 < s < t_2\}$ where $x \in \mathscr{G}(\psi)$ and $t_1 < t_2 < t_1 + p_\psi(x)$. Let E be the simple ψ-time preimage of A based at x which satisfies $E \subseteq (t_1, t_2)$. Then

$$\mu_\psi(A) = m(E)$$
$$= \inf\{m(V) \mid V \text{ is open and } E \subseteq V \subseteq (t_1, t_2)\}$$
$$= \inf\{\mu_\psi(U) \mid U = \psi(V, x), V \text{ is open and } E \subseteq V \subseteq (t_1, t_2)\}.$$

But for each open subset V of (t_1, t_2), V is a countable disjoint union of intervals, say $V = \bigcup_{n=1}^{\infty} V_n$, and if $U_n = \psi(V_n, x)$ for each $n = 1, 2, \ldots$, then $U = \bigcup_{n=1}^{\infty} U_n$, and for each n, we have $\mu_\psi(U_n) \geq \mu_\pi(U_n)$. It follows that if V is an open subset of (t_1, t_2) and $U = \psi(V, x)$, then $\mu_\psi(U) \geq \mu_\pi(U)$. Therefore

$$\mu_\psi(A) \geq \inf\{\mu_\pi(U) \mid U = \psi(V, x), V \text{ is open and } E \subseteq V \subseteq (t_1, t_2)\}$$
$$\geq \mu_\pi(A).$$

Finally, let A be a Borel subset of a non-trivial ψ-orbit. Let x be any point in this orbit, and choose a sequence $\{t_n\}$ $(n = 0, \pm 1, \pm 2, \ldots)$ such that t_n increases to $\frac{1}{2} p_\psi(x)$ as $n \to \infty$, and t_n decreases to $-\frac{1}{2} p_\psi(x)$ as $n \to -\infty$. For each n, let

$$A_n = A \cap \{\psi(s, x) \mid t_n < s < t_{n+1}\},$$

and let

$$B = A \cap \{\psi(t_n, x) \mid n \text{ is an integer}\}.$$

Then since $\mu_\psi(B) = \mu_\pi(B) = 0$, we have

$$\mu_\psi(A) = \mu_\psi(B) + \sum_{n=-\infty}^{\infty} \mu_\psi(A_n) = \sum_{n=-\infty}^{\infty} \mu_\psi(A_n) \geq \sum_{n=-\infty}^{\infty} \mu_\pi(A_n) = \mu_\pi(A).$$

This completes the proof of the theorem.   **QED**

8.3. Theorem. *Let φ be a quasi-flow in a Hausdorff space X, let u and v be nonnegative quasi φ-multipliers, let $\psi = u\varphi$ and let $\pi = v\varphi$.*

Then $\psi \leq \pi$ iff $u(x) \leq v(x)$ for μ_φ-almost all $x \in \mathscr{G}(\varphi)$.

PROOF: Suppose $\psi \leq \pi$. Then since $\mathscr{F}(\pi) \subseteq \mathscr{F}(\psi)$, it follows at once from Theorem 5.41 that $u(x) \leq v(x)$ for μ_φ-almost all $x \in \mathscr{G}(\psi)$. But for every $x \in \mathscr{G}(\varphi) \cap \mathscr{F}(\psi)$, we have $u(x) = 0 \leq v(x)$, and we therefore deduce that $u(x) \leq v(x)$ for μ_φ-almost all $x \in \mathscr{G}(\varphi)$.

Now suppose that $u(x) \leq v(x)$ for μ_φ-almost all $x \in \mathscr{G}(\varphi)$. Then we can show that $\mathscr{G}(\psi) \subseteq \mathscr{G}(\pi)$, for suppose $x \in \mathscr{G}(\psi)$. The integral

$$\int_0^\tau \frac{1}{u\big(\varphi(\sigma, x)\big)} \, d\sigma$$

converges for all sufficiently small τ, and we have $u\big(\varphi(\sigma, x)\big) \leq v\big(\varphi(\sigma,x)\big)$ for almost all $\sigma \in \mathbb{R}$. Therefore, the integral

$$\int_0^\tau \frac{1}{v\big(\varphi(\sigma, x)\big)} \, d\sigma$$

also converges for all sufficiently small τ, and we conclude that $x \in \mathscr{G}(\pi)$. We can now show that $\mu_\pi \leq \mu_\psi$. Let A be any Borel set. If A meets $\mathscr{F}(\psi)$, then

$$\mu_\psi(A) = \infty \geq \mu_\pi(A),$$

and if $A \subseteq \mathscr{G}(\psi)$, then since $A \subseteq \mathscr{G}(\pi)$, we have

$$\mu_\psi(A) = \int_A \frac{1}{u} \, d\mu_\varphi \geq \int_A \frac{1}{v} \, d\mu_\varphi = \mu_\pi(A).$$

This completes the proof.   **QED**

The Relationship between Speed and the Partial Order in $\mathrm{Rep}_q^+(\varphi)$

8.4. Theorem. *Let φ be a quasi-flow in a metric space X, let ψ and π be members of $\mathrm{Rep}_q^+(\varphi)$, and suppose $\psi \leq \pi$.*

Then denoting the upper speeds of ψ and π by $\bar{D}\psi$ and $\bar{D}\pi$, respectively, (see Sections 1.52 and 1.53) we have

$$(\bar{D}\psi)(x) \leq (\bar{D}\pi)(x) \quad \text{for every } x \in X.$$

A fortiori, if for some $x \in X$, $|\dot\psi|(x)$ and $|\dot\pi|(x)$ exist, we have $|\dot\psi|(x) \leq |\dot\pi|(x)$.

PROOF: Choose an isotonic quasi π-reparametrizer f such that $\psi = f$ rep φ, and $|f_x(t)| \leq |t|$ for all $x \in X$ and all $t \in \mathbb{R}$. The inequality $(\overline{D}\psi)(x) \leq (\overline{D}\pi)(x)$ is trivial if $x \in \mathscr{F}(\psi)$. But if $x \in \mathscr{G}(\psi)$ and $t \neq 0$, we have

$$\frac{d\big(\psi(t, x), x\big)}{|t|} = \frac{d\big(\pi(f_x(t), x), x\big)}{|t|} \leq \frac{d\big(\pi(f_x(t), x), x\big)}{|f_x(t)|},$$

and the required inequality follows at once on taking the upper limit as $t \to 0$.   **QED**

8.5. Theorem. *Let φ be a quasi-flow in a metric space X, let $\psi \in \mathrm{Prod}_q^+(\varphi)$, and suppose that $|\dot\varphi|$ exists and is finite and non-zero, μ_φ-almost everywhere in $\mathscr{G}(\psi)$.*
Then $\psi \leq \varphi$ iff $|\dot\psi|(x) \leq |\dot\varphi|(x)$ for μ_φ-almost all $x \in \mathscr{G}(\psi)$.

PROOF: We note first that Theorem 5.48 implies the existence of $|\dot\psi|$ μ_φ-almost everywhere. It therefore follows directly from Theorem 8.4 that if $\psi \leq \varphi$, we have $|\dot\psi| \leq |\dot\varphi|$ μ_φ-almost everywhere in $\mathscr{G}(\psi)$.

Suppose now that $|\dot\psi|(x) \leq |\dot\varphi|(x)$ for μ_φ-almost all $x \in \mathscr{G}(\psi)$. Then the inequality $\psi \leq \varphi$ follows easily from Theorems 5.50 and 8.3. **QED**

8.6. Remark. Theorem 8.5 is no longer true if we omit the assumption that $|\dot\varphi|$ be nonzero μ_φ-almost everywhere in $\mathscr{G}(\psi)$. It is a remarkable fact that there exist flows ψ and φ in $\mathbb{R}$ each of which is an isotonic reparametrization of the other, such that ψ it not slower than φ even though $\dot\varphi$ and $\dot\psi$ exist everywhere in $\mathbb{R}$, are C^∞ functions from $\mathbb{R}$ into $\mathbb{R}$, and satisfy $\dot\varphi(x) = 2\dot\psi(x)$ for every $x \in \mathbb{R}$.

8.7. Example (The Hare and the Tortoise).[3] Let C be the $\frac{1}{3}$-Cantor subset of $[0, 1]$ (see Appendix A) and let w be a C^∞ function from $\mathbb{R}$ into $[0, 1]$, so chosen that

$$C = \{t \in \mathbb{R} \mid w(t) = 0\}, \quad \text{and} \quad w(t) = 1 \text{ whenever } |t| \geq 2.$$

Let

$$f_0(t) = \int\limits_0^t w(s)\,ds, \quad \forall\, t \in \mathbb{R}.$$

---

[3] This Example is known as THE HARE AND THE TORTOISE. If we choose the measure λ so that $\lambda(K) > f_0^{-1}(1)$, then we can recast the example as a paradox after the fable of Aesop: The hare (φ) and the tortoise (ψ) run a race from 0 to 1. At every point of the race course, the velocity of the hare is double the velocity of the tortoise $(\dot\varphi(x) = 2\dot\psi(x))$. Unlike the tale in Aesop, neither the hare nor the tortoise ever sleeps $(\mathscr{F}(\varphi) = \mathscr{F}(\psi) = \square)$. Nonetheless, it again transpires that the tortoise wins:

$$\big(\mu_\varphi([0, 1]) = f_0^{-1}(1) + \lambda(K) > 2f_0^{-1}(1) = \mu_\psi([0, 1])\big).$$

Now (cf. Example 4.53) for each $x \in \mathbb{R}$, let g_x be the function from $\mathbb{R}$ onto $\mathbb{R}$ defined by

$$g_x(\tau) = f_0^{-1}(x + \tau) - f_0^{-1}(x), \qquad \forall \, \tau \in \mathbb{R},$$

and let $f_x = g_x^{-1}$.

Define the function f from $\mathbb{R} \times \mathbb{R}$ into $\mathbb{R}$ by

$$f(x, t) = f_x(t), \qquad \forall \, x \in \mathbb{R} \qquad \forall \, t \in \mathbb{R},$$

and define $\psi = f \operatorname{rep} \mathbf{1}$. We note that $\psi = u\mathbf{1}$, where

$$u(x) = w\big(g_0(x)\big), \qquad \forall \, x \in \mathbb{R},$$

and therefore, defining $K = f_0(C)$, we have $\mu_\psi(K) = 0$, and $\dot{\psi}(x) = 0$ iff $x \in K$.

Choose a finite, non-zero, continuous, nonnegative Borel measure λ on K, and let

$$\mu = \lambda + \frac{1}{2}\,\mu_\psi.$$

If follows at once from Theorem 5.33 that μ is the time measure of a flow $\varphi \in \operatorname{Rep}_q^+(\psi)$. It is clear that $\varphi \leq 2\psi$, and from this, we deduce that $\dot{\varphi}(x) = 0$ for every $x \in K$. Also, since $2\mu_\varphi$ and μ_ψ agree in the open set $\mathbb{R} \setminus K$, it is easy to see that $\dot{\varphi}(x) = 2\dot{\psi}(x)$ for every $x \in \mathbb{R} \setminus K$.

We therefore have $\dot{\varphi}(x) = 2\dot{\psi}(x)$ for every $x \in \mathbb{R}$; but since

$$\mu_\varphi(K) = \lambda(K) > 0 = \mu_\psi(K),$$

we do not have $\psi \leq \varphi$.

The next theorem is a sharper form of Theorem 8.5.

8.8. Theorem. *Let φ be a quasi-flow in a metric space X and let $\psi, \pi \in \operatorname{Rep}_q^+(\varphi)$. Suppose that for μ_π-almost all $x \in \mathscr{G}(\psi)$, $|\dot{\pi}|(x)$ exists and satisfies*

$$0 < |\dot{\pi}|(x) < \infty.$$

Then

(a) $\psi \in \operatorname{Rep}_q^+(\pi)$,

(b) $|\dot{\psi}|$ *exists μ_π-almost everywhere in X,*

and

(c) $\psi \leq \pi$ *iff* $|\dot{\psi}|(x) \leq |\dot{\pi}|(x)$ *for μ_π-almost all $x \in \mathscr{G}(\psi)$.*

PROOF: We show first that $\mathscr{F}(\pi) \subseteq \mathscr{F}(\psi)$. Suppose, to obtain a contradiction, that a point x lies in $\mathscr{F}(\pi) \cap \mathscr{G}(\psi)$. Then since $\mu_\pi(\{x\}) = \infty > 0$, we must have $|\dot{\pi}|(x) > 0$, which is clearly impossible. This shows that $\mathscr{F}(\pi) \subseteq \mathscr{F}(\psi)$, and therefore Theorem 4.14 implies that $\psi \in \mathrm{Rep}_q^+(\pi)$. In view of Theorem 5.48, (b) follows at once from (a).

To prove (c), we note first that if $\psi \leq \pi$, then Theorem 8.4 implies that $|\dot{\psi}|(x) \leq |\dot{\pi}|(x)$ for μ_π-almost all $x \in \mathscr{G}(\psi)$. Suppose now that $|\dot{\psi}|(x) \leq |\dot{\pi}|(x)$ for μ_π-almost all $x \in \mathscr{G}(\psi)$, and choose a quasi π-reparametrizer f such that $\psi = f$ rep π. Denote inv(f) by $\{g_x \mid x \in \mathscr{G}(\psi)\}$. Using Theorem 5.48, it is easy to se that if $x \in \mathscr{G}(\psi)$ then for almost all τ in the domain of g_x, we have

$$g_x'(\tau) = g_{\varphi(\tau, x)}'(0) = \frac{|\dot{\varphi}|\big(\varphi(\tau, x)\big)}{|\dot{\psi}|\big(\varphi(\tau, x)\big)} \geq 1.$$

Consequently, if τ lies in the domain of g_x, we have

$$|g_x(\tau)| \geq \left| \int_0^\tau g_x'(\sigma)\, d\sigma \right| \geq |\tau|,$$

and it follows at once from Theorem 8.2 that $\psi \leq \pi$.   **QED**

Measure Theoretic Preliminaries to the Lattice Operations in $\mathrm{Rep}_q^+(\varphi)$

8.9. Note. Much of the following material on maxima and minima of measures will be known to the reader, but we include it for the sake of completeness. Although many of the results which follow can be extended without much difficulty to handle extended real-valued measures which take only one of the values $-\infty$ and ∞, we shall not bother to do this. All measures considered here will be either finite real measures, or nonnegative extended real-valued measures (*i.e.*, measures which take their values in $[0, \infty]$).

We remind the reader that every finite-valued measure (indeed, every Banach-space valued measure) defined on a σ-algebra is bounded.

8.10. Definition. Let μ and λ be (countably additive)[4] measures on a σ-algebra Σ of subsets of a set S, and suppose that the measures μ and λ are either finite real-valued, or take their values in $[0, \infty]$.

Then the *maximum* $\mu \vee \lambda$ of μ and λ is defined by

$$(\mu \vee \lambda)(A) = \sup\big\{\mu(B) + \lambda(A \smallsetminus B) \mid B \in \Sigma \text{ and } B \subseteq A\big\}, \text{ for all } A \in \Sigma,$$

and the *minimum* $\mu \wedge \lambda$ of μ and λ is defined by

$$(\mu \wedge \lambda)(A) = \inf\big\{\mu(B) + \lambda(A \smallsetminus B) \mid B \in \Sigma \text{ and } B \subseteq A\big\}, \text{ for all } A \in \Sigma.$$

---

[4] Unless explicitly stated otherwise, all measures are countably additive.

8.11. Theorem. *Let μ and λ be measures defined on a σ-algebra Σ of subsets of a set S, and suppose that μ and λ are either finite real-valued, or take their values in $[0, \infty]$.*

Then $\mu \vee \lambda$ and $\mu \wedge \lambda$ are measures defined on Σ, and furthermore,

(a) *If μ and λ are finite, then so are $\mu \vee \lambda$ and $\mu \wedge \lambda$;*

(b) *$\mu \vee \lambda$ and $\mu \wedge \lambda$ are, respectively, the supremum and infimum of μ and λ in the partially ordered family of extended real-valued measures defined on Σ.*

PROOF: The proof of this theorem is elementary, and is omitted. **QED**

8.12. Definition. If μ and λ are measures defined on a σ-algebra Σ of subsets of a set S, then we say that μ and λ are *mutually singular* and write $\mu \perp \lambda$ when there exist mutually disjoint sets A and B in Σ with μ concentrated[5] on A, and λ concentrated on B.

8.13. Theorem. *Let μ and λ be nonnegative, extended real-valued measures defined on a σ-algebra Σ of subsets of a set S.*

Then $\mu \perp \lambda$ iff $\mu \wedge \lambda = 0$.

PROOF: It is obvious that if $\mu \perp \lambda$, then $\mu \wedge \lambda = 0$. Suppose that $\mu \wedge \lambda = 0$, and for each $n = 1, 2, \ldots$, choose $A_n, B_n \in \Sigma$ such that

$$A_n \cap B_n = \square, \qquad A_n \cup B_n = S, \qquad \text{and} \qquad \lambda(A_n) + \mu(B_n) < \frac{1}{2^n}.$$

Let $A = \bigcup_{n=1}^{\infty} \bigcap_{j=n}^{\infty} A_j$ and $B = S \setminus A = \bigcap_{n=1}^{\infty} \bigcup_{j=n}^{\infty} B_j$. It is clear that μ is concentrated on A and λ is concentrated on B, and since $A \cap B = \square$, we see that $\mu \perp \lambda$.   **QED**

8.14. Theorem. *Let μ be a finite real measure defined on a σ-algebra Σ of subsets of a set S, let $\mu^+ = \mu \vee 0$, and let $\mu^- = -(\mu \wedge 0)$.*

Then

(a) *μ^+ and μ^- are nonnegative finite measures defined on Σ, and $\mu = \mu^+ - \mu^-$.*

(b) *Given any nonnegative finite measures μ_1 and μ_2 satisfying $\mu = \mu_1 - \mu_2$, we have*

$$\mu_1 \wedge \mu_2 = 0 \ \textit{iff} \ \mu_1 = \mu^+ \ \textit{and} \ \mu_2 = \mu^-.$$

---

[5] A measure μ is said to be concentrated on a set A if it is zero on every measurable subset of the complement of A.

PROOF: (a): For any $A \in \Sigma$, we have

$$\big(\mu - (\mu \vee 0)\big)(A) = \mu(A) - \sup\{\mu(B) \mid B \in \Sigma \text{ and } B \subseteq A\}$$
$$= \inf\{\mu(A) - \mu(B) \mid B \in \Sigma \text{ and } B \subseteq A\}$$
$$= -(\mu \wedge 0)(A).$$

Thus, $\mu = \mu^+ - \mu^-$, and (a) follows.

(b): Suppose μ_1 and μ_2 are nonnegative finite measures and $\mu = \mu_1 - \mu_2$. For any $A \in \Sigma$, we have

$$\big(\mu_1 - (\mu \vee 0)\big)(A) = \inf\{\mu_1(A) - \mu(B) \mid B \in \Sigma \text{ and } B \subseteq A\}$$
$$= \inf\{\mu_1(A \setminus B) + \mu_2(B) \mid B \in \Sigma \text{ and } B \subseteq A\}$$
$$= (\mu_1 \wedge \mu_2)(A).$$

It follows that $\mu_1 \wedge \mu_2 = \mu_1 - (\mu \vee 0)$, and therefore we have $\mu_1 \wedge \mu_2 = 0$ iff $\mu_1 = \mu \vee 0 = \mu^+$. (b) now follows at once.   **QED**

8.15. Corollary. *If μ is a finite real measure defined on a σ-algebra Σ of subsets of a set S, then μ can be expressed uniquely as a difference*

$$\mu = \mu^+ - \mu^-$$

where μ^+ and μ^- are mutually singular, nonnegative finite measures defined on Σ.

8.16. Theorem. *Let μ and λ be finite real measures defined on a σ-algebra Σ of subsets of a set S.*

Then there exists a set $A \in \Sigma$ such that

(a) whenever $B \in \Sigma$ and $B \subseteq A$ we have $\mu(B) \leq \lambda(B)$,

and

(b) whenever $B \in \Sigma$ and $B \subseteq S \setminus A$ we have $\lambda(B) \leq \mu(B)$.

PROOF: This follows at once by applying Corollary 8.15 to the measure $\mu - \lambda$.   **QED**

8.17. Theorem. *Let μ be a nonnegative measure defined on a σ-algebra Σ of subsets of a set S, and let f_1 and f_2 be nonnegative Σ-measurable functions on S. Let λ_1 and λ_2 be the measures defined on Σ by*

$$\lambda_1(A) = \int_A f_1 \, d\mu \quad \text{and} \quad \lambda_2(A) = \int_A f_2 \, d\mu, \qquad \forall \, A \in \Sigma.$$

Then for every $A \in \Sigma$, *we have*

(a) $(\lambda_1 \vee \lambda_2)(A) = \int\limits_A (f_1 \vee f_2) \, d\mu$

and

(b) $(\lambda_1 \wedge \lambda_2)(A) = \int\limits_A (f_1 \wedge f_2) \, d\mu.$

PROOF: Suppose $A \in \Sigma$. Let $A_1 = \{x \in A \mid f_1(x) \leq f_2(x)\}$ and $A_2 = A \setminus A_1$. Then clearly

$$(\lambda_1 \vee \lambda_2)(A) = (\lambda_1 \vee \lambda_2)(A_1 \cup A_2) = \lambda_2(A_1) + \lambda_1(A_2)$$

$$= \int\limits_{A_1} (f_1 \vee f_2) \, d\mu + \int\limits_{A_2} (f_1 \vee f_2) \, d\mu$$

$$= \int\limits_A (f_1 \vee f_2) \, d\mu.$$

Similarly, we see that

$$(\lambda_1 \wedge \lambda_2)(A) = \lambda_1(A_1) + \lambda_2(A_2) = \int\limits_A (f_1 \wedge f_2) \, d\mu. \qquad \textbf{QED}$$

8.18. Remark and Definition. In this chapter, we shall be concerned with four binary operations in families of nonnegative measures. Among these are $+, \vee$ and $\wedge$; the fourth will be introduced later.

These four operations have certain properties in common, which allow us to prove some theorems for all four of them simultaneously. We call these properties *consistent:*

Let Σ be a σ-algebra of subsets of a set S. A binary operation $\circ$ in the family $\mathcal{M}$ of nonnegative measures defined on Σ is said to be *consistent* if it satisfies the following:

(a) If $\mu_1, \mu_2, \lambda_1, \lambda_2 \in \mathcal{M}$ and $A \in \Sigma$, and if for every measurable subset B of A we have

$$\mu_1(B) = \lambda_1(B) \text{ and } \mu_2(B) = \lambda_2(B),$$

then

$$(\mu_1 \circ \mu_2)(A) = (\lambda_1 \circ \lambda_2)(A).$$

(b) If $\mu_1, \mu_2, \lambda_1, \lambda_1 \in \mathcal{M}$ and $\mu_1 \leq \lambda_1$ and $\mu_2 \leq \lambda_2$, then $\mu_1 \circ \mu_2 \leq \lambda_1 \circ \lambda_2$.

and

(c) If $\mu_1, \mu_2 \in \mathcal{M}$ and $c \in [0, \infty]$, then

$$c(\mu_1 \circ \mu_2) = (c\mu_1) \circ (c\mu_2).$$

8.19. Theorem. *Let* Σ *be a σ-algebra of subsets of a set* S *and let* $\mathscr{M}$ *be the family of nonnegative measures defined on* Σ.

Then $+, \vee$ *and* $\wedge$ *are consistent operations in* $\mathscr{M}$.

PROOF: $+$ and $\vee$ are clearly consistent, and it will be clear that $\wedge$ is consistent when we have shown that, if $\mu_1, \mu_2 \in \mathscr{M}$, then

$$\infty\,(\mu_1 \wedge \mu_2) = (\infty\,\mu_1) \wedge (\infty\,\mu_2).$$

Let $\mu_1, \mu_2 \in \mathscr{M}$ and $A \in \Sigma$. If $(\mu_1 \wedge \mu_2)(A) \neq 0$, it is clear that

$$\big(\infty\,(\mu_1 \wedge \mu_2)\big)(A) = \infty = \big((\infty\,\mu_1) \wedge (\infty\,\mu_2)\big)(A).$$

On the other hand, if $(\mu_1 \wedge \mu_2)(A) = 0$, then by Theorem 8.13, we can choose a measurable subset B of A such that

$$\mu_1(B) = \mu_2(A \setminus B) = 0.$$

Since $(\infty\,\mu_1)(B) = (\infty\,\mu_2)(A \setminus B) = 0$, we have

$$\big((\infty\,\mu_1) \wedge (\infty\,\mu_2)\big)(A) = 0 = \big(\infty\,(\mu_1 \wedge \mu_2)\big)(A). \qquad \textbf{QED}$$

8.20. Theorem. *Let* $\circ$ *be a consistent operation in the family* $\mathscr{M}$ *of nonnegative measures defined on a σ-algebra* Σ *of subsets of a set* S. *Let* $\mu_1, \mu_2 \in \mathscr{M}$ *and let* f *be a nonnegative measurable function defined on* S. *Let* λ_1 *and* λ_2 *be the measures defined on* Σ *by*

$$\lambda_1(A) = \int_A f\,d\mu_1 \ \text{and} \ \lambda_2(A) = \int_A f\,d\mu_2, \qquad \forall\,A \in \Sigma.$$

Then for every $A \in \Sigma$, *we have*

$$(\lambda_1 \circ \lambda_2)(A) = \int_A f\,d(\mu_1 \circ \mu_2).$$

PROOF: We first show that if $A \in \Sigma$, and f takes the constant value c everywhere in A, then

$$(\lambda_1 \circ \lambda_2)(A) = \int_A f\,d(\mu_1 \circ \mu_2).$$

To see this, define $\lambda_1' = c\mu_1$ and $\lambda_2' = c\mu_2$. Then we have

$$(\lambda_1 \circ \lambda_2)(A) = (\lambda_1' \circ \lambda_2')(A) = \big((c\mu_1) \circ (c\mu_2)\big)(A)$$

$$= \big(c(\mu_1 \circ \mu_2)\big)(A) = \int_A f\,d(\mu_1 \circ \mu_2).$$

Next we see that if f is a simple function, say,

$$f = \sum_{j=1}^{n} c_j \chi_{E_j},$$

and $A \in \Sigma$, then

$$(\lambda_1 \circ \lambda_2)(A) = \sum_{j=1}^{n} (\lambda_1 \circ \lambda_2)(A \cap E_j)$$

$$= \sum_{j=1}^{n} \int_{A \cap E_j} f \, d(\mu_1 \circ \mu_2) = \int_A f \, d(\mu_1 \circ \mu_2).$$

Finally, suppose f is any nonnegative measurable function and $A \in \Sigma$. Let

$$A_0 = \{x \in A \mid f(x) = 0\} \quad \text{and} \quad A_\infty = \{x \in A \mid f(x) = \infty\},$$

and for each $k = 1, 2, \ldots$, let

$$A_k = \left\{ x \in A \,\middle|\, \frac{1}{k} \leq f(x) \leq k \right\}.$$

We already know that

$$(\lambda_1 \circ \lambda_2)(A_0) = \int_{A_0} f \, d(\mu_1 \circ \mu_2) \quad \text{and} \quad (\lambda_1 \circ \lambda_2)(A_\infty) = \int_{A_\infty} f \, d(\mu_1 \circ \mu_2).$$

Therefore, since $A_1 \subseteq A_2 \subseteq A_3 \subseteq \ldots$, the proof will be complete as soon as we have shown that, for each $k = 1, 2, \ldots$, we have

$$(\lambda_1 \circ \lambda_2)(A_k) = \int_{A_k} f \, d(\mu_1 \circ \mu_2).$$

Fix k, and choose an increasing sequence $\{f_n\}$ of simple functions which converges uniformly to f on A_k in such a way that, for every $n = 1, 2, \ldots$, and $x \in A_k$, we have

$$f_n(x) \leq f(x) \leq f_n(x) + \frac{1}{n}.$$

For every $n = 1, 2, \ldots$, define $f_n(x) = 0$ for all $x \in S \setminus A_k$. For each $n = 1, 2 \ldots$, define the measures $\underline{\lambda}_1^n, \bar{\lambda}_1^n, \underline{\lambda}_2^n, \bar{\lambda}_2^n$, on Σ by

$$\underline{\lambda}_1^n(E) = \int_E f_n \, d\mu_1, \qquad \bar{\lambda}_1^n(E) = \int_E \left(f_n + \frac{1}{n} \right) d\mu_1,$$

$$\underline{\lambda}_2^n(E) = \int_E f_n \, d\mu_2, \qquad \bar{\lambda}_2^n(E) = \int_E \left(f_n + \frac{1}{n} \right) d\mu_2,$$

for every $E \in \Sigma$.

Then since $\underline{\lambda}_1^n \leq \lambda_1$ and $\underline{\lambda}_2^n \leq \lambda_2$ for every $n = 1, 2, \ldots$, we have

$$(\lambda_1 \circ \lambda_2)\,(A_k) \geq (\underline{\lambda}_1^n \circ \underline{\lambda}_2^n)\,(A_k) = \int_{A_k} f_n\, d(\mu_1 \circ \mu_2)\,.$$

But the Lebesgue Monotone Convergence Theorem implies that

$$\int_{A_k} f_n\, d(\mu_1 \circ \mu_2) \to \int_{A_k} f\, d(\mu_1 \circ \mu_2) \quad \text{as } n \to \infty\,,$$

and we conclude that

$$(\lambda_1 \circ \lambda_2)\,(A_k) \geq \int_{A_k} f\, d(\mu_1 \circ \mu_2)\,.$$

Since $f(x) \geq \frac{1}{k}$ for every $x \in A_k$, the reverse inequality is trivial if $(\mu_1 \circ \mu_2)\,(A_k) = \infty$. Suppose then that $(\mu_1 \circ \mu_2)(A_k) < \infty$. Then the Lebesgue Dominated Convergence Theorem implies that

$$\int_{A_k} \left(f_n + \frac{1}{n}\right) d(\mu_1 \circ \mu_2) \to \int_{A_k} f\, d(\mu_1 \circ \mu_2) \quad \text{as } n \to \infty\,.$$

But for every measurable subset E of A_k, we have

$$\lambda_1(E) \leq \bar{\lambda}_1^n(E) \quad \text{and} \quad \lambda_2(E) \leq \bar{\lambda}_2^n(E)\,,$$

and it follows easily that

$$(\lambda_1 \circ \lambda_2)\,(A_k) \leq (\bar{\lambda}_1^n \circ \bar{\lambda}_2^n)\,(A_k) = \int_{A_k} \left(f_n + \frac{1}{n}\right) d(\mu_1 \circ \mu_2)\,.$$

We conclude that

$$(\lambda_1 \circ \lambda_2)\,(A_k) \leq \int_{A_k} f\, d(\mu_1 \circ \mu_2)\,,$$

and this completes the proof.   **QED**

8.21. Definition. Let φ be a quasi-flow in a Hausdorff space X.

Then $\mathcal{M}_q(\varphi)$ denotes the family of time measures of members of $\mathrm{Rep}_q(\varphi)$.

If φ is a flow, then $\mathcal{M}(\varphi)$ denotes the family of time measures of members of $\mathrm{Rep}(\varphi)$.

8.22. Remark. The next theorem states that Theorem 8.16 holds for measures in any family $\mathcal{M}_q(\varphi)$ and this observation provides us with very simple proofs of theorems such as the distributivity of $\vee$ over $\wedge$. We remark however that many of the results we shall state for time measures are true for all nonnegative measures.

8.23. Theorem. *Let φ be a quasi-flow in a Hausdorff space X, and let $u, \lambda \in \mathcal{M}_q(\varphi)$.*

Then there exists a subset A of X which is $\boldsymbol{\Sigma}(\varphi)$-measurable and satisfies

(i) *whenever $B \in \boldsymbol{\Sigma}(\varphi)$ and $B \subseteq A$ we have $\mu(B) \leq \lambda(B)$,*

and

(ii) *whenever $B \in \boldsymbol{\Sigma}(\varphi)$ and $B \subseteq X \setminus A$ we have $\lambda(B) \leq \mu(B)$.*

PROOF: There is no loss of generality in assuming that φ has only one orbit. Let

$$C = \left\{ x \in X \mid \lambda(\{x\}) = \infty \right\} \text{ and } D = \left\{ x \in X \mid \mu(\{x\}) = \infty \right\}.$$

Then since every total φ-time preimage of C is closed in $\mathbb{R}$, C is a Borel set. In the same way, we see that D is a Borel set. The restrictions of both μ and λ to $X \setminus (C \cup D)$ are σ-finite, and we can therefore express $X \setminus (C \cup D)$ as a disjoint union

$$X \setminus (C \cup D) = \bigcup_{n=1}^{\infty} E_n,$$

where for each $n = 1, 2, \ldots,$ E_n is a Borel set and both $\mu(E_n)$ and $\lambda(E_n)$ are finite. For each $n = 1, 2, \ldots,$ apply Theorem 8.16 to obtain a Borel subset A_n of E_n such that $\mu(B) \leq \lambda(B)$ for all Borel subsets B of A_n, and $\lambda(B) \leq \mu(B)$ for all Borel subsets B of $E_n \setminus A_n$. Let

$$A = C \cup \bigcup_{n=1}^{\infty} A_n.$$

It is clear that A has the required properties.   **QED**

8.24. Theorem. *Let φ be a quasi-flow in a Hausdorff space X and let μ, ν and $\lambda \in \mathcal{M}_q(\varphi)$.*

Then

(a) $\mu \vee (\nu \wedge \lambda) = (\mu \vee \nu) \wedge (\mu \vee \lambda)$;

(b) $\mu \wedge (\nu \vee \lambda) = (\mu \wedge \nu) \vee (\mu \wedge \lambda)$;

(c) $\mu + (\nu \vee \lambda) = (\mu + \nu) \vee (\mu + \lambda)$;

(d) $\mu + (\nu \wedge \lambda) = (\mu + \nu) \wedge (\mu + \lambda)$;

(e) $\mu + \lambda == (\mu \vee \lambda) + (\mu \wedge \lambda)$.

PROOF: The result follows at once if we use Theorem 8.23 to partition X into eight subsets, in each of which the measures μ, ν and λ are comparable.[6]   **QED**

---

[6] Some of the equations $(a), \ldots, (e)$ may be deduced from others by purely algebraic arguments.

Minimum of Two Flows

8.25. Introduction. In this section, we shall show that if φ is a quasi-flow in a Hausdorff space X, then $\text{Rep}_q^+(\varphi)$ and $\text{Prod}_q^+(\varphi)$ are lower semi-lattices. In a later section, we shall show that $\text{Rep}^+(\varphi)$ and $\text{Prod}^+(\varphi)$ are not, in general, lower semi-lattices.

If ψ and π are members of $\text{Rep}_q^+(\varphi)$, and $\vartheta = \psi \wedge \pi$, then it follows from Definition 8.1 that $\mu_\vartheta \geq \mu_\psi \vee \mu_\pi$. The most natural candidate for the office of time measure of $\psi \wedge \pi$ is therefore $\mu_\psi \vee \mu_\pi$. In view of this, the following theorem is important.

8.26. Theorem. *Let φ be a quasi-flow in a Hausdorff space X, and let $\psi, \pi \in \text{Rep}_q^+(\varphi)$.*

Then both $\mu_\psi \vee \mu_\pi$ and $\mu_\psi + \mu_\pi$ belong to $\mathcal{M}_q(\varphi)$.

PROOF: Since $\frac{1}{2}(\mu_\psi + \mu_\pi) \leq \mu_\psi \vee \mu_\pi \leq \mu_\psi + \mu_\pi$, it is sufficient, by Theorem 5.40, to show that $\mu_\psi + \mu_\pi \in \mathcal{M}_q(\varphi)$. To do this, we need to show that $\mu_\psi + \mu_\pi$ satisfies the conditions of Theorem 5.33. But a simple inspection shows that this is the case.   **QED**

8.27. Corollary. *If φ is a quasi-flow in a Hausdorff space X, then $\text{Rep}_q^+(\varphi)$ is a lower semi-lattice, and furthermore, given members $\psi, \pi \in \text{Rep}_q^+(\varphi)$, we have*

(a) $\mu_{\psi \wedge \pi} = \mu_\psi \vee \mu_\pi$,

(b) $\mathcal{O}_{\psi \wedge \pi}(x) = \mathcal{O}_\psi(x) \cap \mathcal{O}_\pi(x)$, $\forall\, x \in X$,

(c) $\mathcal{F}(\psi \wedge \pi) = \mathcal{F}(\psi) \cup \mathcal{F}(\pi)$

and

(d) $\mathcal{G}(\psi \wedge \pi) = \mathcal{G}(\psi) \cap \mathcal{G}(\pi)$.

8.28. Theorem. *Let φ be a quasi-flow in a Hausdorff space X and let u and v be nonnegative quasi φ-multipliers.*

Then $u \wedge v$ is a quasi φ-muliplier and

$$(u \wedge v)\,\varphi = u\varphi \wedge v\varphi.$$

PROOF: It is clear that $u \wedge v$ is a nonnegative $\Sigma(\varphi)$-measurable function which vanishes everywhere in $\mathcal{G}(\varphi) \cap \mathcal{F}(u\varphi \wedge v\varphi)$.

Now suppose A is a $\boldsymbol{\Sigma}(\varphi)$-measurable subset of $\mathscr{G}(u\varphi \wedge v\varphi)$. Then since $A \subseteq \mathscr{G}(u\varphi) \cap \mathscr{G}(v\varphi)$, Theorem 8.17(a) implies that

$$\mu_{(u\varphi \wedge v\varphi)}(A) = (\mu_{u\varphi} \vee \mu_{v\varphi})(A)$$

$$= \int_A \left(\frac{1}{u} \vee \frac{1}{v}\right) d\mu_\varphi = \int_A \left(\frac{1}{u \wedge v}\right) d\mu_\varphi.$$

The theorem now follows immediately from Theorem 5.42.   **QED**

8.29. Remark. If φ is a flow in a Hausdorff space X and u and v are nonnegative φ-multipliers then $u\varphi \wedge v\varphi$ need not be continuous. An example of this will be given later. However, with additional hypotheses, we can make $u\varphi \wedge v\varphi$ continuous. Using Theorem 4.44, we obtain this positive result:

8.30. Theorem. *Let φ be a flow in a Hausdorff space X, and let u and v be nonnegative proper quasi φ-mutipliers. Suppose that u and v are continuous in $\mathscr{G}(\varphi)$, and that for each $x \in \mathscr{F}(\varphi)$, there exists a neighborhood W_x of x such that both u and v are bounded in $\mathscr{G}(\varphi) \cap W_x$.*

Then $u\varphi \wedge v\varphi$ is a flow.

PROOF: It is easy to see that $u \wedge v$ is a proper quasi φ-multiplier which is continuous in $\mathscr{G}(\varphi)$, and bounded in each of the sets $\mathscr{G}(\varphi) \cap W_x$. The result therefore follows at once from Theorem 4.44.   **QED**

8.31. Theorem. *Let φ be a quasi-flow in a Hausdorff space X and let $\psi, \pi \in \mathrm{Rep}_q^+(\varphi)$. Let u be a nonnegative function defined on X, and suppose that u is both a quasi ψ-multiplier and a quasi π-multiplier.*

Then u is a quasi $\psi \wedge \pi$-multiplier, and

$$u(\psi \wedge \pi) = u\psi \wedge u\pi.$$

PROOF: We note first that since $\psi \wedge \pi \in \mathrm{Rep}_q(\psi)$, we have $\boldsymbol{\Sigma}(\psi) \subseteq \boldsymbol{\Sigma}(\psi \wedge \pi)$, and therefore u is $\boldsymbol{\Sigma}(\psi \wedge \pi)$-measurable. Next we note that u vanishes at every point of $\mathscr{G}(\psi \wedge \pi) \cap \mathscr{F}(u\psi \wedge u\pi)$. Now $u\psi \wedge u\pi \in \mathrm{Rep}_q^+(\psi \wedge \pi)$ (by Theorem 4.14), and therefore the result will follow from Theorem 5.42 once we have shown that

$$\mu_{u\psi \wedge u\pi}(A) = \int_A \frac{1}{u} d\mu_{\psi \wedge \pi},$$

for every $\boldsymbol{\Sigma}(\varphi)$-measurable subset A of $\mathscr{G}(u\psi \wedge u\pi)$.

Let A be any $\Sigma(\varphi)$-measurable subset of $\mathscr{G}(u\psi \wedge u\pi)$. Then $A \subseteq \mathscr{G}(u\psi) \cap \mathscr{G}(u\pi)$, and therefore

$$\mu_{u\psi \wedge u\pi}(A) = (\mu_{u\psi} \vee \mu_{u\pi})(A)$$

$$= \int_A \frac{1}{u} \, d(\mu_\psi \vee \mu_\pi), \qquad \text{by Theorems 8.19 and 8.20,}$$

$$= \int_A \frac{1}{u} \, d\mu_{\psi \wedge \pi}.$$

This completes the proof.   **QED**

8.32. Theorem. *Let φ be a quasi-flow in a Hausdorff space X, let $\psi, \pi \in \mathrm{Rep}_q^+(\varphi)$, and suppose that $\psi \wedge \pi$ is a flow. Let u be a non-negative function defined on X which is both a quasi ψ-multiplier and a quasi π-multiplier, and suppose that for at least one of ψ and π, u is a proper quasi-multiplier. Suppose that u is continuous in $\mathscr{G}(\psi) \cap \mathscr{G}(\pi)$ and that for each point $x \in \mathscr{F}(\psi) \cup \mathscr{F}(\pi)$, there is a neighborhood W_x of x such that u is bounded in $W_x \cap \mathscr{G}(\psi) \cap \mathscr{G}(\pi)$.*

Then $u\psi \wedge u\pi$ is a flow.

PROOF: Since u is clearly a proper quasi $\psi \wedge \pi$-multiplier which satisfies the appropriate continuity and boundedness requirements, the result follows from Theorem 4.44.   **QED**

Harmonic Sums

8.33. Definition. For each pair a, b of members of $[0, \infty]$, we define the *harmonic sum* $a \oplus b$ of a and b by

$$a \oplus b = \left(\frac{1}{a} + \frac{1}{b}\right)^{-1}$$

with the convention that

$$\frac{1}{0} = \infty, \qquad \frac{1}{\infty} = 0, \qquad \text{and} \qquad 0 \cdot \infty = \infty \cdot 0 = 0.$$

8.34. Theorem. $\oplus$ *is a commutative, associative binary operation in $[0, \infty]$, and furthermore,*

(a) *For every $a \in [0, \infty]$, we have*

$$a \oplus 0 = 0 \quad \text{and} \quad a \oplus \infty = a;$$

(b) *For every* $a, b \in [0, \infty]$ *we have*

$$\frac{1}{2}(a \wedge b) \leq a \oplus b \leq a \wedge b;$$

(c) *Whenever* $a, b, c \in [0, \infty]$ *and* $a \leq b$, *we have*

$$a \oplus c \leq b \oplus c;$$

(d) *For every* $a, b, c \in [0, \infty]$, *we have*

$$a(b \oplus c) = ab \oplus ac;$$

(e) *For every* $a_1, a_2, b_1, b_2 \in [0, \infty]$, *we have*

$$(a_1 \oplus b_1) + (a_2 \oplus b_2) \leq (a_1 + a_2) \oplus (b_1 + b_2)$$
$$\leq (a_1 \oplus b_1) + (a_2 + b_2).$$

(f) *If* n *is a positive integer, and* $a_1, \ldots, a_n, b_1, \ldots, b_n \in [0, \infty]$, *we have*

$$\left(\sum_{j=1}^{n} a_j\right) \oplus \left(\sum_{j=1}^{n} b_j\right) \geq \sum_{j=1}^{n} (a_j \oplus b_j).$$

PROOF: (a), (b), (c) and (d) are obvious, and (f) follows by induction from the first inequality in (e).

To prove (e), let $a_1, a_2, b_1, b_2 \in [0, \infty]$. The first inequality of (e) clearly holds if any of the numbers a_1, a_2, b_1, b_2 is either 0 or ∞. If a_1, a_2, b_1, b_2 are all nonzero and finite then

$$(a_1 + a_2) \oplus (b_1 + b_2) - [(a_1 \oplus b_1) + (a_2 \oplus b_2)]$$

$$= \frac{(a_1 b_2 - a_2 b_1)^2}{(a_1 + a_2 + b_1 + b_2)(a_1 + b_1)(a_2 + b_2)} \geq 0.$$

This proves the first inequality of (e).

The second inequality of (e) clearly holds if any of the numbers a_1, a_2, b_1, b_2 is ∞, or if $a_1 = 0$ or $b_1 = 0$. If a_1, a_2, b_1, b_2 are all finite and a_1, b_1 are non-zero, then

$$[(a_1 + a_2) \oplus (b_1 + b_2)] - [a_1 \oplus b_1] = \frac{a_1^2 b_2 + a_2 b_1^2 + a_1 a_2 b_2 + a_2 b_1 b_2}{(a_1 + b_1)(a_1 + a_2 + b_1 + b_2)}$$

$$= \frac{(a_1^2 + a_1 a_2) b_2 + (b_1^2 + b_1 b_2) a_2}{(a_1^2 + a_1 a_2) + (b_1^2 + b_1 b_2) + (2a_1 b_1 + a_1 b_2 + a_2 b_1)}$$

$$\leq a_2 + b_2.$$

This completes the proof of (e).   **QED**

Harmonic Sum of Two Flows

8.35. Definition. Let φ be a quasi-flow in a Hausdorff space X and let $\psi, \pi \in \mathrm{Rep}_q^+(\varphi)$. Then the *harmonic sum* $\psi \oplus \pi$ of ψ and π is defined to be that member of $\mathrm{Rep}_q^+(\varphi)$ whose time measure is $\mu_\psi + \mu_\pi$.

8.36. Theorem. *Let φ be a quasi-flow in a Hausdorff space X and let $\psi, \pi \in \mathrm{Rep}_q^+(\varphi)$.*

Then

$$\frac{1}{2}(\psi \wedge \pi) \leq \psi \oplus \pi \leq \psi \wedge \pi.$$

A fortiori, $\psi \wedge \pi$ *and* $\psi \oplus \pi$ *have the same orbits.*

PROOF: The theorem is immediate because

$$\mu_\psi \vee \mu_\pi \leq \mu_\psi + \mu_\pi \leq 2(\mu_\psi \vee \mu_\pi). \qquad \textbf{QED}$$

8.37. Theorem. *Let φ be a quasi-flow in a Hausdorff space X, let $\psi, \pi, \chi \in \mathrm{Rep}_q^+(\varphi)$ and suppose $\psi \leq \chi$.*

Then $\psi \oplus \pi \leq \chi \oplus \pi$.

PROOF: Clear.   **QED**

8.38. Theorem. *Let φ be a quasi-flow in a Hausdorff space X.*

Then $\oplus$ is a commutative, associative binary operation in $\mathrm{Rep}_q^+(\varphi)$, and is distributive over $\wedge$.

PROOF: The theorem follows at once on an examination of the dual statements about time measures. (See Theorem 8.24(c).)   **QED**

8.39. Theorem. *Let φ be a quasi-flow in a Hausdorff space X and let u and v be nonnegative quasi φ-multipliers.*

Then $u \oplus v$ is a quasi φ-multiplier, and

$$(u \oplus v)\,\varphi = u\varphi \oplus v\varphi.$$

PROOF: Since

$$\frac{1}{2}(u \wedge v) \leq u \oplus v \leq u \wedge v,$$

it is clear that $u \oplus v$ is a quasi φ-multiplier, and that $(u \oplus v)\varphi$ and $(u \wedge v)\varphi$ have the same orbits.

Now let $A \in \boldsymbol{\Sigma}(\varphi)$. If A meets $\mathscr{F}(u\varphi \oplus v\varphi)$, then we clearly have

$$(\mu_{u\varphi \oplus v\varphi})(A) = \infty = \mu_{(u \oplus v)\varphi}(A).$$

On the other hand, if $A \subseteq \mathscr{G}(u\varphi \oplus v\varphi)$, we have

$$\mu_{(u\oplus v)\varphi}(A) = \int_A \frac{1}{u \oplus v}\, d\mu_\varphi$$

$$= \int_A \left(\frac{1}{u} + \frac{1}{v}\right) d\mu_\varphi$$

$$= \mu_{u\varphi}(A) + \mu_{v\varphi}(A) = \mu_{u\varphi \oplus v\varphi}(A).$$

Thus, $\mu_{u\varphi \oplus v\varphi} = \mu_{(u\oplus v)\varphi}$, and the proof is complete.   **QED**

8.40. Note. It is easy to see that Theorems 8.30, 8.31 and 8.32 also hold with $\wedge$ replaced by $\oplus$.

Maximum of Two Flows

8.41. Introduction. If φ is a quasi-flow in a Hausdorff space X, then the process of forming maxima in $\mathrm{Rep}_q^+(\varphi)$ is much more pathological than that of forming minima, and $\mathrm{Rep}_q^+(\varphi)$ is not, in general, an upper semi-lattice. $\mathrm{Rep}_q^+(\varphi)$ need not even be upper-directed, *i.e.*, there may exist members ψ and π of $\mathrm{Rep}_q^+(\varphi)$ which have no common upper bound. Furthermore, even when members ψ and π of $\mathrm{Rep}_q^+(\varphi)$ have a common upper bound, a least upper bound need not exist. The process of forming maxima in $\mathrm{Rep}^+(\varphi)$ is even more pathological, as we shall show in a later section.

Now if ψ and π are members of a family $\mathrm{Rep}_q^+(\varphi)$, and $\vartheta = \psi \vee \pi$ exists, then it follows from Definition 8.1 that $\mu_\vartheta \leq \mu_\psi \wedge \mu_\pi$. Therefore, especially in view of what we know about minima, we might expect to have $\mu_\vartheta = \mu_\psi \wedge \mu_\pi$. Unfortunately, this also need not be true. The difficulty occurs in $\mathscr{F}(\psi) \cap \mathscr{F}(\pi)$ where it is quite possible for ϑ to be moving. Outside $\mathscr{F}(\psi) \cap \mathscr{F}(\pi)$, the measures μ_ϑ and $\mu_\psi \wedge \mu_\pi$ agree, as the following theorems will show.

8.42. Theorem. *Let φ be a quasi-flow in a Hausdorff space X and let $\psi, \pi \in \mathrm{Rep}_q^+(\varphi)$. Let $x \in X$ and $-\infty \leq \tau_1 < \tau_2 \leq \infty$, and suppose that $\{\varphi(\tau, x) \mid \tau_1 < \tau < \tau_2\}$ does not meet $\mathscr{F}(\psi) \cap \mathscr{F}(\pi)$.*

Then the restriction of $\mu_\psi \wedge \mu_\pi$ to $\{\varphi(\tau, x) \mid \tau_1 < \tau < \tau_2\}$ is continuous, and is finite on all subsets of the form $\{\varphi(\tau, x) \mid \tau \in C\}$ with C a compact subset of (τ_1, τ_2).

PROOF: It is obvious that $\mu_\psi \wedge \mu_\pi$ is continuous in

$$\{\varphi(\tau, x) \mid \tau_1 < \tau < \tau_2\}.$$

Now let C be a compact subset of (τ_1, τ_2). For each $\tau \in C$, choose a neighborhood U_τ of τ such that $\{\varphi(\sigma, x) \mid \sigma \in U_\tau\}$ is contained in a single orbit of either ψ or π, and by making U_τ smaller if necessary, let it be so chosen that one of the measures μ_ψ, μ_π is finite on $\{\varphi(\sigma, x) \mid \sigma \in U_\tau\}$. Since finitely many of the sets U_τ cover C, it is clear that

$$(\mu_\psi \wedge \mu_\pi)(C) < \infty. \hspace{3em} \textbf{QED}$$

8.43. Theorem. *Let φ be a quasi-flow in a Hausdorff space X, let ψ, $\pi \in \mathrm{Rep}_q^+(\varphi)$ and suppose that ψ and π have a common upper bound $\vartheta \in \mathrm{Rep}_q^+(\varphi)$. Let $x \in X$ and $-\infty \leq \tau_1 < \tau_2 \leq \infty$, and suppose that $\{\varphi(\tau, x) \mid \tau_1 < \tau < \tau_2\}$ does not meet $\mathscr{F}(\psi) \cap \mathscr{F}(\pi)$.*

Then given any non-empty open subset U of (τ_1, τ_2), we have

$$\big(\mu_\psi \wedge \mu_\pi\big)\big(\{\varphi(\tau, x) \mid \tau \in U\}\big) > 0.$$

PROOF: The result is immediate since $\mu_\psi \wedge \mu_\pi \geq \mu_\vartheta$. \hspace{2em} **QED**

8.44. Theorem. *Let φ be a quasi-flow in a Hausdorff space X, let ψ, $\pi \in \mathrm{Rep}_q^+(\varphi)$ and suppose $\psi \vee \pi$ exists.*

Then $\mathscr{F}(\psi \vee \pi) \subseteq \mathscr{F}(\psi) \cap \mathscr{F}(\pi)$, and the measures $\mu_{\psi \vee \pi}$ and $\mu_\psi \wedge \mu_\pi$ agree in the complement of $\big(\mathscr{F}(\psi) \cap \mathscr{F}(\pi)\big) \setminus \mathscr{F}(\psi \vee \pi)$.

PROOF: There is no loss of generality in assuming that there is only one φ-orbit, and it is obvious that $\mathscr{F}(\psi \vee \pi) \subseteq \mathscr{F}(\psi) \cap \mathscr{F}(\pi)$. Now since both $\mu_{\psi \vee \pi}$ and $\mu_\psi \wedge \mu_\pi$ are infinite at each point of $\mathscr{F}(\psi \vee \pi)$, it is clear that these measures agree in $\mathscr{F}(\psi \vee \pi)$. Suppose, to obtain a contradiction, that there exists a Borel set A such that A does not meet $\mathscr{F}(\psi) \cap \mathscr{F}(\pi)$, and

$$\mu_{\psi \vee \pi}(A) \neq (\mu_\psi \wedge \mu_\pi)(A).$$

Then since $\mu_{\psi \vee \pi} \leq \mu_\psi \wedge \mu_\pi$ and $\mu_\psi \wedge \mu_\pi$ is σ-finite on A, we can choose a Borel subset B of A such that

$$\mu_{\psi \vee \pi}(B) < (\mu_\psi \wedge \mu_\pi)(B) < \infty.$$

Let μ be the Borel measure on X defined by

$$\mu(E) = \mu_{\psi \vee \pi}(E \setminus B) + (\mu_\psi \wedge \mu_\pi)(E \cap B),$$

for every Borel set E. It follows from Theorem 5.33 that $\mu \in \mathcal{M}_q(\varphi)$. Let $\vartheta \in \mathrm{Rep}_q^+(\varphi)$ satisfy $\mu_\vartheta = \mu$. Then since $\mu \leq \mu_\psi \wedge \mu_\pi$, we have $\psi \leq \vartheta$ and $\pi \leq \vartheta$. Therefore $\vartheta \geq \psi \vee \pi$, contradicting the fact that $\mu(B) > \mu_{\psi \vee \pi}(B)$.

This completes the proof of the theorem. \hspace{2em} **QED**

8.45. Jumping. Let φ be a quasi-flow in a Hausdorff space X, and let $\psi, \pi \in \text{Rep}_q^+(\varphi)$. A simple way for the measure $\mu_\psi \wedge \mu_\pi$ to fail to be in $\mathscr{M}_q(\varphi)$ is for $\mu_\psi \wedge \mu_\pi$ to vanish on a set of the form $\{\varphi(\tau, x) \mid \tau \in U\}$ where $x \in X$ and U is a non-empty open subinterval of $\mathbb{R}$. When this happens, $\psi \vee \pi$ does not exist because it would be forced to "jump" over the set $\{\varphi(\tau, x) \mid \tau \in U\}$ in zero time. For the same reason, ψ and π have no common upper bound in $\text{Rep}_q^+(\varphi)$. Thus in order for two members ψ, π of $\text{Rep}_q^+(\varphi)$ to have a common upper bound, the measures μ_ψ and μ_π must not have mutually singular restrictions to any subset of a φ-orbit with a non-empty open total φ-time preimage.

The following two examples illustrate the phenomenon of jumping.

8.46. Example. Let C_1 be the usual Cantor subset, (*i.e.*, the $\frac{1}{3}$-Cantor subset) of $[0, 1]$, and let μ_1 be a nonnegative continuous Borel measure on C_1 satisfying $\mu_1(C_1) = \frac{1}{2}$. Let $K_1 = C_1$.

If $C_1, \ldots, C_n, K_1, \ldots, K_n$, and $\mu_1, \ldots, \mu_n$ have been defined, and K_n is compact, define C_{n+1}, K_{n+1}, and μ_{n+1} as follows: For each component I of $[0, 1] \setminus K_n$, let $C_{n+1,I}$ be the $\frac{1}{3}$-Cantor subset of $\bar{I}$, and let $\mu_{n+1,I}$ be a nonnegative continuous Borel measure on $C_{n+1,I}$ which satisfies

$$\mu_{n+1,I}(C_{n+1,I}) = \frac{1}{2^{n+1}} \cdot m(I).$$

Let

$$\mu_{n+1} = \sum_I \mu_{n+1,I},$$

where the sum is taken over all components I of $[0, 1] \setminus K_n$. Let

$$C_{n+1} = \bigcup_I C_{n+1,I},$$

and let $K_{n+1} = K_n \cup C_{n+1}$.

This defines C_n, K_n and μ_n for every positive integer n, and it is easy to see that for each n,

$$\mu_n(C_n) = \frac{1}{2^n}, \quad m(K_n) = 0, \quad K_n = \bigcup_{j=1}^n C_j,$$

and K_n is compact.

Now let μ be the nonnegative Borel measure defined on $\mathbb{R}$ by

$$\mu(A) = m(A \setminus [0, 1]) + \sum_{n=1}^\infty \mu_n(A \cap C_n).$$

It is clear that $\mu \in \mathscr{M}_q(1)$.[7]

---

[7] If we define a Borel measure λ on $\mathbb{R}$ by repeating in each interval $[n, n + 1]$ the procedure used in this example to define μ in $[0, 1]$, we obtain a nonnegative

Now it is clear that

$$\mu\left([0, 1] \setminus \bigcup_{n=1}^{\infty} C_n\right) = m\left(\bigcup_{n=1}^{\infty} C_n\right) = 0,$$

and therefore the restrictions of μ and m to $(0, 1)$ are mutually singular. Consequently, if we let $\psi = 1$ and π be that member of $\text{Rep}_q^+(1)$ whose time measure is μ, then the would-be maximum of ψ and π jumps over $(0, 1)$.

8.47. Example. The above example exhibits two members of $\text{Rep}_q^+(1)$ whose "maximum" jumps. In this example, we employ a similar technique to exhibit two members of $\text{Prod}_q^+(1)$ whose "maximum" jumps.

We shall first partition $[0, 1]$ into two mutually disjoint Borel subsets A and B in such a way that, for every non-empty open subset U of $[0, 1]$, $m(A \cap U) > 0$ and $m(B \cap U) > 0$:

Let C_1 be the δ-Cantor subset of $[0, 1]$, where δ is so chosen that $m(C_1) = \frac{1}{2}$. Let $K_1 = C_1$. If $C_1, ..., C_n, K_1, ..., K_n$ have been defined, and K_n is compact, then for each component I of $[0, 1] \setminus K_n$, define $C_{n+1,I}$ to be the δ-Cantor subset of $\bar{I}$. It is clear that for each I,

$$m(C_{n+1,I}) = \frac{1}{2} m(I).$$

Let

$$C_{n+1} = \bigcup_I C_{n+1,I},$$

where the union is taken over all components I of $[0, 1] \setminus K_n$, and let $K_{n+1} = K_n \cup C_{n+1}$.

This defines C_n and K_n for every positive integer n. Now define $A = \bigcup_{n=1}^{\infty} C_{2n}$ and $B = [0, 1] \setminus A$.

Now using A and B, we shall define two nonnegative 1-multipliers u and v. Define

$$u(x) = \begin{cases} \infty & \text{whenever } x \in A, \\ 1 & \text{whenever } x \in \mathbb{R} \setminus A, \end{cases}$$

and

$$v(x) = \begin{cases} \infty & \text{whenever } x \in B, \\ 1 & \text{whenever } x \in \mathbb{R} \setminus B. \end{cases}$$

---

continuous singular Borel measure which is finite on compact sets. The function f defined by

$$f(x) = \mu\big((0, x)\big), \quad \forall \, x \geq 0,$$
$$= -\mu\big((x, 0)\big), \quad \forall \, x < 0,$$

is a strictly increasing continuous function from $\mathbb{R}$ onto $\mathbb{R}$ whose derivative is zero almost everywhere.

It is easy to see that u and v are **1**-multipliers and that $\mathscr{F}(u\varphi) = \mathscr{F}(v\varphi) = \square$. Further, for every Borel subset E of $\mathbb{R}$, we have

$$\mu_{u\varphi}(E) = \int_E \frac{1}{u}\,dm = m(E \setminus A)$$

and

$$\mu_{v\varphi}(E) = \int_E \frac{1}{v}\,dm = m(E \setminus B),$$

and therefore $\mu_{u\varphi}$ and $\mu_{v\varphi}$ have mutually singular restrictions to $(0, 1)$.

8.48. Remark. Whenever members ψ and π of a family $\mathrm{Rep}_q^+(\varphi)$ have the same fixed points, it is easy to see that both ψ and π are members of $\mathrm{Prod}_q^+(\psi \wedge \pi)$. Therefore to obtain an example of two members of a family $\mathrm{Prod}_q^+(\vartheta)$ which produce jumping, we need not go to the trouble of producing Example 8.47. All we need do is apply this observation to Example 8.46.

Note, however, that if φ is a quasi-flow and $\psi \in \mathrm{Prod}_q^+(\varphi)$, then φ and ψ can never produce jumping.

8.49. Ramming. Let φ be a quasi-flow in a Hausdorff space X and let $\psi, \pi \in \mathrm{Rep}_q^+(\varphi)$. Even if ψ and π do not produce jumping, ψ and π need not have a common upper bound in $\mathrm{Rep}_q^+(\varphi)$. To see how this can happen, note that for any $x \in X$, all three of the measures μ_ψ, μ_π, μ_φ are necessarily infinite on the set $\{\varphi(\tau, x) \mid \tau \geq 0\}$, i.e., all three of the quasi-flows ψ, π, φ take infinite time to reach the ω-end of $\mathcal{O}_\varphi(x)$ (if they reach it at all). However, it is quite possible for $\mu_\psi \wedge \mu_\pi$ to be finite on $\{\varphi(\tau, x) \mid \tau \geq 0\}$, thus causing the "maximum" of ψ and π to "ram" into the ω-end of $\mathcal{O}_\varphi(x)$ in finite time.

The following example exhibits this phenomenon:

8.50. Example. Let u and v be the functions from $\mathbb{R}$ into $(0, \infty)$ defined by:

$u(x) = v(x) = 1 \qquad \forall\, x \leq 1$,

$u(x) = 1$ and $v(x) = n^2$ whenever $n \leq x < n + 1$ and n is an odd positive integer,

and

$u(x) = n^2$ and $v(x) = 1$ whenever $n \leq x < n + 1$ and n is an even positive integer.

It is easy to see that u and v are proper **1**-multipliers. However,

$$(\mu_{u\varphi} \wedge \mu_{v\varphi})\big((0,\infty)\big) = \int_0^\infty \left(\frac{1}{u} \wedge \frac{1}{v}\right) dm \qquad \text{(by Theorem 8.17(b))}$$

$$= 1 + \sum_{n=1}^\infty \frac{1}{n^2} < \infty,$$

and therefore the maximum of $u\mathbf{1}$ and $v\mathbf{1}$ rams the ω-end of the **1**-orbit.

8.51. Definition. Let φ be a quasi-flow in a Hausdorff space X, let $\psi, \pi \in \mathrm{Rep}_q^+(\varphi)$ and let $x \in X$ satisfy $\mathrm{p}_\varphi(x) = \infty$.

We say that the pair $\{\psi, \pi\}$ *rams* $\alpha_\varphi(x)$ if there exists $\tau \in \mathbb{R}$ such that

(i) $\{\sigma < \tau \mid \varphi(\sigma, x) \in \mathscr{F}(\psi) \cap \mathscr{F}(\pi)\}$ has empty interior,

and

(ii) $\mu_\psi \wedge \mu_\pi$ is finite on the set $\{\varphi(\sigma, x) \mid \sigma < \tau\} \setminus \big(\mathscr{F}(\psi) \cap \mathscr{F}(\pi)\big)$.

We say that the pair $\{\psi, \pi\}$ *rams* $\omega_\varphi(x)$ if there exists $\tau \in \mathbb{R}$ such that

(i) $\{\sigma > \tau \mid \varphi(\sigma, x) \in \mathscr{F}(\psi) \cap \mathscr{F}(\pi)\}$ has empty interior,

and

(ii) $\mu_\psi \wedge \mu_\pi$ is finite on the set $\{\varphi(\sigma, x) \mid \sigma > \tau\} \setminus \big(\mathscr{F}(\psi) \cap \mathscr{F}(\pi)\big)$.

8.52. Theorem. *Let φ be a quasi-flow in a Hausdorff space X and let $\psi, \pi \in \mathrm{Rep}_q^+(\varphi)$.*

Then a necessary and sufficient condition for ψ and π to have an upper bound in $\mathrm{Rep}_q^+(\varphi)$ is that

(i) *the pair $\{\psi, \pi\}$ produces no jumping, i.e., whenever $x \in X$ and $\tau_1 < \tau_2$, we have*

$$(\mu_\psi \wedge \mu_\pi)\big(\{\varphi(\tau, x) \mid \tau_1 < \tau < \tau_2\}\big) > 0,$$

(ii) *for every $x \in X$, if the pair $\{\psi, \pi\}$ rams $\alpha_\varphi(x)$, then for every real number τ, the set*

$$\{\sigma < \tau \mid \varphi(\sigma, x) \in \mathscr{F}(\psi) \cap \mathscr{F}(\pi)\}$$

is uncountable,

and

(iii) *for every $x \in X$, if the pair $\{\psi, \pi\}$ rams $\omega_\varphi(x)$, then for every real number τ, the set*

$$\{\sigma > \tau \mid \varphi(\sigma, x) \in \mathscr{F}(\psi) \cap \mathscr{F}(\pi)\}$$

is uncountable.

PROOF: For convenience of notation, let $F = \mathscr{F}(\psi) \cap \mathscr{F}(\pi)$.

To show that the condition is necessary, suppose that ψ and π have a common upper bound $\vartheta \in \mathrm{Rep}_q^+(\varphi)$. Then since $\mu_\vartheta \leq \mu_\psi \wedge \mu_\pi$, (i) is clearly satisfied. Now suppose that (iii) does not hold. Choose $x \in X$ and $\tau \in \mathbb{R}$ such that the pair $\{\psi, \pi\}$ rams $\omega_\varphi(x)$ and the set $\{\sigma > \tau \mid \varphi(\sigma, x) \in F\}$ is countable. We may clearly assume that $\varphi(\tau, x) \in F$, and therefore defining $y = \varphi(\tau, x)$, we have $y \in \mathscr{G}(\vartheta)$. Choose τ_1 and $\tau_2 \in [-\infty, \infty]$ such that

$$\mathcal{O}_\vartheta(y) = \{\varphi(\sigma, y) \mid \tau_1 < \sigma < \tau_2\}.$$

Since the restriction of μ_ϑ to $\mathcal{O}_\vartheta(y)$ is continuous and the set $F \cap \{\varphi(\sigma, y) \mid \sigma > 0\}$ is countable, we have

$$\infty = \mu_\vartheta\big(\{\vartheta(t, y) \mid t \geq 0\}\big) = \mu_\vartheta\big(\{\varphi(\sigma, y) \mid 0 \leq \sigma < \tau_2\}\big)$$
$$= \mu_\vartheta\big(\{\varphi(\sigma, y) \mid 0 \leq \sigma < \tau_2\} \setminus F\big)$$
$$\leq (\mu_\psi \wedge \mu_\pi)\big(\{\varphi(\sigma, x) \mid \sigma \geq \tau\} \setminus F\big) < \infty.$$

This contradiction shows that (iii) holds, and a similar agument can be used to show that (ii) holds. We have therefore shown that the condition is necessary.

To prove that the condition is sufficient, we may assume without loss of generality that there is only one φ-orbit.

CASE 1: The φ-orbit is periodic and $(\mu_\psi \wedge \mu_\pi)(X \setminus F) < \infty$.

Define a Borel measure μ on X by

$$\mu(E) = (\mu_\psi \wedge \mu_\pi)(E \setminus F) + \mu_\varphi(E \cap F),$$

for all Borel subsets E of X. Using Theorem 5.33, we see that $\mu \in \mathcal{M}_q(\varphi)$, and if ϑ is the member of $\mathrm{Rep}_q^+(\varphi)$ for which $\mu = \mu_\vartheta$, we have $\psi \leq \vartheta$ and $\pi \leq \vartheta$.

CASE 2: The φ-orbit is periodic and $\mu_\psi \wedge \mu_\pi(X \setminus F) = \infty$.

Using Theorem 8.42, we see that $F \neq \square$. Choose $x \in F$. Now write the set $U = \big\{\tau \in (0, \mathrm{p}_\varphi(x)) \mid \varphi(\tau, x) \in X \setminus F\big\}$ as a countable disjoint union $\bigcup\limits_{n=1}^{\infty} J_n$, where for each $n = 1, 2, \ldots$, J_n is an interval and $\mu_\psi \wedge \mu_\pi$ is finite on $\varphi(J_n, x)$. (This can be done since each component of U, being an open interval, can be written in this form using Theorem 8.42.) Choose, for each positive integer n, a number $\delta_n \in (0, 1]$ in such a way that

$$\sum_{n=1}^{\infty} \delta_n \cdot (\mu_\psi \wedge \mu_\pi)\big(\varphi(J_n, x)\big) < \infty.$$

Define a Borel measure μ on X by

$$\mu(E) = \mu_\varphi(E \cap F) + \sum_{n=1}^{\infty} \delta_n \cdot (\mu_\psi \wedge \mu_\pi)\left(E \cap \varphi(J_n, x)\right),$$

for all Borel subsets E of X.

As in Case 1, $\mu = \mu_\vartheta$ for some $\vartheta \in \mathrm{Rep}_q^+(\varphi)$ which satisfies $\psi \leq \vartheta$ and $\pi \leq \vartheta$.

There now remain those cases in which the φ-orbit is aperiodic. In each of these cases, we need to define a Borel measure μ on X such that $\mu = \mu_\vartheta$ for some $\vartheta \in \mathrm{Rep}_q^+(\varphi)$ which satisfies $\psi \leq \vartheta$ and $\pi \leq \vartheta$. Choose a point $x \in X \setminus F$. (If this cannot be done the theorem is trivial.) We shall define the measure μ by defining it separately on $\{\varphi(\tau, x) \mid \tau \geq 0\}$ and $\{\varphi(\tau, x) \mid \tau \leq 0\}$. In fact, once we have shown how μ can be defined on $\{\varphi(\tau, x) \mid \tau \geq 0\}$, the method of definition on $\{\varphi(\tau, x) \mid \tau \leq 0\}$ will be clear. We shall therefore confine ourselves to the task of defining μ on $Y = \{\varphi(\tau, x) \mid \tau \geq 0\}$, and for this purpose, we need only consider the following three cases:

CASE 3: $p_\varphi(x) = \infty$ and $(\mu_\psi \wedge \mu_\pi)(Y \setminus F) < \infty$.

In this case, our hypotheses allow us to choose an increasing sequence $\{r_n\}$ of positive integers such that for each $n = 1, 2, \ldots$, the set $\{\tau \in [r_n, r_{n+1}] \mid \varphi(\tau, x) \in F\}$ is uncountable. Using Corollary 7.15, choose for each $n = 1, 2, \ldots$, a subset K_n of $F \cap \{\varphi(\tau, x) \mid r_n \leq \tau \leq r_{n+1}\}$ such that K_n is homeomorphic to the Cantor set, and choose a nonnegative continuous Borel measure λ_n on X such that λ_n is concentrated on K_n and $\lambda_n(K_n) = 1$.

Define a Borel measure μ on Y by

$$\mu(E) = (\mu_\psi \wedge \mu_\pi)(E \setminus F) + \mu_\varphi(E \cap F) + \sum_{n=1}^{\infty} \lambda_n(E \cap F),$$

for every Borel subset E of Y. The reader can readily verify that μ has the required properties.

CASE 4. $p_\varphi(x) = \infty$, $(\mu_\psi \wedge \mu_\pi)(Y \setminus F) = \infty$, and $\{\tau \mid \varphi(\tau, x) \in F\}$ is bounded above.

Let $\tau = \sup\{\sigma \mid \varphi(\sigma, x) \in F\}$. If $\tau < 0$, we can obtain the required measure μ on Y by defining μ to be the restriction of $\mu_\psi \wedge \mu_\pi$ to Y. Suppose now that $\tau > 0$, and write the set

$$U = \left\{\sigma \in [0, \tau + 1) \mid \varphi(\sigma, x) \in Y \setminus F\right\}$$

as a countable disjoint union $\bigcup_{n=1}^{\infty} J_n$, where for each $n = 1, 2, \ldots, J_n$

is an interval and $\mu_\psi \wedge \mu_\pi$ is finite on $\varphi(J_n, x)$. Choose, for each positive integer n, a number $\delta_n \in (0, 1]$ in such a way that

$$\sum_{n=1}^{\infty} \delta_n \cdot (\mu_\psi \wedge \mu_\pi)\big(\varphi(J_n, x)\big) < \infty.$$

Define μ by

$$\mu(E) = \mu_\varphi(E \cap F) + (\mu_\psi \wedge \mu_\pi)\big(E \cap \varphi([\tau + 1, \infty), x)\big)$$

$$+ \sum_{n=1}^{\infty} \delta_n \cdot (\mu_\psi \wedge \mu_\pi)\big(E \cap \varphi(J_n, x)\big).$$

Again, it is easy to see that μ has the required properties.

CASE 5. $\mathrm{p}_\varphi(x) = \infty$, $(\mu_\psi \wedge \mu_\pi)(Y \setminus F) = \infty$, and $\{\tau \mid \varphi(\tau, x) \in F\}$ is not bounded above.

If there exists $\tau > 0$ such that $\mu_\psi \wedge \mu_\pi$ is finite on the set $\{\varphi(\sigma, x) \mid \sigma > \tau\} \setminus F$, then we can use the method of Case 4 to replace $\mu_\psi \wedge \mu_\pi$ by a finite measure μ_1 on $Y \setminus F$, and we can use the method of Case 3 to define a measure λ on $Y \cap F$ (for our assumption clearly implies that sets of the form $\{\varphi(\sigma, x) \mid \sigma \geq \sigma_0\} \cap F$ are uncountable), and we can then define μ by

$$\mu(E) = \mu_1(E \setminus F) + \lambda(E \cap F) + \mu_\varphi(E \cap F),$$

for all Borel subsets E of Y.

We shall therefore assume that for every $\tau > 0$,

$$(\mu_\psi \wedge \mu_\pi)\big(\{\varphi(\sigma, x) \mid \sigma > \tau\} \setminus F\big) = \infty.$$

Write the set $U = \{\sigma \geq 0 \mid \varphi(\sigma, x) \in Y \setminus F\}$ as a countable disjoint union $\bigcup_{n=1}^{\infty} J_n$, where for each $n = 1, 2, \ldots$, J_n is an interval and $\mu_\psi \wedge \mu_\pi$ is finite on $\varphi(J_n, x)$. Choose, for each positive integer n, a number $\delta_n \in (0, 1]$ in such a way that

$$\sum_{n=1}^{\infty} \delta_n \cdot (\mu_\psi \wedge \mu_\pi)\big(\varphi(J_n, x)\big) < \infty.$$

Choose a finite set S_1 of positive integers such that

$$\sum_{n \in S_1} (\mu_\psi \wedge \mu_\pi)\big(\varphi(J_n, x)\big) > 1.$$

Choose a number τ_1 such that $\varphi(\tau_1, x) \in F$ and τ_1 is an upper bound of J_n for every $n \in S_1$. For each positive integer k, if finite sets $S_1, \ldots, S_k$

of positive integers and numbers $\tau_1, \ldots, \tau_k$ have been defined, choose a finite set S_{k+1} of positive integers such that τ_k is a lower bound of J_n for every $n \in S_{k+1}$, and satisfying

$$\sum_{n \in S_{k+1}} (\mu_\psi \wedge \mu_\pi)\big(\varphi(J_n, x)\big) > 1.$$

Now choose τ_{k+1} such that $\varphi(\tau_{k+1}, x) \in F$, and τ_{k+1} is an upper bound of J_n for every $n \in S_{k+1}$.

This defines S_k for every positive integer k. Let

$$S = \bigcup_{k=1}^{\infty} S_k.$$

Define μ by

$$\mu(E) = \mu_\varphi(E \cap F) + \sum_{n \in S} (\mu_\psi \wedge \mu_\pi)\big(E \cap \varphi(J_n, x)\big)$$
$$+ \sum_{n \notin S} \delta_n \cdot (\mu_\psi \wedge \mu_\pi)\big(E \cap \varphi(J_n, x)\big).$$

Once again, μ has the required properties, and this completes the proof of the theorem.   **QED**

8.53. Ramming Points. Let φ be a quasi-flow in a Hausdorff space X and let $\psi, \pi \in \mathrm{Rep}_q^+ (\varphi)$.

A point $x \in \mathscr{F}(\psi) \cap \mathscr{F}(\pi)$ is said to be a *left ramming point* of the pair $\{\psi, \pi\}$ if there exists $\tau < 0$ such that

(i) $\{\sigma \in [\tau, 0] \mid \varphi(\sigma, x) \in \mathscr{F}(\psi) \cap \mathscr{F}(\pi)\}$ has empty interior, and

(ii) $\mu_\psi \wedge \mu_\pi$ is finite on the set $\{\varphi(\sigma, x) \mid \tau \leq \sigma \leq 0\} \setminus \big(\mathscr{F}(\psi) \cap \mathscr{F}(\pi)\big)$.

A point $x \in \mathscr{F}(\psi) \cap \mathscr{F}(\pi)$ is said to be a *right ramming point* of the pair $\{\psi, \pi\}$ if there exists $\tau > 0$ such that

(i) $\{\sigma \in [0, \tau] \mid \varphi(\sigma, x) \in \mathscr{F}(\psi) \cap \mathscr{F}(\pi)\}$ has empty interior, and

(ii) $\mu_\psi \wedge \mu_\pi$ is finite on the set $\{\varphi(\sigma, x) \mid 0 \leq \sigma \leq \tau\} \setminus \big(\mathscr{F}(\psi) \cap \mathscr{F}(\pi)\big)$.

If a point $x \in \mathscr{F}(\psi) \cap \mathscr{F}(\pi)$ is either a left ramming point or a right ramming point of the pair $\{\psi, \pi\}$, then x is said to be a *ramming point* of the pair $\{\psi, \pi\}$.

If a point $x \in \mathscr{F}(\psi) \cap \mathscr{F}(\pi)$ is both a left ramming point and a right ramming point of the pair $\{\psi, \pi\}$, then x is said to be a *removable* ramming point of the pair $\{\psi, \pi\}$.

The set of removable ramming points of the pair $\{\psi, \pi\}$ is denoted by $\mathscr{R}(\psi, \pi)$.

8.54. Example. Let u and v be the functions from $\mathbb{R}$ into $[0, \infty)$ defined as follows:

$$u(x) = v(x) = 1 \text{ whenever } x \leq -1,$$

$$u(x) = v(x) = x \text{ whenever } x \geq 0,$$

$$u(x) = 1 \text{ and } v(x) = \frac{1}{n^2} \text{ whenever } n \text{ is an odd positive integer}$$

$$\text{and} -\frac{1}{n} < x \leq -\frac{1}{n+1},$$

and

$$u(x) = \frac{1}{n^2} \text{ and } v(x) = 1 \text{ whenever } n \text{ is an even positive integer}$$

$$\text{and} -\frac{1}{n} < x \leq -\frac{1}{n+1}.$$

It is easy to see that u and v are proper 1-multipliers, and that 0 is a left ramming point of the pair $\{u1, v1\}$. Note that although Theorem 8.52 implies that $u1$ and $v1$ have a common upper bound in $\text{Rep}_q^+(1)$, it is an elementary consequence of Theorem 8.44 that there is no least upper bound.

8.55. Theorem. *Let φ be a quasi-flow in a Hausdorff space X, let $\psi, \pi \in \text{Rep}_q^+(\varphi)$, and let $F = \mathscr{F}(\psi) \cap \mathscr{F}(\pi)$.*

Then for every $x \in X$, the set

$$\{\tau \in \mathbb{R} \mid \varphi(\tau, x) \in \mathscr{R}(\psi, \pi)\}$$

is totally disconnected, and is relatively open in $\{\tau \in \mathbb{R} \mid \varphi(\tau, x) \in F\}$.

PROOF: Clear.   **QED**

8.56. Theorem. *Let F be a closed subset of $\mathbb{R}$ and let Q be a totally disconnected, relatively open subset of F.*

Then there exist $\psi, \pi \in \text{Rep}_q^+(1)$ such that

$$F = \mathscr{F}(\psi) \cap \mathscr{F}(\pi) \text{ and } Q = \mathscr{R}(\varphi, \pi).$$

PROOF: Each component interval I of the complement of $F \setminus Q$ is a countable union $\bigcup_{n=1}^{\infty} I_n$ of components of $\mathbb{R} \setminus F$, together with a subset of Q. For each n, the techniques of Example 8.54 may be used to find nonnegative continuous measures μ_n and λ_n on I_n which are finite on compact subsets of I_n, non-zero on non-empty open subsets of I_n, and infinite on initial and final segments of I_n. We can require further of μ_n and λ_n that $\mu_n \wedge \lambda_n$ be finite on initial segments of I_n iff I_n has a left endpoint inside the interval I, and that $\mu_n \wedge \lambda_n$ be finite on final segments of I_n iff I_n has a right endpoint inside I. We can also require that $\sum_{n=1}^{\infty} \mu_n \wedge \lambda_n$ be finite on compact subsets of I, and be infinite on all initial segments and all final segments of I.

Let $\mu_I = \sum_{n=1}^{\infty} \mu_n$ and $\lambda_I = \sum_{n=1}^{\infty} \lambda_n$, and repeat this for every component I of the complement of $F \setminus Q$.

Let μ' and λ' be, respectively, the sums of the measures μ_I and λ_I so defined, and let μ and λ be defined as follows

$$\mu(E) = \lambda(E) = \infty, \text{ whenever } E \text{ is a Borel set meeting } F,$$

and

$$\mu(E) = \mu'(E) \text{ and } \lambda(E) = \lambda'(E) \text{ whenever } E \text{ is a Borel set disjoint from } F.$$

It is easy to see that μ and λ are the time measures of members ψ, π of $\mathrm{Rep}_q^+(1)$ which have the required properties.   **QED**

8.57. Definition. Let φ be a quasi-flow in a Hausdorff space X and let $\psi, \pi \in \mathrm{Rep}_q^+(\varphi)$.

ψ and π are said to be *compatible* if

(a) the pair $\{\psi, \pi\}$ produces no jumping,

(b) the pair $\{\psi, \pi\}$ does not ram $\alpha_\varphi(x)$ nor $\omega_\varphi(x)$ for any $x \in X$,

and

(c) every ramming point of $\{\psi, \pi\}$ is removable.

8.58. Theorem. *Let φ be a quasi-flow in a Hausdorff space X and let $\psi, \pi \in \mathrm{Rep}_q^+(\varphi)$.*

Then the following three conditions are equivalent:

(a) *ψ and π are compatible.*

(b) *The measure μ defined by*

$$\mu(E) = (\mu_\psi \wedge \mu_\pi)(E \setminus \mathcal{R}(\psi, \pi)), \qquad \forall \, E \in \mathbf{\Sigma}(\varphi),$$

belongs to $\mathcal{M}_q(\varphi)$.

(c) *There exists a subset P of $\mathcal{F}(\psi) \cap \mathcal{F}(\pi)$ such that the measure μ defined* *by*

$$\mu(E) = (\mu_\psi \wedge \mu_\pi)(E \setminus P), \qquad \forall\, E \in \boldsymbol{\Sigma}(\varphi),$$

belongs to $M_q(\varphi)$.

PROOF: For convenience of notation, let $F = \mathscr{F}(\psi) \cap \mathscr{F}(\pi)$.

Assume that (c) holds, and let P be any subset of F for which the corresponding measure μ belongs to $\mathscr{M}_q(\varphi)$. Let ϑ be the member of $\mathrm{Rep}_q^+(\varphi)$ for which $\mu = \mu_\vartheta$. It is clear that $F \cap \mathscr{G}(\vartheta) = P$.

We claim that every left ramming point of $\{\psi, \pi\}$ must belong to P, for suppose x is a left ramming point of $\{\psi, \pi\}$, and choose $\tau < 0$ such that (i) and (ii) of Definition 8.53 are satisfied. Choose $\tau_1 \in (\tau, 0)$ such that $\varphi(\tau_1, x) \in \mathscr{G}(\vartheta)$, and let $y = \varphi(\tau_1, x)$. Choose $\tau_2 \in (0, \infty]$ such that

$$\{\vartheta(t, y) \mid t > 0\} = \{\varphi(\sigma, y) \mid 0 < \sigma < \tau_2\}.$$

Then since

$$\infty = \mu_\vartheta\big(\{\varphi(\sigma, y) \mid 0 < \sigma < \tau_2\}\big)$$
$$= (\mu_\psi \wedge \mu_\pi)\big(\{\varphi(\sigma, y) \mid 0 < \sigma < \tau_2\} \setminus F\big),$$

the fact that x is a left ramming point of $\{\psi, \pi\}$ implies that $x \in \{\varphi(\sigma, y) \mid 0 < \sigma < \tau_2\}$, and we deduce that

$$x \in F \cap \mathscr{G}(\vartheta) = P.$$

We claim also that every point of P is a left ramming point of $\{\psi, \pi\}$, for suppose $x \in P$, and choose $\tau < 0$ such that $\{\varphi(\sigma, x) \mid \tau \leq \sigma \leq 0\} \subseteq \mathscr{O}_\vartheta(x)$. It is not hard to verify that (i) and (ii) of Definition 8.53 are satisfied.

We have therefore shown that P is the set of left ramming points of $\{\psi, \pi\}$, and a similar argument shows that P is the set of right ramming points of $\{\psi, \pi\}$. From this we deduce at once that $P = \mathscr{R}(\psi, \pi)$ and that every ramming point of $\{\psi, \pi\}$ is removable. It follows that (b) and (c) are equivalent.

Now to prove that (b) implies (a), assume that (b) holds. Then by the above argument, every ramming point of $\{\psi, \pi\}$ is removable. Furthermore, since $\mu \leq \mu_\psi \wedge \mu_\pi$, $\{\psi, \pi\}$ produces no jumping. To obtain a contradiction, assume that $x \in X$ and $\{\psi, \pi\}$ rams $\omega_\varphi(x)$. As before, let ϑ be the member of $\mathrm{Rep}_q^+(\varphi)$ for which $\mu = \mu_\vartheta$. Choose $\tau > 0$ such that $\varphi(\tau, x) \in \mathscr{G}(\vartheta)$ and the measure $\mu_\psi \wedge \mu_\pi$ is finite on the set $\{\varphi(\sigma, x) \mid \sigma > \tau\} \setminus F$. Let $y = \varphi(\tau, x)$ and choose $\tau_1 \in (0, \infty]$ such that

$$\{\vartheta(t, y) \mid t \geq 0\} = \{\varphi(\sigma, y) \mid 0 \leq \sigma < \tau_1\}.$$

Then we have

$$\infty = \mu_\vartheta\big(\{\varphi(\sigma, y) \mid 0 \leq \sigma < \tau_1\}\big)$$

$$= (\mu_\psi \wedge \mu_\pi)\big(\{\varphi(\sigma, y) \mid 0 \leq \sigma < \tau_1\} \setminus F\big)$$

$$\leq (\mu_\psi \wedge \mu_\pi)\big(\{\varphi(\sigma, x) \mid \sigma \geq \tau\} \setminus F\big) < \infty,$$

and this contradiction shows that $\{\psi, \pi\}$ does not ram $\omega_\varphi(x)$ for any $x \in X$. A similar argument shows that $\{\psi, \pi\}$ does not ram $\alpha_\varphi(x)$ for any $x \in X$, and we have therefore shown that (b) implies (a).

Now to prove that (a) $\Rightarrow$ (b), assume that (a) holds, and let μ be defined as in (b). First we shall show that for any $x \in X$ and $\tau_1 < \tau_2$,

$$\mu\big(\{\varphi(\tau, x) \mid \tau_1 < \tau < \tau_2\}\big) > 0.$$

But this is clear, for if $\varphi(\tau, x) \in F \setminus \mathscr{R}(\psi, \pi)$ for some $\tau \in (\tau_1, \tau_2)$, then $\mu\big(\{\varphi(\tau, x)\}\big) = \infty$, and otherwise there must exist a non-empty subinterval (σ_1, σ_2) of (τ_1, τ_2) such that $\varphi(\tau, x) \notin F$, for all $\tau \in (\sigma_1, \sigma_2)$.

We shall complete the proof by showing that μ satisfies the conditions of Theorem 5.33. Certainly, condition (a) of Theorem 5.33 is satisfied. Before proving that condition (b) of Theorem 5.33 is satisfied, we need to show that whenever $x \in X$, $-\infty < \tau_1 \leq \tau_2 < \infty$, and $\{\varphi(\sigma, x) \mid \tau_1 \leq \sigma \leq \tau_2\}$ does not meet $F \setminus \mathscr{R}(\psi, \pi)$, we have

$$\mu\big(\{\varphi(\sigma, x) \mid \tau_1 \leq \sigma \leq \tau_2\}\big) < \infty.$$

We prove this as follows: For every $\sigma \in [\tau_1, \tau_2]$, choose a neighborhood U_σ of σ such that $\mu\big(\varphi(U_\sigma, x)\big) < \infty$. (The choice of U_σ is easy whenever $\varphi(\sigma, x) \notin F$, and when $\varphi(\sigma, x) \in \mathscr{R}(\psi, \pi)$, the choice can still be made because of the way $\mathscr{R}(\psi, \pi)$ is defined.) Using the compactness of $[\tau_1, \tau_2]$, we see at once that $\mu\big(\{\varphi(\sigma, x) \mid \tau_1 \leq \sigma \leq \tau_2\}\big)$ is finite.

Now making use of the fact that for each $x \in X$, $\{\tau \mid \varphi(\tau, x) \in \mathscr{R}(\psi, \pi)\}$ is totally disconnected and $\{\tau \mid \varphi(\tau, x) \in F \setminus \mathscr{R}(\psi, \pi)\}$ is closed, it can be readily verified that for any $x \in X$, x satisfies condition (b)(i) of Theorem 5.33 iff $x \in F \setminus \mathscr{R}(\psi, \pi)$, x satisfies condition (b)(ii) of Theorem 5.33 iff $0 < p_\varphi(x) < \infty$ and $\mathcal{O}_\varphi(x)$ does not meet $F \setminus \mathscr{R}(\psi, \pi)$, and x satisfies condition (b)(iii) of Theorem 5.33 iff $x \notin F \setminus \mathscr{R}(\psi, \pi)$ and either $p_\varphi(x) = \infty$ or $\mathcal{O}_\varphi(x)$ meets $F \setminus \mathscr{R}(\psi, \pi)$.

This shows that (a) $\Rightarrow$ (b), and the proof is therefore complete.   **QED**

8.59. Corollary. *Let φ be a quasi-flow in a Hausdorff space X, let $\psi, \pi \in \mathrm{Rep}_q^+(\varphi)$, and suppose ψ and π are compatible. Let P be a subset of $\mathscr{F}(\psi) \cap \mathscr{F}(\pi)$ and let the measure μ be defined by*

$$\mu(E) = (\mu_\psi \wedge \mu_\pi)(E \setminus P), \qquad \forall \, E \in \mathbf{\Sigma}(\varphi).$$

Then $\mu \in \mathcal{M}_q(\varphi)$ *iff* $P = \mathcal{R}(\psi, \pi)$.

PROOF: This follows from the proof of Theorem 8.58.   **QED**

8.60. Definition. Let φ be a quasi-flow in a Hausdorff space X and let ψ and π be compatible members of $\mathrm{Rep}_q^+(\varphi)$. Let ϑ be the member of $\mathrm{Rep}_q^+(\varphi)$ for which

$$\mu_\vartheta(E) = (\mu_\psi \wedge \mu_\pi)\big(E \setminus \mathcal{R}(\psi, \pi)\big),$$

for every $\Sigma(\varphi)$-measurable subset E of X.

Then we call ϑ the *pseudo-maximum* of ψ and π and we write $\vartheta = \psi \vee \pi$.[8]

8.61. Theorem. *Let φ be a quasi-flow in a Hausdorff space X, let ψ and π be compatible members of $\mathrm{Prod}_q^+(\varphi)$ and let ϑ be their pseudo-maximum.*

Then $\vartheta \in \mathrm{Prod}_q^+(\varphi)$.

PROOF: By Theorem 5.43, we need only show that whenever A is a $\Sigma(\varphi)$-measurable subset of $\mathcal{G}(\vartheta)$ and $\mu_\varphi(A) = 0$ we have $\mu_\vartheta(A) = 0$. Let A be such a set. Then A is the disjoint union $A_1 \cup A_2 \cup A_3$ where $A_1 = A \cap \mathcal{R}(\psi, \pi)$, $A_2 = A \cap \mathcal{G}(\psi)$ and $A_3 = A \cap \mathcal{F}(\psi) \cap \mathcal{G}(\pi)$. It is obvious that $\mu_\vartheta(A_1) = 0$. Furthermore, $\mu_\vartheta(A_2) \leq \mu_\psi(A_2) = 0$, and $\mu_\vartheta(A_3) \leq \mu_\pi(A_3) = 0$.

This completes the proof.   **QED**

8.62. Theorem. *Let φ be a quasi-flow in a Hausdorff space X, let $\psi, \pi \in \mathrm{Rep}_q^+(\varphi)$ and suppose $\psi \vee \pi$ exists.*

Then given any $x \in X$, there does not exist a finite continuous nonnegative non-zero Borel measure λ on the set

$$[\mathcal{O}_\varphi(x) \cap \mathcal{F}(\psi) \cap \mathcal{F}(\pi)] \setminus \mathcal{F}(\psi \vee \pi).$$

A fortiori, $\mu_{\psi \vee \pi}\big([\mathcal{F}(\psi) \cap \mathcal{F}(\pi)] \setminus \mathcal{F}(\psi \vee \pi)\big) = 0$.

PROOF: Suppose that for some $x \in X$, there exists a finite continuous nonnegative non-zero Borel measure λ on

$$[\mathcal{O}_\varphi(x) \cap \mathcal{F}(\psi) \cap \mathcal{F}(\pi)] \setminus \mathcal{F}(\psi \vee \pi).$$

Let $\mu = \lambda + \mu_{\psi \vee \pi}$. It is easy to see that μ satisfies the conditions of Theorem 5.33, and we therefore have $\mu \in \mathcal{M}_q(\varphi)$. Let ϑ be the member

---

[8] We use the bold face $\vee$ to denote the pseudo-maximum, as we shall later use the bold face $+$ to denote the pseudo-sum. As a mnemonic, please remember that the pseudo-maximum, like the pseudo-sum, is in general an imposter, and like all imposters, requires a bold face.

of $\mathrm{Rep}_q^+(\varphi)$ for which $\mu = \mu_\theta$. Then although $\psi \le \vartheta$ and $\pi \le \vartheta$, we do not have $\psi \vee \pi \le \vartheta$, and this contradiction completes the proof of the theorem.   **QED**

8.63. Corollary. *Let φ be a quasi-flow in a Hausdorff space X, let $\psi, \pi \in \mathrm{Rep}_q^+(\varphi)$ and suppose that $\psi \vee \pi$ exists.*

Then ψ and π are compatible and $\psi \vee \pi = \psi \mathbf{V} \pi$.

PROOF: Let $P = \big(\mathscr{F}(\psi) \cap \mathscr{F}(\pi)\big) \setminus \mathscr{F}(\psi \vee \pi)$. Then Theorems 8.44 and 8.62 show that

$$\mu_{\psi \vee \pi}(E) = (\mu_\psi \wedge \mu_\pi)(E \setminus P)$$

for every $\Sigma(\varphi)$-measurable set E. The corollary now follows at once from Theorem 8.58.   **QED**

8.64. Corollary. *Let φ be a quasi-flow in a Hausdorff space X, let $\psi, \pi \in \mathrm{Rep}_q^+(\varphi)$ and suppose that $\psi \vee \pi$ exists.*

Then $\mathscr{R}(\psi, \pi) = \big(\mathscr{F}(\psi) \cap \mathscr{F}(\pi)\big) \setminus \mathscr{F}(\psi \vee \pi)$.

PROOF: Immediate from Corollaries 8.59 and 8.63.   **QED**

8.65. Theorem. *Let φ be a quasi-flow in a Hausdorff space X, let ψ and π be compatible members of $\mathrm{Rep}_q^+(\varphi)$, and let χ be their pseudo-maximum. Let*

$$\mathscr{U} = \big\{ \vartheta \in \mathrm{Rep}_q^+(\varphi) \mid \psi \le \vartheta \text{ and } \pi \le \vartheta \text{ and } \mu_\vartheta\big(\mathscr{R}(\psi, \pi) \cap \mathscr{G}(\vartheta)\big) = 0 \big\}.$$

Then χ is the least member of $\mathscr{U}$.

PROOF: Let $\vartheta \in \mathscr{U}$, and let μ be the time measure of $\chi \wedge \vartheta$. We shall prove the theorem by showing that $\chi \wedge \vartheta = \chi$, and to do this, we must show that $\mu = \mu_\chi$. It is clear that μ and μ_χ agree in $X \setminus \mathscr{R}(\psi, \pi)$, and therefore if $P = \mathscr{R}(\psi, \pi) \cap \mathscr{G}(\vartheta)$, we see that for every $\Sigma(\varphi)$-measurable set E,

$$\mu(E) = (\mu_\psi \wedge \mu_\pi)(E \setminus P).$$

It follows from Corollary 8.59 that $P = \mathscr{R}(\psi, \pi)$, and we therefore have

$$\mu(E) = (\mu_\psi \wedge \mu_\pi)\big(E \setminus \mathscr{R}(\psi, \pi)\big) = \mu_\chi(E),$$

for every $\Sigma(\varphi)$-measurable set E.

This completes the proof of the theorem.   **QED**

8.66. Theorem. *Let φ be a quasi-flow in a Hausdorff space X, let ψ and π be compatible members of $\mathrm{Rep}_q^+(\varphi)$ and suppose that $\mathscr{R}(\psi, \pi)$ has a countable intersection with every φ-orbit.*

Then $\psi \vee \pi$ exists.

PROOF: If ϑ is any common upper bound of ψ and π, we have

$$\mu_\vartheta\big(\mathscr{R}(\psi, \pi) \cap \mathscr{G}(\vartheta)\big) = 0,$$

for given any $x \in X$, the restriction of μ_ϑ to $\mathcal{O}_\varphi(x) \cap \mathscr{R}(\psi, \pi) \cap \mathscr{G}(\vartheta)$ is a continuous measure on a countable set.

It therefore follows from Theorem 8.65 that the pseudo-maximum of ψ and π is the maximum of ψ and π.   **QED**

8.67. Note. It is a remarkable fact that the converse of Theorem 8.66 is true: whenever $\psi \vee \pi$ exists, $\mathscr{R}(\psi, \pi)$ has a countable intersection with every φ-orbit. The key to the proof of this lies in the following lemma.

8.68. Lemma. *Let F_1 and F_2 be closed subsets of $\mathbb{R}$, and suppose $F_1 \setminus F_2$ is uncountable.*

Then there exists a nonnegative finite continuous non-zero Borel measure on $F_1 \setminus F_2$.

PROOF: Choose a decreasing sequence $\{U_n\}$ of open sets such that $F_2 = \bigcap\limits_{n=1}^{\infty} U_n$. Then since

$$F_1 \setminus F_2 = \bigcup_{n=1}^{\infty} (F_1 \setminus U_n),$$

we can choose n such that $F_1 \setminus U_n$ is uncountable. Now $F_1 \setminus U_n$, being an uncountable closed subset of $\mathbb{R}$, contains a homeomorphic image C of the Cantor set, and on C, there exists a nonnegative finite continuous non-zero Borel measure.

This proves the lemma.   **QED**

8.69. Theorem. *Let φ be a quasi-flow in a Hausdorff space X, and let $\psi, \pi \in \mathrm{Rep}_q^+(\varphi)$.*

Then a necessary and sufficient condition for the existence of $\psi \vee \pi$ is that ψ and π be compatible and $\mathscr{R}(\psi, \pi)$ have a countable intersection with every φ-orbit.

PROOF: In view of Corollary 8.63 and Theorem 8.66, we need only show that if $\psi \vee \pi$ exists, then $\mathscr{R}(\psi, \pi)$ has a countable intersection with every φ-orbit.

Suppose that $\psi \vee \pi$ exists, let $x \in X$, and to obtain a contradiction, assume that $\mathscr{R}(\psi, \pi) \cap \mathcal{O}_\varphi(x)$ is uncountable. Then $\{\tau \in \mathbb{R} \mid \varphi(\tau, x) \in \mathscr{R}(\psi, \pi)\}$ is uncountable, and we can therefore choose τ_1 and τ_2 in $\mathbb{R}$ such that $\tau_2 - \tau_1 < \mathrm{p}_\varphi(x)$ and the set

$A = \{\tau \in [\tau_1, \tau_2] \mid \varphi(\tau, x) \in \mathcal{R}(\psi, \pi)\}$ is uncountable. But by Corollary 8.64, A is the difference of the two closed sets

$$\{\tau \in [\tau_1, \tau_2] \mid \varphi(\tau, x) \in \mathcal{F}(\psi) \cap \mathcal{F}(\pi)\}$$

and $\{\tau \in [\tau_1, \tau_2] \mid \varphi(\tau, x) \in \mathcal{F}(\psi \vee \pi)\}.$

Therefore, using Lemma 8.68, we can find a nonnegative continuous finite non-zero Borel measure λ on A.

Let μ be the measure defined by

$$\mu(E) = \lambda\big(\{\tau \in [\tau_1, \tau_2] \mid \varphi(\tau, x) \in E\}\big),$$

for every Borel subset E of $\mathcal{O}_\varphi(x) \cap \mathcal{R}(\psi, \pi)$.

But this contradicts Theorem 8.62, because μ is a nonnegative continuous finite non-zero Borel measure. This completes the proof of the theorem.

QED

8.70. Corollary. *Let φ be a quasi-flow in a Hausdorff space X, let $\psi, \pi \in \mathrm{Rep}_q^+(\varphi)$ and suppose $\mathcal{F}(\psi) \cap \mathcal{F}(\pi) = \mathcal{F}(\varphi)$.*

Then $\psi \vee \pi$ exists iff ψ and π have an upper bound in $\mathrm{Rep}_q^+(\varphi)$.

PROOF: If ψ and π have an upper bound, then Theorem 8.52 implies that ψ and π are compatible, and the existence of $\psi \vee \pi$ follows from Theorem 8.69.   **QED**

8.71. The Distributive Law $(u \vee v)\varphi = u\varphi \vee v\varphi.$

In the following nine sections, we investigate the maximum operation in $\mathrm{Prod}_q^+(\varphi)$. We have already seen in Theorem 8.61 that if ψ and π are compatible members of $\mathrm{Prod}_q^+(\varphi)$, then their pseudo-maximum also lies in $\mathrm{Prod}_q^+(\varphi)$. Consequently, if $\psi \vee \pi$ exists then $\psi \vee \pi$ is also the maximum of ψ and π in the family $\mathrm{Prod}_q^+(\varphi)$. We know that if ψ and π are compatible, then $\psi \vee \pi$ exists iff $\mathcal{R}(\psi, \pi)$ has a countable intersection with every φ-orbit. We shall show that ψ and π have a maximum in $\mathrm{Prod}_q^+(\varphi)$ iff $\mu_\varphi\big(\mathcal{R}(\psi, \pi)\big) = 0$.

8.72. Lemma. *Let φ be a quasi-flow in a Hausdorff space X and let u and v be nonnegative quasi φ-multipliers.*

Then for every $\Sigma(\varphi)$-measurable subset E of $X \setminus \big(\mathcal{F}(u\varphi) \cap \mathcal{F}(v\varphi)\big)$, we have

$$(\mu_{u\varphi} \wedge \mu_{v\varphi})(E) = \int_E \frac{1}{u \vee v}\, d\mu_\varphi.$$

PROOF: Let E be a $\boldsymbol{\Sigma}(\varphi)$-measurable subset of $X \setminus \left(\mathscr{F}(u\varphi) \cap \mathscr{F}(v\varphi)\right)$, and let

$$E_1 = E \cap \mathscr{G}(u\varphi) \cap \mathscr{G}(v\varphi),$$
$$E_2 = E \cap \mathscr{G}(u\varphi) \cap \mathscr{F}(v\varphi),$$
and
$$E_3 = E \cap \mathscr{F}(u\varphi) \cap \mathscr{G}(v\varphi).$$

By Theorem 8.17 (b), we have

$$(\mu_{u\varphi} \wedge \mu_{v\varphi})(E_1) = \int_{E_1} \left(\frac{1}{u} \wedge \frac{1}{v}\right) d\mu_\varphi = \int_{E_1} \frac{1}{u \vee v}\, d\mu_\varphi.$$

Also,

$$(\mu_{u\varphi} \wedge \mu_{v\varphi})(E_2) = \mu_{u\varphi}(E_2) = \int_{E_2} \frac{1}{u}\, d\mu_\varphi = \int_{E_2} \frac{1}{u \vee v}\, d\mu_\varphi.$$

(The latter equality holds since v is zero on E_2 and u is nonnegative.) Similarly, we see that

$$(\mu_{u\varphi} \wedge \mu_{v\varphi})(E_3) = \mu_{v\varphi}(E_3) = \int_{E_3} \frac{1}{u \vee v}\, d\mu_\varphi.$$

From this, the lemma follows. $\hspace{4cm}$ **QED**

8.73. Theorem. *Let φ be a quasi-flow in a Hausdorff space X, let u and v be nonnegative quasi φ-multipliers, and suppose $u\varphi$ and $v\varphi$ are compatible.*

Let $w(x) = u(x) \vee v(x)$, for all $x \in X \setminus \mathscr{R}(u\varphi, v\varphi)$ and $w(x) = \infty$, for all $x \in \mathscr{R}(u\varphi, v\varphi)$.

Then w is a quasi φ-multiplier and

$$w\varphi = u\varphi \vee v\varphi.$$

PROOF: Let ϑ be the pseudo-maximum of $u\varphi$ and $v\varphi$. In view of Theorem 5.42, the theorem will be proved once we have shown that w is zero in $\mathscr{G}(\varphi) \cap \mathscr{F}(\vartheta)$, and that

$$\mu_\vartheta(E) = \int_E \frac{1}{w}\, d\mu_\varphi,$$

for every $\boldsymbol{\Sigma}(\varphi)$-measurable subset E of $\mathscr{G}(\vartheta)$. Since both u and v are zero in $\mathscr{G}(\varphi) \cap \mathscr{F}(\vartheta)$, it is clear that w is zero there. Now let E be a $\boldsymbol{\Sigma}(\varphi)$-measurable subset of $\mathscr{G}(\vartheta)$, and write $E = E_1 \cup E_2$ where $E_1 = E \setminus \left(\mathscr{F}(\psi) \cap \mathscr{F}(\pi)\right)$ and $E_2 = E \cap \mathscr{R}(u\varphi, v\varphi)$. By Lemma 8.72,

$$\mu_\vartheta(E_1) = (\mu_{u\varphi} \wedge \mu_{v\varphi})(E_1) = \int_{E_1} \frac{1}{u \vee v}\, d\mu_\varphi = \int_{E_1} \frac{1}{w}\, d\mu_q.$$

Also,

$$\mu_\vartheta(E_2) = 0 = \int_{E_2} \frac{1}{w} \, d\mu_\varphi.$$

This completes the proof. **QED**

8.74. Theorem. *Let φ be a quasi-flow in a Hausdorff space X, let u and v be nonnegative quasi φ-multipliers, and suppose $u \vee v$ is a quasi φ-multiplier.*

Then

(a) *$u\varphi$ and $v\varphi$ are compatible,*

(b) *$\mu_\varphi\big(\mathscr{R}(u\varphi, v\varphi)\big) = 0$,*

(c) *$(u \vee v)\,\varphi = u\varphi \vee v\varphi$,*

and

(d) *$(u \vee v)\,\varphi$ is the maximum of $u\varphi$ and $v\varphi$ in the family $\mathrm{Prod}_q^+(\varphi)$.*

PROOF: Let $F = \mathscr{F}(u\varphi) \cap \mathscr{F}(v\varphi)$ and $P = F \setminus \mathscr{F}\big((u \vee v)\,\varphi\big)$. We claim that

$$\mu_{(u \vee v)\varphi}(E) = (\mu_{u\varphi} \wedge \mu_{v\varphi})\,(E \setminus P),$$

for every $\boldsymbol{\Sigma}(\varphi)$-measurable set E. Let $E \in \boldsymbol{\Sigma}(\varphi)$ and write $E = E_1 \cup E_2 \cup E_3$, where $E_1 = E \setminus F$, $E_2 = E \cap \mathscr{F}\big((u \vee v)\,\varphi\big)$, and $E_3 = E \cap P$. Then since $\mathscr{F}\big((u \vee v)\,\varphi\big) \subseteq F$, we have

$$\mu_{(u \vee v)\varphi}(E_1) = \int_{E_1} \frac{1}{u \vee v} \, d\mu_\varphi,$$

and Lemma 8.72 therefore implies that

$$\mu_{(u \vee v)\varphi}(E_1) = (\mu_{u\varphi} \wedge \mu_{v\varphi})\,(E_1).$$

It is clear that

$$\mu_{(u \vee v)\varphi}(E_2) = (\mu_{u\varphi} \wedge \mu_{v\varphi})\,(E_2).$$

Thirdly, since $P \subseteq \mathscr{G}\big((u \vee v)\,\varphi\big)$ and $u \vee v$ is zero in P, it follows from Theorem 5.44(c) that $\mu_\varphi(P) = 0$, and therefore Theorem 5.43 implies that $\mu_{(u \vee v)\varphi}(P) = 0$. Thus, $\mu_{(u \vee v)\varphi}(E_3) = 0$, and the above claim is justified.

We now deduce from Theorem 8.58 that $u\varphi$ and $v\varphi$ are compatible, and from Corollary 8.59, we deduce that $P = \mathscr{R}(u\varphi, v\varphi)$. Consequently, $\mu_\varphi\big(\mathscr{R}(u\varphi, v\varphi)\big) = 0$. This proves parts (a), (b) and (c) of the theorem.

To prove (d), suppose that w is a nonnegative quasi φ-multiplier, and that $u\varphi \leq w\varphi$ and $v\varphi \leq w\varphi$. Then by Theorem 8.3, we have $u(x) \leq w(x)$ and $v(x) \leq w(x)$ for μ_φ-almost all $x \in \mathcal{G}(\varphi)$. Therefore $u(x) \vee v(x) \leq w(x)$ for μ_φ-almost all $x \in \mathcal{G}(\varphi)$, and Theorem 8.3 implies that $(u \vee v) \varphi \leq w\varphi$. This proves (d).   **QED**

8.75. Theorem. *Let φ be a quasi-flow in a Hausdorff space X, let u and v be nonnegative quasi φ-multipliers, suppose that $u\varphi$ and $v\varphi$ are compatible and suppose that $\mu_\varphi\big(\mathcal{R}(u\varphi, v\varphi)\big) = 0$.*

Then $u \vee v$ is a quasi φ-multiplier.

PROOF: Let $w(x) = u(x) \vee v(x)$ whenever $x \in X \setminus \mathcal{R}(u\varphi, v\varphi)$, and

$$w(x) = \infty \quad \text{whenever } x \in \mathcal{R}(u\varphi, v\varphi).$$

Then by Theorem 8.73, w is a quasi φ-multiplier.

Since $u \vee v \leq w$, it is clear that $u \vee v$ is zero in $\mathcal{G}(\varphi) \cap \mathcal{F}(w\varphi)$. Since $\mu_\varphi\big(\mathcal{R}(u\varphi, v\varphi)\big) = 0$, the functions $u \vee v$ and w agree μ_φ-almost everywhere. It now follows at once from Theorem 5.42 that $u \vee v$ is a quasi φ-multiplier.   **QED**

8.76. Theorem. *Let φ be a quasi-flow in a Hausdorff space X, let u and v be nonnegative quasi φ-multipliers, and suppose $u\varphi$ and $v\varphi$ have a maximum in the family $\mathrm{Prod}_q^+(\varphi)$.*

Then $u \vee v$ is a quasi φ-multiplier.

PROOF: Let $w\varphi$ be the maximum of $u\varphi$ and $v\varphi$ in $\mathrm{Prod}_q^+(\varphi)$. Since $\mathcal{F}(w\varphi) \subseteq \mathcal{F}(u\varphi) \cap \mathcal{F}(v\varphi)$, we see that $u \vee v$ is zero in $\mathcal{G}(\varphi) \cap \mathcal{F}(w\varphi)$. Therefore the theorem will follow from Theorem 5.42 once we have shown that $u \vee v$ and w agree μ_φ-almost everywhere in $\mathcal{G}(\varphi)$.

Suppose, to obtain a contradiction, that $u \vee v$ and w are different on a set of positive μ_φ-measure in $\mathcal{G}(\varphi)$. Then we can choose a $\Sigma(\varphi)$-measurable subset A of $\mathcal{G}(\varphi)$ such that $\mu_\varphi(A) > 0$ and $(u \vee v)(x) < w(x)$, for every $x \in A$. Choose $\delta \in (0, 1)$ and a $\Sigma(\varphi)$-measurable subset B of A such that $\mu_\varphi(B) > 0$ and $(u \vee v)(x) < \delta \cdot w(x)$, for every $x \in B$. Define $\tilde{w}$ by

$$\tilde{w}(x) = \begin{cases} w(x) & \text{if } x \in X \setminus B, \\ \delta \cdot w(x) & \text{if } x \in B. \end{cases}$$

Since $\delta w \leq \tilde{w} \leq w$, $\tilde{w}$ is clearly a quasi φ-multiplier. But $\tilde{w}\varphi$ is strictly slower than $w\varphi$, in spite of the fact that $\tilde{w}\varphi$ is an upper bound of $u\varphi$ and $v\varphi$, and this contradicts the choice of $w\varphi$. Thus, $u \vee v$ and w agree μ_φ-almost everywhere in $\mathcal{G}(\varphi)$, and the proof is complete.   **QED**

8.77. Example. Let $0 < \delta \le \frac{1}{3}$, and let C be the δ-Cantor subset of $[0, 1]$. Using the techniques of Example 8.54 and the proof of Theorem 8.56, choose two nonnegative 1-multipliers u and v such that $u\mathbf{1}$ and $v\mathbf{1}$ are compatible and

$$\mathscr{F}(u\mathbf{1}) = \mathscr{F}(v\mathbf{1}) = \mathscr{R}(u\mathbf{1}, v\mathbf{1}) = C.$$

Then since C is uncountable, Theorem 8.69 implies that $u\mathbf{1} \vee v\mathbf{1}$ does not exist, *i.e.*, $u\mathbf{1}$ and $v\mathbf{1}$ do not have a least upper bound in $\mathrm{Rep}_\mathrm{q}^+(\mathbf{1})$. However, Theorems 8.74, 8,75 and 8.76 show that $u\mathbf{1}$ and $v\mathbf{1}$ have a maximum in $\mathrm{Prod}_\mathrm{q}^+(\mathbf{1})$ iff $\delta = \frac{1}{3}$.

In other words, if ϑ is the pseudo-maximum of $u\mathbf{1}$ and $v\mathbf{1}$, then $\vartheta \in \mathrm{Prod}_\mathrm{q}^+(\mathbf{1})$, and $\mathrm{Rep}_\mathrm{q}^+(\mathbf{1})$ has members that are strictly smaller than ϑ while still being upper bounds of $u\mathbf{1}$ and $v\mathbf{1}$. $\mathrm{Prod}_\mathrm{q}^+(\mathbf{1})$ has members strictly smaller than ϑ while still being upper bounds of $u\mathbf{1}$ and $v\mathbf{1}$, precisely when $\delta < \frac{1}{3}$.

8.78. Theorem. *Let φ be a quasi-flow in a Hausdorff space X, let u and v be nonnegative quasi φ-multipliers and suppose $u\varphi \vee v\varphi$ exists.*

Then $(u \vee v)\varphi = u\varphi \vee v\varphi$.

PROOF: Clear.   **QED**

8.79. Theorem. *Let φ be a flow in a Hausdorff space X, and let u and v be nonnegative proper quasi φ-multipliers. Suppose that u and v are continuous in $\mathscr{G}(\varphi)$, and that for each $x \in \mathscr{F}(\varphi)$, there exists a neighborhood W_x of x such that both u and v are bounded in $\mathscr{G}(\varphi) \cap W_x$. Suppose further than $u\varphi$ and $v\varphi$ are compatible, and $\mathscr{R}(u\varphi, v\varphi) = \square$.*

Then $u\varphi \vee v\varphi$ is a flow.

PROOF: Since $u\varphi$ and $v\varphi$ are compatible and $\mathscr{R}(u\varphi, v\varphi) = \square$, Theorem 8.69 implies the existence of $u\varphi \vee v\varphi$. Therefore, since $\mathscr{R}(u\varphi, v\varphi) = \square$, it is clear that $u \vee v$ is a proper quasi φ-multiplier. Further, $u \vee v$ is continuous in $\mathscr{G}(\varphi)$ and bounded in each of the sets $\mathscr{G}(\varphi) \cap W_x$. The result therefore follows from Theorem 4.44.   **QED**

8.80. The Distributive Law $u(\psi \vee \pi) = u\psi \vee u\pi$.

Suppose ψ and π are members of a family $\mathrm{Rep}_\mathrm{q}^+(\varphi)$, that u is a nonnegative function and that $u\psi$, $u\pi$ and $\psi \vee \pi$ all exist. Then it is of interest to ask what relationship there might be between the condition that u be a quasi $\psi \vee \pi$-multiplier and the condition that $u\psi$ and $u\pi$ be compatible.

In the following ten sections, we examine this situation, and show that the equality

$$u\,(\psi \vee \pi) = u\psi \vee u\pi$$

holds whenever it is meaningful.

8.81. Lemma. *Let φ be a quasi-flow in a Hausdorff space X and let $\psi, \pi \in \mathrm{Rep}_q^+(\varphi)$. Let u be a nonnegative function defined on X, and suppose that u is both a quasi ψ-multiplier and a quasi π-multiplier.*

Then for every $\Sigma(\varphi)$-measurable subset E of $X \setminus \big(\mathscr{F}(u\psi) \cap \mathscr{F}(u\pi)\big)$, we have

$$(\mu_{u\psi} \wedge \mu_{u\pi})\,(E) = \int_E \frac{1}{u}\,d(\mu_\psi \wedge \mu_\pi).$$

PROOF: Let E be a $\Sigma(\varphi)$-measurable subset of $X \setminus \big(\mathscr{F}(u\psi) \cap \mathscr{F}(u\pi)\big)$ and let

$$E_1 = E \cap \mathscr{G}(u\psi) \cap \mathscr{G}(u\pi),$$

$$E_2 = E \cap \mathscr{G}(u\psi) \cap \mathscr{F}(u\pi) \cap \mathscr{G}(\pi),$$

$$E_3 = E \cap \mathscr{G}(u\psi) \cap \mathscr{F}(\pi),$$

$$E_4 = E \cap \mathscr{F}(u\psi) \cap \mathscr{G}(\psi) \cap \mathscr{G}(u\pi),$$

and $\qquad E_5 = E \cap \mathscr{F}(\psi) \cap \mathscr{G}(u\pi).$

Theorem 5.41, 8.19 and 8.20 imply that

$$(\mu_{u\psi} \wedge \mu_{u\pi})\,(E_1) = \int_{E_1} \frac{1}{u}\,d(\mu_\psi \wedge \mu_\pi).$$

Now since $E_2 \subseteq \mathscr{F}(u\pi) \cap \mathscr{G}(\pi)$, we have $u(x) = 0$ for every $x \in E_2$, and therefore Theorem 5.44 (c) implies that $\mu_\psi(E_2) = 0$. Therefore $\mu_{u\psi}(E_2) = 0$, and it follows that

$$(\mu_{u\psi} \wedge \mu_{u\pi})\,(E_2) = 0 = \int_{E_2} \frac{1}{u}\,d(\mu_\psi \wedge \mu_\pi).$$

We see similarly that

$$(\mu_{u\psi} \wedge \mu_{u\pi})\,(E_4) = 0 = \int_{E_4} \frac{1}{u}\,d(\mu_\psi \wedge \mu_\pi).$$

Since $\mu_\pi(\{x\}) = \mu_{u\pi}(\{x\}) = \infty$ for every $x \in E_3$, Theorem 5.41 implies that

$$(\mu_{u\psi} \wedge \mu_{u\pi})\,(E_3) = \mu_{u\psi}(E_3) = \int_{E_3} \frac{1}{u}\,d\mu_\psi = \int_{E_3} \frac{1}{u}\,d(\mu_\psi \wedge \mu_\pi).$$

We see similarly that

$$(\mu_{u\psi} \wedge \mu_{u\pi})\,(E_5) = \int\limits_{E_5} \frac{1}{u}\, d(\mu_\psi \wedge \mu_\pi).$$

From this, the lemma follows at once.   **QED**

8.82. Theorem. *Let φ be a quasi-flow in a Hausdorff space X, and let $\psi, \pi \in \mathrm{Rep}_q^+(\varphi)$. Let u be a nonnegative function defined on X, and suppose that u is both a quasi ψ-multiplier and a quasi π-multiplier. Suppose that ψ and π are compatible, and that $u\psi$ and $u\pi$ are compatible. Let w be the function defined by*

$$w(x) = \begin{cases} \infty, & \forall\, x \in \mathscr{R}(u\psi, u\pi) \\ 0, & \forall\, x \in \mathscr{R}(\psi, \pi) \setminus \mathscr{R}(u\psi, u\pi) \\ u(x), & \text{otherwise.} \end{cases}$$

Then a necessary and sufficient condition for w to be a quasi $(\psi \vee \pi)$-multiplier is that

$$\mathscr{F}(\psi \vee \pi) \subseteq \mathscr{F}(u\psi \vee u\pi).$$

Furthermore, when this condition holds, we have

$$w(\psi \vee \pi) = u\psi \vee u\pi.$$

PROOF: Suppose that w is a quasi $\psi \vee \pi$-multiplier. Since u and w agree in $\mathscr{G}(u\psi)$ and $\mu_{\psi\psi\pi} \leq \mu_\psi$, it follows easily from Theorem 5.41 that $u\psi \leq w(\psi \vee \pi)$. We see similarly that $u\pi \leq w(\psi \vee \pi)$. Since $w(x) = \infty$ for every $x \in \mathscr{R}(u\psi, u\pi)$, Theorem 5.41 implies that

$$\mu_{w(\psi\psi\pi)}\big(\mathscr{R}(u\psi, u\pi) \cap \mathscr{G}(w(\psi \vee \pi))\big) = 0,$$

and it follows from Theorem 8.65, that $u\psi \vee u\pi \leq w(\psi \vee \pi)$. It follows that

$$\mathscr{F}(\psi \vee \pi) \subseteq \mathscr{F}\big(w(\psi \vee \pi)\big) \subseteq \mathscr{F}(u\psi \vee u\pi).$$

This shows that the condition is necessary.

Suppose that $\mathscr{F}(\psi \vee \pi) \subseteq \mathscr{F}(u\psi \vee u\pi)$. Then by Theorem 4.14, we have $u\psi \vee u\pi \in \mathrm{Rep}_q^+(\psi \vee \pi)$. We claim that w is zero in the set $\mathscr{F}(u\psi \vee u\pi) \cap \mathscr{G}(\psi \vee \pi)$, for if $x \in \mathscr{G}(\psi) \cap \mathscr{F}(u\psi \vee u\pi)$, then $x \in \mathscr{G}(\psi) \cap \mathscr{F}(u\psi)$, and $w(x) = u(x) = 0$; and if $x \in \mathscr{G}(\pi) \cap \mathscr{F}(u\psi \vee u\pi)$, then $x \in \mathscr{G}(\pi) \cap \mathscr{F}(u\pi)$, and $w(x) = u(x) = 0$; and lastly, if $x \in \mathscr{R}(\psi, \pi) \cap \mathscr{F}(u\psi \vee u\pi)$, then $x \in \mathscr{R}(\psi, \pi) \setminus \mathscr{R}(u\psi, u\pi)$, and $w(x) = 0$ by hypothesis.

Therefore Theorem 5.42 will imply that w is a quasi $\psi \vee \pi$-multiplier and that $w(\psi \vee \pi) = u\psi \vee u\pi$ when we have shown that for every $\Sigma(\varphi)$-measurable subset A of $\mathscr{G}(u\psi \vee u\pi)$, we have

$$\mu_{u\psi\vee u\pi}(A) = \int_A \frac{1}{w} \, d\mu_{\psi\vee\pi}.$$

To show this, let $A_1 = A \setminus \big(\mathscr{F}(u\psi) \cap \mathscr{F}(u\pi)\big)$ and let

$$A_2 = A \setminus A_1 = A \cap \mathscr{R}(u\psi, u\pi).$$

Then using Lemma 8.81, we have

$$\mu_{u\psi\vee u\pi}(A_1) = (\mu_{u\psi} \wedge \mu_{u\pi})(A_1)$$

$$= \int_{A_1} \frac{1}{u} \, d(\mu_\psi \wedge \mu_\pi) = \int_{A_1} \frac{1}{w} \, d\mu_{\psi\vee\pi}.$$

Also, $\mu_{u\psi\vee u\pi}(A_2) = 0$, and since $w(x) = \infty$ for every $x \in A_2$, we have

$$\int_{A_2} \frac{1}{w} \, d\mu_{\psi\vee\pi} = 0.$$

This completes the proof of the theorem.   **QED**

8.83. Theorem. *Let φ be a quasi-flow in a Hausdorff space X, and let ψ and π be compatible members of $\mathrm{Rep}_q^+(\varphi)$. Let u be a nonnegative function defined on X and suppose that u is a quasi ψ-multiplier and a quasi π-multiplier and a quasi $\psi \vee \pi$-multiplier.*

Then

(a) *$u\psi$ and $u\pi$ are compatible,*

(b) *$\mu_{\psi\vee\pi}\big(\mathscr{R}(u\psi, u\pi)\big) = 0$,*

(c) *$\mathscr{F}(\psi \vee \pi) \subseteq \mathscr{F}(u\psi \vee u\pi)$,*

(d) *$u(x) = 0$ for every $x \in \mathscr{R}(\psi, \pi) \setminus \mathscr{R}(u\psi, u\pi)$,*

and

(e) *$u(\psi \vee \pi) = u\psi \vee u\pi$.*

PROOF: Let $P = \mathscr{F}(u\psi) \cap \mathscr{F}(u\pi) \cap \mathscr{G}\big(u(\psi \vee \pi)\big)$ and let

$$\lambda(A) = (\mu_{u\psi} \wedge \mu_{u\pi})(A \setminus P),$$

for every $\Sigma(\varphi)$-measurable set A. We shall prove that λ is the time measure of $u(\psi \vee \pi)$.

Let
$$P_1 = P \cap \mathscr{G}(\psi) \cap \mathscr{F}(\pi),$$

$$P_2 = P \cap \mathscr{G}(\pi),$$

and
$$P_3 = P \cap \mathscr{F}(\psi) \cap \mathscr{F}(\pi).$$

Since $P_1 \subseteq \mathscr{G}(\psi) \cap \mathscr{F}(u\psi)$, u is zero on P_1 and since $P_1 \subseteq \mathscr{G}\big(u(\psi \vee \pi)\big)$, Theorem 5.44(c) implies that $\mu_{\psi \vee \pi}(P_1) = 0$. Therefore

$$\mu_{u(\psi \vee \pi)}(P_1) = 0.$$

We see similarly that

$$\mu_{u(\psi \vee \pi)}(P_2) = \mu_{\psi \vee \pi}(P_2) = 0.$$

Since $P_3 \subseteq \mathscr{G}\big(u(\psi \vee \pi)\big) \subseteq \mathscr{G}(\psi \vee \pi)$, we have $P_3 \subseteq \mathscr{R}(\psi, \pi)$, and therefore $\mu_{\psi \vee \pi}(P_3) = 0$. Thus

$$\mu_{u(\psi \vee \pi)}(P_3) = 0.$$

This shows that $\mu_{u(\psi \vee \pi)}$ and λ agree in P.

Let A be a $\mathbf{\Sigma}(\varphi)$-measurable set and let

$$A_1 = A \cap P,$$

$$A_2 = A \cap \mathscr{F}\big(u(\psi \vee \pi)\big),$$

and

$$A_3 = A \smallsetminus \big(\mathscr{F}(u\psi) \cap \mathscr{F}(u\pi)\big).$$

It is clear that the sets A_1, A_2 and A_3 partition A, and we already know that

$$\mu_{u(\psi \vee \pi)}(A_1) = \lambda(A_1).$$

It is clear that

$$\mu_{u(\psi \vee \pi)}(A_2) = \lambda(A_2),$$

and using Lemma 8.81, we see that

$$\lambda(A_3) = (\mu_{u\psi} \wedge \mu_{u\pi})(A_3) = \int_{A_3} \frac{1}{u}\, d(\mu_\psi \wedge \mu_\pi)$$

$$= \int_{A_3} \frac{1}{u}\, d\mu_{\psi \vee \pi} = \mu_{u(\psi \vee \pi)}(A_3).$$

Thus, λ is the time measure of $u(\psi \vee \pi)$.

Therefore, by Theorem 8.58, $u\psi$ and $u\pi$ are compatible, and by Corollary 8.59, we have $P = \mathscr{R}(u\psi, u\pi)$. Furthermore, $u(\psi \vee \pi) = u\psi \vee u\pi$.

This proves parts (a) and (e) of the theorem, and since the above argument shows that $\mu_{\psi \vee \pi}(P) = 0$, (b) is also proved. (c) follows at once from (e). Since

$$\mathscr{R}(\psi, \pi) \setminus \mathscr{R}(u\psi, u\pi) \subseteq \mathscr{G}(\psi \vee \pi) \cap \mathscr{F}\big(u(\psi \vee \pi)\big),$$

(d) is also proved.   **QED**

8.84. Theorem. *Let φ be a quasi-flow in a Hausdorff space X, and let $\psi, \pi \in \mathrm{Rep}_q^+(\varphi)$. Let u be a nonnegative function defined on X, and suppose that u is both a quasi ψ-multiplier and a quasi π-multiplier. Suppose that ψ and π are compatible, and that $u\psi$ and $u\pi$ are compatible. Suppose also that*

$$\mu_{\psi \vee \pi}\big(\mathscr{R}(u\psi, u\pi)\big) = 0.$$

Then u is a quasi $\psi \vee \pi$-multiplier iff the following conditions hold:

(a) $\mathscr{F}(\psi \vee \pi) \subseteq \mathscr{F}(u\psi \vee u\pi)$

and

(b) $u(x) = 0$ *for every* $x \in \mathscr{R}(\psi, \pi) \setminus \mathscr{R}(u\psi, u\pi)$.

PROOF: It follows at once from Theorem 8.83 that if u is a quasi $\psi \vee \pi$-multiplier, then conditions (a) and (b) hold.

Suppose that conditions (a) and (b) hold and let w be defined as in Theorem 8.82. By Theorem 8.82, w is a quasi $\psi \vee \pi$-multiplier. Furthermore, since $\mu_{\psi \vee \pi}\big(\mathscr{R}(u\psi, u\pi)\big) = 0$, u and w agree $\mu_{\psi \vee \pi}$ almost everywhere. Therefore, since u is clearly zero in the set $\mathscr{F}(u\psi \vee u\pi) \cap \mathscr{G}(\pi)$, Theorem 5.42 implies that u is a quasi $\psi \vee \pi$-multiplier.

This completes the proof of the theorem.   **QED**

8.85. Corollary. *Let φ be a quasi-flow in a Hausdorff space X, let $\psi, \pi \in \mathrm{Rep}_q^+(\varphi)$, and let u be a nonnegative function defined on X. Suppose that*

(i) *u is both a quasi ψ-multiplier and a quasi π-multiplier,*

(ii) *both $\psi \vee \pi$ and $u\psi \vee u\pi$ exist,*

(iii) *$\mathscr{F}(\psi \vee \pi) \subseteq \mathscr{F}(u\psi \vee u\pi)$,*

and

(iv) *u is zero in $\mathscr{R}(\psi, \pi) \setminus \mathscr{R}(u\psi, u\pi)$.*

Then u is a quasi $\psi \vee \pi$-multiplier, and

$$u(\psi \vee \pi) = u\psi \vee u\pi.$$

PROOF: The result will follow from Theorems 8.83 and 8.84 as soon as we have shown that

$$\mu_{\psi\vee\pi}\big(\mathscr{R}(u\psi,\,u\pi)\big) = 0.$$

But since $\mathscr{R}(u\psi,\,u\pi) \subseteq \mathscr{G}(\psi\vee\pi)$, the restriction of $\mu_{\psi\vee\pi}$ to $\mathscr{R}(u\psi,\,u\pi)$ is continuous, and since $u\psi\vee u\pi$ exists, the intersection of $\mathscr{R}(u\psi,\,u\pi)$ with every φ-orbit is countable. Thus, the desired equality holds, and the proof is complete. **QED**

8.86. Remark. If ψ and π are members of a family $\mathrm{Rep}_q^+(\varphi)$, u is a nonnegative function, and $\psi\vee\pi$ and $u\psi\vee u\pi$ exist, we cannot deduce that u is a quasi $\psi\vee\pi$-multiplier. It is quite possible for the set

$$\mathscr{F}(\psi\vee\pi) \setminus \mathscr{F}(u\psi\vee u\pi)$$

to be non-empty. We illustrate this in the following example.

8.87. Example. Let sequences $\{a_n\}$, $\{b_n\}$, $\{c_n\}$ be chosen such that

$$-1 = a_1 < b_1 < c_1 < a_2 < b_2 < c_2 < \ldots$$

and such that

$$\lim_{n\to\infty} a_n = \lim_{n\to\infty} b_n = \lim_{n\to\infty} c_n = 0.$$

Define $v(x) = 1$ for $x \leq -1$,

$v(x) = 1$ for $a_n < x \leq b_n$, $n = 1, 2, \ldots$,

$v(x) = c_n - b_n$ for $b_n < x \leq c_n$, $n = 1, 2, \ldots$,

$v(x) = a_{n+1} - c_n$ for $c_n < x \leq a_{n+1}$, $n = 1, 2, \ldots$,

$v(0) = 0$,

and $v(x) = v(-x)$ for $x > 0$.

It is clear that v is a **1**-multiplier. Let $\psi = v\mathbf{1}$.

Define $w(x) = 1$ for $x \leq -1$,

$w(x) = b_n - a_n$ for $a_n < x \leq b_n$, $n = 1, 2, \ldots$,

$w(x) = 1$ for $b_n < x \leq c_n$, $n = 1, 2, \ldots$,

$w(x) = a_{n+1} - c_n$ for $c_n < x \leq a_{n+1}$, $n = 1, 2, \ldots$,

$w(0) = 0$,

and $w(x) = w(-x)$ for $x > 0$.

It is clear that w is a **1**-multiplier. Let $\pi = w\mathbf{1}$.

Define $u(x) = 1$ for $x \leq -1$,

$\qquad u(x) = 1$ for $a_n < x \leq c_n$, $n = 1, 2, \ldots$,

$\qquad u(x) = \dfrac{1}{a_{n+1} - c_n}$ for $c_n < x \leq a_{n+1}$, $n = 1, 2, \ldots$,

$\qquad u(0) = 0$,

and $\qquad u(x) = u(-x)$ for $x > 0$.

It is easy to see that u is both a ψ-multiplier and a π-multiplier, and that both $\psi \vee \pi$ and $u\psi \vee u\pi$ exist. However, it is clear that

$$\mathscr{F}(\psi \vee \pi) \setminus \mathscr{F}(u\psi \vee u\pi) = \{0\} \neq \square.$$

8.88. Remark. If ψ and π are compatible members of a family $\mathrm{Rep}_q^+(\varphi)$ and if u is a quasi ψ-multiplier and a quasi π-multiplier and a quasi $\psi \vee \pi$-multiplier, then Theorem 8.83 implies that $u\psi$ and $u\pi$ are compatible. We cannot however deduce that $u\psi \vee u\pi$ exists even if $\psi \vee \pi$ is known to exist, because $\mathscr{R}(u\psi, u\pi)$ can still be too large. The following lemma allows us to construct an example which illustrates this pathology.

8.89. Lemma. *Let α and β be real numbers with $\alpha < \beta$.*

Then there exist functions u, v and w from $[\alpha, \beta)$ into $(0, \infty)$ such that

(a) $\displaystyle \int_\alpha^\beta \frac{1}{v}\, dm \leq \beta - \alpha$ *and* $\displaystyle \int_\alpha^\beta \frac{1}{w}\, dm \leq \beta - \alpha$,

(b) $\displaystyle \int_\alpha^\beta \frac{1}{u(v \vee w)}\, dm \leq \beta - \alpha$,

(c) $\displaystyle \int_\alpha^\delta \frac{1}{uv}\, dm < \infty$ *and* $\displaystyle \int_\alpha^\delta \frac{1}{uw}\, dm < \infty$, *for every $\delta \in [\alpha, \beta)$,*

and

(d) $\displaystyle \int_\alpha^\beta \frac{1}{uv}\, dm = \infty = \int_\alpha^\beta \frac{1}{uw}\, dm$.

PROOF: Choose sequences $\{a_n\}$ and $\{b_n\}$ such that

$$\alpha = a_1 < b_1 < a_2 < b_2 < \cdots,$$

$$\lim_{n \to \infty} a_n = \lim_{n \to \infty} b_n = \beta,$$

and $b_n - a_n < 1$ and $a_{n+1} - b_n < 1$ for every $n = 1, 2, \ldots$.

Let $u(x) = \begin{cases} b_n - a_n & \text{for } a_n \le x < b_n, & n = 1, 2, \ldots, \\ a_{n+1} - b_n & \text{for } b_n \le x < a_{n+1}, & n = 1, 2, \ldots; \end{cases}$

$v(x) = \begin{cases} \dfrac{1}{b_n - a_n} & \text{for } a_n \le x < b_n, & n = 1, 2, \ldots, \\ 1 & \text{for } b_n \le x < a_{n+1}, & n = 1, 2, \ldots; \end{cases}$

and $w(x) = \begin{cases} 1 & \text{for } a_n \le x < b_n, & n = 1, 2, \ldots, \\ \dfrac{1}{a_{n+1} - b_n} & \text{for } b_n \le x < a_{n+1}, & n = 1, 2, \ldots. \end{cases}$

It is clear that these functions have the required properties.   **QED**

8.90. Example. Let C be the $\frac{1}{3}$-Cantor subset of $[-1, 1]$, and for convenience, denote the family of components of $[-1, 1] \setminus C$ by $\{(a_n - \delta_n, a_n + \delta_n) \mid n = 1, 2, \ldots\}$.

Let $u(x) = 0$ and $v(x) = w(x) = 1$ for $x \in C$.

For each $n = 1, 2, \ldots$, define u, v and w in $[a_n, a_n + \delta_n)$ in such a way that they satisfy the conditions of Lemma 8.89 with α and β replaced by a_n and $a_n + \delta_n$. Then define u, v and w in $(a_n - \delta_n, a_n)$ by the equations

$$u(a_n - x) = u(a_n + x),$$
$$v(a_n - x) = v(a_n + x),$$
and $$w(a_n - x) = w(a_n + x),$$

for every $x \in (0, \delta_n)$.

Define u, v and w in $[-2, -1)$ in such a way that they satisfy the conditions of Lemma 8.89, with α and β replaced by -2 and -1. Then define u, v and w in $(1, 2]$ by the equations

$$u(x) = u(-x), \quad v(x) = v(-x), \quad \text{and} \quad w(x) = w(-x),$$

for every $x \in (1, 2]$.

Finally, define $u(x) = v(x) = w(x) = 1$, whenever $|x| \ge 2$.

It is easy to see that u is a $(v1 \vee w1)$-multiplier and that $\mathscr{R}(uv1, uw1) = C$. Since C is uncountable, $uv1 \vee uw1$ does not exist.

8.91. Maximum of Finitely many Flows. The problem of finding the maximum of finitely many members $\psi_1, \psi_2, \ldots, \psi_n$ of a family $\text{Rep}_q^+(\varphi)$ is similar to the case $n = 2$ which we have considered in Sections 8.41

to 8.90. Given finitely many members $\psi_1, \psi_2, \ldots, \psi_n$ of a family $\text{Rep}_q^+(\varphi)$, it is clear how to define jumping, ramming, ramming points and compatibility of the family $\{\psi_1, \psi_2, \ldots, \psi_n\}$. When the family $\{\psi_1, \psi_2, \ldots, \psi_n\}$ is compatible, we can define the pseudo-maximum $\overset{n}{\underset{j=1}{V}} \psi_j$ to be that member of $\text{Rep}_q^+(\varphi)$ whose time measure μ is defined by

$$\mu(E) = (\mu_{\psi_1} \wedge \mu_{\psi_2} \wedge \ldots \wedge \mu_{\psi_n})(E \setminus \mathcal{R}(\psi_1, \psi_2, \ldots, \psi_n)),$$

for every $\Sigma(\varphi)$-measurable set E. (Of course, $\mathcal{R}(\psi_1, \psi_2, \ldots, \psi_n)$ is the set of removable ramming points of $\{\psi_1, \psi_2, \ldots, \psi_n\}$.) We also write $\overset{n}{\underset{j=1}{V}} \psi_j$ as $\psi_1 V \psi_2 V \ldots V \psi_n$.

With these definitions, we obtain analogues of all the results in Sections 8.41 to 8.90, and among other things we know that if $\psi_1, \psi_2, \ldots, \psi_n \in \text{Rep}_q^+(\varphi)$, then $\{\psi_1, \psi_2, \ldots, \psi_n\}$ has a least upper bound (*i.e.*, a maximum) if and only if the following two conditions hold:

(a) $\{\psi_1, \psi_2, \ldots, \psi_n\}$ is compatible;

and

(b) $\mathcal{R}(\psi_1, \psi_2, \ldots, \psi_n)$ has a countable intersection with every φ-orbit.

When $\{\psi_1, \psi_2, \ldots, \psi_n\}$ has a least upper bound, we denote it by $\overset{n}{\underset{j=1}{V}} \psi_j$, and also by $\psi_1 \vee \psi_2 \vee \ldots \vee \psi_n$.

8.92. Theorem. *Let φ be a quasi-flow in a Hausdorff space X and let $\psi_1, \psi_2, \psi_3 \in \text{Rep}_q^+(\varphi)$.*

If $(\psi_1 \vee \psi_2) \vee \psi_3$ exists, we have

$$(\psi_1 \vee \psi_2) \vee \psi_3 = \psi_1 \vee \psi_2 \vee \psi_3.$$

A fortiori, if $(\psi_1 \vee \psi_2) \vee \psi_3$ and $\psi_1 \vee (\psi_2 \vee \psi_3)$ both exist, they are equal.

PROOF: Trivial.   **QED**

8.93. Remarks.

(a) If ψ_1, ψ_2, ψ_3 belong to a family $\text{Rep}_q^+(\varphi)$, then the existence of $(\psi_1 \vee \psi_2) \vee \psi_3$ does not imply the existence of $\psi_1 \vee (\psi_2 \vee \psi_3)$.

(b) If ψ_1, ψ_2, ψ_3 belong to a family $\text{Rep}_q^+(\varphi)$, then the existence of $\psi_1 \vee \psi_2 \vee \psi_3$ does not imply the existence of any of the expressions $(\psi_1 \vee \psi_2) \vee \psi_3$, $\psi_1 \vee (\psi_2 \vee \psi_3)$, $(\psi_1 \vee \psi_3) \vee \psi_2$.

We leave it to the reader to construct examples to prove (a) and (b). Such examples may be constructed by the usual technique, *i.e.*, making

ψ_1, ψ_2 and ψ_3 have the correct behavior in a neighborhood of a prospective ramming point.

Note, however, that if $\psi_1 \vee \psi_2 \vee \psi_3$ exists and $\psi_1 \vee \psi_2$ exists, then $(\psi_1 \vee \psi_2) \vee \psi_3$ clearly exists.

We remark finally

(c) Theorem 8.92 is no longer true if $\vee$ is replaced by V.

8.94. Theorem. The Distributive Law $\psi_1 \vee (\psi_2 \wedge \psi_3) = (\psi_1 \vee \psi_2) \wedge (\psi_1 \vee \psi_3)$. *Let φ be a quasi-flow in a Hausdorff space X, let $\psi_1, \psi_2, \psi_3 \in \mathrm{Rep}_q^+(\varphi)$, and suppose that $\psi_1 \vee \psi_2$ and $\psi_1 \vee \psi_3$ both exist.*

Then $\psi_1 \vee (\psi_2 \wedge \psi_3) = (\psi_1 \vee \psi_2) \wedge (\psi_1 \vee \psi_3)$.

PROOF: Denote the time measures of $\psi_1, \psi_2, \psi_3, \psi_1 \vee \psi_2, \psi_1 \vee \psi_3$ and $(\psi_1 \vee \psi_2) \wedge (\psi_1 \vee \psi_3)$ by $\mu_1, \mu_2, \mu_3, \mu_{12}, \mu_{13}$ and μ respectively. We clearly have

$$\mu = \mu_{12} \vee \mu_{13}.$$

Since μ_{12} agrees with $\mu_1 \wedge \mu_2$ in $X \setminus \mathscr{R}(\psi_1, \psi_2)$ and μ_{13} agrees with $\mu_1 \wedge \mu_3$ in $X \setminus \mathscr{R}(\psi_1, \psi_3)$, it follows from Theorem 8.24 (b) that μ agrees with $\mu_1 \wedge (\mu_2 \vee \mu_3)$ in $X \setminus [\mathscr{R}(\psi_1, \psi_2) \cup \mathscr{R}(\psi_1, \psi_3)]$. Now

$$\mathscr{R}(\psi_1, \psi_2) \cup \mathscr{R}(\psi_1, \psi_3) \subseteq \mathscr{F}(\psi_1) \cap \mathscr{F}(\psi_2 \wedge \psi_3)$$

and therefore

$$\big(\mu_1 \wedge (\mu_2 \vee \mu_3)\big)(\{x\}) = \infty, \qquad \forall\, x \in \mathscr{R}(\psi_1, \psi_2) \cup \mathscr{R}(\psi_1, \psi_3),$$

and consequently μ agrees with $\mu_1 \wedge (\mu_2 \vee \mu_3)$ in the set

$$\{x \in \mathscr{R}(\psi_1, \psi_2) \cup \mathscr{R}(\psi_1, \psi_3) \mid \mu(\{x\}) = \infty\}.$$

Let

$$P = \{x \in \mathscr{R}(\psi_1, \psi_2) \cup \mathscr{R}(\psi_1, \psi_3) \mid \mu(\{x\}) = 0\}.$$

Then since P has a countable intersection with every φ-orbit, $\mu(P) = 0$, and we therefore have

$$\mu(E) = \mu_1 \wedge (\mu_2 \vee \mu_3)(E \setminus P)$$

for every $\mathbf{\Sigma}(\varphi)$-measurable set E. Therefore, from Corollary 8.59 and Theorem 8.66 we see that $\psi_1 \vee (\psi_2 \wedge \psi_3)$ exists and equals $(\psi_1 \vee \psi_2) \wedge (\psi_1 \vee \psi_3)$.
QED

8.95. Remark. In the proof of Theorem 8.94, the fact that P has a countable intersection with every φ-orbit is essential. If, in the hypotheses of the theorem, we replace $\psi_1 \vee \psi_2$ and $\psi_1 \vee \psi_3$ by $\psi_1 \mathsf{V} \psi_2$ and $\psi_1 \mathsf{V} \psi_3$, then we cannot conclude that $\psi_1 \mathsf{V} (\psi_2 \wedge \psi_3)$ exists. Furthermore, even if all the expressions $\psi_1 \vee \psi_2$, $\psi_1 \vee \psi_3$ and $\psi_1 \vee (\psi_2 \wedge \psi_3)$ exist, we need not have

$$\psi_1 \vee (\psi_2 \wedge \psi_3) = (\psi_1 \vee \psi_2) \wedge (\psi_1 \vee \psi_3).$$

Examples justifying these assertions can readily be obtained.

The next theorem gives an analogous result in $\mathrm{Prod}_q^+ (\varphi)$.

8.96. Theorem. *Let φ be a quasi-flow in a Hausdorff space X, let ψ_1, ψ_2, $\psi_3 \in \mathrm{Prod}_q^+ (\varphi)$, and suppose that both $\{\psi_1, \psi_2\}$ and $\{\psi_1, \psi_3\}$ have maxima in the family $\mathrm{Prod}_q^+ (\varphi)$.*

Then $\{\psi_1, \psi_2 \wedge \psi_3\}$ has a maximum in $\mathrm{Prod}_q^+ (\varphi)$ and

$$\psi_1 \vee (\psi_2 \wedge \psi_3) = (\psi_1 \vee \psi_2) \wedge (\psi_1 \vee \psi_3).$$

PROOF: Denote the time measures of $\psi_1, \psi_2, \psi_3, \psi_1 \vee \psi_2, \psi_1 \vee \psi_3$ and $(\psi_1 \vee \psi_2) \wedge (\psi_1 \vee \psi_3)$ by $\mu_1, \mu_2, \mu_3, \mu_{12}, \mu_{13}$ and μ, respectively, and as in the proof of Theorem 8.94, let

$$P = \{x \in \mathscr{R}(\psi_1, \psi_2) \cup \mathscr{R}(\psi_1, \psi_3) \mid \mu(\{x\}) = 0\}.$$

As in the proof of Theorem 8.94, we can see that μ agrees with $\mu_1 \wedge (\mu_2 \vee \mu_3)$ in $X \setminus P$. The result will therefore follow at once from Corollary 8.59 and Theorems 8.75 and 8.74 once we have shown that

$$\mu(P) = \mu_\varphi(P) = 0.$$

Now since $P \subseteq \mathscr{R}(\psi_1, \psi_2) \cup \mathscr{R}(\psi_1, \psi_3)$, we have

$$\mu_\varphi(P) \le \mu_\varphi\big(\mathscr{R}(\psi_1, \psi_2)\big) + \mu_\varphi\big(\mathscr{R}(\psi_1, \psi_3)\big) = 0.$$

Therefore since Theorem 8.28 implies that $(\psi_1 \vee \psi_2) \wedge (\psi_1 \vee \psi_3) \in \mathrm{Prod}_q^+ (\varphi)$ and

$$P \subseteq \mathscr{G}\big((\psi_1 \vee \psi_2) \wedge (\psi_1 \vee \psi_3)\big),$$

it follows from Theorem 5.43 (b) that $\mu(P) = 0$.

This completes the proof. **QED**

8.97. Theorem. The Distributive Law $\psi_1 \wedge (\psi_2 \vee \psi_3) = (\psi_1 \wedge \psi_2) \vee (\psi_1 \wedge \psi_3)$.
Let φ be a quasi-flow in a Hausdorff space X and let ψ_1, ψ_2, $\psi_3 \in \mathrm{Rep}_q^+ (\varphi)$.

Then

(a) *If $\psi_2 \vee \psi_3$ exists, we have*

$$\psi_1 \wedge (\psi_2 \vee \psi_3) = (\psi_1 \wedge \psi_2) \vee (\psi_1 \wedge \psi_3).$$

(b) *If* $\psi_1, \psi_2, \psi_3 \in \mathrm{Prod}_q^+(\varphi)$ *and* $\{\psi_2, \psi_3\}$ *has a maximum in* $\mathrm{Prod}_q^+(\varphi)$, *then so does* $\{\psi_1 \wedge \psi_2, \psi_1 \wedge \psi_3\}$, *and we have*

$$\psi_1 \wedge (\psi_2 \vee \psi_3) = (\psi_1 \wedge \psi_2) \vee (\psi_1 \wedge \psi_3).$$

PROOF: Denote the time measures of $\psi_1, \psi_2, \psi_3, \psi_2 \vee \psi_3$ and $\psi_1 \wedge (\psi_2 \vee \psi_3)$ by $\mu_1, \mu_2, \mu_3, \mu_{23}$, and μ, respectively. Then since $\mu = \mu_1 \vee \mu_{23}$, it is easy to see that, if

$$P = \mathscr{R}(\psi_2, \psi_3) \cap \mathscr{G}(\psi_1),$$

then μ agrees with $\mu_1 \vee (\mu_2 \wedge \mu_3)$ in $X \smallsetminus P$. Therefore, by Theorem 8.24(a), μ agrees with $(\mu_1 \vee \mu_2) \wedge (\mu_1 \vee \mu_3)$ in $X \smallsetminus P$, and since $P \subseteq \mathscr{F}(\psi_1 \wedge \psi_2) \cap \mathscr{F}(\psi_1 \wedge \psi_3)$, the theorem will follow as soon as we have shown that $\mu(P) = 0$.

Suppose (a) holds. Then since $P \subseteq R(\psi_2, \psi_3)$, P has a countable intersection with every φ-orbit, and the fact that μ is continuous on P implies that $\mu(P) = 0$.

Suppose (b) holds. Then since $P \subseteq \mathscr{R}(\psi_2, \psi_3)$, $\mu_\varphi(P) = 0$, and Theorem 5.43(b) implies that $\mu_1(P) = 0$. Therefore, since $\mu_{23}(P) = 0$, we have $\mu(P) = 0$.
This completes the proof. **QED**

8.98. Theorem. The Distributive Law $\psi_1 \oplus (\psi_2 \vee \psi_3) = (\psi_1 \oplus \psi_2) \vee (\psi_1 \oplus \psi_3)$. *Let* φ *be a quasi-flow in a Hausdorff space* X *and let* $\psi_1, \psi_2, \psi_3 \in \mathrm{Rep}_q^+(\varphi)$.

Then

(a) *If* $\psi_2 \vee \psi_3$ *exists, we have*

$$\psi_1 \oplus (\psi_2 \vee \psi_3) = (\psi_1 \oplus \psi_2) \vee (\psi_1 \oplus \psi_3).$$

(b) *If* $\psi_1, \psi_2, \psi_3 \in \mathrm{Prod}_q^+(\varphi)$ *and* $\{\psi_2, \psi_3\}$ *has a maximum in* $\mathrm{Prod}_q^+(\varphi)$, *then so does* $\{\psi_1 \oplus \psi_2, \psi_1 \oplus \psi_3\}$, *and we have*

$$\psi_1 \oplus (\psi_2 \vee \psi_3) = (\psi_1 \oplus \psi_2) \vee (\psi_1 \oplus \psi_3).$$

PROOF: The proof is similar to that of Theorem 8.97, and is omitted.
QED

8.99. Theorem. *Let* φ *be a quasi-flow in a Hausdorff space* X *and let* $\psi_1, \psi_2, \psi_3 \in \mathrm{Rep}_q^+(\varphi)$.

Then $\psi_1 \oplus (\psi_2 \wedge \psi_3) = (\psi_1 \oplus \psi_2) \wedge (\psi_1 \oplus \psi_3)$.

PROOF: The result follows at once from Theorem 8.24(c). **QED**

8.100. Theorem. *Let φ be a quasi-flow in a Hausdorff space X, and let ψ_1 and ψ_2 be compatible members of* $\mathrm{Rep}_q^+(\varphi)$.

Then $\psi_1 \oplus \psi_2 = (\psi_1 \wedge \psi_2) \oplus (\psi_1 \vee \psi_2)$.

PROOF: Even though $\mu_{\psi_1 \vee \psi_2}$ may differ from $\mu_{\psi_1} \wedge \mu_{\psi_2}$, it is easy to see that the time measure of $(\psi_1 \wedge \psi_2) \oplus (\psi_1 \vee \psi_2)$ is $(\mu_{\psi_1} \vee \mu_{\psi_2}) + (\mu_{\psi_1} \wedge \mu_{\psi_2})$. The result now follows at once from Theorem 8.24 (e).   **QED**

CHAPTER NINE

ALGEBRAIC COMBINATIONS OF FLOWS II

Sum of Two Flows: Motivation

9.1. Introduction. There is an important difference between the theory of maxima and minima of members of a family $\mathrm{Rep_q^+}\,(\varphi)$, and the theory of harmonic sums and sums. In dealing with the maximum or minimum of two members ψ, π of a family $\mathrm{Rep_q^+}\,(\varphi)$, we had only to ask when $\psi \vee \pi$ and $\psi \wedge \pi$ could exist and how they behaved. The *meaning* of the symbols $\psi \vee \pi$ and $\psi \wedge \pi$ was preassigned by the partial order $\leq$ in $\mathrm{Rep_q^+}\,(\varphi)$. However, the harmonic sum and sum need to be *defined* in such a way that they have the behavior we expect of them. In the case of harmonic sums, the definition given in Section 8.35 was "justified" by the equation

$$(u \oplus v)\, \varphi = u\varphi \oplus v\varphi.$$

At the outset, we expect the sum in a family $\mathrm{Rep_q^+}\,(\varphi)$ to have the following three properties:

(a) If $\psi, \pi \in \mathrm{Rep_q^+}\,(\varphi)$ and $\psi + \pi$ exists, then $\psi + \pi$ is an upper bound of $\{\psi, \pi\}$.

(b) For every $\psi \in \mathrm{Rep_q^+}\,(\varphi)$ and nonnegative real numbers a, b,

$$(a + b)\, \psi = a\psi + b\psi.$$

(c) If $\psi, \pi \in \mathrm{Rep_q^+}\,(\varphi)$ and $\psi + \pi$ exists, then given any upper bound ϑ of $\{\psi, \pi\}$, we have

$$\psi + \pi \leq 2\vartheta.$$

9.2. Theorem. *Let φ be a quasi-flow in a Hausdorff space X, let $\psi, \pi, \chi \in \mathrm{Rep_q^+}\,(\varphi)$, suppose that χ is an upper bound of ψ and π and suppose that, for every upper bound ϑ of ψ and π we have $\chi \leq 2\vartheta$.*

Then $\psi \vee \pi$ exists and

$$\psi \vee \pi \leq \chi \leq 2(\psi \vee \pi).$$

PROOF: Let $F = \mathscr{F}(\psi) \cap \mathscr{F}(\pi)$. We shall show first that

$$\frac{1}{2}(\mu_\psi \wedge \mu_\pi)(E) \leq \mu_\chi(E),$$

for every $\Sigma(\varphi)$ -measurable subset E of $X \setminus F$.

Assume that this is false. Then we can choose a $\Sigma(\varphi)$ -mesurable subset A of $X \setminus F$ such that

$$\mu_\chi(A) < \frac{1}{2}(\mu_\psi \wedge \mu_\pi)(A) < \infty.$$

Define the measure μ by

$$\mu(E) = (\mu_\psi \wedge \mu_\pi)(E \cap A) + \mu_\chi(E \setminus A),$$

for every $\Sigma(\varphi)$ -measurable set E . It is clear that $\mu \in \mathscr{M}_q(\varphi)$, and that $\mu \leq \mu_\psi \wedge \mu_\pi$. But if ϑ is the member of $\mathrm{Rep}^+(\varphi)$ for which $\mu = \mu_\vartheta$, then ϑ is an upper bound of ψ and π which does not satisfy the inequality $\chi \leq 2\vartheta$. This contradiction shows that $\frac{1}{2}(\mu_\psi \wedge \mu_\pi) \leq \mu_\chi$ in $X \setminus F$. Now define the measure λ by

$$\lambda(E) = (\mu_\psi \wedge \mu_\pi)(E \setminus F) + 2\mu_\chi(E \cap F),$$

for every $\Sigma(\varphi)$ -measurable set E . It is clear that

$$\mu_\chi \leq \lambda \leq 2\mu_\chi,$$

and therefore Theorem 5.40 implies that λ is the time measure of some member ϑ of $\mathrm{Rep}_q^+(\varphi)$. ϑ is clearly an upper bound of ψ and π . But if $\tilde{\vartheta}$ is any upper bound of ψ and π , then since $\mu_{\tilde{\vartheta}} \leq \mu_\psi \wedge \mu_\pi$ and $\mu_{\tilde{\vartheta}} \leq 2\mu_\chi$, we have $\mu_{\tilde{\vartheta}} \leq \lambda$, *i.e.*, $\vartheta \leq \tilde{\vartheta}$. Thus $\vartheta = \psi \vee \pi$.

We note finally that the inequality

$$\psi \vee \pi \leq \chi \leq 2(\psi \vee \pi)$$

is obvious.   QED

9.3. Motivation of the Definition. In view of Theorem 9.2, the sum of two members ψ and π of a family $\mathrm{Rep}_q^+(\varphi)$ can only be defined when $\psi \vee \pi$ exists, and the inequality

$$\psi \vee \pi \leq \psi + \pi \leq 2(\psi \vee \pi)$$

implies that $\mu_{\psi \vee \pi}$ and $\mu_{\psi+\pi}$ must agree in $\mathscr{F}(\psi) \cap \mathscr{F}(\pi)$.

Thus, the problem of defining $\psi + \pi$ reduces to the problem of finding a measure $\mu \in \mathcal{M}_q(\varphi)$ which agrees with $\mu_{\psi \vee \pi}$ in $\mathcal{F}(\psi) \cap \mathcal{F}(\pi)$, and whose behavior in $X \setminus \big(\mathcal{F}(\psi) \cap \mathcal{F}(\pi)\big)$ gives us the properties we expect of a sum.

Let us first consider the simplest case of all: Let φ be a quasi-flow in a Hausdorff space X, let a and b be positive real numbers, and let $\psi = a\varphi$ and $\pi = b\varphi$. Then $\mathcal{F}(\psi) = \mathcal{F}(\pi) = \mathcal{F}(\varphi)$ and for every $\Sigma(\varphi)$-measurable subset E of $\mathcal{G}(\varphi)$,

$$\mu_{(a+b)\varphi}(E) = \int\limits_E \frac{1}{a+b}\, d\mu_\varphi = \mu_{a\varphi}(E) \oplus \mu_{b\varphi}(E).$$

However, we cannot in general define the time measure of the sum of ψ and π by

$$\mu_{\psi+\pi}(E) = \mu_\psi(E) \oplus \mu_\pi(E),$$

for all $\Sigma(\varphi)$-measurable subsets E of $X \setminus \big(\mathcal{F}(\psi) \cap \mathcal{F}(\pi)\big)$, because this definition of $\mu_{\psi+\pi}$ does not yield a measure.

Suppose next that u and v are nonnegative simple functions which are both quasi φ-multipliers, and that $u + v$ is also a quasi φ-multiplier. Write

$$u = \sum_{j=1}^{n} a_j \chi_{A_j} \text{ and } v = \sum_{j=1}^{n} b_j \chi_{A_j}.$$

If E is a $\Sigma(\varphi)$-measurable subset of $\mathcal{G}(u\varphi) \cap \mathcal{G}(v\varphi)$, we have

$$\mu_{(u+v)\varphi}(E) = \sum_{j=1}^{n} \mu_{(u+v)\varphi}(E \cap A_j)$$

$$= \sum_{j=1}^{n} \int\limits_{E \cap A_j} \frac{1}{a_j + b_j}\, d\mu_\varphi$$

$$= \sum_{j=1}^{n} \mu_{u\varphi}(E \cap A_j) \oplus \mu_{v\varphi}(E \cap A_j).$$

Also, bearing in mind that u and v vanish respectively in $\mathcal{F}(u\varphi) \cap \mathcal{G}(\varphi)$ and $\mathcal{F}(v\varphi) \cap \mathcal{G}(\varphi)$, it is easy to see that the equation

$$\mu_{(u+v)\varphi}(E) = \sum_{j=1}^{n} \mu_{u\varphi}(E \cap A_j) \oplus \mu_{v\varphi}(E \cap A_j)$$

holds for $E \subseteq \mathcal{F}(u\varphi) \cap \mathcal{G}(v\varphi)$ and also for $E \subseteq \mathcal{G}(u\varphi) \cap \mathcal{F}(v\varphi)$. Therefore this equation holds for every $E \subseteq X \setminus \big(\mathcal{F}(u\varphi) \cap \mathcal{F}(v\varphi)\big)$.

This suggests that in the general case, given a $\Sigma(\varphi)$-measurable subset E of $X \setminus \big(\mathscr{F}(\psi) \cap \mathscr{F}(\pi)\big)$, we should define $\mu_{\psi+\pi}(E)$ to be a limit of sums of the form

$$\sum_{j=1}^{n} \mu_\psi(E_j) \oplus \mu_\pi(E_j),$$

where $\{E_j \mid = 1, \ldots, n\}$ partitions E, and the partition is, in some sense, a "fine" one. Fortunately, we do not have to deal with the awkward question as to which partitions of E are fine, and which are not. For suppose the partition $\{E_j \mid j = 1, \ldots, n\}$ is refined to produce the partition $\{E_{jk} \mid j = 1, \ldots, n \; ; \; k = 1, \ldots, n_j\}$, where for each $j = 1, \ldots, n$

$$E_j = \bigcup_{k=1}^{n_j} E_{jk}.$$

Then

$$\sum_{j=1}^{n} \sum_{k=1}^{n_j} \big(\mu_\psi(E_{jk}) \oplus \mu_\pi(E_{jk})\big) \leq \sum_{j=1}^{n} \left(\sum_{k=1}^{n_j} \mu_\psi(E_{jk})\right) \oplus \left(\sum_{k=1}^{n_j} \mu_\pi(E_{jk})\right)$$

$$\text{(by Theorem 8.34(f))}$$

$$= \sum_{j=1}^{n} \mu_\psi(E_j) \oplus \mu_\pi(E_j).$$

This at once suggests the definition

$$\mu_{\psi+\pi}(E) = \inf\left\{ \sum_{j=1}^{n} \mu_\psi(E_j) \oplus \mu_\pi(E_j) \;\middle|\; E = \bigcup_{j=1}^{n} E_j \right\},$$

where the infimum is taken over all partitions of E into finitely many $\Sigma(\varphi)$-measurable subsets.

Harmonic Sums of Measures

9.4. Definition. Let μ and λ be nonnegative measures on a σ-algebra Σ of subsets of a set S. Then the *harmonic sum* $\mu \oplus \lambda$ of μ and λ is defined by

$$(\mu \oplus \lambda)(A) = \inf\left\{ \sum_{j=1}^{n} \mu(A_j) \oplus \lambda(A_j) \;\middle|\; A = \bigcup_{j=1}^{n} A_j \right\},$$

for every measurable set A, where the infimum is taken over all partitions of A into finitely many measurable sets.

9.5. Theorem. *Let μ and λ be nonnegative measures defined on a σ-algebra Σ of subsets of a set S.*

Then $\mu \oplus \lambda$ *is a measure, and*

$$\frac{1}{2}(\mu \wedge \lambda) \leq \mu \oplus \lambda \leq \mu \wedge \lambda.$$

PROOF: Let $A \in \boldsymbol{\Sigma}$. Then for any partition $\{A_j \mid j = 1, \ldots, n\}$ of A, we have

$$\frac{1}{2}(\mu \wedge \lambda)(A) = \sum_{j=1}^{n} \frac{1}{2}(\mu \wedge \lambda)(A_j)$$

$$= \sum_{j=1}^{n} (\mu \wedge \lambda)(A_j) \oplus (\mu \wedge \lambda)(A_j)$$

$$\leq \sum_{j=1}^{n} \mu(A_j) \oplus \lambda(A_j) \qquad \text{(by Theorem 8.34 (c))}.$$

It follows that $\frac{1}{2}(\mu \wedge \lambda)(A) \leq (\mu \oplus \lambda)(A)$. Also, for any measurable subset B of A,

$$\mu(B) + \lambda(A \setminus B) \geq [\mu(B) \oplus \lambda(B)] + [\mu(A \setminus B) \oplus \lambda(A \setminus B)]$$
$$\geq (\mu \oplus \lambda)(A).$$

It follows that $(\mu \wedge \lambda)(A) \geq (\mu \oplus \lambda)(A)$.

This shows that $\frac{1}{2}(\mu \wedge \lambda) \leq \mu \oplus \lambda \leq \mu \wedge \lambda$. Because of this inequality, it will be clear that $\mu \oplus \lambda$ is a measure as soon as we have shown that it is finitely additive.

Let A and B be mutually disjoint measurable sets. Then for any partition $\{E_j \mid j = 1, \ldots, n\}$ of $A \cup B$, we have

$$\sum_{j=1}^{n} \mu(E_j) \oplus \lambda(E_j) = \sum_{j=1}^{n} [(\mu(E_j \cap A) + \mu(E_j \cap B)) \oplus (\lambda(E_j \cap A) + \lambda(E_j \cap B))]$$

$$\geq \sum_{j=1}^{n} [(\mu(E_j \cap A) \oplus \lambda(E_j \cap A)) + (\mu(E_j \cap B) \oplus \lambda(E_j \cap B))]$$
$$\text{(by Theorem 8.34 (f))}$$

$$\geq (\mu \oplus \lambda)(A) + (\mu \oplus \lambda)(B).$$

If follows that

$$(\mu \oplus \lambda)(A \cup B) \geq (\mu \oplus \lambda)(A) + (\mu \oplus \lambda)(B).$$

Also, for any partitions $\{A_j \mid j = 1, \ldots, k\}$ and $\{B_j \mid j = 1, \ldots, n\}$ of A and B respectively, we have

$$\sum_{j=1}^{k} \mu(A_j) \oplus \lambda(A_j) + \sum_{j=1}^{n} \mu(B_j) \oplus \lambda(B_j) \geq (\mu \oplus \lambda)(A \cup B).$$

It follows that

$$(\mu \oplus \lambda)(A \cup B) \leq (\mu \oplus \lambda)(A) + (\mu \oplus \lambda)(B).$$

Thus $\mu \oplus \lambda$ is additive, and the proof is complete.   **QED**

9.6. Theorem. *Let Σ be a σ-algebra of subsets of a set S and let $\mathcal{M}$ be the family of nonnegative measures defined on Σ.*

Then $\oplus$ is a consistent operation in $\mathcal{M}$.

PROOF: It is clear that conditions (a) and (b) of Definition 8.18 are satisfied by $\oplus$. Using the inequality

$$\frac{1}{2}(\mu \wedge \lambda) \leq \mu \oplus \lambda \leq \mu \wedge \lambda,$$

condition (c) may be proved by the techniques of the proof of Theorem 8.19.   **QED**

9.7. Theorem. *Let μ be a nonnegative measure defined on a σ-algebra Σ of subsets of a set S, and let f and g be nonnegative measurable functions defined on S.*

Then for any measurable set A, we have

$$\int_A f \, d\mu \oplus \int_A g \, d\mu \geq \int_A (f \oplus g) \, d\mu.$$

PROOF: If f and g are simple functions, the result follows at once from Theorem 8.34 (f).

In general, choose increasing sequences $\{f_n\}$ and $\{g_n\}$ of simple functions, which converge to f and g respectively. Then since $f_n \oplus g_n$ increases to $f \oplus g$, we have

$$\int_A (f \oplus g) \, d\mu = \lim_{n \to \infty} \int_A (f_n \oplus g_n) \, d\mu$$

$$\leq \lim_{n \to \infty} \left(\int_A f_n \, d\mu \oplus \int_A g_n \, d\mu \right)$$

$$= \left(\lim_{n \to \infty} \int_A f_n \, d\mu \right) \oplus \left(\lim_{n \to \infty} \int_A g_n \, d\mu \right)$$

$$= \int_A f \, d\mu \oplus \int_A g \, d\mu.$$ **QED**

9.8. Note. The following theorem will allow us to sharpen Theorem 9.7 substantially.

9.9. Theorem. *Let μ and λ be finite, nonnegative measures defined on a σ-algebra Σ of subsets of a set S. Let $\{\mu_n\}$ and $\{\lambda_n\}$ be increasing sequences of nonnegative measures defined on Σ and suppose that*

$$\mu_n(A) \to \mu(A) \quad and \quad \lambda_n(A) \to \lambda(A) \quad as \quad n \to \infty, \qquad \forall \; A \in \Sigma.$$

Then for every $A \in \Sigma$, we have

$$(\mu_n \oplus \lambda_n)(A) \to (\mu \oplus \lambda)(A).$$

PROOF: Let $A \in \Sigma$. We clearly have

$$(\mu \oplus \lambda)(A) \geq \lim_{n \to \infty} (\mu_n \oplus \lambda_n)(A).$$

Now given any partition $\{A_j \mid j = 1, \ldots, k\}$ of A, and any $n = 1, 2, \ldots,$ we have

$$(\mu \oplus \lambda)(A) \leq \sum_{j=1}^{k} \mu(A_j) \oplus \lambda(A_j)$$

$$= \sum_{j=1}^{k} \left[\left(\mu_n(A_j) + (\mu - \mu_n)(A_j) \right) \oplus \left(\lambda_n(A_j) + (\lambda - \lambda_n)(A_j) \right) \right]$$

$$\leq \sum_{j=1}^{k} \left[\left(\mu_n(A_j) \oplus \lambda_n(A_j) \right) + \left((\mu - \mu_n)(A_j) + (\lambda - \lambda_n)(A_j) \right) \right]$$

$$\text{(by Theorem 8.34(e))}$$

$$= \left[\sum_{j=1}^{k} \left(\mu_n(A_j) \oplus \lambda_n(A_j) \right) \right] + (\mu - \mu_n)(A) + (\lambda - \lambda_n)(A).$$

Therefore, for every $n = 1, 2, \ldots,$ we have

$$(\mu \oplus \lambda)(A) \leq (\mu_n \oplus \lambda_n)(A) + (\mu - \mu_n)(A) + (\lambda - \lambda_n)(A),$$

and we deduce that

$$(\mu \oplus \lambda)(A) \leq \lim_{n \to \infty} (\mu_n \oplus \lambda_n)(A). \qquad \qquad \textbf{QED}$$

9.10. Theorem.[1] *Let ν be a nonnegative measure defined on a σ-algebra Σ of subsets of a set S. Let f and g be nonnegative measurable functions defined on S, and suppose the measures μ and λ are defined by the equations*

$$\mu(A) = \int_A f \, d\nu \quad and \quad \lambda(A) = \int_A g \, d\nu,$$

for every measurable set A.

---

[1] cf. Theorem 8.17.

Then for every measurable set A, we have

$$(\mu \oplus \lambda)(A) = \int_A (f \oplus g) \, d\nu.$$

PROOF: Suppose $A \in \Sigma$ and both f and g are finite and constant in A. Write $f(s) = a$ and $g(s) = b$ for all $s \in A$. Then

$$(\mu \oplus \lambda)(A) = \inf \left\{ \sum_{j=1}^{k} \left(a\nu(A_j) \oplus b\nu(A_j) \right) \,\Big|\, A = \bigcup_{j=1}^{k} A_j \right\}$$

$$= \inf \left\{ \sum_{j=1}^{k} (a \oplus b) \left(\nu(A_j) \right) \,\Big|\, A = \bigcup_{j=1}^{k} A_j \right\}$$

$$= (a \oplus b) \left(\nu(A) \right) = \int_A (f \oplus g) \, d\nu.$$

Suppose $A \in \Sigma$, and either f or g is constantly zero in A. Then we clearly have

$$(\mu \oplus \lambda)(A) = \int_A (f \oplus g) \, d\nu.$$

Suppose $A \in \Sigma$, and either f or g (say f) is constantly ∞ in A. Then for every measurable subset B of A, we clearly have

$$\mu(B) \oplus \lambda(B) = \lambda(B),$$

and therefore

$$(\mu \oplus \lambda)(A) = \lambda(A).$$

But $f \oplus g = g$, and therefore

$$(\mu \oplus \lambda)(A) = \int_A g \, d\nu = \int_A (f \oplus g) \, d\nu.$$

From these observations, we prove the following special case of the theorem.

CASE 1: f and g are simple functions.

Write $f = \sum_{j=1}^{n} a_j \chi_{A_j}$ and $g = \sum_{j=1}^{n} b_j \chi_{A_j}$. Then for any measurable set A,

$$\int_A (f \oplus g) \, d\nu = \sum_{j=1}^{n} \int_{A \cap A_j} (f \oplus g) \, d\nu$$

$$= \sum_{j=1}^{n} (\mu \oplus \lambda)(A \cap A_j) = (\mu \oplus \lambda)(A).$$

CASE 2: $\mu(S) < \infty$ and $\lambda(S) < \infty$.

Choose increasing sequences $\{f_n\}$ and $\{g_n\}$ of nonnegative simple functions which converge to f and g respectively, and for each $n = 1, 2, \ldots$, define

$$\mu_n(A) = \int_A f_n \, d\nu \quad \text{and} \quad \lambda_n(A) = \int_A g_n \, d\nu, \qquad \forall \, A \, \epsilon \, \boldsymbol{\Sigma}.$$

Then since $\mu_n(A) \to \mu(A)$ and $\lambda_n(A) \to \lambda(A)$, $\forall \, A \, \epsilon \, \boldsymbol{\Sigma}$, Theorem 9.9 implies that $(\mu_n \oplus \lambda_n)(A) \to (\mu \oplus \lambda)(A)$, $\forall \, A \, \epsilon \, \boldsymbol{\Sigma}$. Therefore

$$(\mu \oplus \lambda)(A) = \lim_{n \to \infty} (\mu_n \oplus \lambda_n)(A)$$

$$= \lim_{n \to \infty} \int_A (f_n \oplus g_n) \, d\nu = \int_A (f \oplus g) \, d\nu.$$

THE GENERAL CASE:

Let $A \, \epsilon \, \boldsymbol{\Sigma}$, and define

$$A_0 = \{s \, \epsilon \, A \mid f(s) = 0 \text{ or } g(s) = 0 \text{ or } f(s) = \infty \text{ or } g(s) = \infty\}.$$

For each $n = 1, 2, \ldots$, define

$$A_n = \left\{ s \, \epsilon \, A \,\middle|\, \frac{1}{n} \leq f(s) \leq n \text{ and } \frac{1}{n} \leq g(s) \leq n \right\}.$$

Then the above cases show that

$$(\mu \oplus \lambda)(A_0) = \int_{A_0} (f \oplus g) \, d\nu,$$

and therefore since A_n increases to $A \setminus A_0$ as $n \to \infty$, the proof will be complete when we have shown that

$$(\mu \oplus \lambda)(A_n) = \int_{A_n} (f \oplus g) \, d\nu$$

for every $n = 1, 2, \ldots$. But given any $n = 1, 2, \ldots$, the required equality follows from Case 2 if $\nu(A_n) < \infty$, and if $\nu(A_n) = \infty$, it follows from

$$\int_{A_n} (f \oplus g) \, d\nu \geq \int_{A_n} \frac{1}{2n} \, d\nu = \infty,$$

and

$$(\mu \oplus \lambda)(A_n) \geq \left(\frac{1}{n} \nu \oplus \frac{1}{n} \nu \right)(A_n) = \frac{1}{2n} \nu(A_n) = \infty.$$

This completes the proof.   **QED**

9.11. Theorem. *Let $\mathcal{M}$ be the family of nonnegative measures defined on a σ-algebra Σ of subsets of a set S.*

Then $\oplus$ is a commutative associative binary operation on $\mathcal{M}$. Furthermore, for every $A \in \Sigma$ and $\mu_1, \mu_2, \ldots, \mu_k \in \mathcal{M}$, we have

$$(\mu_1 \oplus \mu_2 \oplus \ldots \oplus \mu_k)(E)$$

$$= \inf \left\{ \sum_{j=1}^{n} \left(\mu_1(A_j) \oplus \mu_2(A_j) \oplus \ldots \oplus \mu_k(A_j) \right) \middle| A = \bigcup_{j=1}^{k} A_j \right\}$$

$$= \inf \left\{ \sum_{j=1}^{\infty} \left(\mu_1(A_j) \oplus \mu_2(A_j) \oplus \ldots \oplus \mu_k(A_j) \right) \middle| A = \overset{\infty}{\underset{j=1}{\cup}} A_j \right\},$$

where the first infimum is taken over all finite measurable partitions of A, and the second infimum is taken over all countable measurable partitions of A.

PROOF: The commutativity and associativity of $\oplus$ and the first equality stated are all clear. The second equality is an immediate consequence of the countable additivity of $\mu_1 \oplus \mu_2 \oplus \ldots \oplus \mu_k$.   **QED**

9.12. Theorem. *Let φ be a quasi-flow in a Hausdorff space X and let $\mu_1, \mu_2, \lambda_1, \lambda_2, \in \mathcal{M}_q(\varphi)$.*
Then

$$\text{(a)} \quad \mu_1 \oplus \mu_2 = (\mu_1 \wedge \mu_2) \oplus (\mu_1 \vee \mu_2),$$

$$\text{(b)} \quad \mu_1 \oplus (\lambda_1 \vee \lambda_2) = (\mu_1 \oplus \lambda_1) \vee (\mu_1 \oplus \lambda_2),$$

$$\text{(c)} \quad \mu_1 \oplus (\lambda_1 \wedge \lambda_2) = (\mu_1 \oplus \lambda_1) \wedge (\mu_1 \oplus \lambda_2),$$

and

$$\text{(d)} \quad (\mu_1 + \mu_2) \oplus (\lambda_1 + \lambda_2) \geq (\mu_1 \oplus \lambda_1) + (\mu_2 \oplus \lambda_2).$$

PROOF: (a), (b) and (c) may be proved by the method of proof of Theorem 8.24.

Now, to prove (d), let $A \in \Sigma(\varphi)$. Then given a $\Sigma(\varphi)$-measurable partition $\{A_j \mid j = 1, \ldots, n\}$ of A, we have

$$\sum_{j=1}^{n} (\mu_1 + \mu_2)(A_j) \oplus (\lambda_1 + \lambda_2)(A_j)$$

$$\geq \sum_{j=1}^{n} \left[\left(\mu_1(A_j) \oplus \lambda_1(A_j) \right) + \left(\mu_2(A_j) \oplus \lambda_2(A_j) \right) \right] \text{ (by Theorem 8.34(e))},$$

$$\geq (\mu_1 \oplus \lambda_1)(A) + (\mu_2 \oplus \lambda_2)(A).$$

From this, (d) follows.   **QED**

Sum of Two Flows

We begin by stating two theorems that follow at once from Theorem 8.58 and Corollary 8.59, and Theorem 9.5.

9.13. Theorem. *Let φ be a quasi-flow in a Hausdorff space X and let $\psi, \pi \in \mathrm{Rep}_q^+(\varphi)$.*

Then the following three conditions are equivalent:

(a) *ψ and π are compatible.*

(b) *The measure μ defined by*

$$\mu(E) = (\mu_\psi \oplus \mu_\pi)\big(E \smallsetminus \mathscr{R}(\psi, \pi)\big), \qquad \forall\, E \in \boldsymbol{\Sigma}(\varphi),$$

belongs to $\mathscr{M}_q(\varphi)$.

(c) *There exists a subset P of $\mathscr{F}(\psi) \cap \mathscr{F}(\pi)$ such that the measure μ defined by*

$$\mu(E) = (\mu_\psi \oplus \mu_\pi)(E \smallsetminus P), \qquad \forall\, E \in \boldsymbol{\Sigma}(\varphi),$$

belongs to $\mathscr{M}_q(\varphi)$.

9.14. Theorem. *Let φ be a quasi-flow in a Hausdorff space X, let $\psi, \pi \in \mathrm{Rep}_q^+(\varphi)$, and suppose ψ and π are compatible. Let P be a subset of $\mathscr{F}(\psi) \cap \mathscr{F}(\pi)$ and let the measure μ be defined by*

$$\mu(E) = (\mu_\psi \oplus \mu_\pi)(E \smallsetminus P), \qquad \forall\, E \in \boldsymbol{\Sigma}(\varphi).$$

Then $\mu \in \mathscr{M}_q(\varphi)$ iff $P = \mathscr{R}(\psi, \pi)$.

9.15. Definition. Let φ be a quasi-flow in a Hausdorff space X and let ψ and π be compatible members of $\mathrm{Rep}_q^+(\varphi)$. Let ϑ be the member of $\mathrm{Rep}_q^+(\varphi)$ for which

$$\mu_\vartheta(E) = (\mu_\psi \oplus \mu_\pi)\big(E \smallsetminus R(\psi, \pi)\big),$$

for every $\boldsymbol{\Sigma}(\varphi)$-measurable subset E of X.

Then we call ϑ the *pseudo-sum* of ψ and π, and we write $\vartheta = \psi \dotplus \pi$.

9.16. Definition. Let φ be a quasi-flow in a Hausdorff space X, let $\psi, \pi \in \mathrm{Rep}_q^+(\varphi)$, and suppose $\psi \vee \pi$ exists.

Then $\psi \dotplus \pi$ is called the *sum* of ψ and π, and is denoted by $\psi + \pi$.

9.17. Definition. Let φ be a quasi-flow in a Hausdorff space X and let $\psi_1, \psi_2, \ldots, \psi_n$ be members of $\mathrm{Rep}_q^+(\varphi)$. If $\{\psi_1, \psi_2, \ldots, \psi_n\}$ is compatible, then the *pseudo-sum* $\psi_1 + \psi_2 + \ldots + \psi_n$ of $\psi_1, \psi_2, \ldots, \psi_n$ is defined to be that member of $\mathrm{Rep}_q^+(\varphi)$ whose time measure μ is defined by

$$\mu(E) = (\mu_{\psi_1} \oplus \mu_{\psi_2} \oplus \ldots \oplus \mu_{\psi_n})\left(E \setminus \mathscr{R}(\psi_1, \psi_2, \ldots, \psi_n)\right),$$

for every $\Sigma(\varphi)$-measurable set E.

When $\bigvee\limits_{j=1}^{n} \psi_j$ exists, we call $\psi_1 + \psi_2 + \ldots + \psi_n$ the *sum* of $\psi_1, \psi_2, \ldots, \psi_n$, and denote it by $\psi_1 + \psi_2 + \ldots + \psi_n$, or by $\sum\limits_{j=1}^{n} \psi_j$.

9.18. Theorem. *Let φ be a quasi-flow in a Hausdorff space X and let $\psi_1, \psi_2, \ldots, \psi_n \in \mathrm{Rep}_q^+(\varphi)$.*

Then a necessary and sufficient condition for the existence of $\psi_1 + \ldots + \psi_n$ is that $\{\psi_1, \ldots, \psi_n\}$ be compatible and $\mathscr{R}(\psi_1, \ldots, \psi_n)$ have a countable intersection with every φ-orbit.

PROOF: Clear. (See Section 8.91.)   **QED**

9.19. Theorem. *Let φ be a quasi-flow in a Hausdorff space X and let ψ, π be compatible members of $\mathrm{Rep}_q^+(\varphi)$.*

$$\text{Then } \psi + \pi = \pi + \psi \text{ and } \psi \vee \pi \leq \psi + \pi \leq 2(\psi \vee \pi).$$

PROOF: Clear.   **QED**

9.20. Theorem. *Let φ be a quasi-flow in a Hausdorff space X, let $\psi, \psi_2, \pi_1, \pi_2 \in \mathrm{Rep}_q^+(\varphi)$, suppose $\psi_1 \leq \psi_2$ and $\pi_1 \leq \pi_2$, and suppose that $\psi_1 + \pi_1$ and $\psi_2 + \pi_2$ exist.*

Then $\psi_1 + \pi_1 \leq \psi_2 + \pi_2$.

PROOF: Let μ_1, μ_2, λ_1, and λ_2 be the time measures of ψ_1, ψ_2, π_1 and π_2, respectively. Then since $\mu_1 \geq \mu_2$ and $\lambda_1 \geq \lambda_2$, we have $\mu_1 \oplus \lambda_1 \geq \mu_2 \oplus \lambda_2$. Therefore, to complete the proof, we need only show that

$$\mu_{\psi_2 + \pi_2}\left(\mathscr{R}(\psi_1, \pi_1)\right) = 0.$$

But this clearly holds, because

$$\mu_{\psi_2 + \pi_2}\left(\mathscr{R}(\psi_1, \pi_1)\right) \leq \mu_{\psi_2 \vee \pi_2}\left(\mathscr{R}(\psi_1, \pi_1)\right) \leq \mu_{\psi_1 \vee \pi_1}\left(\mathscr{R}(\psi_1, \pi_1)\right) = 0. \quad \textbf{QED}$$

9.21. Remark. If ψ_1 and π_1 are compatible members of a family $\mathrm{Rep}_q^+(\varphi)$ and $\psi_1 + \pi_1$ does not exist, then Theorem 9.2 implies the existence of compatible members ψ_2, π_2 of $\mathrm{Rep}_q^+(\varphi)$ such that $\psi_1 \leq \psi_2$ and $\pi_1 \leq \pi_2$, and for which nonetheless, the inequality

$$\psi_1 + \pi_1 \leq \psi_2 + \pi_2$$

is false.

One can also construct an example to show that if $\psi_1, \psi_2, \pi \in \mathrm{Rep}_q^+ (\varphi)$, $\psi_1 \leq \psi_2$, and the pairs $\{\psi_1, \pi\}$ and $\{\psi_2, \pi\}$ are both compatible, then we need NOT have

$$\psi_1 + \pi \leq \psi_2 + \pi.$$

9.22. Theorem. *Let φ be a quasi-flow in a Hausdorff space X and let $\psi_1, \psi_2, \psi_3 \in \mathrm{Rep}_q^+ (\varphi)$.*

Then $\psi_1 + (\psi_2 + \psi_3)$ exists iff $\psi_1 \vee (\psi_2 \vee \psi_3)$ exists, and when $\psi_1 + (\psi_2 + \psi_3)$ exists, we have

$$\psi_1 + (\psi_2 + \psi_3) = \psi_1 + \psi_2 + \psi_3.$$

A fortiori, *if $\psi_1 + (\psi_2 + \psi_3)$ and $(\psi_1 + \psi_2) + \psi_3$ both exist, they are equal.*

PROOF: Let μ_1, μ_2 and μ_3 be the time measures of ψ_1, ψ_2 and ψ_3 respectively. Since

$$\frac{1}{2} (\mu_2 \wedge \mu_3) \leq \mu_2 \oplus \mu_3 \leq \mu_2 \wedge \mu_3,$$

we have

$$\frac{1}{2} (\mu_1 \wedge \mu_2 \wedge \mu_3) \leq \mu_1 \wedge (\mu_2 \oplus \mu_3) \leq \mu_1 \wedge \mu_2 \wedge \mu_3,$$

and it follows easily that $\psi_1 \vee (\psi_2 \vee \psi_3)$ exists precisely when $\psi_1 \vee (\psi_2 + \psi_3)$ exists. Therefore $\psi_1 \vee (\psi_2 \vee \psi_3)$ exists precisely when $\psi_1 + (\psi_2 + \psi_3)$ exists.

Now suppose that $\psi_1 + (\psi_2 + \psi_3)$ exists, and denote its time measure by μ. Let $F = \mathscr{F}(\psi_1) \cap \mathscr{F}(\psi_2) \cap \mathscr{F}(\psi_3)$, and denote the time measure of $\psi_2 + \psi_3$ by μ_{23}. Given any $\Sigma(\varphi)$-measurable set E, we have

$$\mu(E) = (\mu_1 \oplus \mu_{23})\left(E \smallsetminus \mathscr{R}(\psi_1, \psi_2 + \psi_3)\right)$$

$$= (\mu_1 \oplus \mu_2 \oplus \mu_3)\left[E \smallsetminus [\mathscr{R}(\psi_1, \psi_2 + \psi_3) \cup \mathscr{R}(\psi_2, \psi_3)]\right]$$

$$= (\mu_1 \oplus \mu_2 \oplus \mu_3)(E \smallsetminus P)$$

where

$$P = [\mathscr{R}(\psi_1, \psi_2 + \psi_3) \cup \mathscr{R}(\psi_2, \psi_3)] \cap F.$$

(The last equality holds because $[\mathscr{R}(\psi_1, \psi_2 + \psi_3) \cup \mathscr{R}(\psi_2, \psi_3)] \smallsetminus F$ is a subset of $\mathscr{G}(\psi_1)$ which meets each φ-orbit in a countable set.)
It follows that $P = \mathscr{R}(\psi_1, \psi_2, \psi_3)$ and that μ is the time measure of $\psi_1 + \psi_2 + \psi_3$. This completes the proof.   **QED**

9.23. Remark. Theorem 9.22 is no longer true if $+$ is replaced by $\dotplus$. But using the method of proof of Theorem 8.96, one can show easily that if

$\psi_1, \psi_2, \psi_3 \in \text{Prod}_q^+(\varphi)$ and if both the pairs $\{\psi_2, \psi_3\}$ and $\{\psi_1, \psi_2 \vee \psi_3\}$ have maxima in $\text{Prod}_q^+(\varphi)$, then

$$\psi_1 \dotplus (\psi_2 \dotplus \psi_3) = \psi_1 \dotplus \psi_2 \dotplus \psi_3.$$

9.24. Theorem. *Let φ be a quasi-flow in a Hausdorff space X, and let ψ and π be compatible members of $\text{Rep}_q^+(\varphi)$.*

Then $\psi \dotplus \pi = (\psi \vee \pi) \dotplus (\psi \wedge \pi)$.

PROOF: It is clear that $\psi \vee \pi$ and $\psi \dotplus \pi$ exist, and the inequality

$$\psi \wedge \pi \leq \psi \vee \pi$$

implies at once that $\mathscr{R}(\psi \wedge \pi, \psi \vee \pi) = \square$ and that $(\psi \vee \pi) \dotplus (\psi \wedge \pi)$ exists. Let μ be the time measure of $(\psi \vee \pi) \dotplus (\psi \wedge \pi)$. Then given any $\boldsymbol{\Sigma}(\varphi)$-measurable set E, we have

$$\begin{aligned}
\mu(E) &= (\mu_{\psi \wedge \pi} \oplus \mu_{\psi \wedge \pi})(E) \\
&= \left((\mu_\psi \wedge \mu_\pi) \oplus (\mu_\psi \vee \mu_\pi)\right)\left(E \setminus \mathscr{R}(\psi, \pi)\right) \\
&= (\mu_\psi \oplus \mu_\pi)\left(E \setminus \mathscr{R}(\psi, \pi)\right) \qquad \text{(by Theorem 9.12(a)).}
\end{aligned}$$

This completes the proof.   **QED**

9.25. Theorem. The Distributive Law $\psi_1 \dotplus (\psi_2 \wedge \psi_3) = (\psi_1 \dotplus \psi_2) \wedge (\psi_1 \dotplus \psi_3)$. *Let φ be a quasi-flow in a Hausdorff space X, let $\psi_1, \psi_2, \psi_3 \in \text{Rep}_q^+(\varphi)$ and suppose that $\psi_1 \dotplus \psi_2$ and $\psi_1 \dotplus \psi_3$ both exist.*

Then $\psi_1 \dotplus (\psi_2 \wedge \psi_3) = (\psi_1 \dotplus \psi_2) \wedge (\psi_1 \dotplus \psi_3)$.

PROOF: The proof uses Theorem 9.12(b), and is essentially similar to that of Theorem 8.94.   **QED**

9.26. Remark. Theorem 9.25 is no longer true if we replace $\dotplus$ by $\dotplus$. However, as we see in the next theorem, the analogue in $\text{Prod}_q^+(\varphi)$ is true.

9.27. Theorem. *Let φ be a quasi-flow in a Hausdorff space X, let $\psi_1, \psi_2, \psi_3 \in \text{Prod}_q^+(\varphi)$, and suppose that both $\{\psi_1, \psi_2\}$ and $\{\psi_1, \psi_3\}$ have maxima in $\text{Prod}_q^+(\varphi)$.*

Then $\psi_1 \dotplus (\psi_2 \wedge \psi_3) = (\psi_1 \dotpm \psi_2) \wedge (\psi_1 \dotplus \psi_3)$.

PROOF: Essentially similar to Theorem 8.96.   **QED**

9.28. Theorem. The Distributive Law $\psi_1 \dotplus (\psi_2 \vee \psi_3) = (\psi_1 \dotplus \psi_2) \vee (\psi_1 \dotplus \psi_3)$. *Let φ be a quasi-flow in a Hausdorff space X, let $\psi_1, \psi_2, \psi_3 \in \text{Rep}_q^+(\varphi)$*

and suppose that $\psi_1 \vee \psi_2$, $\psi_1 \vee \psi_3$, $\psi_2 \vee \psi_3$ *and* $\psi_1 \vee \psi_2 \vee \psi_3$ *all exist.*

Then $\psi_1 + (\psi_2 \vee \psi_3) = (\psi_1 + \psi_2) \vee (\psi_1 + \psi_3)$.

PROOF: It is easy to see that both $\psi_1 + (\psi_2 \vee \psi_3)$ and $(\psi_1 + \psi_2) \vee (\psi_1 + \psi_3)$ exist, and have the same orbits as $\psi_1 \vee \psi_2 \vee \psi_3$. Furthermore, it is clear that

$$\mathscr{R}(\psi_1, \psi_2 \vee \psi_3) \subseteq \mathscr{R}(\psi_1, \psi_2, \psi_3) \text{ and } \mathscr{R}(\psi_1 + \psi_2, \psi_1 + \psi_3) \subseteq \mathscr{R}(\psi_1, \psi_2, \psi_3).$$

Let μ and λ be the time measures of $\psi_1 + (\psi_2 \vee \psi_3)$ and $(\psi_1 + \psi_2) \vee (\psi_1 + \psi_3)$ respectively. Then since μ and λ are continuous on the sets $\mathscr{R}(\psi_1, \psi_2)$, $\mathscr{R}(\psi_1, \psi_3)$, $\mathscr{R}(\psi_2, \psi_3)$, $\mathscr{R}(\psi_1, \psi_2, \psi_3)$, and each of these sets meets every φ-orbit in a countable set, μ and λ are zero on the sets $\mathscr{R}(\psi_1, \psi_2)$, $\mathscr{R}(\psi_1, \psi_3)$, $\mathscr{R}(\psi_2, \psi_3)$ and $\mathscr{R}(\psi_1, \psi_2, \psi_3)$. Therefore, to prove the theorem, we need only show that μ and λ agree in the complement Y of the union of these four sets.

Let E be a $\boldsymbol{\Sigma}(\varphi)$-measurable subset of Y. Then

$$
\begin{aligned}
\lambda(E) &= \big((\mu_{\psi_1} \oplus \mu_{\psi_2}) \wedge (\mu_{\psi_1} \oplus \mu_{\psi_3})\big)(E) \\
&= \big(\mu_{\psi_1} \oplus (\mu_{\psi_2} \wedge \mu_{\psi_3})\big)(E) \qquad \text{(by Theorem 9.12(c))} \\
&= \mu(E).
\end{aligned}
$$

This completes the proof.   **QED**

9.29. Remark. $+$ and $\vee$ cannot be replaced by $\dotplus$ and V in Theorem 9.28, but, as usual, the analogue in $\mathrm{Prod}_q^+(\varphi)$ is true.

9.30. Theorem. *Let* φ *be a quasi-flow in a Hausdorff space* X, *and let* $\psi_1, \psi_2, \pi_1, \pi_2 \in \mathrm{Rep}_q^+(\varphi)$.

Then if

$$(\psi_1 \oplus \psi_2) + (\pi_1 \oplus \pi_2) \quad \text{and} \quad (\psi_1 + \pi_1) \oplus (\psi_2 + \pi_2)$$

both exist, we have

$$(\psi_1 \oplus \psi_2) + (\pi_1 \oplus \pi_2) \leq (\psi_1 + \pi_1) \oplus (\psi_2 + \pi_2).$$

PROOF: It is easy to see that

$$\psi_1 \oplus \psi_2) \vee (\pi_1 \oplus \pi_2) \leq (\psi_1 + \pi_1) \wedge (\psi_2 + \pi_2),$$

and it follows from this that

$$\mathscr{G}\big((\psi_1 \oplus \psi_2) + (\pi_1 \oplus \pi_2)\big) \subseteq \mathscr{G}\big((\psi_1 + \pi_1) \oplus (\psi_2 + \pi_2)\big).$$

Let μ and λ be the time measures of $(\psi_1 \oplus \psi_2) + (\pi_1 \oplus \pi_2)$ and $(\psi_1 + \pi_1) \oplus (\psi_2 + \pi_2)$ respectively. We need to show that $\lambda \leq \mu$. But since $\mu(\{x\}) = \infty$ for every $x \in \mathcal{F}\big((\psi_1 \oplus \psi_2) + (\pi_1 \oplus \pi_2)\big)$, we certainly have $\lambda \leq \mu$ in $\mathcal{F}\big((\psi_1 \oplus \psi_2) + (\pi_1 \oplus \pi_2)\big)$. Also, since $\mathcal{R}(\psi_1 \oplus \psi_2, \pi_1 \oplus \pi_2)$ meets every φ-orbit in a countable set, and λ is continuous on $\mathcal{R}(\psi_1 \oplus \psi_2, \pi_1 \oplus \pi_2)$, we have

$$\lambda\big(\mathcal{R}(\psi_1 \oplus \psi_2, \pi_1 \oplus \pi_2)\big) = 0.$$

But if a $\Sigma(\varphi)$-measurable set E does not meet

$$\mathcal{F}\big(\psi_1 \oplus \psi_2) + (\pi_1 \oplus \pi_2)\big) \cup \mathcal{R}(\psi_1 \oplus \psi_2, \pi_1 \oplus \pi_2),$$

then

$$E \subseteq \mathcal{G}(\psi_1 \oplus \psi_2) \cup \mathcal{G}(\pi_1 \oplus \pi_2)$$

$$\subseteq [\mathcal{G}(\psi_1) \cup \mathcal{G}(\pi_1)] \cap [\mathcal{G}(\psi_2) \cup \mathcal{G}(\pi_2)],$$

and therefore

$$\lambda(E) = (\mu_{\psi_1+\pi_1} + \mu_{\psi_2+\pi_2})(E)$$

$$= \big((\mu_{\psi_1} \oplus \mu_{\pi_1}) + (\mu_{\psi_2} \oplus \mu_{\pi_2})\big)(E)$$

$$\leq \big((\mu_{\psi_1} + \mu_{\psi_2}) \oplus (\mu_{\pi_1} + \mu_{\pi_2})\big)(E) \qquad \text{(by Theorem 9.12(d))}$$

$$= \mu(E).$$

This completes the proof.   **QED**

9.31. Remark. An example can be constructed to show that $+$ cannot be replaced by $\dotplus$ in Theorem 9.30. But as usual, the analogue in $\text{Prod}_q^+(\varphi)$ is true. Furthermore, it is necessary to assume the existence of *both* $(\psi_1 \oplus \psi_2) + (\pi_1 \oplus \pi_2)$ *and* $(\psi_1 + \pi_1) \oplus (\psi_2 + \pi_2)$, for it is possible for either one to exist without the other.

9.32. The Distributive Law $(u + v)\varphi = u\varphi + v\varphi$. We have seen that many theorems on sums remain true for pseudo-sums of members of a family $\text{Prod}_q^+(\varphi)$ as long as we only take the pseudo-sum of two members of $\text{Prod}_q^+(\varphi)$ when they have a maximum in $\text{Prod}_q^+(\varphi)$. This is not surprising, in view of the following considerations:

(a) It is clear that if ψ and π are compatible members of $\text{Prod}_q^+(\varphi)$, then $\psi \dotplus \pi \in \text{Prod}_q^+(\varphi)$.

(b) The argument used at the beginning of the chapter (Sections 9.1, 9.2 and 9.3) can be modified to show that if $\psi, \pi \in \text{Prod}_q^+(\varphi)$ then ψ and π should have a sum in $\text{Prod}_q^+(\varphi)$ precisely when they have a maximum in $\text{Prod}_q^+(\varphi)$.

(c) If u and v are nonnegative quasi φ-multipliers, then the inequality

$$u \vee v \leq u + v \leq 2(u \vee v)$$

implies that $u + v$ is a quasi φ-multiplier iff $u \vee v$ is a quasi φ-multiplier, and by Theorems 8.74 and 8.76, this happens precisely when $u\varphi$ and $v\varphi$ have a maximum in $\mathrm{Prod}_q^+ (\varphi)$.

The following theorems may be proved by the same techniques that were used in Sections 8.71 to 8.79.

9.33. Theorem. *Let φ be a quasi-flow in a Hausdorff space X, let u and v be nonnegative quasi φ-multipliers, and suppose $u + v$ is a quasi φ-multiplier.*

Then

(a) *$u\varphi$ and $v\varphi$ are compatible,*

(b) *$\mu_\varphi\big(\mathcal{R}(u\varphi, v\varphi)\big) = 0$,*

and

(c) *$(u + v)\varphi = u\varphi + v\varphi$.*

PROOF: The proof is similar to the proof of Theorem 8.74 except that, instead of using Lemma 8.72, we use a similar lemma based on Theorem 9.10. **QED**

9.34. Theorem. *Let φ be a quasi-flow in a Hausdorff space X, let u and v be nonnegative quasi φ-multipliers, suppose that $u\varphi$ and $v\varphi$ are compatible and suppose that $\mu_\varphi\big(\mathcal{R}(u\varphi, v\varphi)\big) = 0$.*

Then $u + v$ is a quasi φ-multiplier.

PROOF: Immediate from Theorem 8.75. **QED**

9.35. Theorem. *Let φ be a quasi-flow in a Hausdorff space X, let u and v be nonnegative quasi φ-multipliers, and suppose $u\varphi + v\varphi$ exists.*

Then $(u + v)\varphi = u\varphi + v\varphi$.

PROOF: Clear. **QED**

9.36. Theorem. *Let φ be a flow in a Hausdorff space X and let u and v be nonnegative proper quasi φ-multipliers. Suppose that u and v are continuous in $\mathcal{G}(\varphi)$, and that for each $x \in \mathcal{F}(\varphi)$, there exists a neighborhood W_x of x such that both u and v are bounded in $\mathcal{G}(\varphi) \cap W_x$. Suppose further that $u\varphi$ and $v\varphi$ are compatible and $\mathcal{R}(u\varphi, v\varphi) = \square$.*

Then $u\varphi + v\varphi$ is a flow.

PROOF: Similar to the proof of Theorem 8.79. **QED**

9.37. The Distributive Law $u(\psi + \pi) = u\psi + u\pi$

Since the theory is essentially the same as the theory developed in Sections 8.80 to 8.90, it will suffice to state the main theorems without proof. The only significant difference between the proof of the following theorem and that of Theorem 8.83 is that, instead of using Lemma 8.81, we use a similar lemma which is based on Theorem 8.21 and the fact that $\oplus$ is a consistent operation in a family of measures.

Note that, if ψ and π are compatible members of a family $\mathrm{Rep}_q^+(\varphi)$ and u is a nonnegative function, then u is a quasi $(\psi \vee \pi)$-multiplier iff u is a quasi $(\psi + \pi)$-multiplier.

9.38. Theorem. *Let φ be a quasi-flow in a Hausdorff space X, and let ψ and π be compatible members of $\mathrm{Rep}_q^+(\varphi)$. Let u be a nonnegative function defined on X and suppose that u is a quasi ψ-multiplier and a quasi π-multiplier and a quasi $(\psi + \pi)$-multiplier.*

Then

(a) *$u\psi$ and $u\pi$ are compatible,*

(b) $\mu_{\psi + \pi}\big(\mathscr{R}(u\psi, u\pi)\big) = 0$,

(c) $\mathscr{F}(\psi + \pi) \subseteq \mathscr{F}(u\psi + u\pi)$,

(d) $u(x) = 0$ *for every* $x \in \mathscr{R}(\psi, \pi) \setminus \mathscr{R}(u\psi, u\pi)$,

and

(e) $u(\psi + \pi) = u\psi + u\pi$.

9.39. Theorem. *Let φ be a quasi-flow in a Hausdorff space X, and let $\psi, \pi \in \mathrm{Rep}_q^+(\varphi)$. Let u be a nonnegative function defined on X, and suppose that u is both a quasi ψ-multiplier and a quasi π-multiplier. Suppose that each of the pairs $\{\psi, \pi\}$ and $\{u\psi, u\pi\}$ is compatible, and suppose that*

$$\mu_{\psi + \pi}\big(\mathscr{R}(u\psi, u\pi)\big) = 0.$$

Then u is a quasi $(\psi + \pi)$-multiplier iff the following conditions hold:

(a) $\mathscr{F}(\psi + \pi) \subseteq \mathscr{F}(u\psi + u\pi)$;

and

(b) $u(x) = 0$ *for every* $x \in \mathscr{R}(\psi, \pi) \setminus \mathscr{R}(u\psi, u\pi)$.

9.40. Corollary. *Let φ be a quasi-flow in a Hausdorff space X, let $\psi, \pi \in \mathrm{Rep}_q^+(\varphi)$, and let u be a nonnegative function defined on X. Suppose that*

(i) *u is both a quasi ψ-multiplier and a quasi π-multiplier,*

(ii) *both $\psi + \pi$ and $u\psi + u\pi$ exist,*

(iii) $\mathscr{F}(\psi + \pi) \subseteq \mathscr{F}(u\psi + u\pi)$,

and

(iv) *u is zero in* $\mathscr{R}(\psi, \pi) \setminus \mathscr{R}(u\psi, u\pi)$.

Then u is a quasi $\psi + \pi$-*multiplier, and*

$$u(\psi + \pi) = u\psi + u\pi.$$

The Relationship between Speed and the Algebraic Combinations

9.41. Lemma. *Let* φ *be a quasi-flow in a Hausdorff space* X, *let* f *be an isotonic quasi* φ-*reparametrizer, let* $\psi = f$ rep φ, *and let* inv $(f) = \{g_x \mid x \in \mathscr{G}(\psi)\}$. *Let* λ_1 *and* λ_2 *be the measures defined on* $\mathscr{G}(\psi)$ *by*

$$\lambda_1(A) = \int_A g_x'(0) \, d\mu_\varphi(x), \quad and \quad \lambda_2(A) = \mu_\psi(A) - \lambda_1(A),$$

for all $\Sigma(\varphi)$-*measurable subsets* A *of* $\mathscr{G}(\psi)$.

Then λ_1 *and* λ_2 *are nonnegative measures on* $\mathscr{G}(\psi)$. *Furthermore, in* $\mathscr{G}(\psi)$, *we have* $\lambda_1 \ll \mu_\varphi$ *and* $\lambda_2 \perp \mu_\varphi$, *i.e.,* $\lambda_1 + \lambda_2$ *is the Lebesgue decomposition of* μ_ψ *with respect to* μ_φ.

PROOF: Let $x \in \mathscr{G}(\psi)$ and denote the range of f_x by (α_x, ω_x). Let ν be the Borel measure defined on (α_x, ω_x) by

$$\nu(E) = m(g_x(E)),$$

for all Borel subsets E of (α_x, ω_x). Then the Lebesgue decomposition of ν with respect to m is $\nu_1 + \nu_2$, where

$$\nu_1(E) = \int_E g_x'(\tau) \, d\tau = \int_E g_{\varphi(\tau, x)}'(0) \, d\tau,$$

and

$$\nu_2(E) = \nu(E) - \nu_1(E),$$

for all Borel subsets E of (α_x, ω_x). (For the properties of the Lebesgue decomposition used here, see Rudin [1], Theorem 8.6.)

Now suppose A is a Borel subset of $\mathscr{O}_\psi(x)$, and using Theorem 5.31, choose a simple φ-time preimage E of A, such that $E \subseteq (\alpha_x, \omega_x)$. Then

$$\mu_\psi(A) = \nu(E) \qquad \text{(by Theorem 5.31 (d))}$$
$$= \nu_1(E) + \nu_2(E)$$
$$= \int_E g_{\varphi(\tau, x)}'(0) \, d\tau + \nu_2(E)$$
$$= \int_A g_y'(0) \, d\mu_\varphi(y) + \nu_2(E) \qquad \text{(by Theorem 5.29)}$$
$$= \lambda_1(A) + \nu_2(E).$$

It follows that $\lambda_2(A) = \nu_2(E)$.

Using these observations, it is easy to complete the proof of the lemma.

QED

9.42. Lemma. *Let λ, ν and μ be nonnegative measures defined on a σ-algebra Σ of subsets of a set S. Suppose that λ and ν have Lebesgue decompositions $\lambda_1 + \lambda_2$ and $\nu_1 + \nu_2$, respectively, with respect to μ, where $\lambda_1 \ll \mu$, $\lambda_2 \perp \mu$, $\nu_1 \ll \mu$ and $\nu_2 \perp \mu$.*

Let $\circ$ be any one of the operations $+, \vee, \wedge, \oplus$.

Then the Lebesgue decomposition of $\lambda \circ \nu$ with respect to μ is $(\lambda_1 \circ \nu_1) + (\lambda_2 \circ \nu_2)$.

PROOF: Using the fact that λ_2 and ν_2 are μ-singular, choose measurable subsets A and B of S such that $\mu(A) = \mu(B) = 0$, λ_2 is concentrated on A and ν_2 is concentrated on B. Now we clearly have $\lambda_1 \circ \nu_1 \ll \mu$, and therefore all the measures $\lambda_1, \nu_1, \lambda_1 \circ \nu_1$ and μ vanish in $A \cup B$. Therefore, in $A \cup B$, we have

$$\lambda \circ \nu = \lambda_2 \circ \nu_2 = (\lambda_1 \circ \nu_1) + (\lambda_2 \circ \nu_2).$$

But it is also clear that $\lambda_2 \circ \nu_2$ is concentrated on $A \cup B$, and in particular $\lambda_2 \circ \nu_2$ is μ-singular. Therefore, in the complement of $A \cup B$, we have

$$\lambda \circ \nu = \lambda_1 \circ \nu_1 = (\lambda_1 \circ \nu_1) + (\lambda_2 \circ \nu_2).$$

This completes the proof.   **QED**

9.43. Theorem. *Let φ be a quasi-flow in a metric space X, and suppose that $|\dot\varphi|$ exists and is non-zero in a $\Sigma(\varphi)$-measurable subset A of $\mathscr{G}(\varphi)$. Let $\circ$ be any one of the operations $+, \vee, \wedge, \oplus$, let $\psi, \pi, \vartheta \in \mathrm{Rep}_q^+(\varphi)$, and suppose*

$$\vartheta = \psi \circ \pi.$$

Then we have

$$|\dot\vartheta|(x) = |\dot\psi|(x) \circ |\dot\pi|(x),$$

for μ_φ-almost all $x \in A$.

PROOF: Choose isotonic quasi φ-reparametrizers f_ψ, f_π and f_ϑ such that $\psi = f_\psi \operatorname{rep} \varphi$, $\pi = f_\pi \operatorname{rep} \varphi$ and $\vartheta = f_\vartheta \operatorname{rep} \varphi$, and let $\operatorname{inv}(f_\psi) = \{g_{\psi,x} \mid x \in \mathscr{G}(\psi)\}$, $\operatorname{inv}(f_\pi) = \{g_{\pi,x} \mid x \in \mathscr{G}(\pi)\}$ and $\operatorname{inv}(f_\vartheta) = \{g_{\vartheta,x} \mid x \in \mathscr{G}(\vartheta)\}$.

Inside $\mathscr{G}(\psi)$, denote the Lebesgue decomposition of μ_ψ with respect to μ_φ by $\mu_{\psi,1} + \mu_{\psi,2}$, with $\mu_{\psi,1} \ll \mu_\varphi$. Then for each $\Sigma(\varphi)$-measurable subset B of $\mathscr{G}(\psi)$, we have

$$\mu_{\psi,1}(B) = \int_B g'_{\psi,x}(0)\, d\mu_\varphi(x) \qquad \text{(by Lemma 9.41).}$$

Inside $\mathscr{G}(\pi)$, denote the Lebesgue decomposition of μ_π with respect to μ_φ by $\mu_{\pi,1} + \mu_{\pi,2}$, with $\mu_{\pi,1} \ll \mu_\varphi$. Then for each $\Sigma(\varphi)$-measurable subset B of $\mathscr{G}(\pi)$, we have

$$\mu_{\pi,1}(B) = \int_B g'_{\pi,x}(0) \, d\mu_\varphi(x) .$$

In the same way, let $\mu_{\vartheta,1} + \mu_{\vartheta,2}$ be the Lebesgue decomposition of μ_ϑ with respect to μ_φ inside $\mathscr{G}(\vartheta)$. Then for each $\Sigma(\varphi)$-measurable subset B of $\mathscr{G}(\vartheta)$, we have

$$\mu_{\vartheta,1}(B) = \int_B g'_{\vartheta,x}(0) \, d\mu_\varphi(x) .$$

CASE 1: $\circ = \wedge$.

Since the equation

$$|\dot{\vartheta}|(x) = |\dot{\psi}|(x) \wedge |\dot{\pi}|(x)$$

clearly holds for every $x \in \mathscr{F}(\psi) \cup \mathscr{F}(\pi)$, there is no loss of generality in assuming that $A \subseteq \mathscr{G}(\psi) \cap \mathscr{G}(\pi)$. Now Theorem 8.17(a) implies that for every $\Sigma(\varphi)$-measurable subset B of A, we have

$$(\mu_{\psi,1} \vee \mu_{\pi,1})(B) = \int_B \left(g'_{\psi,x}(0) \vee g'_{\pi,x}(0)\right) d\mu_\varphi(x) ,$$

and from Lemma 9.42, it follows that

$$\mu_{\vartheta,1}(B) = \int_B \left(g'_{\psi,x}(0) \vee g'_{\pi,x}(0)\right) d\mu_\varphi(x) .$$

Therefore for μ_φ-almost all $x \in A$, we have

$$g'_{\vartheta,x}(0) = g'_{\psi,x}(0) \vee g'_{\pi,x}(0) ,$$

i.e.,

$$f'_{\vartheta,x}(0) = f'_{\psi,x}(0) \wedge f'_{\pi,x}(0) ,$$

and it follows from Theorem 5.48(a) that for μ_φ-almost all $x \in A$, we have

$$\begin{aligned}
|\dot{\vartheta}|(x) &= |\dot{\varphi}|(x) \cdot f'_{\vartheta,x}(0) \\
&= |\dot{\varphi}|(x)\left(f'_{\psi,x}(0) \wedge f'_{\pi,x}(0)\right) \\
&= |\dot{\varphi}|(x) f'_{\psi,x}(0) \wedge |\dot{\varphi}|(x) f'_{\pi,x}(0) \\
&= |\dot{\psi}|(x) \wedge |\dot{\pi}|(x) .
\end{aligned}$$

CASE 2: $\circ = \oplus$.

This case is similar to Case 1, and is omitted.

CASE 3: $\circ = \vee$.

Using an argument similar to the one in Case 1 (making use of Theorem 8.17(b) instead of 8.17(a)), we can see that the equation

$$|\dot{\vartheta}|(x) = |\dot{\psi}|(x) \vee |\dot{\pi}|(x)$$

holds for μ_φ-almost all $x \in A \cap \mathscr{G}(\psi) \cap \mathscr{G}(\pi)$. Furthermore, this equation clearly holds in $\mathscr{F}(\vartheta)$, and $\mu_\varphi(\mathscr{R}(\psi, \pi)) = 0$. Therefore, to complete the proof, it is sufficient to assume that $A \subseteq \mathscr{G}(\psi) \cap \mathscr{F}(\pi)$. With this assumption, it is clear that μ_ψ and μ_θ agree in A, and we therefore have

$$\int_B g'_{\psi,x}(0) \, d\mu_\varphi(x) = \int_B g'_{\vartheta,x}(0) \, d\mu_\varphi(x)$$

for every $\Sigma(\varphi)$-measurable subset B of A. Therefore, for μ_φ-almost all $x \in A$, we have

$$g'_{\psi,x}(0) = g'_{\vartheta,x}(0), \quad i.e., \quad f'_{\psi,x}(0) = f'_{\vartheta,x}(0),$$

and using Theorem 5.48(a), we see that, for μ_φ-almost all $x \in A$,

$$|\dot{\vartheta}|(x) = |\dot{\varphi}|(x) \cdot f'_{\vartheta,x}(0) = |\dot{\varphi}|(x) \cdot f'_{\psi,x}(0)$$
$$= |\dot{\psi}|(x) = |\dot{\psi}|(x) \vee |\dot{\pi}|(x).$$

CASE 4: $\circ = +$.

Using an argument similar to the one in Case 1 (making use of Theorem 9.10 instead of Theorem 8.17(a)), we can see that the equation

$$|\dot{\varphi}|(x) = |\dot{\psi}|(x) + |\dot{\pi}|(x)$$

holds for μ_φ-almost all $x \in A \cap \mathscr{G}(\psi) \cap \mathscr{G}(\pi)$.

Having seen this, we can complete the proof using the method of Case 3.

QED

9.44. Example. In this section, we demonstrate how the behavior of speed can be inconsistent with the algebraic operations. We shall show that there exist two members ψ, π of $\mathrm{Prod}_q^+(1)$ such that all the flows $\psi, \pi, \psi \wedge \pi, \psi \vee \pi, \psi \oplus \pi, \psi + \pi$ have bounded non-vanishing speeds, but do not satisfy any of the equations linking the speeds and the combinations $\wedge, \vee, \oplus, +$.

If $a_n = 1$ for every even positive integer n, and $a_n = 2$ for every odd positive integer n, and if

$$x_n = \sum_{j=1}^n \frac{a_j}{j^2}, \quad \text{for every } n = 1, 2, \ldots,$$

then $\{x_n\}$ increases to a number x, and the following conditions hold:

$$\text{(i)} \quad \lim_{n \to \infty} \left(\frac{x_{2n+1} - x_{2n}}{x_{2n} - x_{2n-1}} \right) = 2,$$

$$\text{(ii)} \quad \lim_{n \to \infty} \left(\frac{x_{2n+1} - x_{2n}}{x_{2n-1} - x_{2n-2}} \right) = 1,$$

and

$$\text{(iii)} \quad \lim_{n \to \infty} \left(\frac{x_{n+1} - x_n}{\displaystyle\sum_{j=n}^{\infty} (x_{j+1} - x_j)} \right) = 0.$$

Now having established the existence of a strictly increasing convergent sequence $\{x_n\}$ which satisfies conditions (i), (ii) and (iii), choose such a sequence, and denote its limit by x.

Choose sequences $\{p_n\}$ and $\{q_n\}$ such that, for each $n = 1, 2, \ldots$,

$$p_n < x_n < q_n < p_{n+1},$$

and such that

$$\lim_{n \to \infty} \left(\frac{q_n - p_n}{x_{n+1} - x_n} \right) = 0.$$

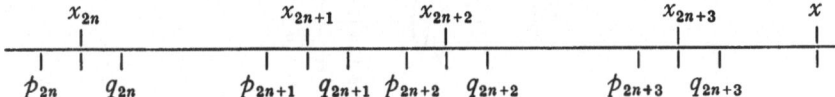

For each $n = 1, 2, \ldots$, define

$$a_n = \sum_{j=n}^{\infty} (x_{2j+2} - x_{2j+1}), \qquad b_n = \sum_{j=n}^{\infty} (p_{2j+2} - q_{2j+1}),$$

$$c_n = \sum_{j=n}^{\infty} (q_{2j+2} - p_{2j+1}), \qquad d_n = \sum_{j=n}^{\infty} (x_{2j+1} - x_{2j}),$$

$$e_n = \sum_{j=n}^{\infty} (p_{2j+1} - q_{2j}), \quad \text{and} \quad f_n = \sum_{j=n}^{\infty} (q_{2j+1} - p_{2j}).$$

Given any integer k, it can readily be seen that as $n \to \infty$,

$$\frac{b_{n+k}}{a_n} \to 1, \quad \frac{c_{n+k}}{a_n} \to 1, \quad \frac{d_{n+k}}{a_n} \to 2, \quad \frac{e_{n+k}}{a_n} \to 2 \quad \text{and} \quad \frac{f_{n+k}}{a_n} \to 2.$$

Choose real numbers δ and η satisfying $0 < \delta < \eta$. Define continuous functions u and v from $\mathbb{R} \setminus \{x\}$ into $[\delta, \eta]$ which have the following properties:

$$u(y) = \delta \text{ and } v(y) = \eta, \quad \text{whenever } q_{2n} \leq y \leq p_{2n+1}, \quad n = 1, 2, \ldots,$$

$$u(y) = \eta \text{ and } v(y) = \delta, \quad \text{whenever } q_{2n-1} \leq y \leq p_{2n}, \quad n = 1, 2, \ldots,$$

and

$$u(x + y) = u(x - y) \text{ and } v(x + y) = v(x - y), \text{ for all } y > 0.$$

It is clear that u and v are quasi 1-multipliers. Let $\psi = u\mathbf{1}$ and $\pi = v\mathbf{1}$. It is clear that $\psi \vee \pi$ exists, and that all the flows $\psi, \pi, \psi \wedge \pi, \psi \vee \pi, \psi \oplus \pi, \psi + \pi$ have speeds which are continuous, bounded, and bounded away from 0 in $\mathbb{R} \setminus \{x\}$.

To find $|\dot\psi|(x)$, observe that, for $x_{2n} \leq y \leq x_{2n+2}$, we have

$$\frac{x - y}{\mu_\psi((y, x))} \leq \frac{a_n + d_n}{\frac{1}{\eta} c_{n+1} + \frac{1}{\delta} e_{n+1}} = \frac{1 + \dfrac{d_n}{a_n}}{\dfrac{1}{\eta} \dfrac{c_{n+1}}{a_n} + \dfrac{1}{\delta} \dfrac{e_{n+1}}{a_n}},$$

and that consequently

$$\varlimsup_{y \uparrow x} \left(\frac{x - y}{\mu_\psi((y, x))} \right) \leq \frac{3}{\dfrac{1}{\eta} + \dfrac{2}{\delta}}.$$

Observe also that for $x_{2n} \leq y \leq x_{2n+2}$, we have

$$\frac{x - y}{\mu_\psi((y, x))} \geq \frac{a_{n+1} + d_{n+1}}{\frac{1}{\delta} f_n + \frac{1}{\eta} b_n} = \frac{1 + \dfrac{d_{n+1}}{a_{n+1}}}{\dfrac{1}{\delta} \dfrac{f_n}{a_{n+1}} + \dfrac{1}{\eta} \dfrac{b_n}{a_{n+1}}},$$

and that consequently

$$\varliminf_{y \uparrow x} \left(\frac{x - y}{\mu_\psi((y, x))} \right) \geq \frac{3}{\dfrac{2}{\delta} + \dfrac{1}{\eta}}.$$

We therefore have

$$\lim_{y \uparrow x} \left(\frac{x - y}{\mu_\psi((y, x))} \right) = \frac{3}{\dfrac{2}{\delta} + \dfrac{1}{\eta}},$$

and a similar argument shows that

$$\lim_{v \downarrow x} \left(\frac{y - x}{\mu_\varphi((x, y))} \right) = \frac{3}{\frac{2}{\delta} + \frac{1}{\eta}} \, .$$

From this, we deduce that

$$|\dot\psi|(x) = \frac{3}{\frac{2}{\delta} + \frac{1}{\eta}} \, ,$$

and similar arguments can be used to show that[2]

$$|\dot\pi|(x) = \frac{3}{\frac{2}{\eta} + \frac{1}{\delta}} \, , \quad |(\psi \overset{\vee}{\vphantom{v}} \pi)|(x) = \eta, \quad |(\psi \overset{+}{\vphantom{+}} \pi)|(x) = \delta + \eta,$$

$$|(\psi \overset{\wedge}{\vphantom{\wedge}} \pi)|(x) = \delta, \text{ and } |(\psi \overset{\cdot}{\oplus} \pi)|(x) = \delta \oplus \eta.$$

We clearly have

$$|(\psi \overset{\vee}{\vphantom{v}} \pi)|(x) \neq |\dot\psi|(x) \vee |\dot\pi|(x),$$

$$|(\psi \overset{+}{\vphantom{+}} \pi)|(x) \neq |\dot\psi|(x) + |\dot\pi|(x),$$

$$|(\psi \overset{\wedge}{\vphantom{\wedge}} \pi)|(x) \neq |\dot\psi|(x) \wedge |\dot\pi|(x),$$

and

$$|(\psi \overset{\cdot}{\oplus} \pi)|(x) \neq |\dot\psi|(x) \oplus |\dot\pi|(x).$$

Operations in the Family $\mathrm{Rep}^+(\varphi)$

9.45. Introduction. In the preceding material, we have discussed the various algebraic combinations of two members ψ, π of a family $\mathrm{Rep}_q^+(\varphi)$, and we have shown that, in several important circumstances, if ψ and π are *flows*, then so are their combinations. Theorem 8.30, Note 8.40, and Theorems 8.79 and 9.36 assert the continuity of $\psi \wedge \pi$, $\psi \oplus \pi$, $\psi \vee \pi$ and $\psi + \pi$ when ψ and π belong to a family $\mathrm{Rep}^+(\varphi)$, and certain other necessary conditions are satisfied. It can be noted, however, that the proofs of all these theorems employ essentially the same technique, namely an appeal to Theorem 4.44.

In the next few sections, we shall demonstrate that for the operations $\wedge$, $\vee$ and $+$, a great deal of pathology can occur in the absence of some

---

[2] The notation $|(\psi \overset{\vee}{\vphantom{v}} \pi)|$ is awkward. It denotes, of course, the speed of $\psi \vee \pi$, and $|(\psi \overset{+}{\vphantom{+}} \pi)|$, $|(\psi \overset{\wedge}{\vphantom{\wedge}} \pi)|$ and $|(\psi \overset{\cdot}{\oplus} \pi)|$ are interpreted similarly.

rather restrictive hypotheses, and to this end, we shall construct a sequence $\{g_n\}$ of strictly increasing functions from $\mathbb{R}$ onto $\mathbb{R}$, and a sequence $\{\mu_n\}$ of nonnegative Borel measures defined on $\mathbb{R}$, whose behavior is sufficiently nasty to suit our purposes.

9.46. Construction. Fix a number η in $(0, 1)$.

We shall define sequences $\{C_n\}$, $\{K_n\}$, $\{V_n\}$, $\{\mu_n\}$, $\{g_n\}$, $\{P_n\}$ recursively.

Let C_1 be the $\frac{\eta}{3}$-Cantor subset of $[0, 1]$, let $K_1 = C_1$, and let $V_1 = [0, 1] \setminus K_1$. We clearly have $m(V_1) = \eta$. Let μ_1 be the Borel measure defined on $\mathbb{R}$ as follows:

$$\mu_1(A) = m(A \setminus [0, 1]) + \frac{1}{\eta} m(A \cap V_1),$$

for every Borel subset A of $\mathbb{R}$. Let g_1 be the strictly increasing continuous function from $\mathbb{R}$ onto $\mathbb{R}$ defined by

$$g_1(\tau) = \begin{cases} \mu_1([0, \tau]) & \text{if } \tau \geq 0, \\ -\mu_1([\tau, 0]) & \text{if } \tau < 0. \end{cases}$$

Let P_1 be the set of endpoints of components of V_1. Note that P_1 is a countable subset of C_1, and that P_1 meets every subinterval of $[0, 1]$ whose length exceeds $\frac{\eta}{3}$.

For any positive integer n, if $C_1, C_2, \ldots, C_n$, $K_1, K_2, \ldots, K_n$, $V_1, V_2, \ldots, V_n$, $\mu_1, \mu_2, \ldots, \mu_n$, $g_1, g_2, \ldots, g_n$, and $P_1, P_2, \ldots, P_n$, have all been defined, and V_n is open, we define $C_{n+1}, K_{n+1}, V_{n+1}, \mu_{n+1}, g_{n+1}$ and P_{n+1} by the following procedure:

For each component I of V_n, Let $C_{n+1,I}$ be the $\eta/3$-Cantor subset of $\bar{I}$. Define

$$C_{n+1} = \bigcup_I C_{n+1,I},$$

where the union is taken over all components I of V_n. Now let $K_{n+1} = K_n \cup C_{n+1}$, and let $V_{n+1} = [0, 1] \setminus K_{n+1}$. Let μ_{n+1} be the Borel measure defined on $\mathbb{R}$ as follows:

$$\mu_{n+1}(A) = m(A \setminus [0, 1]) + \frac{1}{\eta^{n+1}} m(A \cap V_{n+1}),$$

for every Borel subset A of $\mathbb{R}$. Let

$$g_{n+1}(\tau) = \begin{cases} \mu_{n+1}([0, \tau]) & \text{if } \tau \geq 0, \\ -\mu_{n+1}([\tau, 0]) & \text{if } \tau < 0, \end{cases}$$

and finally, let P_{n+1} be the set of endpoints of V_{n+1}.

Using an induction argument, we see at once that, for each $n = 1, 2, \ldots$, V_n is an open dense subset of $[0, 1]$, $m(V_n) = \eta^n$, g_n is a strictly increasing, continuous function from $\mathbb{R}$ onto $\mathbb{R}$, P_n is a countable subset of C_n, and P_n meets every subinterval of $[0, 1]$ whose length exceeds $\left(\dfrac{\eta}{3}\right)^n$.

Define $P = \bigcup\limits_{n=1}^{\infty} P_n$. It is clear that P is a countable dense subset of $[0, 1]$.

9.47. Lemma. *If n is any positive integer and $\tau \in P_n$, we have*

$$g_n(\tau) = g_{n+k}(\tau), \quad \text{for every } k = 1, 2, \ldots$$

PROOF: The result follows at once from the observation that if I is a component of V_n, then for any $k = 1, 2, \ldots$,

$$\mu_{n+k}(I) = \frac{1}{\eta^n}\, m(I) = \mu_n(I). \qquad \textbf{QED}$$

9.48. The function g. It follows from Lemma 9.47 that the sequence $\{g_n\}$ converges pointwise on P. We define the function g on P by

$$g(\tau) = \lim_{n \to \infty} g_n(\tau), \quad \text{for all } \tau \in P.$$

9.49. Lemma. *g is a strictly increasing function from P onto a dense subset of $[0, 1]$.*

PROOF: To show that g is strictly increasing, suppose that $\tau_1, \tau_2 \in P$ and $\tau_1 < \tau_2$. Choose n_0 so large that P_{n_0} contains two distinct points σ_1, σ_2 in the interval (τ_1, τ_2).[3] We may assume that $\sigma_1 < \sigma_2$. Now for any $n \geq n_0$ we have

$$g_n(\tau_2) - g_n(\tau_1) \geq g_n(\sigma_2) - g_n(\sigma_1)$$
$$= g_{n_0}(\sigma_2) - g_{n_0}(\sigma_1), \qquad \text{(by Lemma 9.47)},$$

and since $g_{n_0}(\sigma_2) - g_{n_0}(\sigma_1) > 0$, we must therefore have $g(\tau_2) - g(\tau_1) > 0$. This shows that g is strictly increasing.

To show that the range of g is dense in $[0, 1]$, suppose $0 \leq t_1 < t_2 \leq 1$. Choose n such that

$$\frac{1}{3^n} < t_2 - t_1,$$

---

[3] Explicitly, choose n_0 such that $\left(\dfrac{\eta}{3}\right)^{r_0} < \dfrac{1}{2}\,(\tau_2 - \tau_1)$. This is sufficient, by the remarks made at the end of Section 9.46.

let $\tau_1 = g_n^{-1}(t_1)$ and let $\tau_2 = g_n^{-1}(t_2)$. Then since

$$g_n(\tau_2) - g_n(\tau_1) \le \frac{1}{\eta^n}(\tau_2 - \tau_1),$$

we must have

$$\tau_2 - \tau_1 \ge \eta^n(t_2 - t_1) > \left(\frac{\eta}{3}\right)^n.$$

Therefore P_n meets (τ_1, τ_2), and it follows that (t_1, t_2) meets the range of g.

QED

9.50. Corollary. *g has an extension to a strictly increasing continuous function from* $[0, 1]$ *onto* $[0, 1]$.

9.51. Extension of the function g. Using Corollary 9.50 and the fact that if $\tau \in \mathbb{R} \setminus (0, 1)$, then

$$g_n(\tau) = \tau, \text{ for every } n = 1, 2, \ldots,$$

we can now extend g to a strictly increasing continuous function from $\mathbb{R}$ onto $\mathbb{R}$ that satisfies

$$g(\tau) = \lim_{n \to \infty} g_n(\tau), \qquad \forall\ \tau \in \mathbb{R}.$$

Let μ be the Borel measure defined on $\mathbb{R}$ by

$$\mu(A) = m\big(g(A)\big), \text{ for every Borel subset } A \text{ of } \mathbb{R}.$$

9.52. Lemma. *If* $\{k_n\}$ *is a sequence of positive integers, and* $k_n \to \infty$ *and* $\tau_n \to \tau$, *as* $n \to \infty$, *we have*

$$g_{k_n}(\tau_n) \to g(\tau), \text{ as } n \to \infty.$$

PROOF: Suppose first that $0 < \tau < 1$, and let $\varepsilon > 0$. Choose n_0 so large that $3^{-n_0} < \frac{\varepsilon}{4}$, and choose $\sigma_1, \sigma_2 \in P_{n_0}$ such that $\sigma_1 < \tau < \sigma_2$ and such that

$$\sigma_2 - \tau < 2\left(\frac{\eta}{3}\right)^{n_0} \quad \text{and} \quad \tau - \sigma_1 < 2\left(\frac{\eta}{3}\right)^{n_0}.$$

Clearly,

$$g_{n_0}(\sigma_1) = g(\sigma_1) < g(\tau) < g(\sigma_2) = g_{n_0}(\sigma_2),$$

and we have

$$g_{n_0}(\sigma_2) - g_{n_0}(\sigma_1) \le \frac{1}{\eta^{n_0}}(\sigma_2 - \sigma_1) < \frac{4}{3^{n_0}} < \varepsilon.$$

Choose N such that $\tau_n \in (\sigma_1, \sigma_2)$ and $k_n \geq n_0$ whenever $n \geq N$. Then for $n \geq N$, we have

$$g_{n_0}(\sigma_1) = g_{k_n}(\sigma_1) < g_{k_n}(\tau_n) < g_{k_n}(\sigma_2) = g_{n_0}(\sigma_2)$$

and consequently

$$|g_{k_n}(\tau_n) - g(\tau)| \leq |g_{n_0}(\sigma_2) - g_{n_0}(\sigma_1)| < \varepsilon.$$

Therefore $g_{k_n}(\tau_n) \to g(\tau)$, as required.

If $\tau = 0$ or $\tau = 1$, the proof that $g_{k_n}(\tau_n) \to g(\tau)$ is similar to the above proof for the case $0 < \tau < 1$, and if $\tau \in \mathbb{R} \setminus [0, 1]$, the result is trivial.   **QED**

9.53. Corollary. *If $\{k_n\}$ is a sequence of positive integers, and $k_n \to \infty$ and $t_n \to t$ as $n \to \infty$, we have*

$$g_{k_n}^{-1}(t_n) \to g^{-1}(t), \text{ as } n \to \infty.$$

PROOF: The result is trivial if $t \notin [0, 1]$, and we therefore suppose that $t \in [0, 1]$. Let $\tau_n = g_{k_n}^{-1}(t_n)$, for each $n = 1, 2, \ldots$ Then since $\{\tau_n\}$ is bounded, we need only show that $g^{-1}(t)$ is the only finite cluster-point of $\{\tau_n\}$. But if τ is a finite cluster-point of $\{\tau_n\}$, then by Lemma 9.52, $g(\tau)$ is a cluster-point of $\{g_{k_n}(\tau_n)\}$, and this implies that

$$g(\tau) = \lim_{n \to \infty} t_n = t, \text{ as required.}   \textbf{QED}$$

9.54. Example. We now describe the flow which will provide us with the counter-examples we are seeking.

For each $n = 1, 2, \ldots$, let $X_n = \{\langle x, \frac{1}{n}\rangle \mid x \in \mathbb{R}\}$, and let $X_0 = \{\langle x, 0\rangle \mid x \in \mathbb{R}\}$. Let

$$X = \bigcup_{n=0}^{\infty} X_n.$$

Let ψ be the flow in X defined by

$$\psi\left(t, \langle x, \tfrac{1}{n}\rangle\right) = \langle g_n^{-1}\left(t + g_n(x)\right), \tfrac{1}{n}\rangle \qquad (n = 1, 2, \ldots)$$

and

$$\psi(t, \langle x, 0\rangle) = \langle g^{-1}(t + g(x)), 0\rangle,$$

for all $t \in \mathbb{R}$ and all $x \in \mathbb{R}$. Note that, if A is a Borel subset of $\mathbb{R}$, then

$$\mu_\psi\left(\{\langle x, \tfrac{1}{n}\rangle \mid x \in A\}\right) = \mu_n(A), \text{ for all } n = 1, 2, \ldots,$$

and

$$\mu_\psi(\{\langle x, 0\rangle \mid x \in A\} = \mu(A).$$

It follows easily from Lemma 9.52 and Corollary 9.53 that ψ is a flow (not merely a quasi-flow) in X.

In the next example, we describe a simple modification of this flow.

9.55. Example. Let $X_0, X_1, \ldots$, and X be as in Example 9.54, and let π be the flow in X defined by

$$\pi\left(t, \langle x, \tfrac{1}{n}\rangle\right) = \langle g_{n+1}^{-1}\left(t + g_{n+1}(x)\right), \tfrac{1}{n}\rangle, \qquad \text{for } n \text{ even,}$$

$$\pi\left(t, \langle x, \tfrac{1}{n}\rangle\right) = \langle g_n^{-1}\left(t + g_n(x)\right), \tfrac{1}{n}\rangle, \qquad \text{for } n \text{ odd,}$$

and

$$\pi(t, \langle x, 0\rangle) = \langle g^{-1}(t + g(x)), 0\rangle,$$

for all $t \in \mathbb{R}$ and all $x \in \mathbb{R}$.

Note that, if A is a Borel subset of $\mathbb{R}$, then

$$\mu_\pi\left(\{\langle x, \tfrac{1}{n}\rangle \mid x \in A\}\right) = \mu_{n+1}(A), \qquad \text{if } n \text{ is even,}$$

$$\mu_\pi\left(\{\langle x, \tfrac{1}{n}\rangle \mid x \in A\}\right) = \mu_n(A), \qquad \text{if } n \text{ is odd,}$$

and

$$\mu_\pi(\{\langle x, 0\rangle \mid x \in A\}) = \mu(A).$$

Once again, Lemma 9.52 and Corollary 9.53 imply that π is a flow.

9.56. Theorem. *Let $X_0, X_1, \ldots$, and X be as in Examples 9.54 and 9.55 and let φ be the flow in X defined by*

$$\varphi(t, \langle x, y\rangle) = \langle x + t, y\rangle, \qquad \forall\, t \in \mathbb{R}, \qquad \forall\, \langle x, y\rangle \in X.$$

Then the flows ψ and π defined in Examples 9.54 and 9.55 are members of $\mathrm{Rep}^+(\varphi)$, and $\psi \vee \pi$ exists.

However, none of the quasi-flows $\psi \wedge \pi$, $\psi \vee \pi$ and $\psi + \pi$ is continuous, and in fact, ψ and π have neither a maximum nor a minimum in the family $\mathrm{Rep}^+(\varphi)$.

PROOF: We have already seen that ψ and π belong to $\mathrm{Rep}^+(\varphi)$, and it is clear that the measures μ_ψ and μ_π agree in

$$X_0 \cup \bigcup_{n=1}^{\infty} X_{2n-1}.$$

It is also easy to see that if n is even, and A is a Borel subset of $\mathbb{R}$, then

$$(\mu_\psi \vee \mu_\pi)\left(\left\{\langle x, \tfrac{1}{n}\rangle \mid x \in A\right\}\right)$$

$$= m(A \setminus [0, 1]) + \frac{1}{\eta^{n+1}} m(A \cap V_{n+1}) + \frac{1}{\eta^n} m\left(A \cap (V_n \setminus V_{n+1})\right),$$

$$(\mu_\psi + \mu_\pi)\left(\left\{\langle x, \tfrac{1}{n}\rangle \mid x \in A\right\}\right)$$

$$= 2m(A \setminus [0, 1]) + \left(\frac{1}{\eta^{n+1}} + \frac{1}{\eta^n}\right) m(A \cap V_{n+1})$$

$$+ \frac{1}{\eta^n} m\left(A \cap (V_n \setminus V_{n+1})\right),$$

$$(\mu_\psi \wedge \mu_\pi)\left(\left\{\langle x, \tfrac{1}{n}\rangle \mid x \in A\right\}\right) = m(A \setminus [0, 1]) + \frac{1}{\eta^n} m(A \cap V_{n+1}),$$

and

$$(\mu_\psi \oplus \mu_\pi)\left(\left\{\langle x, \tfrac{1}{n}\rangle \mid x \in A\right\}\right) = \frac{1}{2} m(A \setminus [0, 1])$$

$$+ \left(\frac{1}{\eta^{n+1} + \eta^n}\right) m(A \cap V_{n+1}).$$

From this, it follows easily that $\psi \vee \pi$ exists, and we deduce also that if n is odd, we have

$$(\mu_\psi \vee \mu_\pi)\left(\left\{\langle x, \tfrac{1}{n}\rangle \mid x \in [0, 1]\right\}\right) = 1,$$

$$(\mu_\psi + \mu_\pi)\left(\left\{\langle x, \tfrac{1}{n}\rangle \mid x \in [0, 1]\right\}\right) = 2,$$

$$(\mu_\psi \wedge \mu_\pi)\left(\left\{\langle x, \tfrac{1}{n}\rangle \mid x \in [0, 1]\right\}\right) = 1$$

and

$$(\mu_\psi \oplus \mu_\pi)\left(\left\{\langle x, \tfrac{1}{n}\rangle \mid x \in [0, 1]\right\}\right) = \frac{1}{2},$$

and if n is even, we have

$$(\mu_\psi \vee \mu_\pi)\left(\left\{\langle x, \tfrac{1}{n}\rangle \mid x \in [0, 1]\right\}\right) = 2 - \eta,$$

$$(\mu_\psi + \mu_\pi)\left(\left\{\langle x, \tfrac{1}{n}\rangle \mid x \in [0, 1]\right\}\right) = 2,$$

$$(\mu_\psi \wedge \mu_\pi)\left(\left\{\langle x, \tfrac{1}{n}\rangle \mid x \in [0, 1]\right\}\right) = \eta$$

and

$$(\mu_\psi \oplus \mu_\pi)\left(\left\{\langle x, \tfrac{1}{n}\rangle \mid x \in [0, 1]\right\}\right) = \frac{\eta}{1 + \eta}.$$

We shall now show that $\psi \wedge \pi$ is not a flow. To obtain a contradiction, suppose that $\psi \wedge \pi$ is a flow, and let h be the φ-reparametrizer for which $\psi \wedge \pi = h \operatorname{rep} \varphi$. Then Theorem 4.23 implies that

$$h^{-1}_{\langle 0, \frac{1}{n} \rangle}(1) \to h^{-1}_{\langle 0, 0 \rangle}(1) \quad \text{as} \quad n \to \infty.$$

But if n is odd,

$$h^{-1}_{\langle 0, \frac{1}{n} \rangle}(1) = (\mu_\psi \vee \mu_\pi)\left(\left\{\langle x, \tfrac{1}{n} \rangle \mid x \in [0, 1]\right\}\right) = 1,$$

and if n is even,

$$h^{-1}_{\langle 0, \frac{1}{n} \rangle}(1) = (\mu_\psi \vee \mu_\pi)\left(\left\{\langle x, \tfrac{1}{n} \rangle \mid x \in [0, 1]\right\}\right) = 2 - \eta.$$

Therefore $\left\{ h^{-1}_{\langle 0, \frac{1}{n} \rangle}(1) \right\}$ fails to converge, and we have a contradiction.

We have therefore shown that $\psi \wedge \pi$ is not a flow; in fact, we have shown more than this. The argument just presented shows that the restriction of $\psi \wedge \pi$ to $\bigcup_{n=1}^{\infty} X_n$ has no extension to a flow in X. From this we can deduce that ψ and π have no minimum in the family $\operatorname{Rep}^+(\varphi)$, for if ϑ were such a minimum, then ϑ would have to agree with $\psi \wedge \pi$ in each of the sets X_n $(n = 1, 2, \ldots)$.

A similar argument can be used to show that neither $\psi \vee \pi$ nor $\psi + \pi$ is a flow, and that ψ and π do not have a maximum in the family $\operatorname{Rep}^+(\varphi)$.

QED

9.57. Note. We note that (with the notation of Theorem 9.56)

$$(\mu_\psi + \mu_\pi)\left(\left\{\langle x, \tfrac{1}{n} \rangle \mid x \in [0, 1]\right\}\right) = 2,$$

regardless of whether n is even or odd. Therefore we cannot use the techniques of the proof of Theorem 9.56 to show that $\psi \oplus \pi$ is not a flow. As a matter of fact, $\psi \oplus \pi$ *is* a flow, as we shall show in Theorem 9.59.

In order to establish the continuity of the harmonic sum of two flows, we shall prove the following theorem, which is an analogue of Theorem 4.32.

9.58. Theorem. *Let φ be a quasi-flow in a Hausdorff space X, let f be a quasi φ-reparametrizer, let $\psi = f \operatorname{rep} \varphi$, and let $\operatorname{inv}(f) = \{g_x \mid x \in \mathcal{G}(\psi)\}$. Suppose that $\mathcal{G}(\psi)$ is an open subset of X.*

Then the following are equivalent:

(a) *f is a continuous function from the subspace $\mathscr{G}(\psi) \times \mathbb{R}$ of $X \times \mathbb{R}$ into $\mathbb{R}$.*

(b) *Whenever $\{\langle x_\alpha, \tau_\alpha \rangle \mid \alpha \in A\}$ is a net of points in $X \times \mathbb{R}$, $x_\alpha \to x \in \mathscr{G}(\psi)$ and $\tau_\alpha \to \tau \in f_x(\mathbb{R})$, we have*

$$g_{x_\alpha}(\tau_\alpha) \to g_x(\tau).$$

PROOF: The proof that (a) $\Rightarrow$ (b) is identical to the proof of Theorem 4.23 and we refer the reader to that theorem.

Suppose now that (b) holds. Let $x \in \mathscr{G}(\psi)$, $t \in \mathbb{R}$, and let $\{\langle x_\alpha, t_\alpha \rangle \mid \alpha \in A\}$ be a net in $\mathscr{G}(\psi) \times \mathbb{R}$ such that $x_\alpha \to x$ and $t_\alpha \to t$. Write $\tau = f_x(t)$, and let $\varepsilon > 0$. By making ε smaller, if necessary, we may suppose that $[\tau - \varepsilon, \tau + \varepsilon] \subseteq f_x(\mathbb{R})$. Then since $t = g_x(\tau) \in \big(g_x(\tau - \varepsilon), g_x(\tau + \varepsilon) \big)$, and since $t_\alpha \to t$ and $g_{x_\alpha}(\tau - \varepsilon) \to g_x(\tau - \varepsilon)$ and $g_{x_\alpha}(\tau + \varepsilon) \to g_x(\tau + \varepsilon)$, we must have

$$t_\alpha \in \big(g_{x_\alpha}(\tau - \varepsilon), g_{x_\alpha}(\tau + \varepsilon) \big)$$

for all sufficiently large α, *i.e.*,

$$f_{x_\alpha}(t_\alpha) \in (\tau - \varepsilon, \tau + \varepsilon)$$

for all sufficiently large α.

We have therefore shown that $f_{x_\alpha}(t_\alpha) \to \tau$, and the proof is complete.

QED

9.59. Theorem. *Let φ be a flow in a Hausdorff space X and let $\psi, \pi \in \mathrm{Rep}^+(\varphi)$.*

Then $\psi \oplus \pi$ is a flow.

PROOF: Choose isotonic quasi φ-reparametrizers f, f_ψ and f_π such that

$$\psi \oplus \pi = f \text{ rep } \varphi, \quad \psi = f_\psi \text{ rep } \varphi \text{ and } \pi = f_\pi \text{ rep } \varphi,$$

and let $\mathrm{inv}(f) = \{g_x \mid x \in \mathscr{G}(\psi \oplus \pi)\}$, $\mathrm{inv}(f_\psi) = \{g_{\psi,x} \mid x \in \mathscr{G}(\psi)\}$, and $\mathrm{inv}(f_\pi) = \{g_{\pi,x} \mid x \in \mathscr{G}(\pi)\}$.

Suppose $x \in \mathscr{G}(\psi \oplus \pi)$. Then since $\mathscr{O}_{\psi \oplus \pi}(x) = \mathscr{O}_\psi(x) \cap \mathscr{O}_\pi(x)$, we see that $f_x(\mathbb{R}) = f_{\psi,x}(\mathbb{R}) \cap f_{\pi,x}(\mathbb{R})$. Therefore, whenever $\tau \in f_x(\mathbb{R})$ and $|\tau| < \mathrm{p}_\varphi(x)$, we have

$$|g_x(\tau)| = \mu_{\psi \oplus \pi}\big(\{\varphi(\sigma, x) \mid \sigma \in [0, \tau]\}\big)$$

$$= \mu_\psi\big(\{\varphi(\sigma, x) \mid \sigma \in [0, \tau]\}\big) + \mu_\pi\big(\{\varphi(\sigma, x) \mid \sigma \in [0, \tau]\}\big)$$

$$= |g_{\psi,x}(\tau)| + |g_{\pi,x}(\tau)|,$$

and therefore, since the numbers $g_x(\tau)$, $g_{\psi,x}(\tau)$ and $g_{\pi,x}(\tau)$ have the same sign, we have

$$g_x(\tau) = g_{\psi,x}(\tau) + g_{\pi,x}(\tau).$$

It follows easily that whenever $\tau \in f_x(\mathbb{R})$, we have

$$g_x(\tau) = g_{\psi,x}(\tau) + g_{\pi,x}(\tau).$$

Now let $\{\langle x_\alpha, \tau_\alpha \rangle \mid \alpha \in A\}$ be a net in $X \times \mathbb{R}$, and suppose $x_\alpha \to x \in \mathscr{G}(\psi \oplus \pi)$ and $\tau_\alpha \to \tau \in f_x(\mathbb{R})$. Then since $\mathscr{G}(\psi \oplus \pi) = \mathscr{G}(\psi) \cap \mathscr{G}(\pi)$ is open, we see at once from Theorem 4.23 and the fact that both ψ and π are flows that

$$g_{x_\alpha}(\tau_\alpha) = g_{\psi,x_\alpha}(\tau_\alpha) + g_{\pi,x_\alpha}(\tau_\alpha) \to g_{\psi,x}(\tau) + g_{\pi,x}(\tau) = g_x(\tau).$$

It follows immediately from Theorem 9.58 that f is a continuous function from $\mathscr{G}(\psi \oplus \pi) \times \mathbb{R}$ into $\mathbb{R}$.

To show that $\psi \oplus \pi$ is continuous, let $x \in X$, $t \in \mathbb{R}$ and let $\{\langle x_\alpha, t_\alpha \rangle \mid \alpha \in A\}$ be a net in $X \times \mathbb{R}$ such that $x_\alpha \to x$ and $t_\alpha \to t$. If $x \in \mathscr{G}(\psi \oplus \pi)$, we have $x_\alpha \in \mathscr{G}(\psi \oplus \pi)$ for all sufficiently large α, and therefore

$$(\psi \oplus \pi)(t_\alpha, x_\alpha) = \varphi(f_{x_\alpha}(t_\alpha), x_\alpha) \to \varphi(f_x(t), x) = (\psi \oplus \pi)(t, x).$$

If $x \in \mathscr{F}(\psi \oplus \pi)$, then either $x \in \mathscr{F}(\psi)$ or $x \in \mathscr{F}(\pi)$. Suppose, to be precise, that $x \in \mathscr{F}(\psi)$. Then since $\psi \oplus \pi \leq \psi$, we can choose for each $\alpha \in A$, a number $s_\alpha \in [0, t_\alpha]$ such that

$$(\psi \oplus \pi)(t_\alpha, x_\alpha) = \psi(s_\alpha, x_\alpha).$$

Since the net $\{s_\alpha \mid \alpha \in A\}$ is bounded, and $x \in \mathscr{F}(\psi)$, we have

$$\psi(s_\alpha, x_\alpha) \to x,$$

and we conclude that

$$(\psi \oplus \pi)(t_\alpha, x_\alpha) \to x.$$

We have therefore shown that $\psi \oplus \pi$ is continuous, and the proof is complete.   **QED**

9.60. Corollary. *Let φ be a flow in a Hausdorff space X and let u and v be nonnegative φ-multipliers.*

Then $u \oplus v$ is a φ-multiplier.

PROOF: Immediate from Theorems 8.39 and 9.59.   **QED**

Notes and Remarks to Chapters 8 and 9

The problem of defining the sum of two flows was first investigated by the author and Aryeh Dvoretsky[4] who designed a procedure for summing two members of the family $\mathscr{F}$ of all flows which move from left to right along the real line, and have no fixed points.

In order to give a brief description of this procedure, let ψ and π be members of $\mathscr{F}$, and let g_ψ and g_π be the continuous, strictly increasing functions from $\mathbb{R}$ into $\mathbb{R}$ defined by:

$\psi(g_\psi(\tau), 0) = \tau$ and $\pi(g_\pi(\tau), 0) = \tau$, $\forall \tau \in \mathbb{R}$. Define a continuous, non-decreasing function g from $\mathbb{R}$ into $\mathbb{R}$ as follows:

Fix $\tau > 0$. (The procedure for $\tau < 0$ is similar.)

Given a partition $P = \{\tau_0, \tau_1, \ldots, \tau_k\}$ of $[0, \tau]$ which satisfies

$$0 = \tau_0 < \tau_1 < \ldots < \tau_k = \tau,$$

let

$$g_P(\tau) = \sum_{j=1}^{k} \left(\frac{1}{g_\psi(\tau_j) - g_\psi(\tau_{j-1})} + \frac{1}{g_\pi(\tau_j) - g_\pi(\tau_{j-1})} \right)^{-1}.$$

Then define $g(\tau)$ to be the limit $\lim\limits_{n \to \infty} g_{P_n}(\tau)$, where $\{P_n\}$ is any sequence of partitions of $[0, \tau]$ which satisfies

(i) $P_1 \subseteq P_2 \subseteq P_3 \ldots$,

and

(ii) The mesh of P_n tends to 0 as $n \to \infty$.

It can be shown that $g(\tau)$ is uniquely defined (independent of the particular sequence chosen) for every $\tau \in \mathbb{R}$, and that g is continuous and nondecreasing.

In those cases in which g turned out to be a *strictly* increasing function from $\mathbb{R}$ *onto* $\mathbb{R}$, $\psi + \pi$ was then defined to be that member of $\mathscr{F}$ which satisfies

$$(\psi + \pi)(g(\tau), 0) = \tau, \qquad \forall \tau \in \mathbb{R}.$$

In order to investigate the conditions under which the sum could be defined, we introduced the relation $\leq$ in $\mathscr{F}$ (using a condition similar to Theorem 8.2(b)), and showed that if ψ and π are any members of $\mathscr{F}$, then $\psi + \pi$ exists iff ψ and π have a common upper bound in $\mathscr{F}$ iff ψ and π have a maximum in $\mathscr{F}$.

The theory as it appears in Chapters 8 and 9 is a generalization by the Lewins of this joint work with Dvoretsky. By introducing the concept of

---

[4] This work has not been previously published.

time measure, the Lewins were able to simplify considerably the manipulation involved in taking the sum, maximum or minimum of two flows. This in turn allowed them to extend the definitions to include flows with fixed points, and also to make a close examination of the algebraic structure. Also, by means of the reparametrization theory developed in Chapters 4 and 5, the Lewins were able to discuss algebraic combinations in a much wider setting than $\mathrm{Rep}^+(1)$. The ability to deal with flows globally, instead of having to consider one orbit at a time, made it possible to prove the various theorems in Chapter 8 and 9, which assert that unter the right conditions, algebraic combinations of flows are continuous.

Differences. Dvoretsky and the author have shown that if ψ and π are members of the family $\mathscr{F}$, and $\psi < \pi$, then there does not necessarily exist a member ϑ of $\mathscr{F}$ such that $\psi + \vartheta = \pi$,

and also that if $\psi_1, \psi_2, \pi \in \mathscr{F}$ and

$$\psi_1 + \pi = \psi_2 + \pi,$$

then we do not necessarily have $\psi_1 = \psi_2$.

The Lewins are currently working on the problem of forming differences in a family $\mathrm{Rep}_\mathrm{q}^+(\varphi)$. In the same way that the sum in $\mathrm{Rep}_\mathrm{q}^+(\varphi)$ corresponds to the harmonic sum in $\mathscr{M}_q(\varphi)$, the difference $\psi - \pi$ of two members $\psi, \pi \in \mathrm{Rep}_\mathrm{q}^+(\varphi)$, where $\pi \leq \psi$, will have for its time measure the "harmonic difference" $\mu_\psi \ominus \mu_\pi$ in much of the space. The harmonic difference of two measures is defined as follows:

Let λ and μ be nonnegative measures defined on a σ-algebra Σ of subsets of a set S, and suppose that $\mu \leq \lambda$. Then the *harmonic difference* $\mu \ominus \lambda$ of μ and λ is defined by

$$(\mu \ominus \lambda)(A) = \sup \left\{ \sum_{j=1}^n \left(\frac{1}{\mu(A_j)} - \frac{1}{\lambda(A_j)} \right)^{-1} \middle| A = \bigcup_{j=1}^n A_j \right\},$$

for every measureable set A, where the supremum is taken over all partitions of A into finitely many measurable sets.

It can be shown that if λ and μ are nonnegative measures and $\mu \leq \lambda$, then $\mu \ominus \lambda$ is a countably additive nonnegative measure, and furthermore, if $\mu, \lambda \in \mathscr{M}_q(\varphi)$, we have $(\mu \ominus \lambda) \oplus \lambda = \mu$.

It can therefore be shown that if ψ, π are members of a family $\mathrm{Rep}_\mathrm{q}^+(\varphi)$ and $\psi - \pi$ exists, then

$$(\psi - \pi) + \pi = \psi.$$

Chapters 8 and 9 constitute the major portion of the doctoral dissertation "Algebraic Combinations of Continuous Flows" of Mirit Hope Lewin (University of Wisconsin, 1970).

CHAPTER TEN

FINE STRUCTURE IN $\mathscr{G}_r(\varphi)$

10.1. Introduction. We now undertake an examination of $\mathscr{G}_r(\varphi)$ in greater detail. In Chapter 3, we specify the form of the components of $\mathscr{G}_r(\varphi)$. We will now look more carefully at the actual action of the flow φ in each such component. It will be our purpose to build "standard models" for these component flows, *i.e.* for each component, we will construct a flow ϕ in a specified annulus A (say the annulus $\{z \mid 1 < |z| < 2\}$), which is conjugate to the flow φ (the concept of conjugacy of flows will be set forth in this chapter). The flow ϕ will, in certain subannuli of A, be the spiral flows set forth in Examples 3.14(b) and 3.14(c), and in the rest of A, ϕ will be an ordinary rotation flow, counterclockwise on some orbits, clockwise on others. The meaning of a conjugacy is quite simple: There exist a homeomorphism h of the component Ω onto A, and a ϕ-multiplier u defined from A into $\mathbb{R} \setminus \{0\}$ and continuous in $\mathscr{G}(\phi)$ so that $h[\varphi] = u\phi$. In this sense, then, we will be able to say that we know all about the action of the flow φ in $\mathscr{G}_r(\varphi)$.

We will begin by working with the "spiral flows" and characterizing their actions.

10.2. Theorem. *Let Ω be the annulus bounded by two periodic orbits $\mathcal{O}_\varphi(y^-)$ and $\mathcal{O}_\varphi(y^+)$, and satisfying $\mathrm{p}_\varphi(x) = \infty$, $\forall\, x \in \Omega$.*

Then for every $\varepsilon > 0$, $x_0 \in \Omega$, we can find a curve γ in Ω satisfying

(i) *the endpoints of γ belong to $\mathcal{O}_\varphi(x_0)$,*

(ii) *no other point of γ belongs to $\mathcal{O}_\varphi(x_0)$,*

(iii) *$\forall\, x \in \Omega \setminus \mathcal{O}_\varphi(x_0)$, $\mathcal{O}_\varphi(x)$ intersects γ in exactly one point,*

(iv) *γ lies within ε of γ^+.*

PROOF: The hypotheses allow no fixed points, and thus no stagnation points, in Ω. Thus, for $x_1, x_2 \in \Omega$, $\alpha_\varphi(x_1) = \alpha_\varphi(x_2)$ and $\omega_\varphi(x_1) = \omega_\varphi(x_2)$.

Let us assume that $\mathcal{O}_\varphi(y^+) = \omega_\varphi(x)$, $\forall\, x \in \Omega$, and by choosing an appropriate point to play the role of ∞, we can have $\mathcal{O}_\varphi(y^+)$ be the *inner* boundary of Ω.

Let $\varepsilon_1 > 0$ be chosen with $\mathrm{cl}\big(N_d(y^+, \varepsilon_1)\big) \subseteq N(y^+, \varepsilon)$. Choose $\delta_1 < 0$ so small that $0 < \delta_1 < \frac{1}{4}\, p_\varphi(y^+)$, and $d\big(\varphi(t, y^+), y^+\big) < \varepsilon_1$, $\forall\, |t| \leq \delta_1$. Choose $\varepsilon_2 > 0$ so small that $\varepsilon_2 < \varepsilon_1$, and $d\big(\varphi(t, x), y^+\big) < \varepsilon_1$, $\forall\, |t| \leq \delta_1$, $\forall\, x \in N_d(y^+, \varepsilon_2)$.

Now choose any $x_0 \in \Omega$, and let $\langle \tau^-, \tau^+, x^-, x^+, J, L, T \rangle$ be a φ-gate construction for the pair $\langle x_0, y^+ \rangle$ with the property that $L \subseteq N_d(y^+, \varepsilon_2)$ (cf. Definition 2.6). Theorem 2.3 assures us of the existence of an integer $k \geq 2$ and a positive δ_2 satisfying $\delta_1 = k\delta_2$ and other properties set forth in that theorem. It is clear from the statements of the Gate Theorems (Theorems 2.3 and 2.4) that since $\mathcal{O}_\varphi(y^-) \subseteq \mathrm{outs}\,(J)$, we have $\mathcal{O}_\varphi(y^+) \subseteq \mathrm{ins}\,(J)$ and $\mathrm{ins}\,\varphi\big((n+1)\,\delta_2, J\big) \subseteq \mathrm{ins}\,\varphi(n\delta_2, J)$, $\forall\, n = 0, \pm 1, \pm 2, \dots$. It is also clear that $\forall\, x \in \Omega$, $\mathcal{O}_\varphi(x) \cap J \neq \square$.

By Theorem 2.4 (b), we know that $\mathcal{O}_\varphi(x_0) \cap J = \{\varphi(t, x_0) \mid \tau^- \leq t \leq \tau^+\}$. We now wish to modify L so that $\mathcal{O}_\varphi(x) \cap J$ will be either a curve (closed arc) or a point for each $x \in \Omega$. Theorem 2.3 (vi) states that for each $x \in L$, the curve $\{\varphi(t, x) \mid |t| \leq \delta_2\}$ does not intersect either $\varphi(-\delta_1, L)$ or $\varphi(\delta_1, L)$. We will show that if $x \in L \setminus \{x^-, x^+\}$ and $\varphi(t, x) \in L$, we must have $|t| < \delta_1$. Indeed, suppose $t \geq \delta_1$ and that $\varphi(t, x) \in L$. Set $t = m\delta_2 + s$, where $0 \leq s < \delta_2$. Clearly $m \geq k$. If $m = k$, $\varphi(m\delta_2, x) \in \varphi(\delta_1, L)$, and $\varphi(m\delta_2 + s, x) \in L$, which is impossible since $|s| < \delta_2$. If $m > k$, then $\varphi(m\delta_2, x) \in \mathrm{ins}\,\varphi(\delta_1, J)$ and thus $\varphi(\tau, x) \in \varphi(\delta_1, J)$ for some $m\delta_2 < \tau < t$, since $\varphi(t, x) \in \mathrm{outs}\,\varphi(\delta_1, J)$. Since $x \notin \mathcal{O}_\varphi(x_0)$, $\varphi(\tau, x) \in \varphi(\delta_1, L)$, which is again impossible since $t - \tau < t - m\delta_2 = s < \delta_2$.

The case for $t \leq -\delta_1$ follows from this one, and thus x and $\varphi(t, x)$ cannot both belong to $L \setminus \{x^-, x^+\}$ if $|t| \geq \delta_1$.

We now define for each $x \in L \setminus \{x^-, x^+\}$ the number $\bar{d}(x) = \mathrm{diam}\big(\mathcal{O}_\varphi(x) \cap J\big)$. By the compactness of $L \times [0, \delta_1]$ and the continuity of φ, we know that there is a point $x_1 \in L \setminus \{x^-, x^+\}$ for which $\bar{d}(x)$ achieves its maximum. Note that this is proved by applying the standard compactness arguments, while noting that for points lying close to x^- or x^+, $\bar{d}(x)$ is very small. If there is more than one point x maximizing $\bar{d}(x)$, we avoid the use of the Axiom of Choice by selecting the one closest to x^-, again using the continuity of φ to assure the existence of a closest point.

We have seen that for $x \in L \setminus \{x^-, x^+\}$, $\mathcal{O}_\varphi(x) \cap J$ contains no points of the form $\varphi(t, x)$, where $|t| \geq \delta_1$. The same analysis shows that if $\varphi(t_1, x)$ and $\varphi(t_2, x)$ are both in L, then $|t_1| < \delta_1$ and $|t_2| < \delta_1$, so that $|t| < \delta_1$ for all $t_1 \leq t \leq t_2$. Now let x_1^- and x_1^+ be the points of $\mathcal{O}_\varphi(x_1) \cap L$ most distant from each other, and label them so that $x_1^- = \varphi(\tau_1^-, x_1)$ and $x_1^+ = \varphi(\tau_1^+, x_1)$, with $\tau_1^- < \tau_1^+$. We eliminate $[x_1^-, x_1^+]$ from L, replacing it by the arc $\Lambda_1 = \{\varphi(t, x_1) \mid \tau_1^- \leq t \leq \tau_1^+\}$. We assume WoLOG that $x_1^- = x_1$. Then $\tau_1^- = 0$ and we define $\tau_1 = \tau_1^+$. We now note that $L \setminus [x_1^-, x_1^+]$ consists of two intervals, and that for any $x \notin \mathcal{O}_\varphi(x_0)$, $\mathcal{O}_\varphi(x)$ can intersect at most one of these intervals. We now choose x_2 in a way similar to x_1, so as to maximize $\text{diam}\left(\mathcal{O}_\varphi(x_2) \cap (L \setminus [x_1, x_1^+])\right)$. We continue, each time choosing x_i so that $\text{diam}\left(\mathcal{O}_\varphi(x_i) \cap \left(L \setminus \bigcup_{j=1}^{i-1} [x_j, x_j^+]\right)\right)$ is maximal, defining x_i^- and x_i^+, τ_i^- and τ_i^+ as with x_1, and then setting $x_i = x_i^-$, and $\tau_i = \tau_i^+$, all WoLOG, as before. We continue with this process until it is no longer possible to continue. At each stage, we define the curve

$$\Lambda_i = \{\varphi(t, x_i) \mid 0 \leq t \leq \tau_i\}$$

and replace the interval $[x_i, x_i^+]$ in L by Λ_i. Note that for each $j > 0$, $\left(L \setminus \bigcup_{i=1}^{j} [x_i, x_i^+]\right) \cup \bigcup_{i=1}^{j} \Lambda_i$ is an arc joining x^- to x^+, and that each $\Lambda_i \subseteq \{\varphi(t, x_i) \mid |t| < \delta_1\}$, so that $\Lambda_i \subseteq N_d(y^+, \varepsilon_1)$. We also have

$$\mathscr{L}([x_i, x_i^+]) \geq \mathscr{L}([x_{i+1}, x_{i+1}^+]), \qquad \forall \, i \geq 1,$$

and

$$\sum_i \mathscr{L}([x_i, x_i^+]) \leq \mathscr{L}([x^-, x^+]),$$

so that the process ends in at most a countable number of steps, that is, once through the integers.

We wish to show that $\gamma_0 = \left(L \setminus \bigcup_i [x_i, x_i^+]\right) \cup \bigcup_i \Lambda_i$ is a curve. To do this, we must show that if there are infinitely many Λ_i, then $\text{diam}(\Lambda_i) \to 0$ as $i \to \infty$. We show first that $\tau_i \to 0$. If not, then there is a subsequence $\{\tau_{i_j}\}$ with $\tau_{i_j} > \delta_3$, $\exists \, \delta_3 > 0$. Since $0 < \tau_i < \delta_1$, $\forall i$, there must be a sub-subsequence of $\{\tau_i\}$ which converges to $\bar{\delta}$, $\exists \, \delta_3 \leq \bar{\delta} \leq \delta_1$. WoLOG we take $\{\tau_{i_j}\}$ to be that sub-subsequence. Now the sequence $\{x_{i_j}\}$ must have a cluster point, and we assume WoLOG that $x_{i_j} \to \bar{x}$. Then also, $x_{i_j}^+ \to \bar{x}$, since $d(x_i, x_i^+) \to 0$. It follows then that $\varphi(\bar{\delta}, \bar{x}) = \bar{x}$. However, since $p_\varphi(x) = \infty$, $\forall \, x \in \Omega$, we cannot have $\varphi(\bar{\delta}, \bar{x}) = \bar{x}$, so that the assertion $\tau_i \to 0$ is proved. Since $\tau_i \to 0$ and L is compact, we have $\text{diam}(\Lambda_i) \to 0$ by the continuity of φ.

The curve γ_0 now has the property that $\gamma_0 \cap \mathcal{O}_\varphi(x)$ is either a single point or a curve for each $x \in \Omega \setminus \mathcal{O}_\varphi(x_0)$. We now wish to modify it so that all these intersections are points. We observe first that there are at most finitely many values of i for which we have $\frac{1}{2}\delta_1 < \tau_i \leq \delta_1$. WoLOG we can assume that these are the numbers $1, \ldots, k_1$. We will now choose a collection $N_1, \ldots, N_{k_1}$ of open subarcs of γ_0 which have the following properties:

(i) The N_i are pairwise disjoint, $\quad \forall\, i = 1, \ldots, k_1$.

(ii) No N_i contains x^- or x^+.

(iii) The N_i are disjoint from the Λ_j, $\forall\, i = 1, \ldots, k_1$, $\forall\, j = 1, \ldots, k_1$.

(iv) x_i is an endpoint of N_i, $\forall\, i = 1, \ldots, k_1$.

(v) $\forall\, x \in N_i \quad \forall\, 0 \leq t \leq \tau_i, \quad d\big(\varphi(t, x), y^+\big) < \varepsilon_1$.

That we can satisfy (i)—(iv) is obvious, and (v) follows from the continuity of φ, combined with the fact that

$$d\big(\varphi(t, x_i), y^+\big) < \varepsilon_1, \quad \forall\, 0 \leq t \leq \tau_i, \quad \forall\, i.$$

Next, we define a nonnegative continuous function f_1 in γ_0 satisfying:

(i) $f_1(x) = 0$ unless $x \in \bigcup_{i=1}^{k_1} (N_i \cup \Lambda_i)$.

(ii) $0 \leq f_1(x) \leq \tau_i, \quad \forall\, x \in N_i, \quad \forall\, i = 1, \ldots, k_1$.

(iii) $\forall\, x \in \Lambda_i, \ f_1(x) = \tau_i - t$, where $x = \varphi(t, x_i)$, $\quad \forall\, i = 1, \ldots, k_1$.

(iv) f_1 is constant on each Λ_i, $\quad \forall\, i \neq 1, \ldots, k_1$.

To show that such a function f_1 exists, it is sufficient to show that in each of the N_i, $i = 1, \ldots, k_1$, there is a function which increases along that open subarc from 0 at one end to τ_i at the other, and is constant on each subarc Λ_j, $j \neq 1, \ldots, k_1$. To show this, we can use the same technique as in showing the existence of a mapping of Λ onto $[0, 1]$ which is constant on each of the Λ_i, and this is done below, in the construction of the function g_1, which is also required to be monotonic.

We now map γ_0 onto a curve γ_1 by a mapping $h_1 : h_1(x) = \varphi(f_1(x), x)$. Note that $h_1(\Lambda_i) = \{x_i^+\}$, $\forall\, i = 1, \ldots, k_1$, and otherwise h_1 is one-to-one.

We continue by iterating the above process. There are at most finitely many values of i for which $\frac{1}{4}\delta_1 < \tau_i \leq \frac{1}{2}\delta_1$. WoLOG, we take these to be the numbers between k_1 and $k_2 : k_1 < i \leq k_2$. We now define a collection N_i of open subarcs of γ_0 for *these* values of i, satisfying conditions similar to the previous ones:

(i) The N_i are pairwise disjoint, $\forall\, k_1 < i \leq k_2$.

(ii) No N_i contains x^- or x^+.

(iii) The N_i are disjoint from the Λ_j, $\bigvee k_1 < i \leq k_2$, $\bigvee 1 \leq j \leq k_2$.

(iv) x_i is an endpoint of N_i, $\bigvee k_1 < i \leq k_2$.

(v) $\bigvee x \in N_i$, $\bigvee 0 \leq t \leq \tau_i$, $d\big(\varphi(f_1(x) + t, x), y^+\big) < \varepsilon_1$.

And again, we define a nonnegative continuous function in γ_0, this time called f_2, satisfying:

(i) $f_2(x) = 0$ unless $x \in \bigcup_{k_1 < i \leq k_2} (N_i \cup \Lambda_i)$.

(ii) $0 \leq f_2(x) \leq \tau_i$, $\bigvee x \in N_i$, $\bigvee k_1 < i \leq k_2$.

(iii) $\bigvee x \in \Lambda_i$, $f_2(x) = \tau_i - t$, where $x = \varphi(t, x_i)$, $\bigvee k_1 < i \leq k_2$.

(iv) f_2 is constant on each Λ_i, $\bigvee i \neq k_1 + 1, \ldots, k_2$.

Having defined f_2, we now define a mapping h_2 from γ_1 onto an arc γ_2 by mapping $h_1(x)$ onto $h_2\big(h_1(x)\big) = \varphi\big(f_2(x), h_1(x)\big)$, which is the same as $\varphi\big(f_1(x) + f_2(x), x\big)$. Note that

$$h_2\big(h_1(\Lambda_i)\big) = x_i^+, \quad \bigvee 1 \leq i \leq k_1$$

and

$$h_2\big(h_1(\Lambda_i)\big) = h_1(x_i^+), \quad \bigvee k_1 < i \leq k_2.$$

We continue in this same manner: for each $n > 0$, there are at most finitely many values of i for which $2^{-n}\delta_1 < \tau_i \leq 2^{-n+1}\delta_1$. WoLOG we number these as $k_{n-1} < i \leq k_n$. Define the N_i, $\bigvee k_{n-1} < i \leq k_n$, as open subarcs of γ_0 satisfying

(i) The N_i are pairwise disjoint, $\bigvee k_{n-1} < i \leq k_n$,

(ii) No N_i contains x^- or x^+,

(iii) The N_i are disjoint from the Λ_j, $\bigvee k_{n-1} < i \leq k_n$, $\bigvee 1 \leq j \leq k_n$,

(iv) x_i is an endpoint of N_i, $\bigvee k_{n-1} < i \leq k_n$,

(v) $\bigvee x \in N_i$, $\bigvee 0 \leq t \leq \tau_i$, $d\big(\varphi(f_1(x) + \ldots + f_{n-1}(x) + t, x), y^+\big) < \varepsilon_1$,
$$\bigvee k_{n-1} < i \leq k_n.$$

Then we define f_n on γ_0 so that it is nonnegative and continuous and

(i) $f_n(x) = 0$ unless $x \in \bigcup_{k_{n-1} < i \leq k_n} (N_i \cup \Lambda_i)$,

(ii) $0 \leq f_n(x) \leq \tau_i$, $\bigvee k_{n-1} < i \leq k_n$, $\bigvee x \in N_i$,

(iii) $\bigvee x \in \Lambda_i$, $f_n(x) = \tau_i - t$, where $x = \varphi(t, x_i)$, $\bigvee k_{n-1} < i \leq k_n$.

(iv) f_n is constant on each Λ_i, $\bigvee i \neq k_{n-1} + 1, \ldots, k_n$.

Once f_n is defined, we obtain h_n from γ_{n-1} to a curve γ_n by setting

$$h_n\big(h_{n-1}(\ldots(h_1(x))\ldots)\big) = \varphi\big(f_1(x) + \ldots + f_n(x), x\big), \quad \bigvee x \in \gamma_0.$$

This definition of h_n completes the induction step of our construction.

Once the induction is complete, we define $f(x) = \sum_{n=1}^{\infty} f_n(x)$, $\forall \, x \in \gamma_0$, which converges since $0 < f_n(x) \leq 2^{-n+1} \delta_1$, $\forall \, n > 0$. Our construction assures that $\varphi(f(x), x) = \lim_{m \to \infty} \left(\varphi \left(\sum_{n=1}^{m} f_n(x), x \right) \right)$, $\forall \, x \in \gamma_0$, and since these last all lie within ε_1 of y^+, we have

$$d\left(\varphi(f(x), x), y^+ \right) \leq \varepsilon_1$$

and thus $\varrho\left(\varphi(f(x), x) \, y^+ \right) < \varepsilon$. Set $h(x) = \varphi(f(x), x)$, $\forall \, x \in \Omega$ and define $\gamma = \{ h(x) \mid x \in \gamma_0 \}$. We already know that for each $x \in \Omega \setminus \mathcal{O}_\varphi(x_0)$, $\mathcal{O}_\varphi(x)$ intersects γ_0 in an arc or a point. If a point, then $\mathcal{O}_\varphi(x)$ intersects γ in a point, namely $\varphi(f(\tilde{x}), \tilde{x})$, where $\{ \tilde{x} \} = \mathcal{O}_\varphi(x) \cap \gamma_0$. If $\mathcal{O}_\varphi(x)$ intersects γ_0 in an arc, then for some n (which depends on the φ-time measure of that arc), $\mathcal{O}_\varphi(x)$ intersects γ_n in a point, and thus $\mathcal{O}_\varphi(x)$ intersects γ in a single point, namely $\varphi(f(x_i), x_i)$ for an appropriately chosen x_i.

All that now remains is to show that γ is a curve. We will exhibit a homeomorphism g of $[0, 1]$ onto γ. First, we define an order-preserving continuous map g_0 of L onto $[0, 1]$. Assume WoLOG that $\mathscr{L}(L) = 1$, and define $g_0(x^-) = 0$ and $g_0(x^+) = 1$. Let $x_{1/2}$ be the midpoint of L, and define $g_0(x_{1/2}) = \frac{1}{2}$. If it should happen that $x_{1/2} \in [x_i, x_i^+]$, $\exists \, i$, then set $g_0(x) = \frac{1}{2}$, $\forall \, x \in [x_i, x_i^+]$. The removal of all points assigned the values 0, 1, or $\frac{1}{2}$ leaves in L two open intervals, each of length no more than $\frac{1}{2}$. Call the midpoints of these intervals $x_{1/4}$ and $x_{3/4}$ in such a way as to preserve the order, define $g_0(x_{1/4}) = \frac{1}{4}$ and $g_0(x_{3/4}) = \frac{3}{4}$, again extending the definition as before if either $x_{1/4}$ or $x_{3/4}$ should lie in one of the intervals $[x_i, x_i^+]$. The removal from L of the points where the values of g_0 have already been assigned leaves four open intervals, each having length no more then $\frac{1}{4}$. We continue in this manner, assigning dyadic rational values for g_0 to points or intervals chosen in the way specified. Note that every interval $[x_i, x_i^+]$ is eventually mapped to some dyadic interval, and in fact if $\mathscr{L}[x_i, x_i^+] \geq 2^{-n}$, then

$$g_0(x) = \frac{p}{2^n}, \quad \forall \, x \in [x_i, x_i^+], \quad \exists \, 1 \leq p \leq 2^n. \quad \text{The process described}$$

defines g_0 on a dense set in L and the function is order-preserving and uniformly continuous on that set. We can now extend g_0 to all of L by continuity.

Our next step is to define a function g_1 on γ_0. For each $x \in \gamma_0$, if $\mathcal{O}_\varphi(x) \cap \gamma_0 = \{ x \}$, then $x \in L$, and we set $g_1(x) = g_0(x)$. Otherwise, we have $x \in \Lambda_i$, $\exists \, i$, and we define $g_1(x) = g_0(x_i)$. It is immediate that g_1 is continuous and order-preserving, and takes constant values on each of the Λ_i. We now define $g(t) = h\big(g_1^{-1}(t)\big)$, $\forall\ 0 \leq t \leq 1$. We see at once that $\sup\limits_{p} \text{diam}\left(g_1^{-1}\left(\left[\dfrac{p}{2^n}, \dfrac{p+1}{2^n}\right]\right)\right)$ converges to 0 as $n \to \infty$. Since h is continuous, g is also continuous, and g maps $[0, 1]$ one-to-one onto γ. Thus g is a homeomorphism, and this theorem is proved. **QED**

10.3. Corollary. *Under the same hypotheses, there exists a Jordan curve J in Ω so that $\mathcal{O}_\varphi(x) \cap J$ is a single point for each $x \in \Omega$ (i.e. J is a cross-section for Ω.)*

PROOF: Let g be the homeomorphism of $[0, 1]$ onto the arc γ defined in the proof of Theorem 10.2. By the definition of g, we have $g(0) = x^-$ and $g(1) = x^+$, where $x^- = \varphi(\tau^-, x_0)$ and $x^+ = \varphi(\tau^+, x_0)$. We now define $f(t) = \varphi\big(t(\tau^- - \tau^+),\, g(t)\big)$, $\forall\ 0 \leq t \leq 1$. Let J be the image of $[0, 1]$ under f. Since $t(\tau^- - \tau^+)$ and $g(t)$ are continuous functions of t, arïd φ is continuous in its two variables, f is a continuous function. For each $x \in \Omega \setminus \mathcal{O}_\varphi(x_0)$, $\mathcal{O}_\varphi(x) \cap \gamma$ consists of exactly one point, and therefore the same is true of $\mathcal{O}_\varphi(x) \cap J$. Since $f(0) = f(1)$, $\mathcal{O}_\varphi(x) \cap J$ also consists of just one point. Since f is one-to-one except that $f(0) = f(1)$, J is a Jordon curve. **QED**

10.4. Corollary. *Let φ and Ω be as in Theorem 10.2 and Corollary 10.3. Let $\hat{\varphi}$ be the flow defined in $\mathbb{R} \times \Gamma_1$ by $\hat{\varphi}(t, \langle \tau, k \rangle) = \langle t + \tau, k \rangle$.*

Then φ is homeomorphic with $\hat{\varphi}$.

PROOF: Let $\bar{f}$ be the homeomorphism of Γ_1 onto the Jordan curve J of Corollary 10.3 defined by $\bar{f}(e^{2\pi i t}) = f(t)$, $\forall\ 0 \leq t \leq 1$, where f is the function defined in the proof of that corollary. Then for each $x \in \Omega$, $\mathcal{O}_\varphi(x) \cap J$ is a single point, which we denote as $f_1(x)$, and there is thus a unique real number, call it $t(x)$, such that $x = \varphi\big(t(x), f_1(x)\big)$. Define

$$F(x) = \big\langle t(x), \bar{f}^{-1}\big(f_1(x)\big)\big\rangle, \quad \forall\ x \in \Omega.$$

It is clear that F is a one-to-one mapping of Ω onto $\mathbb{R} \times \Gamma_1$, and that F^{-1} is continuous. Thus, F^{-1} is a homeomorphism on each compact cylinder $[-n, n] \times \Gamma_1$. Since the interiors of the images of these cylinders exhaust Ω, F is a homeomorphism. **QED**

As a result of the last corollary, we have established homeomorphism of all flows in spiral regions. It is not the case, however, that the flows on the spiral regions, *including their boundaries*, are homeomorphic. To begin with, it is obvious that the period of a periodic orbit is an invariant of homeomorphisms. Thus, if two flows defined in spiral regions have different periods on the boundaries, then they cannot be homeomorphic. However, even if the periods agree, there are still difficulties, as will be seen from the next theorem.

10.5. Theorem. *The flows constructed in Examples* 3.14 (b) *and* 3.14 (c) *are not homeomorphic.*

PROOF: Suppose that φ_1 and φ_2 are two flows defined on the annulus $A = \{z \mid 1 \leq |z| \leq 2\}$ with φ_1 of the type of Example 3.14 (b) and φ_2 of the type of Example 3.14 (c). Assume that h is a homeomorphism of A onto A satisfying $\varphi_2 = h[\varphi_1]$. Let z_1 and z_2 be chosen with $|z_1| = 1$, $|z_2| = 2$. The identification of $\varphi_1(t, z_1)$ with $\varphi_1(t, z_2)$ for all values of t maps A onto a torus T, and this mapping is one-to-one except for the indicated identification. Let the mapping of A onto T be denoted by f. Then $g = fh^{-1}$ is a continuous mapping of A onto T which is one-to-one except that

$$f\big(\varphi_1(t, z_1)\big) = f\big(\varphi_1(t, z_2)\big), \qquad \forall\, t \in \mathbb{R}.$$

Since $h(z_1)$ and $h(z_2)$ lie in opposite boundaries of A, one of the boundaries, say $\mathcal{O}_{\varphi_2}(h(z_1))$, moves clockwise with increasing values of t, while the other moves counter-clockwise. Thus the identification of $g\big(\varphi_2(t, h(z_1))\big)$ with $g\big(\varphi_2(t, h(z_2))\big)$, $\forall\, t \in \mathbb{R}$, gives us a Klein bottle, not a torus, *i.e.* $g(A) \neq f(A)$. It follows that there is no such homeomorphism h.

QED

In connection with the previous theorem, it is interesting to note that φ_1 and φ_2 are homeomorphic when restricted to int (A) or to either boundary of A, and that both flows are continuous to the boundary, but they are nonetheless not homeomorphic on A; in some sense, the boundaries are not properly attached. It is natural to conjecture that every pair of orientation-preserving spirals with the same periods on the boundaries are homeomorphic, but this too is false, as the next example shows.

10.6. Example. We begin with Example 3.14 (b), and we use the notation of that example, with $c_1 = 1$ and $c_2 = 0$. The flow φ constructed there, and a modification φ_1 of that flow, will have the properties that both are orientation-preserving spiral flows in the same annulus, that both have period 2π on the boundaries of the annulus, but they will not be homeomorphic. To construct φ_1, we first re-examine the flow ψ and modify it to yield a flow ψ_1 in the same strip. Let a continuously differentiable function g be defined in $\mathbb{R}$ so that

(i) $0 < g'(t) < 1$, $\qquad \forall\, t \in \mathbb{R}$,

(ii) $g(-t) = -g(t)$, $\qquad \forall\, t \in \mathbb{R}$,

(iii) $g(t) \to \infty$ as $t \to \infty$,

(iv) $g'(t) \downarrow 0$ as $t \to \infty$.

It is clear that such functions exist. For each point $i\sigma$ lying on the imaginary axis, we define $\psi_1(t, i\sigma) = \tan^{-1}(t + g(t)) + i(\sigma + t + g(t))$. At each point $\tau + i\sigma$ in the strip, the flow ψ is moving so that its velocity in the imaginary direction is 1, while the corresponding velocity for ψ_1 is $1 + g'(\tan(\tau))$. It follows directly that $\psi_1 = v\psi$, where v is a continuous function which is bounded and bounded away from 0. Thus, *a fortiori*, ψ_1 is a flow. The function v takes the value 1 on the boundaries of the strip, and so ψ_1 is the same as ψ there. The flow φ_1 will now be defined as $f[\psi_1]$. It is clear that $\varphi_1 = u\varphi$ for some continuous φ-multiplier u, and that φ_1 is an orientation-preserving spiral flow which has the required behavior on the boundaries. It is an interesting fact that although the flows ψ and ψ_1 are homeomorphic, their images $\varphi = f[\psi]$ and $\varphi_1 = f[\psi_1]$ are not. We will prove only the latter statement; the former is a routine exercise and also irrelevant to our example. In order to see that the two flows are not homeomorphic, let us select a point $i\sigma_0$ on the imaginary axis, and for each $t > 0$, denote $\psi_1(t, i\sigma_0)$ as $\tau_t + i\sigma_t$. Then, for each $t_0, t > 0$, we have

$$t_0\left(1 + g'(t + t_0)\right) < \sigma_{t+t_0} - \sigma_t < t_0\left(1 + g'(t)\right)$$

by the monotonicity of g'. Thus, the sequence $\{\mathrm{Im}(\psi_1(t + 2n\pi, i\sigma_0))\}$ is dense on the interval $[0, 2\pi]$ (mod 2π) for positive values of n, and also for negative values. It follows that for each point z of the annulus, the sequence $\{\varphi_1(2n\pi, z)\}$ has its cluster points dense in one boundary of the annulus as $n \to +\infty$ and in the other as $n \to -\infty$. By contrast, the sequence $\{\varphi(2n\pi, z)\}$ has a single limit point as $n \to +\infty$ and also as $n \to -\infty$, for each such z. Since these are properties which are invariant under homeomorphism, the example is established.

Example 10.6 still leaves the possibility that for any orientation-preserving spiral flow φ_1 in an annulus Ω, there is a homeomorphism h of $\bar{\Omega}$ onto A *and a φ-multiplier u* such that $h[\varphi_1] = u\varphi$, where φ is the flow of example 3.14(b) defined on the closed annulus A. This conjecture is, in fact, true, as the next series of theorems will establish.

10.7. Theorem. *Let Ω be an annulus bounded by two periodic φ-orbits $\mathcal{O}_\varphi(y^-)$ and $\mathcal{O}_\varphi(y^+)$, and satisfying $\mathrm{p}_\varphi(x) = \infty$ and $\omega_\varphi(x) = \mathcal{C}_\varphi(y^+)$, $\forall x \in \Omega$.*

Then for every ε and δ satisfying $\varepsilon > 0$ and $0 < \delta < \frac{1}{4} \mathrm{p}_\varphi(y^+)$, there exists an arc $\gamma \subseteq \Omega$ lying within ε of y^+ and having y^+ for its sole endpoint *at one end, which meets every orbit in Ω, and with the property that for every $x \in \gamma$ there is precisely one $t \in \mathbb{R}$ satisfying $\delta < t < 2p_\varphi(y^+) - \delta$ such that $\varphi(t, x) \in \gamma$. Furthermore, that value t must satisfy*

$$p_\varphi(y^+) - \delta < t < p_\varphi(y^+) + \delta.$$

PROOF: We choose x_0, ε_1, δ_1, ε_2, δ_2, *etc.* as in the proof of Theorem 10.2, but we set additional conditions on δ_1 and ε_2. We require that $\delta_1 \leq \tfrac{1}{6}\delta$, and define $B_1 = \{\varphi(t, y^+) \mid \delta_1 \leq t \leq p_\varphi(y^+) - \delta_1\}$. Then we choose ε_2 so that we never have both $-\delta_1 < t_1 < t_2 < t_3 < \delta_1$ and also $d(\varphi(t_1, x), y^+) < \varepsilon_2$, $d(\varphi(t_3, x), y^+) < \varepsilon_2$, and $d(\varphi(t_2, x), B_1) < \varepsilon_2$. This is certainly possible for ε_2 small enough, since the opposite implies $y^+ \in B_1$. Now choose ε_3 so small that $d(\varphi(t, x), \varphi(t, y)) < \tfrac{1}{2}\varepsilon_2$ for all x satisfying $d(x, y) < \varepsilon_3$, $y \in \mathcal{O}_\varphi(y^+)$, and $|t| < 2p_\varphi(y^+)$.

Now let J be such a Jordan curve as is specified in Corollary 10.3, and let Ω_n denote the annulus bounded by $\mathcal{O}_\varphi(y^+)$ and $\varphi(n, J)$. We note that $\overline{\Omega_n}$ is compact, while $\underset{n}{\cap} \overline{\Omega_n} = \mathcal{O}_\varphi(y^+)$, so that for n large enough, $\overline{\Omega_n}$ contains no point x for which $d(x, \mathcal{O}_\varphi(y^+)) \geq \varepsilon_3$. Thus, for each point x belonging to such an Ω_n, and for each positive interval $[a, b]$ whose length is at least $p_\varphi(y^+)$, the set $\{\varphi(t, x) \mid a \leq t \leq b\}$ contains at least one point lying within $\tfrac{1}{2}\varepsilon_2$ of y^+. Similarly, if $b - a > p_\varphi(y^+) + 2\delta_1$, then there is, in fact, a sub-interval $[c - \delta_1, c + \delta_1]$ of $[a, b]$ such that $d(\varphi(t, y^+), \varphi(c + t, x)) < \tfrac{1}{2}\varepsilon_2$, $\forall |t| \leq \delta_1$.

Assume now that we are working with a fixed value n_0 of n, meeting the above criteria, and consider Ω_{n_0}. WoLOG, we may assume that $n_0 = 0$. Define $\varepsilon_4 = d(J, \mathcal{O}_\varphi(y^+))$ and apply Theorem 10.2 to obtain a curve γ_0 lying within ε_4 of y^+ and having the required properties. It is clear that $\varepsilon_4 < \varepsilon_3$, and so $\gamma_0 \subseteq \Omega_0$. It will now be our purpose to produce continuous functions f_i, $\forall i = 1, 2, \ldots$, defined on γ_0 into $\mathbb{R}$ such that $p_\varphi(y^+) - \delta < f_i(x) - f_{i-1}(x) < p_\varphi(y^+) + \delta$, $\forall x \in \gamma_0$, $\forall i = 1, 2, \ldots$ in such a way that the image curves γ_i defined by

$$\gamma_i = \{\varphi(f_i(x), x) \mid x \in \gamma_0\}, \qquad \forall i = 1, 2, \ldots$$

join end to end to produce a half-open arc which has y^+ for an endpoint.[1] For each $x \in \gamma_0$, we observe that $\varphi(p_\varphi(y^+), x)$ lies within $\tfrac{1}{2}\varepsilon_2$ of y^+, and thus the minimum value of $d(\varphi(t, x), y^+)$ for values of t between $p_\varphi(y^+) - \delta_1$ and $p_\varphi(y^+) + \delta_1$ is less than $\tfrac{1}{2}\varepsilon_2$. Since for all other values of t between δ_1 and $2p_\varphi(y^+) - \delta_1$, $d(\varphi(t, x), y^+) > \varepsilon_2$, this minimum applies as well for the longer time interval. We denote the minimum

---

[1] We define $f_0(x) = 0$, $\forall x \in \gamma_0$.

distance as $m_1(x)$, and define the number $d_1(x)$ as the least value of t which exceeds δ_1 for which $d(\varphi(d_1(x), x), y^+) = m_1(x)$. Then, as we have noted, $\mathrm{p}_\varphi(y^+) - \delta_1 < d_1(x) < \mathrm{p}_\varphi(y^+) + \delta_1$, $\forall\ x\ \epsilon\ \gamma_0$. We also define $a_1(x)$ as the greatest value of t less than $d_1(x)$ for which $d(\varphi(a_1(x), x), y^+) = \varepsilon_2$. As above, $\mathrm{p}_\varphi(y^+) - \delta_1 < a_1(x)$, $\forall\ x\ \epsilon\ \gamma_0$. We now define $\ \ W_1 = \{\varphi(t, x) \mid x\ \epsilon\ \gamma_0,\ \ \mathrm{p}_\varphi(y^+) - \delta_1 \leq t \leq \mathrm{p}_\varphi(y^+) + \delta_1\}$, $V_1 = \{\varphi(t, x) \mid x\ \epsilon\ \gamma_0,\ \mathrm{p}_\varphi(y^+) - \delta_1 \leq t < d_1(x)\}$, and $U_1 = \{\varphi(t, x) \mid x\ \epsilon\ \gamma_0,\ a_1(x) < t \leq \mathrm{p}_\varphi(y^+) + \delta_1\}$. It is clear that m_1 is a continuous function, so that if $d(\varphi(t, x), y^+) > m_1(x)$ for a particular x and all t satisfying $\mathrm{p}_\varphi(y^+) - \delta_1 \leq t \leq t_0$, then the same must hold for nearby values of x and t_0. That is, V_1 is open in W_1. By a similar argument, U_1 is open in W_1.

We now define the "strip" $\hat{U} = \mathbb{R} \times \gamma_0$, and the flow $\hat{\phi}$ in $\hat{U}$ by $\hat{\phi}(t, \langle\tau, x\rangle) = \langle t + \tau, x\rangle$, $\forall\ x\ \epsilon\ \gamma_0$, $\forall\ t, \tau\ \epsilon\ \mathbb{R}$. We observe that φ maps $\hat{U}$ continuously into $\mathbb{S}^2$, so that the sets $\{\langle t, x\rangle \mid x\ \epsilon\ \gamma_0, t \geq d_1(x)\}$ and $\{\langle t, x\rangle \mid x\ \epsilon\ \gamma_0, t \leq a_1(x)\}$ are closed in $\hat{U}$. Since $\hat{U}$ is a metrizable space, we can select a continuous $\hat{\phi}$-multiplier $\hat{u}_1$ defined on $\hat{U}$ so that $\hat{u}_1(\langle t, x\rangle) = 1$ whenever $t \leq a_1(x)$, $\hat{u}_1(\langle t, x\rangle) = 0$ whenever $t \geq d_1(x)$, and $0 < \hat{u}_1(\langle t, x\rangle) < 1$ otherwise. We define $\hat{\psi}_1 = \hat{u}_1\hat{\phi}$.

We now define $m_1(\gamma_0) = \min \{m_1(x) \mid x\ \epsilon\ \gamma_0\}$, and denote by the symbol $\partial(V_1)$ the boundary of V_1 in W_1. For each $\ \ x\ \epsilon\ \gamma_0$, we define $b_1(x)$ as the greatest real number less than $d_1(x)$ for which $d(\varphi(b_1(x), x), \partial(V_1)) = m_1(\gamma_0)$. By the same reasoning used previously, we see that the set $U_{0,1} = \{\varphi(t, x) \mid x\ \epsilon\ \gamma_0,\ b_1(x) < t < \mathrm{p}_\varphi(y^+) + \delta_1\}$ is open in W_1. By the definition of $\hat{u}_1$, there must be a positive number r_1 such that $\{\varphi(t, x) \mid 0 < \hat{u}_1(\langle t, x\rangle) < r_1\} \subseteq U_{0,1}$. It follows immediately from the definitions of $\hat{u}_1$ and $\hat{\psi}_1$ that whenever

$$\tau_1 > \mathrm{p}_\varphi(y^+) - \delta_1 + \frac{2\delta_1}{r_1}, \quad \text{then} \quad b_1(x) < c_1(x) < d_1(x), \quad \forall\ x\ \epsilon\ \gamma_0,$$

where $c_1(x)$ is defined by $\hat{\psi}_1(\tau_1, \langle 0, x\rangle) = \langle c_1(x), x\rangle$, $\forall\ x\ \epsilon\ \gamma_0$. We choose such a τ_1. Then the curve $\gamma_{01} = \{\varphi(c_1(x), x) \mid x\ \epsilon\ \gamma_0\}$ lies within $m_1(\gamma_0)$ of $\partial(V_1)$, which in turn lies rather close to y^+ (we shall see how close in a moment).

By the additional condition which we imposed on ε_2, we will now show that no point which lies within ε_2 of B_1 can be in $\partial(V_1)$. Indeed, assume that $\varphi(\bar{t}, \bar{x})$ is such a point, where $\bar{x}\ \epsilon\ \gamma_0$. Then $d_1(\bar{x}) \leq \bar{t} \leq \mathrm{p}_\varphi(y^+) + \delta_1$, since $\varphi(\bar{t}, \bar{x}) \notin V_1$. We have $d(\varphi(\bar{t}, \bar{x}), B_1) < \varepsilon_2$ and $d(\varphi(d_1(x), x), y^+) = m_1(x) < \frac{1}{2}\varepsilon_2$, $\forall\ x\ \epsilon\ \gamma_0$. We can find $\tilde{x}\ \epsilon\ \gamma_0$ and $\tilde{t}$ satisfying $\mathrm{p}_\varphi(y^+) - \delta_1 \leq \tilde{t} < d_1(\tilde{x})$ so that $\varphi(\tilde{t}, \tilde{x})$ is as close as we please to $\varphi(\bar{t}, \bar{x})$, while still lying in V_1. We choose it close enough so that $d(\varphi(\tilde{t}, \tilde{x}), B_1) < \varepsilon_2$ and also $d(\varphi(\tilde{t} - \bar{t} + d_1(\tilde{x}), \tilde{x}), y^+) < \varepsilon_2$. Since $\mathrm{p}_\varphi(y^+) - \delta_1 < \tilde{t} - \bar{t} + d_1(\tilde{x}) < \tilde{t} < d_1(\tilde{x}) < \mathrm{p}_\varphi(y^+) + \delta_1$ and

$d\big(\varphi(d_1(\tilde{x}), \tilde{x}), y^+\big) < \varepsilon_2$, we have derived a contradiction to the defining property of ε_2. Now we can see that no point of γ_{01} lies within $\frac{1}{2}\varepsilon_2$ of B_1, since each such point lies within $\frac{1}{2}\varepsilon_2$ of $\partial(V_1)$. Thus, each point of γ_{01} lies within ε_3 of a point $\varphi(t, y^+)$ where $|t| < \delta_1$, since each point of A lies within ε_3 of *some* point of $\mathcal{O}_\varphi(y^+)$, and $\varepsilon_3 < \frac{1}{2}\varepsilon_2$.

Our next step is to repeat the construction. Since $\varphi(c_1(x), x)$ lies within ε_2 of some point $\varphi(t, y^+)$ with $|t| \leq \delta_1$, we define $m_2(x)$ as the minimum distance between $\varphi(c_1(x) + t, x)$ and y^+, where $p_\varphi(y^+) - 2\delta_1 < t < p_\varphi(y^+) + 2\delta_1$, $\forall\, x \in \gamma_0$. Then we define $d_2(x)$ as the least real number greater than $p_\varphi(y^+) - 2\delta_1$ for which $d\big(\varphi(c_1(x) + d_2(x), x), y^+\big) = m_2(x)$. Proceeding in a like manner, we define $a_2(x)$, $b_2(x)$, and $c_2(x)$, and in the process, we define $m_2(\gamma_0)$ and τ_2. In the definition of τ_2, however, we add another condition. We note that the points $\varphi(d_1(x^+), x^+)$ and $\varphi(c_1(x^-) + d_2(x^-), x^-)$ are in fact the same point, so that if $c_2(x^-)$ is large enough, then $\varphi(c_1(x^-) + c_2(x^-), x^-)$ will be a point of the arc $\{\varphi(t, x^+) \mid c_1(x^+) < t < d_1(x^+)\}$. We choose τ_2 large enough so that this condition on $c_2(x^-)$ is also satisfied, and define the curve γ_{02} by $\gamma_{02} = \{\varphi(c_1(x) + c_2(x), x) \mid x \in \gamma_0\}$. Now we can define f_1 to be any nonnegative continuous function satisfying $f_1(x^-) = c_1(x^-)$, $c_1(x) \leq f_1(x) < d_1(x)$, $\forall\, x \in \gamma_0$, and $\varphi(f_1(x^+), x^+) = \varphi(c_1(x^-) + c_2(x^-), x^-)$. We define γ_1 by $\gamma_1 = \{\varphi(f_1(x), x) \mid x \in \gamma_0\}$. Observe once again that every point of γ_{02} lies within ε_2 of $\varphi(t, y^+)$, $\exists\, |t| \leq \delta_1$.

When we have carried through the nth iteration of this process, we will have defined the continuous function c_n defined on γ_0, and we will have the curve $\gamma_{0n} = \{\varphi(t, x) \mid x \in \gamma_0,\ t = c_1(x) + \ldots + c_n(x)\}$ and this curve will lie within ε_2 of $\{\varphi(t, y^+) \mid |t| \leq \delta_1\}$. Now, we can increase $c_1 + \ldots + c_{n-1}$ to a continuous function f_{n-1} satisfying

$$f_{n-1}(x^-) = c_1(x^-) + \ldots + c_{n-1}(x^-),$$

$$c_1(x) + \ldots + c_{n-1}(x) \leq f_{n-1}(x) \leq c_1(x) + \ldots + c_{n-2}(x) + d_{n-1}(x), \quad \forall\, x \in \gamma_0,$$

and

$$\varphi(f_{n-1}(x^+), x^+) = \varphi(c_1(x^-) + \ldots + c_{n-1}(x^-), x^-).$$

Now again, we define $m_{n+1}(x)$ as the minimum distance between y^+ and $\varphi(c_1(x) + \ldots + c_n(x) + t, x)$ for $p_\varphi(y^+) - 2\delta_1 \leq t \leq p_\varphi(y^+) + 2\delta_1$, $\forall\, x \in \gamma_0$. We then define d_{n+1}, c_{n+1}, etc. as before, and continue inductively. For each n, we obtain a continuous function f_n and use it to define the curve γ_n. The γ_n link up, end to end. We will now show that their union is an arc joining $\varphi(f_1(x^-), x^-)$ to y^+.

To begin with, we observe that $\gamma_1 \subseteq \operatorname{ins} \varphi(p_\varphi(y^+) - \delta_1, J)$ and thus, *a fortiori*, $\gamma_1 \subseteq \operatorname{ins} \varphi(\frac{1}{2} p_\varphi(y^+), J)$. In the same way, $\gamma_2 \subseteq \operatorname{ins} \varphi(p_\varphi(y^+), J)$, and in general, $\gamma_i \subseteq \operatorname{ins} \varphi(\tfrac{1}{2} i \, \mathrm{p}_\varphi(y^+), J)$, $\forall \; i = 1, 2, \ldots$. It follows immediately that $m_i(\gamma_0) \to 0$ as $i \to \infty$. Thus we see that y^+ is a limit point of the half-open arc γ. We now assume that there were another point $\bar{y}$ which could be a limit point of γ at the same end as y^+. We see at once that $\bar{y} \in \mathcal{O}_\varphi(y^+)$. Let $\bar{y} = \varphi(t_0, y^+)$, where t_0 is so chosen that $|t_0| < \delta_1$.

We choose $\varepsilon_5 < d(\bar{y}, y^+)$, and so small that we can never find an $x \in \Omega$ and numbers t_1, t_2, and t_3 satisfying $-\delta_1 < t_1 < t_2 < t_3 < \delta_1$ for which $d(\varphi(t_1, x), y^+) < \varepsilon_5$, $d(\varphi(t_3, x), y^+) < \varepsilon_5$, but $d(\varphi(t_2, x), \bar{y}) < \varepsilon_5$. Then we define ε_6 so small that for all $y \in \mathcal{O}_\varphi(y^+)$, and all t satisfying $|t| < \mathrm{p}_\varphi(y^+)$, we have $d(\varphi(t, y), \varphi(t, x)) < \tfrac{1}{2}\varepsilon_5$ if $d(x, y) < \varepsilon_6$. We now observe that for i large enough, say $i > N$, we have $d(x, \mathcal{O}_\varphi(y^+)) < \varepsilon_6$, $\forall \; x \in \gamma_i$. Since $\bar{y}$ is, by assumption, an endpoint of γ, we choose a point from some γ_j, where $j > N$, which lies within ε_6 of $\bar{y}$. Let us denote this point as $\varphi(\bar{t}, \bar{x})$, where $\bar{x} \in \gamma_j$. Then $y^+ = \varphi(-t_0, \bar{y})$, so that $\varphi(\bar{t} - t_0, \bar{x})$ lies within $\tfrac{1}{2}\varepsilon_5$ of y^+. Our assumption on ε_5 assures us that no point of $\partial(V_j)$ lies within ε_5 of $\bar{y}$, in the same way that the similar assumption on ε_2 assured that $\partial(V_i)$ lies far from B_1. Also, we know that $m_j(\gamma_0) < \tfrac{1}{2}\varepsilon_5$. However, our assumptions on f_j are such that $d(\varphi(f_j(\bar{x}), \bar{x}), \partial(V_j)) < m_j(\gamma_0) < \tfrac{1}{2}\varepsilon_5$, while $d(\varphi(\bar{t}, \bar{x}), \bar{y}) < \varepsilon_6 < \tfrac{1}{2}\varepsilon_5$. Since $f_j(\bar{x}) = \bar{t}$, we have $d(y, \partial(V_j)) < \varepsilon_5$, which is a contradiction. Thus we see that γ joins $\varphi(f_1(x^-), x^-)$ to y^+ alone. We now observe that for each i, we have

$$\mathrm{p}_\varphi(y^+) - 2\delta_1 \leq c_i(x) < f_i(x) - \big(c_1(x) + \ldots + c_{i-1}(x)\big)$$

$$< d_i(x) \leq \mathrm{p}_\varphi(y^+) + 2\delta_1, \qquad \forall \; x \in \gamma_0.$$

Thus, $\mathrm{p}_\varphi(y^+) - 6\delta_1 \leq c_i(x) - \big(d_{i-1}(x) - c_{i-1}(x)\big) < f_i(x) - f_{i-1}(x)$

$$< d_i(x) \leq \mathrm{p}_\varphi(y^+) + 2\delta_1,$$

and this completes the proof, since $6\delta_1 < \delta$.   **QED**

10.8. Lemma. *Let $\varepsilon > 0$, and let γ be any curve satisfying Theorem 10.7. For each $x \in \gamma$, let $f(x)$ be the smallest positive number t satisfying $\varphi(t, x) \in \gamma$.*

Then f is continuous in γ, and in particular, $f(x) \to \mathrm{p}_\varphi(y^+)$ as $x \to y^+$ in γ.

PROOF: Suppose that $x_i \to x_0 \neq y^+$ in γ. Then x_0 divides γ into two parts, and we can assume WoLOG that all the x_i lie in one of these parts. We see immediately that all but finitely many of the x_i belong to one of the γ_k, and we can assume again that all the x_i are in γ_j, $\forall \; i = 1, 2, \ldots$. Then for each x_i, we have an $\bar{x}_i \in \gamma_0$ such that $x_i = \varphi(f_j(\bar{x}_i), \bar{x}_i)$,

$\forall\, i = 0, 1, 2, \ldots$. Since the mapping $x \to \varphi\big(f_j(x), x\big)$ is one-to-one and continuous from γ_0 (which is compact) onto γ_j, it is a homeomorphism, so that $\tilde{x}_i \to \tilde{x}_0$, and $f_k(\tilde{x}_i) \to f_k(\tilde{x}_0)$, $\forall\, k = 1, 2, \ldots$. Specifically, this convergence holds for $k = j, j+1$, so that

$$f(x_i) = f_{j+1}(\tilde{x}_i) - f_j(\tilde{x}_i) \to f_{j+1}(\tilde{x}_0) - f_j(\tilde{x}_0) = f(x_0).$$

If $x_i \to y^+$ in γ, and $f(x_i) \to p_\varphi(y^+)$, then there is a subsequence such that $f(x_{ij}) \to t \neq p_\varphi(y^+)$. Since in any case,

$$p_\varphi(y^+) - \delta < f(x_{ij}) < p_\varphi(y^+) + \delta,$$

we have $|t - p_\varphi(y^+)| \leq \delta$. Then $\varphi\big(f(x_{ij}), x_{ij}\big) \to \varphi(t, y^+) \neq y^+$, which is impossible since $\varphi\big(f(x_{ij}), x_{ij}\big) \in \gamma$ and $\varphi(t, y^+) \in \mathcal{O}_\varphi(y^+)$.   **QED**

10.9. Lemma. *Let φ and Ω be as in Theorem 10.2 and let $\hat{\varphi}$ be the flow defined in Example 3.14 (b). Let $r_1 < r < r_2$, and define $\hat{A}_r = \{z \mid r \leq |z| \leq r_2\}$.*

Then there is a $\hat{\varphi}$-multiplier $\hat{u}$, a Jordan curve J in Ω, and a homeomorphism h from the closed annulus Ω_r lying between J and $\mathcal{O}_\varphi(y^+)$ onto $\hat{A}_r$ such that

$$h\big(\varphi(t, x)\big) = \hat{u}\hat{\varphi}\big(t, h(x)\big), \quad \forall\, t \geq 0, \quad \forall\, x \in \Omega_r.$$

PROOF: Let J_0 be the Jordan curve constructed in Corollary 10.3 and let t_0 be chosen so that $\varphi(t_0, J_0) \cap \gamma \neq \square$, where γ is some curve satisfying Theorem 10.7. The portion $\bar{\gamma}$ of γ which is the minimal continuum from $\mathcal{O}_\varphi(y^+)$ to $J = \varphi(t_0, J_0)$ has the same properties as γ, and we assume WoLOG that $\bar{\gamma} = \gamma$. We define the homeomorphism g of γ into itself by $g(x) = \varphi\big(f(x), x\big)$, $\forall\, x \in \gamma$, where f is the function defined in Theorem 10.7. We denote as y_0 the intersection of γ and J, and set $y_n = g^n(y_0)$. The closed arc of γ between y_{n-1} and y_n we denote as γ_n, $\forall\, n = 1, 2, \ldots$ In a similar way, we denote the interval $[r, r_2]$ as $\hat{\gamma}$, and the interval $\big[\hat{\varphi}(2\pi(n-1), r), \hat{\varphi}(2\pi n, r)\big]$ as $\hat{\gamma}_n$. Let the homeomorphism h be defined from J onto Γ_r so that $h(y_0) = r$, and so that the orientation is correct, in a sense to be defined later. For each $x \in J$, let $\bar{i}(x)$ be the least nonnegative real number satisfying $\varphi\big(\bar{i}(x), x\big) \in \gamma$, and set $H(x) = \varphi\big(\bar{i}(x), x\big)$. Similarly, for all $\tilde{x}$ of the form $\varphi(t, x)$, $\forall\, x \in J$ $0 \leq t \leq \bar{i}(x)$, we set $\bar{i}(\tilde{x}) = \bar{i}(x) - t$, and $H(\tilde{x}) = H(x)$. For all other $x \in \Omega_r \setminus \mathcal{O}_\varphi(y^+)$, we can write $x = \varphi\big(t(x), G(x)\big)$, where $G(x) \in \gamma$ and $0 \leq t(x) < f\big(G(x)\big)$. For each $x \in \gamma_1$, other than y_1, we have an image for x in $\hat{\gamma}_1$, defined in the following way: Let $x' \in J$ be chosen so that $x = \varphi\big(\bar{i}(x'), x'\big)$, map x' into Γ_r by h, and then map that image into $\hat{\varphi}\big(\hat{i}(x'), h(x')\big)$, where $\hat{i}(x')$ is the least nonnegative number s for which $\hat{\varphi}\big(s, h(x')\big)$ lies in $\mathbb{R}^+$. The orientation of h should be so chosen that for $x \in J$, $\hat{\imath}(x)$ and $\bar{\imath}(x)$ are close to 0 on the same side of y_0. Next we extend h to all the points $\varphi(t, x)$, where $x \in J$, and $0 < t < \bar{\imath}(x)$. We define $h\big(\varphi(t, x)\big) = \hat{\phi}\left(t \dfrac{\hat{\imath}(x)}{\bar{\imath}(x)}, h(x)\right)$.

We now define h on the remainder of γ by defining it on each γ_n, $n = 2, 3, \ldots$. We see that g^{-n} maps γ_{n+1} onto γ_1. Thus, we define $h(x) = \hat{\phi}\big(2n\pi, h(g^{-n}(x))\big)$, $\forall\, x \in \gamma_{n+1}$, $\forall\, n = 1, 2, \ldots$. Now for all the remaining points of $\Omega_r \setminus \mathcal{O}_\varphi(y^+)$,

$$h(x) = \hat{\phi}\left(\frac{2\pi t(x)}{\bar{\imath}(G(x))}, h(G(x))\right).$$

For the point $x \in \mathcal{O}_\varphi(y^+)$, we write $x = \varphi\big(t(x), y^+\big)$ and then define $h(x) = \hat{\phi}\left(\dfrac{2\pi t(x)}{\mathrm{p}_\varphi(y^+)}, r_2\right)$. We have now defined h for all the points of Ω_r. Despite the complexity of the definition, reference to a sketch will make it abundantly clear what the definition of h is and also that h is a homeomorphism of Ω_r onto $\hat{A}_r$.

We now define $\hat{u}$. For each z with $r_1 \leq |z| < r$, we set $\hat{u}(z) = 1$. For each point $h\big(\varphi(t, x)\big)$ with $x \in J$ and $0 \leq t < \bar{\imath}(x)$, we set $\hat{u}\big(h(\varphi(t, x))\big) = \hat{\imath}(x) \,/\, \bar{\imath}(x)$. For each point $h\big(\varphi(t, x)\big)$ with $x \in \gamma$ and $0 \leq t < \bar{\imath}(x)$, we set $\hat{u}\big(h(\varphi(t, x))\big) = 2\pi/\bar{\imath}(x)$. For $|z| = r_2$, we set $\hat{u}(z) = \dfrac{2\pi}{\mathrm{p}_\varphi(y^+)}$.

An elementary calculation will show that

$$h\big(\varphi(t, x)\big) = \hat{u}\hat{\phi}\big(t, h(x)\big), \quad \forall\, t \geq 0, \ \forall\, x \in \Omega_r. \qquad \textbf{QED}$$

10.10. Lemma. *The function $\bar{\imath}$ defined in the proof of Lemma* 10.9 *is continuous in* $J \setminus \{y_0\}$.

PROOF: Let $x_i \to x_0 \neq y_0$, where $x_i \in J \setminus \{y_0\}$, $\forall\, i = 1, 2, \ldots$. We want to show that $\bar{\imath}(x_i) \to \bar{\imath}(x_0)$. We note first that $\{\bar{\imath}(x_i)\}$ is bounded, for otherwise $\big\{\varphi(\bar{\imath}(x_i), x_i)\big\}$ would have a cluster point in $\partial(\Omega)$. It is now clear that $\bar{\imath}(x_i) \to \bar{\imath}(x_0)$, for it not, $\{\bar{\imath}(x_i)\}$ would have a cluster point $s \neq \bar{\imath}(x_0)$, and we would thus obtain two different points $\varphi(s, x_0)$ and $\varphi\big(\bar{\imath}(x_0), x_0\big)$ in $\gamma_1 \cap \mathcal{O}_\varphi(x_0)$. $\qquad$ **QED**

Our next theorem involves the replacement of the $\hat{\phi}$-multiplier $\hat{u}$ in Lemma 10.9 by a *continuous* $\hat{\phi}$-multiplier. To accomplish this, we will have use of a generalization of part of the Implicit Function Theorem:

10.11. Theorem. *Let* $f(x_1, \ldots, x_n)$ *be a function continuous in its n real arguments together. Assume that* $f(x_1, \ldots, x_n)$ *is strictly monotonic in* x_1

for each choice of $x_2, \ldots, x_n$ *(a special case is when* $\partial f/\partial x_1$ *exists and is positive everywhere).*

Then for each $n-1$ *tuple* $(x_2, \ldots, x_n)$, *there is at most one value of* x_1 *for which* $f(x_1, x_2, \ldots, x_n) = 0$, *and setting* $x_1 = g(x_2, \ldots, x_n)$, *we obtain that g is continuous wherever it is defined.*

PROOF: We assume WoLOG that f is strictly *increasing* in x_1 for each $x_2, \ldots, x_n$. If the theorem were false, we could find an $\varepsilon > 0$ and a sequence $(x_{i,2}, \ldots, x_{i,n})$ of $n-1$ tuples so that $x_{i,j} \to x_j, \; \forall \; j = 2, \ldots, n$ as $i \to \infty$, while

$$|g(x_{i,2}, \ldots, x_{i,n}) - g(x_2, \ldots, x_n)| > \varepsilon, \qquad \forall \; i = 1, 2, \ldots.$$

Set $g(x_{i,2}, \ldots, x_{i,n}) = x_{i,1}, \;\; \forall \; i = 1, 2, \ldots,$ and $g(x_2, \ldots, x_n) = x_1$. WoLOG, we assume that $x_{i,1} > x_1 + \varepsilon, \;\; \forall \; i = 1, 2, \ldots.$ Let $f(x_1 + \varepsilon, x_2, \ldots, x_n) - f(x_1, x_2, \ldots, x_n) = \eta > 0$, and let $\delta > 0$ be so chosen that the conditions

$$|x_j - \xi_j| < \delta, \qquad \forall \; j = 2, \ldots, n$$

and

$$|x_1 + \varepsilon - \xi_1| < \delta$$

together imply that

$$|f(x_1 + \varepsilon, x_2, x_3, \ldots, x_n) - f(\xi_1, \xi_2, \ldots, \xi_n)| < \tfrac{1}{2}\eta.$$

Now choose i so large that $|x_{i,j} - x_j| < \delta, \; \forall \; j = 2, \ldots, n$. We then have

$$0 = f(x_{i,1}, x_{i,2}, \ldots, x_{i,n})$$
$$\geq f(x_1 + \varepsilon, x_{i,2}, \ldots, x_{i,n})$$
$$> f(x_1 + \varepsilon, x_2, \ldots, x_n) - \tfrac{1}{2}\eta$$
$$= f(x_1, x_2, \ldots, x_n) + \eta - \tfrac{1}{2}\eta = \tfrac{1}{2}\eta > 0,$$

since $f(x_1, x_2, \ldots, x_n) = 0$, but this is a contradiction.   **QED**

10.12. Theorem. *The $\hat{\varphi}$-multiplier $\hat{u}$ in Theorem* 10.9 *can be replaced by a $\hat{\varphi}$-multiplier $\hat{v}$ continuous in $\hat{A}_r$.*

PROOF: We begin by modifying J, h and $\hat{u}$ so that the modified function, $\hat{u}_1$ will take a constant value on the new image $h'(J')$ of the modification J' of J. We want that value to be $\hat{u}(h(y_0)) = \hat{u}(r)$, and so we define J' by mapping each point $x \in J$ onto the point $E(x) = \varphi(\bar{i}(x) - (\bar{i}(x)/\hat{u}(r)), x)$. Then J' is a continuous one-to-one image of J and the φ-time taken for any point $E(x)$ of J' to reach γ is exactly $\big(\hat{t}(x)/\hat{u}(r)\big)$. Let us now modify h to h' in the following way. $E(x)$ is mapped onto $h(x)$, and $\varphi\big((t/\hat{u}(r)),E(x)\big)$ is mapped onto $\hat{\varphi}\big(t,h(x)\big)$, $\forall\, 0\le t\le \hat{t}(x)$, $\forall\, x\in J$. Let $\hat{u}_1$ be the same as $\hat{u}$ except that for all $x\in J$, $\hat{u}_1\big(\hat{\varphi}(t,h(x))\big)=\hat{u}(r)$, $\forall\, 0\le t\le \hat{t}(x)$. Let $\Omega_r{}'$ denote the closed annulus between J' and $\mathcal{O}_\varphi(y^+)$. Define $\hat{u}_1(z)=1$, $\forall\, z\in\Omega\setminus\Omega_r{}'$. Then we see easily that with this new definition,

$$h'\big(\varphi(t,x)\big)=\hat{u}_1\hat{\varphi}\big(t,h'(x)\big),\qquad \forall\, t\ge 0,\qquad \forall\, x\in\Omega_r{}'.$$

Furthermore, $\hat{u}_1$ is continuous in $\hat{A}_r$ except possibly along the positive real axis, but in any case, is continuous at r and at r_2. We begin by altering $\hat{u}_1$ to be continuous in $\hat{A}_r$, and to have certain other properties mentioned below. For $z\in\hat{\gamma}\setminus\hat{\gamma}_1$, $\hat{u}\big(\hat{\varphi}(t,z)\big)$ takes one value for $-2\pi\le t<0$ and another for $0\le t<2\pi$. We call this first value $c(z)$ and the second $d(z)$. For $z\in\hat{\gamma}_1$, we set $c(z)=\hat{u}(y_0)$ and $d(z)$ is the value that $\hat{u}\big(\hat{\varphi}(t,z)\big)$ assumes for $0\le t<2\pi$. In fact, $c(z)$ and $d(z)$ are the values which $\hat{u}_1$ assumes along $\mathcal{O}_{\hat{\varphi}}(z)$ in the vicinity of z. We will first generate a function $\hat{v}_1$ by altering $\hat{u}_1$ only in a neighborhood of $\hat{\gamma}$ in $\hat{A}_r$, and even in $\hat{A}_r$ we will deal principally with those points where $c(z)\ne d(z)$.

For each value of z satisfying the inequality, we will change $\hat{u}_1$ to $\hat{v}_1$ in the arc $\{\hat{\varphi}(t,z)\mid -a(z)<t<b(z)\}$, where the functions a and b are yet to be defined, in such a way that $\hat{v}_1$ is continuous and also

$$\int\limits_{-a(z)}^{b(z)}\frac{dt}{\hat{v}_1\big(\hat{\varphi}(t,z)\big)}=\int\limits_{-a(z)}^{b(z)}\frac{dt}{\hat{u}_1\big(\hat{\varphi}(t,z)\big)}.$$

Refering to the proof of Theorem 10.9, we note that $\forall\, x\in\gamma$, $\mathrm{p}_\varphi(y^+)-\delta_1\le f(x)\le \mathrm{p}_\varphi(y^+)+\delta_1$, so that

$$\frac{2\pi}{\mathrm{p}_\varphi(y^+)+\delta_1}\le \hat{u}_1(z)\le \frac{2\pi}{\mathrm{p}_\varphi(y^+)-\delta_1},\qquad \forall\, z\in\hat{A}_r.$$

We now choose any neighborhood N of γ in $\Omega_r{}'$ contained in the set $\{\varphi(t,x)\mid x\in\gamma,|t|<\delta_1\}$. There is an $\hat{\varepsilon}>0$ so small that the $\hat{\varepsilon}$-neighborhood of $\hat{\gamma}$ in $\hat{A}_r$ is contained in $h'(N)$. If $z\in\hat{A}_r\cap\mathbb{R}^+$ and $|t|<\dfrac{\hat{\varepsilon}}{r_2}$, then $\hat{\varphi}(t,z)\in N_d(\hat{\varepsilon},\hat{A}_r\cap\mathbb{R}^+)\subseteq h'(N)$. Thus, if $|t|<\varepsilon_2=\dfrac{\hat{\varepsilon}}{r_2}\cdot\dfrac{\mathrm{p}_\varphi(y^+)-\delta_1}{2\pi}$, then $\hat{u}\hat{\varphi}(t,z)\in h'(N)$, $\forall\, z\in\hat{A}_r\cap\mathbb{R}^+$. We choose a and b to be continuous nonnegative real functions defined on $\hat{p}$, both of which are bounded by $|d - c|$ and ε_2 with both of them 0 exactly at r and at r_2 and at all the points where c and d are equal, and finally, $a(z)$ is always small enough so that $\hat{\varphi}\big(-a(z), z\big) \in \hat{A}_r$.

We observe that the $\hat{u}\hat{\varphi}$-time measure of the arc

$$\gamma(z) = \{\hat{\varphi}(t, z) \mid -a(z) \le t \le b(z)\}$$

is

$$\frac{a(z)}{c(z)} + \frac{b(z)}{d(z)}.$$

We define $w(\beta, t, a, b, c, d) = c + (d - c)\left(\dfrac{t + a}{b + a}\right)^{\beta}$ for all positive values of β, a, b, c, and d, and all values of t between $-a$ and b. By choosing values of β close enough to 0, we can make $w(\beta, t, a, b, c, d)$ so close to d for so many values of $t \in [a, b]$ that $\displaystyle\int_{-a}^{b} \frac{dt}{w(\beta, t, a, b, c, d)}$ is as close to $\dfrac{b + a}{d}$ as we please. Similarly, sufficiently large values of β will make the same integral as close as we like to $\dfrac{b + a}{c}$. In any case $w(\beta, t, a, b, c, d)$ lies between c and d for all allowable values of the arguments, and when $c \neq d$, the inequality is proper. We define

$$F(\beta, a, b, c, d) = \int_{-a}^{b} \frac{dt}{w(\beta, t, a, b, c, d)} - \left(\frac{a}{c} + \frac{b}{d}\right).$$

Then F is continuous in all its variables, and monotonic in β, increasing in β if $d > c$ and decreasing if $d < c$. The set

$$D = \{\hat{\varphi}(t, z) \mid c(z) \neq d(z), -a(z) < t < b(z)\}$$

has for its components open sets in which $d(z) > c(z)$ throughout or $d(z) < c(z)$ throughout. For each component D_0 of D, we can imagine F as defined in the domain $\beta \in \mathbb{R}^+$, $a = a(z)$, $b = b(z)$, $c = c(z)$, $d = d(z)$, for all $z \in D_0 \cap \mathbb{R}^+$. In that domain, F is continuous in its variables and monotonic in β. Thus the relation $F(\beta, a, b, c, d) = 0$ has at most one solution $\beta = \beta(a, b, c, d)$ for each quadruple, and we observe that there is always a solution, since $\dfrac{b + a}{d} < \dfrac{a}{c} + \dfrac{b}{d} < \dfrac{b + a}{c}$ if $d > c$, and the opposite inequality holds if $d < c$. We define $\beta(z) = \beta\big(a(z), b(z), c(z), d(z)\big)$ and observe thereby, using Theorem 10.11, that $\beta(z)$ is a continuous function of z.

We now set

$$\hat{v}_1\big(\hat{\phi}(t, z)\big) = w\big(\beta(z), t, a(z), b(z), c(z), d(z)\big), \quad \forall - a(z) \leq t \leq b(z),$$

and observe that the condition $f\big(\beta(z), a(z), b(z), c(z), d(z)\big) = 0$ means exactly that $\hat{u}_1\hat{\phi}$ and $\hat{v}_1\hat{\phi}$ require the same time to cover $\gamma(z)$ for each z in $D_0 \cap \mathbb{R}^+$. Thus the equality of times holds in all of $D \cap \mathbb{R}^+$. We see that $\hat{v}_1$ is continuous in t and z and is thus a continuous function of $\hat{\phi}(t, z)$. The modification of $\hat{u}_1$ to $\hat{v}_1$ requires a corresponding modification h_1 of h' which is made in the obvious way.

Our problem would now be solved, except for two things: The homeomorphism h_1 maps Ω_r', not Ω_r, onto $\hat{A}_r$, and the boundary value is not 1 but $\hat{u}(y_0)$. To make up this deficiency, we shall expand our mapping to a larger pair of annuli, which we will then map into the annuli we want. For n large enough, $\varphi\big(-n p_\varphi(y^+), J\big)$ does not intersect Ω_r', and we denote by J_2 the Jordan curve $\varphi\big(- (n + 1) p_\varphi(y^+), J\big)$. We shall make correspond to J_2 the image $\hat{\phi}\big(-2\pi(n + 1), \Gamma_r\big)$, which is $\Gamma_{r'}$, where $r' = \hat{\phi}\big(-2\pi(n + 1), r\big)$. For each point $x \in J$, let us denote as $\tilde{\gamma}(x)$ the arc of $\mathcal{O}_\varphi(x)$ lying between $E(x)$ and $\varphi\big(- (n + 1) p_\varphi(y^+), x\big)$. We will want to map this arc in a properly continuous fashion onto the arc between $h(x)$ and $\hat{\phi}\big(- 2\pi(n + 1), h(x)\big)$, so as to give the equation we are seeking for an appropriate extension $\hat{u}_2$ of $\hat{u}_1$.

For each real number β, each $x \in J$, and each $0 \leq t \leq 2\pi(n + 1)$, we define $v\big(\beta, \hat{\phi}(-t, h(x))\big)$ to be equal to $\big((1 - t) \hat{u}(y_0) + t\big) e^{-t\beta}$ when $0 \leq t \leq 1$, and equal to $e^{-\beta \frac{2\pi(n+1)-t}{2\pi(n+1)-1}}$ when $1 \leq t \leq 2\pi(n + 1)$. We note that for x fixed, $v\big(\beta, \hat{\phi}(-t, h(x))\big)$ takes values very close to 0 for most values of t when β is large and positive, and takes very large values for most values of t when β is large and negative. In fact, as β decreases from $+\infty$ to $-\infty$, the integral

$$\int_{-2\pi(n+1)}^{0} \frac{dt}{v\big(\beta, \hat{\phi}(t, h(x))\big)}$$

decreases from ∞ to 0 in a strictly monotonic fashion. We define $\beta(x)$ so that the integral taken with that value for β is exactly $\mu_\varphi\big(\tilde{\gamma}(x)\big)$. As before, $\beta(x)$ is a continuous function of x, and we now define $\hat{v}_2(z)$ to be $\hat{v}_1(z)$ if $z \in \hat{A}_r$ and $v\big(\beta(x), \hat{\phi}(-t, h(x))\big)$ if $z = \hat{\phi}\big(-t, h(x)\big)$ for some $x \in J$ and $0 \leq t \leq 2\pi(n + 1)$, i.e. if $z \in \hat{A}_{r'} \setminus \hat{A}_r$. We now extend h' to h_2 in the unique manner which will make

$$h_2\big(\varphi(t, x)\big) = \hat{v}_2\hat{\phi}\big(t, h_2(x)\big), \quad \forall t \geq 0, \quad \forall x \in \varphi\big(- (n + 1) p_\varphi(y^+), \Omega_r\big).$$

However, h_2 is a homeomorphism of $\varphi\left(-(n+1)\,\mathrm{p}_\varphi(y^+),\,\Omega_r\right)$ onto $\hat\varphi\left(-2\pi(n+1),\,\hat A_r\right)$. We define h_3 by

$$h_3(x) = \hat\varphi\left(2\pi(n+1),\, h_2\left(\varphi\left(-(n+1)\,\mathrm{p}_\varphi(y^+),\, x\right)\right)\right), \qquad \forall\, x \in \Omega_r,$$

and define $\hat v(z) = \hat v_2\left(\hat\varphi(-2\pi(n+1),\, z)\right)$. An elementary calculation now shows that h_3 and $\hat v$ give us the required relationship, and that $\hat v$ is continuous and takes the value 1 on Γ_r. **QED**

10.13. Theorem. *Let φ be a flow in the closed annulus $\bar\Omega$, and assume that the boundaries of Ω are periodic orbits $\mathcal{O}_\varphi(y^-)$ and $\mathcal{O}_\varphi(y^+)$. Let $\hat A$ be the closed annulus $\{z \mid r_1 \le |z| \le r_2\}$, and let $\hat\varphi$ be the flow defined in $\hat A$ by Example 3.14(b). Assume that $\mathrm{p}_\varphi(x) = \infty$, $\forall\, x \in \Omega$, and that the flow φ is orientation preserving.*

Then there exist a homeomorphism h from $\bar\Omega$ onto $\hat A$ and a continuous $\hat\varphi$-multiplier $\hat u$ such that $h[\varphi] = \hat u\hat\varphi$.

PROOF: As in the proof of Lemma 10.9, we start with a Jordan curve J_0 which intersects each orbit $\mathcal{O}_\varphi(x)$ in a single point, for each $x \in \Omega$. The Jordan curve J used in that lemma is of the form $\varphi(t_0, J_0)$, where $t_0 \ge 0$. Since φ_{t_0} is a homeomorphism, we see at once that the results of Lemma 10.9, and thus also those of Theorem 10.12, apply to the annulus lying between J_0 and $\mathcal{O}_\varphi(y^+)$ (as before, we assume WoLOG that $\mathcal{O}_\varphi(y^+) = \omega_\varphi(x)$, $\forall\, x \in \Omega$), which we now denote as Ω^+. We call the homeomorphism involved h^+ and the $\hat\varphi$-multiplier $\hat u^+$. In the same way, we can find a homeomorphism h^- of the closed annulus Ω^- lying between J_0 and $\mathcal{O}_\varphi(y^-)$ onto the closed annulus $\hat A^- = \{z \mid r_1 \le |z| \le r\}$ and a $\hat\varphi$-multiplier $\hat u^-$ continuous in that annulus such that

$$h^-\left(\varphi(t, x)\right) = \hat u^- \cdot \hat\varphi\left(t, h^-(x)\right), \qquad \forall\, t \le 0, \qquad \forall\, x \in \Omega^-.$$

We now define a homeomorphism h_0 of Ω onto $\hat A = \{z \mid r_1 \le |z| \le r_2\}$ to be h^+ in the domain of h^+, and h^- in the domain of h^-. The two domains overlap in the curve J, and we must choose the homeomorphisms h^+ and h^- to be the same on J, as we can do. We have not pointed out that the orientation making the proper homeomorphism h^+ and that for h^- are the same, but this is true, since φ and $\hat\varphi$ are both orientation-preserving. Reference to a sketch will show this fact. We define a $\hat\varphi$-multiplier $\hat u_0$ by

$$\begin{aligned}
\hat u_0(z) &= \hat u^+(z), \qquad \forall\, r \le |z| \le r_2 \\
&= \hat u^-(z), \qquad \forall\, r_1 \le |z| < r.
\end{aligned}$$

Note that $\hat u_0$ is continuous except possibly on Γ_r.

However since $\hat u^+$ and $\hat u^-$ both take the value 1 there, the function $\hat u_0$ must be continuous there as well. **QED**

10.14. Theorem. *The same theorem holds if φ is an orientation-reversing flow in $\bar{\Omega}$ and $\hat{\varphi}$ is the flow in Example* 3.14(c).

PROOF: Same proof, *mutatis mutandis.*   **QED**

The question immediately arises whether the proof of Theorem 10.13 doesn't show as well that an orientation-reversing spiral flow can be mapped onto the flow of 3.14(b) as well as onto 3.14(c). The answer is "no", as we shall see from the following material, but this answer is hardly believable, for it hardly appears that we have made any use of the hypothesis of the orientation-preserving character of the flow. The use of the hypothesis lies in the order chosen on the Jordan curve J, *i.e.* the orientation of the homeomorphism h of J onto Γ_r. It seems a slender reed, and yet it is essential. To clinch this nebulous argument, we will show that there is no mapping of the kind indicated between any orientation-preserving flow and any orientation-reversing flow.

10.15. Theorem. *Let φ^+ be an orientation-preserving spiral flow in an annulus A^+, and let φ^- be an orientation-reversing spiral flow in an annulus A^-. Then there do not exist a homeomorphism h from A^+ onto A^- and a φ^--multiplier u such that $h[\varphi^+] = u\varphi^-$.*

PROOF: Let $\mathcal{O}_{\varphi^+}(x^+)$ and $\mathcal{O}_{\varphi^+}(y^+)$ be the boundaries of A^+, and let $\mathcal{O}_{\varphi^-}(x^-)$ and $\mathcal{O}_{\varphi^-}(y^-)$ be the boundaries of A^-, so named that $\omega_{\varphi^+}(x) = \mathcal{O}_{\varphi^+}(y^+)$ and $\omega_{\varphi^-}(y) = \mathcal{O}_{\varphi^-}(y^-)$ for all $x \in A^+$ and $y \in A^-$. Then we can map A^+ onto a torus by identifying $\varphi^+(t p_{\varphi^+}(x^+), x^+)$ with $\varphi^+(t p_{\varphi^+}(y^+), y^+)$, $\forall\, t \in \mathbb{R}$. If u is a φ^--multiplier, it cannot change sign on any orbit, and thus it has the same sign (*i.e.* it is nonnegative or it is nonpositive) on $\mathcal{O}_{\varphi^-}(x^-)$ and $\mathcal{O}_{\varphi^-}(y^-)$ (and thus, in fact, on all of A^-, though that is not needed in the proof). Thus, $\mathcal{O}_{u\varphi^-}(x^-)$ and $\mathcal{O}_{u\varphi^-}(y^-)$ are oppositely directed. It follows that any identification of the points of $\mathcal{O}_{u\varphi^-}(x^-)$ and those of $\mathcal{O}_{u\varphi^-}(y^-)$ which preserves their order gives not a torus but a Klein bottle. This shows that the equation $h[\varphi^+] = u\varphi^-$ is not solved for any h and u.   **QED**

10.16. Conjugacy of Flows. Theorem 10.14 inspires a definition of *conjugacy* for continuous flows. We recognize that homeomorphism of flows (Definition 1.34) is too strict a condition to act as a meaningful working equivalence between flows. There are too few instances. By contrast, the remaining work of this chapter shows the value of the following less stringent similarity condition:

If φ is a flow in a topological space X and ψ is a flow in a topological space Y, then φ is *conjugate to* ψ if there exist

(i) a homeomorphism h from X onto Y, and

(ii) a ψ-multiplier u continuous in $\mathscr{G}(\psi)$ and never equal to either 0 or $\pm\infty$ there,

such that
$$h[\varphi] = u\psi.$$

If, in addition, u is positive, then φ is *positively conjugate to* ψ.

10.17. Theorem. *If* φ *is (positively) conjugate to* ψ, *then* ψ *is (positively) conjugate to* φ.

PROOF: Let $h[\varphi] = u\psi$. Since ψ and $u\psi$ are flows with the same orbits, we can apply Theorem 5.46 to obtain a $u\psi$-multiplier v such that $\psi = v(u\psi)$, and $v(y)\,u(y) = 1$, $\forall\, y \in \mathscr{G}(\psi)$. Therefore $\psi = v(h[\varphi]) = h[w\varphi]$, where $w(x) = v(h(x))$, $\forall\, x \in X$. Then
$$h^{-1}[\psi] = h^{-1}\big[h[w\varphi]\big] = w\varphi,$$

so that ψ is conjugate to φ.

If u is positive, then so are v and, consequently, w.   **QED**

10.18. Theorem. *If* φ *is (positively) conjugate to* ψ *and* ψ *is (positively) conjugate to* π, *then* φ *is (positively) conjugate to* π.

PROOF: Let $h_1[\varphi] = u\psi$ and $h_2[\psi] = v\pi$. Then $\psi = h_2^{-1}[v\pi]$, so that $u\psi = u h_2^{-1}[v\pi] = h_2^{-1}[w(v\pi)]$, where $w(z) = u\big(h_2^{-1}(z)\big)$, $\forall\, z \in \mathscr{G}(\pi)$. Then
$$h_1[\varphi] = h_2^{-1}[w(v\pi)]$$

so that $h_2\big[h_1[\varphi]\big] = w(v\pi) = (wv)\pi$, by Theorem 5.46, i.e. $(h_2 h_1)\varphi = (wv)\pi$, so that φ is conjugate to π. It is clear that if u and v are positive, then so are w and, consequently, wv.   **QED**

10.19. Theorem. *Conjugacy and positive conjugacy are equivalence relations in the class of all flows in topological spaces.*

PROOF: Reflexivity is obvious. Symmetry follows from Theorem 10.17, transitivity from Theorem 10.18.   **QED**

10.20. Definition. If the flow φ is conjugate to the flow ψ, then we alter our usage to say that φ is conjugate *with* ψ or φ *and* ψ *are conjugate flows*, and write $\varphi \equiv \psi$. If the conjugacy is positive, we write $\varphi \equiv^+ \psi$.

10.21. Theorem. *Let* Ω *be a component of* $\mathscr{G}_r(\varphi)$ *and let* y^- *and* $y^+ \in \Omega$ *satisfy* $\mathrm{p}_\varphi(y^-) < \infty$, $\mathrm{p}_\varphi(y^+) < \infty$. *Let* Ω_0 *be the closed annulus lying between* $\mathcal{O}_\varphi(y^-)$ *and* $\mathcal{O}_\varphi(y^+)$.

Then we can find a simple curve $\gamma \subseteq \Omega_0$ *such that* $\gamma \cap \mathcal{O}_\varphi(x)$ *is a single point for each* $x \in \Omega_0$ *satisfying* $\mathrm{p}_\varphi(x) < \infty$.

PROOF: We can assume WoLOG that $d(y^-, y^+) = d(\mathcal{O}_\varphi(y^-),\ \mathcal{O}_\varphi(y^+))$. Let L be the closed interval $[y^-, y^+]$. We then replace subintervals of L, as in Theorem 10.2, by arcs of the periodic orbits so that each periodic orbit meets the modified arc in a single point or a closed arc. When substituting arcs of the periodic orbits for subintervals of L, we always choose the arc of the lesser φ-time measure. By continuity, these measures, and the diameters of the corresponding arcs, converge to 0. Thus, the final modification is an arc. Now we continue, by the method of Theorem 10.2, to further modify the arc so that each of the closed arcs is shrunk to a point, while each orbit which intersects in a point continues to do so. The method is exactly the same.   **QED**

10.22. Theorem. *Let* φ, Ω, y^-, *and* y^+ *be as before, and define* $K(\Omega_0) = \{x \in \Omega_0 \mid \mathrm{p}_\varphi(x) < \infty\}$. *Let* γ *be the curve constructed in Theorem* 10.21, *and let* h_0 *be a homeomorphism of* γ *onto* $[1, 2]$. *Let* $B_0 = h_0(\gamma \cap K(\Omega_0))$ *and* $B_1 = \{z \mid |z| \in B_0\}$. *Define* ψ_1 *in* B_1 *by* $\psi_1(t, z) = ze^{it}$, $\forall\ z \in B_1$, $\forall\ t \in \mathbb{R}$, *and* φ_1 *in* $K(\Omega_0)$ *as the restriction of* φ *there.*
Then $\varphi_1 \equiv^+ \psi_1$.

PROOF: The function p_φ is continuous in $K(\Omega_0)$ by Theorem 2.15. For $x \in K(\Omega_0)$, write $x = \varphi(t, \tilde{x})$, where $\tilde{x} \in \gamma$, and define

$$h(x) = e^{\frac{2\pi i t}{\mathrm{p}_\varphi(x)}} h_0(\tilde{x}).$$

Note that the definition of $h(x)$ is not dependent on the choice of t. Now if we set $u(h(x)) = \dfrac{2\pi}{\mathrm{p}_\varphi(x)} > 0$, $\forall\ x \in K(\Omega_0)$, we see at once that $h[\varphi_1] = u\psi_1$.   **QED**

10.23. Orientation-Preserving Spirals. In addition to the flow denoted as φ in Example 3.14(b), we can consider three other orientation-preserving spiral flows in the same annulus, all very similar to φ. We consider the flow $-\varphi = -1 \cdot \varphi$, the flow $\bar{\varphi}$, which is the image of φ under the homeomorphism of complex conjugation $(z \to \bar{z})$, and the flow $-\bar{\varphi}$, which is $-1 \cdot \bar{\varphi}$. All the four flows are homeomorphic, but in combinations, the differences are important. Consider a flow in the annulus $\{z \mid 1 \leq |z| \leq 3\}$ which is an orientation-preserving spiral flow in the annulus $\{z \mid 1 \leq |z| \leq 2\}$ and also in the annulus $\{z \mid 2 \leq |z| \leq 3\}$. Then there are eight possible ways of combining flows of the type of Example 3.14(b) to meet these criteria, according to the choice of the orientation of the flow (two choices) and then a choice of a flow in each subannulus consistent with that orientation. These eight flows fall into two classes under conjugacy, and three classes under positive conjugacy. Thus, when we combine flows in a manner similar to this, we must take account of these differences.

10.24. Definition. Let φ be a flow of the type of Example 3.14(b). We refer to it as a *positive expansive standard orientation-preserving spiral flow*. $\bar{\varphi}$ is called a *negative expansive standard orientation-preserving spiral flow*; $-\varphi$ is called *negative contractive*, and $-\bar{\varphi}$ is called *positive contractive*.

Referring to Example 3.14(c), we call φ a *positive expansive standard orientation-reversing spiral flow*; $\bar{\varphi}$ is called *negative expansive*, $-\varphi$ is called *negative contractive*, and $-\bar{\varphi}$ is called *positive contractive*.

10.25. Theorem. *Let φ, Ω_0, y^- and y^+ be as in Theorem 10.21 and assume that in every spiral region contained in Ω_0, φ is an orientation-preserving spiral flow. Let A be the annulus $\{z \mid 1 \leq |z| \leq 2\}$ and let the closures of the component annuli of $A \setminus B_1$ be denoted as A_i, where B_1 is as defined in Theorem 10.22, and for each i, $A_i = \{z \mid r_{i,1} \leq |z| \leq r_{i,2}\}$. In each A_i, let $\hat{\varphi}_i$ be a positive standard orientation-preserving spiral flow which is expansive or contractive.*

Then if a proper assignment of these flows as expansive or contractive is made, the flow ψ defined by $\psi = \psi_1 \cup \bigcup_i \hat{\varphi}_i$ will be positively conjugate with the flow φ_{Ω_0} which is the restriction of φ to Ω_0: $\psi \equiv^+ \varphi_{\Omega_0}$.

PROOF: Note first that the flow ψ_1 and the flow $\hat{\varphi}_i$ agree on the circles $\Gamma_{r_{i,1}}$ and $\Gamma_{r_{i,2}}$, $\forall i$. Now let γ be the arc defined in Theorem 10.21, and let h_0 be the mapping of γ onto $[1, 2]$ which gives rise to B_0 and B_1. We assume WoLOG that $h_0(y^-) = 1$ and $h_0(y^+) = 2$. For each i, we define as Ω_i that component annulus of $\Omega_0 \setminus K(\Omega_0)$ which contains the open arc $h_0^{-1}(r_{i,1}, r_{i,2})$, and define $y_i^- = h_0^{-1}(r_{i,1})$ and $y_i^+ = h_0^{-1}(r_{i,2})$. Let us imagine a flow $\hat{\varphi}_i$ in A_i which is expansive or contractive according as $\omega_\varphi(x)$ is $\mathcal{O}_\varphi(y_i^+)$ or $\mathcal{O}_\varphi(y_i^-)$ for $x \in \Omega_i$, and which is generated in the manner of Theorem 10.13. Let h_i be the homeomorphism constructed there, and u_i the multiplier. Assume further that $h_i(y_i^-) = r_{i,1}$ and $h_i(y_i^+) = r_{i,2}$.[2] Then the mapping h_i in Theorem 10.13 agrees with the

---

[2] This involves no loss of generality. In Theorm 10.9, y^+ is chosen arbitrarily, and if t_0 is taken as a multiple of $p_\varphi(y^+)$, then the construction would give, among other things, $h(y^+) = r_2$. In a similar way, the application of the same method to the same Jordan curve J and to y^- would give $h(y^-) = r_1$, where y^- is also arbitrary. In Theorem 10.12 the application of the indicated method to each half of the annulus gives $h(y^+) = r_2$ and $h(y^-) = r_1$. Thus we can produce the conjugacy mapping any desired points of the boundaries into r_1 and r_2. We use that additional fact here.

mapping h of Theorem 10.22 on both $\mathcal{O}_\varphi(y_i^-)$ and $\mathcal{C}_\varphi(y_i^+)$ and u_i agrees with u there as well. Thus, we can construct a one-to-one mapping of Ω_0 onto A by combining h with the h_i, and the flows ψ_1 and the $\dot\varphi_i$ combine to give an algebraic flow $h[\varphi]$ in A, where we give the name h to the conjunction of the previously defined h with the h_i. Even without making any special demands on the h_i, we see that $h[\varphi]$ has many of the attributes of continuity, although it is not, in general, continuous. If $\{x_i\}$ is chosen convergent to $x_0 \in \Omega_0$, and $t_i \to t_0$, then we have $h[\varphi](t_i, x_i) \to h[\varphi](t_0, x_0)$ in all the following cases: when $p_\varphi(x_i) < \infty$, $\forall\, i$, when all the x_i (or all but finitely many) belong to the same $\overline{\Omega}_j$, including as a subcase the case when $x_0 \in \Omega_j, \exists\, j$. Thus, the only possibility of non-continuity occurs with $p_\varphi(x_0) < \infty$, $p_\varphi(x_i) = \infty$, $\forall\, i \geq 1$, and with the x_i chosen from infinitely many different sub-annuli Ω_j. WoLOG, we can assume that all the x_i lie in the same side of $\mathcal{O}_\varphi(x_0)$, that the Ω_j are numbered so that $x_i \in \Omega_i$, $\forall\, i \geq 1$, and so that Ω_i separates the x_j, $1 \leq j < i$ from the other x_j. If the h_i are adjusted so that *every* sequence $\{x_i\}$ with these properties gives the desired convergence, then it will follow that the desired continuity of $h[\varphi]$ will have been achieved, and h will be seen to be a homeomorphism.

We shall now construct the h_i to produce the desired effect. Using the techniques of Theorems 10.2 and 10.7, and the fact that the maximum distance from a point of Ω_i to a point of $\partial(\Omega_i)$ converges to 0 as $i \to \infty$, we see that for every $\varepsilon > 0$, and for all but finitely many of the Ω_i, we can construct an arc γ_i in Ω_i joining y_i^- to y_i^+ so that

(i) $\operatorname{diam}(\gamma_i) < \varepsilon$,

(ii) for every $x \in \gamma_i$, $\varphi(t, x) \in \gamma_i$, for all $t \in \mathbb{R}$ satisfying

$$0 < |t| \leq \tfrac{1}{2} \min_{\bar{x} \in \Omega_0} p_\varphi(\bar{x}).$$

We choose any sequence $\varepsilon_n \downarrow 0$, and for each Ω_i, we construct γ_i so that $\operatorname{diam}(\gamma_i) < \varepsilon_n$ for the largest n possible. This assigns finitely many Ω_i to each ε_n, and leaves finitely many without assignment. For these last, we choose γ_i merely satisfying condition (ii). It is immediate that $\gamma = (\gamma_0 \cap K(\Omega_0)) \cup \bigcup_i \gamma_i$ is a closed arc joining y^- to y^+, and that for each $x \in \gamma$, we have $\varphi(t, x) \notin \gamma$ whenever $0 < |t| \leq \tfrac{1}{2} \min_{\bar{x} \in \Omega_0} p_\varphi(\bar{x})$.

Now, for each $x \in \gamma$, we define $g(x)$ as the minimum positive number such that $\varphi(g(x), x) \in \gamma$. Then g is continuous.

We define the mapping h of Ω_0 onto A by requiring first that h map γ onto $[1, 2]$, that h coincide with h_0 on $\gamma \cap K(\Omega_0)$, that

$h\big(\varphi(g(x), x)\big) = \hat{\varphi}_i\big(2\pi, h(x)\big)$ for all $x \in \gamma_i$, and finally that

$$h\big(\varphi(t, x)\big) = \hat{\varphi}_i\left(\frac{2\pi t}{g(x)}, h(x)\right), \qquad \text{for all} \quad x \in \gamma_i \quad \text{and} \quad 0 \leq t < g(x),$$

$$= h(x) \cdot e^{2\pi i t/\mathrm{p}_\varphi(x)} \qquad \text{for all} \quad x \in K(\Omega_0).$$

We wish to show that h as so defined is a homeomorphism of $\overline{\Omega}_0$ onto A. Since h maps $\overline{\Omega}_0$ 1-1 onto A, it will suffice to show that h is continuous. We know that h is continuous except in the one case discussed earlier, i.e. when we have a sequence $\{x_n\}$ in Ω_0 converging to $x_0 \in \overline{\Omega}_0$ with $x_0 \in K(\Omega_0)$ and each of the x_n in a different Ω_i, which we renumber so that $x_n \in \Omega_n$. We choose t_n and $\bar{x}_n$ for each $n \geq 0$ so that $\bar{x}_n \in \gamma$, $0 \leq t_n < g(\bar{x}_n)$, and $x_n = \varphi(t_n, \bar{x}_n)$. Consider the sequence $\{y_n^-\}$. This is a monotonic sequence of points of γ which converges to a point y_0 of γ which must belong to $K(\Omega_0)$. Since $\mathrm{diam}(\gamma_n) \to 0$, we have $\bar{x}_n \to y_0$ also. We see that $\mathcal{O}_\varphi(x_0)$ cannot lie in $\mathrm{ins}\big(\mathcal{O}_\varphi(y_0)\big)$, nor can $\mathcal{O}_\varphi(y_0)$ lie in $\mathrm{ins}\big(\mathcal{O}_\varphi(x_0)\big)$, since neither y_0 nor x_0 can be cut off from the x_n by the orbit of the other. Thus $\mathcal{O}_\varphi(x_0) = \mathcal{O}_\varphi(y_0)$, and we must have $\bar{x}_0 = y_0$. Now if $t_n \to t_0$, the definition of h will assure that $h(x_n) \to h(x_0)$, since g is a continuous function, and $g(x) = \mathrm{p}_\varphi(x)$, $\forall\, x \in \gamma \cap K(\Omega_0)$. If we do not have this convergence, then the sequence $\{t_n\}$, which is bounded by $\sup\limits_{x \in \gamma} g(x)$, must have a subsequence convergent to some limit $\bar{t} \neq t_0$, and we assume WoLOG that $t_n \to \bar{t}$. By continuity, then, $x_n = \varphi(t_n, \bar{x}_n) \to \varphi(\bar{t}, \bar{x}_0) \neq x_0$, except in the case that $t_0 = 0$, $\bar{t} = \mathrm{p}_\varphi(\bar{x}_0)$. (We use the fact that $0 \leq t_n < g(x_n)$.) However, in this case, we have $h(x_n) \to h(x_0)$. Thus, h is a homeomorphism of Ω_0 onto A. We now define u in A by setting $u\big(h(\varphi(t, x))\big) = 2\pi/g(x)$, $\forall\, x \in \gamma$, $\forall\, 0 \leq t < g(x)$. It is clear that $h[\varphi] = u\psi$, and u is positive but not, in general, continuous. However, the method of Theorem 10.12 will enable us to replace h and u by $\bar{h}$ and $\bar{u}$ so that $\bar{u}$ is both positive and continuous.   **QED**

10.26. Definition. Let $A = \{z \,|\, r_1 < |z| < r_2\}$ and let φ be a flow in $\tilde{A}$, where $\tilde{A}$ is the annulus A, or its closure, or its union with one of its boundaries. Then φ is said to be an *annular flow of standard type* if

(i) $\forall\, z \in \tilde{A}$, $\mathrm{p}_\varphi(z) > 0$,

(ii) $\forall\, z \in \tilde{A}$, $\mathrm{p}(z) < \infty \Rightarrow \varphi(t, z) = z e^{it}$ or $\varphi(t, z) = z e^{-it}$, $\forall\, t \in \mathbb{R}$,

(iii) $K(\tilde{A}) = \{z \,|\, \mathrm{p}_\varphi(z) < \infty\}$ is closed in $\tilde{A}$,

(iv) the supremum and infimum respectively of the radii of the circles in $K(\tilde{A})$ are r_1 and r_2,

(v) for each component annulus A_i of $\tilde{A} \setminus K(\tilde{A})$, the flow φ_{A_i} which is the restriction of φ to $\overline{A}_i$, is a positive or negative expansive or contractive standard order-preserving or order-reversing spiral flow,

(vi) for every compact $C \subseteq A$, C intersects at most finitely many A_i for which φ_{A_i} is order-reversing.

Note that the definition is redundant. Conditions (iii) and (vi) follow from the others.

10.27. Definition. If Ω is an annulus, and φ is a flow in $\overline{\Omega}$, then let us consider the flow φ_Ω which is the restriction of φ to $\Omega \cup K(\overline{\Omega})$, where $K(\overline{\Omega}) = \{x \mid 0 < p_\varphi(x) < \infty\}$. If φ_Ω is homeomorphic to a flow which is an annular flow of standard type, then we say that φ *has a homeomorphic standard annular model,* and that the indicated standard annular flow is that model.

If there exists a continuous positive φ-multiplier u which is constant on periodic φ-orbits such that $u\varphi$ has a homeomorphic standard annular model, then we say that φ *has a standard annular model,* and that the indicated standard annular flow is that model.

10.28. Theorem. *Let Ω be an annulus and let φ be a flow in $\overline{\Omega}$ such that $p_\varphi(x) > 0$, $\forall x \in \Omega$ and the boundaries of Ω are $\mathcal{O}_\varphi(y^-)$ and $\mathcal{O}_\varphi(y^+)$. Define $K(\Omega) = \{x \in \Omega \mid p_\varphi(x) < \infty\}$. Suppose that in every component annulus of $\Omega \setminus K(\Omega)$, φ is an orientation-preserving spiral flow.*

Then φ has a standard annular model.

PROOF: This follows from the proof of Theorem 10.25. **QED**

10.29. Theorem. *Let Ω be an annulus, and let φ be a flow in $\overline{\Omega}$ such that $p_\varphi(x) > 0$, $\forall x \in \Omega$, and the boundaries of Ω are $\mathcal{O}_\varphi(y^-)$ and $\mathcal{O}_\varphi(y^+)$.*

Then φ has a standard annular model.

PROOF: Define $K(\Omega) = \{x \in \Omega \mid p_\varphi(x) < \infty\}$. We see at once that of the component annuli of $\Omega \setminus K(\Omega)$, only finitely many have φ as an orientation-reversing spiral flow, by Theorem 3.16(f). We number these as Ω_k, $1 \leq k \leq n$, $\exists\, n \geq 0$, where the numbering is so chosen that Ω_k is on the y^- side of Ω_j, $\forall k < j$. Denote the boundaries of Ω_k as $\mathcal{O}_\varphi(y_{k-1}^+)$ and $\mathcal{O}_\varphi(y_k^-)$, where $y_0^- = y^-$ is in the y_{k-1}^+ side of Ω_k and $y_n^+ = y^+$ is in the y_k^- side of Ω_k. For each value of k between 0 and n, either $\mathcal{O}_\varphi(y_k^-) = \mathcal{O}_\varphi(y_k^+)$ or the closed annulus $\overline{\Omega}_k$ lying between $\mathcal{O}_\varphi(y_k^-)$ and $\mathcal{O}_\varphi(y_k^+)$ satisfies the hypotheses of Theorem 10.28. Let $0 < r_0^- \leq r_0^+ < r_1^- \leq r_1^+ < \ldots < r_n^- \leq r_n^+ < \infty$ be chosen, with $r_k^- = r_k^+$ iff $\mathcal{O}_\varphi(y_k^-) = \mathcal{O}_\varphi(y_k^+)$. For each $k = 0, 1, \ldots, n$ for which we have $\mathcal{O}_\varphi(y_k^-) \neq \mathcal{O}_\varphi(y_k^+)$, we put into the annulus $\{z \mid r_k^- \leq |z| \leq r_k^+\}$ a standard annular model ψ_k for the flow φ_k which is the restriction of the flow φ to the closed annulus $\tilde{\Omega}_k$. We do this in such a way that the image of $\mathcal{O}_\varphi(y_k^-)$ is $\Gamma_{r_k^-}$ and so that

$$\psi_k(t, r_k^-) = \begin{cases} r_k^- e^{it} & \text{if } i \text{ is even} \\ r_k^- e^{-it} & \text{if } i \text{ is odd.} \end{cases}$$

Now, for each $k = 1, \ldots, n$, we put into the annulus $A_k = \{z \mid r_{k-1}^+ \leq |z| \leq r_k^-\}$ an orientation-reversing standard spiral flow which is positive or negative according as k is even or odd, and expansive or contractive according as $\omega_\varphi(x) = \mathcal{C}_\varphi(y_k^-)$ or $\omega_\varphi(x) = \mathcal{O}_\varphi(y_{k-1}^+)$, $\forall x \in \Omega_k$. It is easy to see that the flow ψ defined in $A = \{z \mid r_0^- \leq |z| \leq r_n^+\}$ as the conjunction of all these flows is an annular flow of standard type, and that ψ is a standard annular image of φ.   **QED**

10.30. Theorem. *Let Ω be an annulus, and let φ be a flow in $\bar{\Omega}$. Assume that one of the boundaries of Ω is a periodic orbit $\mathcal{O}_\varphi(y_0)$, and that the other boundary contains fixed points of φ. Assume further that $\Omega \subseteq \mathscr{G}_r(\varphi)$.*

Then the restriction φ_Ω of φ to $\Omega \cup \mathcal{O}_\varphi(y_0)$ has a standard annular model.

PROOF: Denote the boundary of Ω which is not $\mathcal{O}_\varphi(y_0)$ as B. Since no orbit lying in Ω stagnates in B, there must be a sequence $\{y_k\}$ converging to a point $\bar{y} \in B$ such that $y_k \in \Omega$ and $p_\varphi(y_k) < \infty$, $\forall k = 1, 2, \ldots$. We assume WoLOG that $y_0 \in \text{ins}\big(\mathcal{O}_\varphi(y_k)\big) \subseteq \text{ins}\big(\mathcal{O}_\varphi(y_{k+1})\big)$, $\forall k = 1, 2, \ldots$. Denote as Ω_k the annulus between $\mathcal{O}_\varphi(y_{k-1})$ and $\mathcal{O}_\varphi(y_k)$, $\forall k = 1, 2, \ldots$, and note that for all k, φ_k has a standard annular model, by Theorem 10.29, where φ_k is the restriction of φ to $\overline{\Omega}_k$. Let a sequence $\{r_k\}$ be chosen with $0 < r_0 < r_k < r_{k+1} < r_\infty < \infty$, $\forall k = 1, 2, \ldots$. For each k, let a_k be the number of orientation-reversing spirals in Ω_k. Let the flow ψ_k be defined in the annulus $A_k = \{z \mid r_{k-1} \leq |z| \leq r_k\}$ to be an annular flow of standard type which is positively conjugate with φ_k and so chosen that $\arg\big(\psi_k(t, r_{k-1})\big)$ increases with increasing t iff $\sum_{j < k} a_j$ is even. Let ψ be the conjunction of the flows ψ_k. Then ψ is an annular flow of standard type in the annulus $A = \{z \mid r_0 \leq |z| < r_\infty\}$, and it is clear that ψ is a standard annular model for φ_Ω.   **QED**

10.31. Theorem. *Let Ω be an annulus, and let φ be a flow in $\bar{\Omega}$ such that each boundary of Ω contains fixed points of φ and $\Omega = \mathscr{G}_r(\varphi)$.*

Then the restriction φ_Ω of φ to Ω has a standard annular model.

PROOF: It is clear that $\exists y \in \Omega : p_\varphi(y) < \infty$. Then $\mathcal{O}_\varphi(y)$ divides Ω into two half-closed annuli, Ω^- and Ω^+, and the hypotheses of Theorem

10.30 are satisfied for the restrictions φ^- and φ^+ of φ to Ω^- and Ω^+ respectively. Choose $0 < r^- < r < r^+ < \infty$, and let ψ^- be a standard annular model for φ^- in $A^- = \{z \mid r^- < |z| \leq r\}$ and ψ^+ a standard annular model for φ^+ in $A^+ = \{z \mid r \leq |z| < r^+\}$, both chosen so that the image of $\mathcal{O}_\varphi(y)$ is Γ_r and so that $\psi^-(t, r) = \psi^+(t, r) = re^{it}$, $\forall\ t \in \mathbb{R}$. Let $\psi = \psi^- \cup \psi^+$. Then clearly ψ is a standard annular model for φ_Ω. **QED**

10.32. Theorem. *Let φ be a flow in $\mathbf{S}^2$, and $\bar{\Omega}$ be a component of $\mathcal{G}_r(\varphi)$.*

Then the flow φ_Ω which is the restriction of φ to $\bar{\Omega}$ has a standard annular model.

PROOF: This theorem is proved by applying Theorems 10.29, 10.30 or 10.31 according as $\mathcal{F}(\varphi)$ intersects none, one, or two of the boundaries of Ω.   **QED**

The next theorems characterize the standard annular flows.

10.33. Theorem. *Let $0 < r^- < r^+ < \infty$, and $A = \{z \mid r^- \leq |z| \leq r^+\}$. Let K be a closed set in A which is the union of circles about 0, including the circles of radius r^- and r^+. Let the components of $A \setminus K$ be denoted as $\{A_k\}$. Label finitely many of the A_k with the letter R, the rest with the letter P, and label each of the A_k with either the letter E or C.*

Let the annuli A_k be renumbered so that those with the letter R are taken first, i.e. A_k is labelled R if $1 \leq k \leq n$, $\exists\ n \geq 0$, and A_k is labelled P, $\forall\ k > n$. Furthermore, let the A_k, $1 \leq k \leq n$ be so numbered that A_k lies in the r^- side of A_j, $\forall\ 1 \leq k < j \leq n$. For each point $z \in A \setminus \bigcup\limits_{j=1}^{n} A_j$, we assign to z the integer $l = l(z)$ which is 0 if z lies in the r^- side of A_1 or if $n = 0$, is k if z lies in the r^+ side of A_k and the r^- side of A_{k+1}, and is n if z lies in the r^+ side of A_n. For each $z \in K$, define $\varphi(t, z) = ze^{it}$ or ze^{-it} according as $l(z)$ is even or odd. For each A_k, $k > n$, let φ in A_k be defined as a standard orientation-preserving spiral flow which is positive or negative according as $l(z)$ is even or odd for $z \in A_k$, and expansive or contractive according as A_k is labelled E or C. For each A_k, $k \leq n$, we let φ in A_k be defined as a standard orientation-reversing spiral flow which is positive or negative according as k is even or odd, and expansive or contractive according as A_k is labelled E or C.

Then φ defined in this way is a standard annular flow in the closed annulus A, and every such flow in that annulus is so constructed, up to reflection in the real axis.

PROOF: It is clear that a flow constructed in the way indicated is a standard annular flow in A. The preceding theorems show that every such flow is so constructed. Indeed, if φ is a standard annular flow in A, we take $K = \{z \mid p_\varphi(z) < \infty\}$, and label each annulus of $A \setminus K$ with R or P, E or C, according as the flow φ is an orientation-reversing or orientation-preserving flow, expansive or contractive in that annulus. If we then carry out the indicated construction, the resulting flow will be φ if $\arg\big(\varphi(t, r^-)\big)$ is isotonic in t, and $\bar\varphi$ if $\arg\big(\varphi(t, r^-)\big)$ is antitonic.

QED

10.34. Theorem. *Let* $0 < r^- < r^+ \leq \infty$, *and let* $A = \{z \mid r^- \leq |z| < r^+\}$. *Let* K *be a closed set in* A *which is the union of circles about* 0, *whose radii include* r^- *and have* r^+ *for their supremum. Let the components of* $A \setminus K$ *be denoted as* $\{A_k\}$. *Label countably many of the* A_k *with the letter* R *in such a way that the union of the annuli so labelled is closed in* A, *and label the rest with the letter* P. *Label each of the annuli with either the letter* E *or* C.

Let the A_k *be relabelled as* $\{A_k^R\}$ *and* $\{A_j^P\}$ *according as they are labelled* R *or* P, *with the* A_k^R *so labelled that* A_k^R *lies in the* r^- *side of* A_j^R, $\forall\ 1 \leq k < j$. *For each point* $z \in A \setminus \bigcup_k A_k^R$, *assign to* z *the integer* $l = l(z)$ *so that* z *lies on the* r^- *side of* A_{l+1}^R *and on the* r^+ *side of* A_l^R, *assigning* $l = 0$ *if* z *lies in the* r^- *side of* A_1^R *or if none of the* A_k *are labelled* R, *and assigning* $l = n$ *if the* A_k^R *are only* n *in number and* z *lies in the* r^+ *side of* A_n^R. *For each* $z \in K$, *define* $\varphi(t, z) = z e^{it}$ *or* $z e^{-it}$ *according as* $l(z)$ *is even or odd. For each* A_j^P, *let* φ *in* A_j^P *be defined as a standard orientation-preserving spiral which is positive or negative according as* $l(z)$ *is even or odd for* $z \in A_j^P$, *and expansive or contractive according as we have labelled it* E *or* C. *For each* A_k^R, *let* φ *in* A_k^R *be defined as a standard orientation-reversing spiral flow which is positive or negative according as* k *is even or odd, and expansive or contractive according as* A_k^R *is labelled* E *or* C.

Then φ *defined in this way is a standard annular flow in the half-open annulus* A, *and every such flow in that annulus is so constructed, up to reflections in the real axis.*

PROOF: Same as for Theorem 10.33.   QED

10.35. Theorem. *Let* $0 \leq r^- < r^+ \leq \infty$, *and* $A = \{z \mid r^- < |z| < r^+\}$. *Let* K *be a closed set in* A *which is the union of circles about* 0 *whose radii have* r^- *and* r^+ *for their infimum and supremum respectively. Let the components of* $A \setminus K$ *be denoted as* $\{A_k\}$. *Label countably many of the* A_k *with the letter* R *in such a way that the union of the annuli so labelled is closed in* A, *and label the rest with the letter* P. *Label each of the annuli with the letter* E *or* C.

Let the A_k *be relabelled as* $\{A_j^R\}$ *and* $\{A_j^P\}$ *according as they are labelled* R *or* P, *with the* A_k^R *indexed over a finite or infinite interval in the set of (positive and negative) integers in such a way that* A_k^R *lies in the* r^- *side of*

A_j^R whenever $k < j$. For each point $z \in A \setminus \bigcup\limits_k A_k^R$, assign to z the integer $l = l(z)$ so chosen that z lies in the r^- side of A_{l+1}^R and in the r^+ side of A_l^R, making the obvious modifications for $l + 1$ the least or l the greatest integer in the index interval. For each $z \in K$, define $\varphi(t, z) = z e^{it}$ or $z e^{-it}$ according as $l(z)$ is even or odd. For each A_j^P, let φ in A_j^P be defined as a standard orientation-preserving spiral flow which is positive or negative according as $l(z)$ is even or odd for $z \in A_j^P$, and expansive or contractive according as we have labelled it E or C. For each A_k^R, let φ in A_k^R be defined as a standard orientation-reversing spiral flow which is positive or negative according as k is even or odd, and expansive or contractive according as A_k^R is labelled E or C.

Then φ defined in this way is a standard annular flow in the open annulus A, and every such flow in that annulus is so constructed, up to reflections in the real axis.

PROOF: Same as for Theorem 10.33. **QED**

10.36. Conclusion. Let φ be a flow in $\mathbb{R}^2$, and let Ω be a component of $\mathscr{G}_r(\varphi)$. Then the flow φ in $\bar{\Omega}$ has a standard annular model, and this model is one of the flows described in Theorem 10.33, Theorem 10.34, or Theorem 10.35 according as the number of boundaries of Ω containing fixed points is zero, one, or two respectively.

Regular Annulus Diagrams

10.37. Definition. For each component Ω of $\mathscr{G}_r(\varphi)$, we will now define a topological invariant which will identify the conjugacy class of the flow φ_Ω which is φ restricted to Ω. We begin with a flow ψ in the annulus $A = \{z \mid 1 < |z| < 2\}$ which is a standard annular model for φ_Ω. Let $K(A) = \{z \mid p_\psi(z) < \infty\}$, and $\check{K} = \big(\partial(K(A)) \cap [1, 2]\big) \times [0, 1]$. For

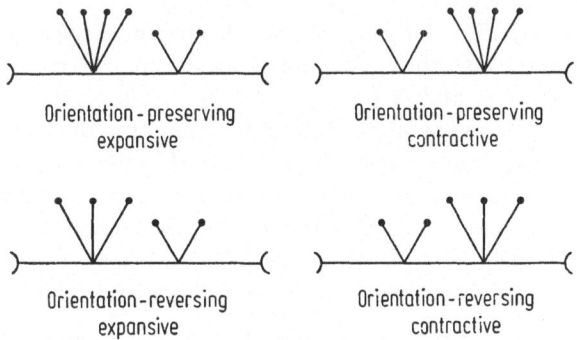

| Orientation-preserving expansive | Orientation-preserving contractive |
| Orientation-reversing expansive | Orientation-reversing contractive |

Fig. 10.1

each annular component A_k of $A \setminus K(A)$, we append to $A_k \cap [1, 2]$ the intervals necessary to produce one of the configurations seen in Fig. 10.1, according as ψ in A_k is an orientation-reversing or orientation-preserving expansive or contractive spiral flow, in such a way that the augmented interval has the same diameter as the unaugmented one. Now we append to $\bar{K}$ and the indicated configurations the interval $[0, 3]$. The closure of the union of all these configurations is a tree-like continuum, which will be called a *regular annulus diagram* for φ_Ω.

10.38. Theorem. *Let φ and ψ be standard annular flows.*

Then the following are equivalent:

(i) *φ and ψ are homeomorphic.*

(ii) *φ and ψ are positively conjugate.*

(iii) *The regular annulus diagrams of φ and ψ are homeomorphic.*

PROOF: Assume WoLOG that φ and ψ are both defined in $A = \{z \mid 1 < |z| < 2\}$. It is clear that (i) $\Rightarrow$ (ii). We will now show the converse. Suppose $h[\varphi] = u\psi$. Then h maps $K(\varphi)$ onto $K(\psi)$, where $K(\varphi) = \{z \mid \mathrm{p}_\varphi(z) = 2\pi\}$, and $K(\psi) = \{z \mid \mathrm{p}_\psi(z) = 2\pi\}$. For each component annulus A_k of $A \setminus K(\varphi)$, we let $\bar{A}_k$ be its image under h. We will now define a new homeomorphism h_1 of A onto itself. For $z \in K(\varphi)$, we define $h_1(z)$ so that $|h_1(z)| = |h(z)|$ and so that $h_1(z)$ is real when z is real, and positive when z is positive, and so that $h_1[\varphi] = \psi$ on $\mathcal{O}_\psi(h(z))$. For each A_k, let $f_{\varphi,k}$ be the mapping of the strip $\left\{ z \,\middle|\, -\dfrac{\pi}{2} \leq \mathrm{Re}(z) \leq \dfrac{\pi}{2} \right\}$ onto A_k which carries the flow called ψ in Example 3.14 (b) or (c) onto the flow φ in A_k, and let $f_{\psi,k}$ be the mapping which carries the same strip onto $\bar{A}_k$ to generate ψ there. Then for each z belonging to that strip, let $h_1(f_{\varphi,k}(z)) = f_{\psi,k}(z)$. It is now easily seen that h_1 is a homeomorphism, and that $h_1[\varphi] = \psi$. Note that h_1 need not be the same as h.

To see that (i) $\Rightarrow$ (iii), let h be a homeomorphism satisfying $h[\varphi] = \psi$. Using the method just above, we can replace h by a homeomorphism h_1 such that $h_1[\varphi] = \psi$ and $h_1([1, 2]) = [1, 2]$. Then $h_1(1)$ is either 1 or 2. In the former case, the symbols assigned to the intervals intersecting the A_k are the same as those assigned to the $\bar{A}_k$, and in the same order, so that the regular annulus diagrams are homeomorphic. In the latter case, the symbols are the same, but applied in the opposite order, since h then takes expansive spiral flows into contractive ones, and *vice versa*. Thus, in this case also, the regular annulus diagrams are homeomorphic.

To show that (iii) $\Rightarrow$ (ii), we observe that if h is a homeomorphism of the regular annulus diagram of φ onto the regular annulus diagram of ψ, then $h\big(K(\varphi) \cap \mathbb{R}\big) = K(\psi) \cap \mathbb{R}$. Also, the assignment of flows as orientation-reversing or orientation-preserving, expansive, or contractive, is such that each of these flows is a standard annular model of the other. Thus, they are positively conjugate.   **QED**

10.39. Theorem. *Let φ and ψ be flows in $\mathbb{S}^2$. Let Ω_φ be a component of $\mathcal{G}_r(\varphi)$ and Ω_ψ a component of $\mathcal{G}_r(\psi)$. Let the restrictions of φ and ψ to Ω_φ and Ω_ψ respectively be φ_1 and ψ_1.*

Then φ_1 and ψ_1 are positively conjugate if and only if they have homeomorphic regular annulus diagrams.

PROOF: If φ_1 and ψ_1 are positively conjugate, then their standard annular models are also, and by Theorem 10.38, their regular annulus diagrams are homeomorphic.

If the regular annulus diagrams are homeomorphic, then the standard annular models are positively conjugate, and since each is positively conjugate to its standard annular model, they are positively conjugate to each other.

10.40. Corollary. *On the same hypotheses, φ_1 and ψ_1 are conjugate if the regular annulus diagram of φ_1 is homeomorphic to that of either ψ_1 or $-\psi_1$.*

PROOF: If $h[\varphi_1] = u\psi_1$, then u must be positive or negative throughout Ω_{ψ_1}, which is connected. Thus, φ_1 is positively conjugate to ψ_1 or $-\psi_1$. Then we apply Theorem 10.39.   **QED**

CHAPTER ELEVEN

FINE STRUCTURE IN $\mathscr{G}_s(\varphi)$ I

11.1. Introduction. These next two chapters deal with work which is current as the manuscript is being written, and thus will have a greater share of discussion, examples, and sketches of ideas, and proportionately fewer theorems than the previous chapters. The singular part of the flow has a richer and more complex structure than the regular part, and this causes the theory to be much harder. In spite of this, we will set forth some theory, though not nearly so complete a presentation as for $\mathscr{G}_r(\varphi)$.

The typical component of $\mathscr{G}_s(\varphi)$ is a multiply-connected set Ω, some of whose boundaries lie in $\mathscr{F}(\varphi)$, and some of which are periodic orbits. These have grossly different relations to the structure of $\mathscr{G}_s(\varphi)$, as we shall see. The boundaries which lie in $\mathscr{F}(\varphi)$ are in $\mathscr{F}_s(\varphi)$, and this portion of $\partial(\Omega)$ is not empty. The periodic orbits in $\partial(\Omega)$ are regular on the side away from Ω. It is possible that Ω contains a periodic orbit, J, in its interior. If so, we see at once that J divides Ω into two subsets, which we could treat separately. Indeed, the flow in one of these subsets could be completely changed, and so long as it remained continuous to the same flow on the boundary J, it would not affect the behavior on the other side of J. Furthermore, we could make as small change in the structure of the flow φ (in a sense which we will not make precise), and "replace" J by an annular region in which the new flow is regular, without changing the essential structure of φ on either side.

Each of the sides of J must include, in its boundary, a subset of $\mathscr{F}_s(\varphi)$ and thus, in the case where $\mathscr{S}(\varphi)$ has finitely many components, we can reduce the study of the components of $\mathscr{G}_s(\varphi)$ to the case where Ω contains no periodic orbits (Of course, we are not excluding the periodic orbits in the boundary).

Our aim is to reduce the study to the simplest cases possible, and then to use these as building blocks, combining them in some way to give all configurations. We can only do this to a small degree, compared with

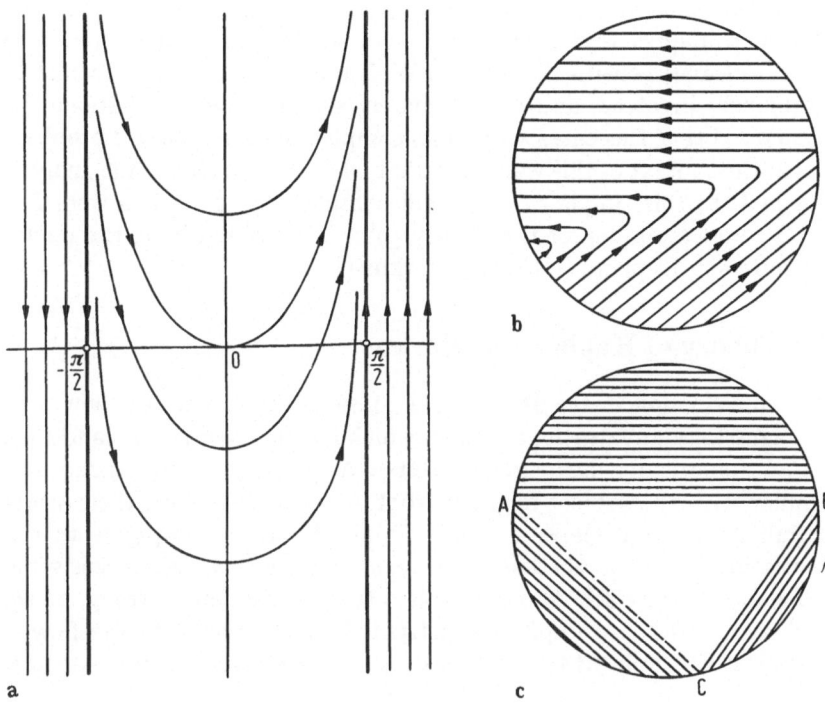

Fig. 11.1

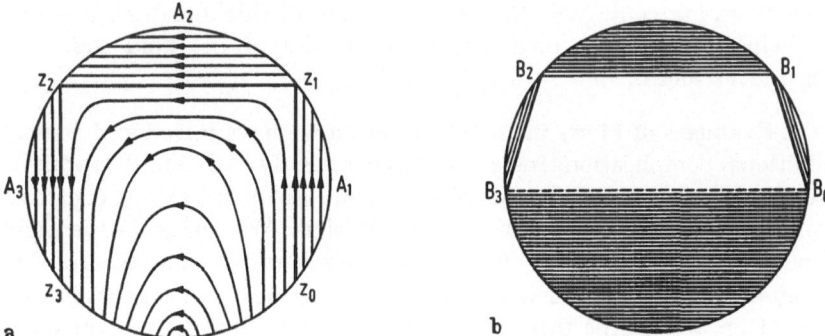

Fig. 11.2

the previous chapter. The simplest case of all is Ω a disc. Each orbit begins and ends on the boundary, and it would not even simplify the analysis to require that $\mathscr{S}(\varphi) \cap \partial(\Omega)$ be a single point. The reduction to this case leaves us with the work of W. Kaplan and L. Markus, which is enormously involved, and will be merely sketched below. It should be understood that this case, with all its complications, is a solved problem, within the context of this work, and that when we can reduce the analysis of φ in Ω to this simple case, we will consider that solved as well. For a fuller understanding of this theory than is presented here, the reader is referred to Kaplan [1], [2] and Markus [1].

The Theory of Kaplan and Markus

11.2. Introduction to the Theory. We consider the orbits of a flow φ in the plane (or equivalently, a disc Ω) which has no fixed points (and thus also, automatically, no periodic orbits). A quick guess by readers unfamiliar with the subject might be that all such flows have their orbits "parallel" in some sense, as, for example, by being conjugate to the translation flow ψ_y in $\mathbb{R}^2$ defined by $\psi_y(t, x) = ty + x$, where y is some non-zero vector in $\mathbb{R}^2$. Actually, this intuitive reaction is wrong, as we shall see. An easy example is constructed by starting with the flow ψ defined in Example 3.14 (c) and with $c_1 = 1$ and $c_2 = 0$. We extend ψ beyond the strip $-\dfrac{\pi}{2} \leq \operatorname{Re}(z) \leq \dfrac{\pi}{2}$ by having it move vertically with unit speed in such a way as to be continuous (see Fig. 11.1 a). We can build a flow conjugate to this one in the disc Ω, and we see an illustration of that in Fig. 11.1 b. For each flow in the disc Ω, we would like to develop some "normal form", which will then tell us whether two such flows are conjugate. Before we develop in this direction, it would be helpful to explore further the pathologies of the kind shown in Figs. 11.1 a and b.

11.3. Examples of Flows in Ω. In Fig. 11.2a we have introduced a small additional complication. Instead of having merely two "limiting orbits", we now have three. In like manner, we could easily extend the construction for any finite number of "limiting orbits". We can go further, and construct flows for which there are infinitely many limiting orbits in many ways, two of which are illustrated in Figs. 11.3 a and 11.4. In Fig. 11.2a, we see the three chords $[z_0, z_1]$, $[z_1, z_2]$, and $[z_2, z_3]$ which cut off the three segments which we denote as A_1, A_2, and A_3, respectively. In each of these three segments, we define the flow to follow parallel orbits, with some motion which will make the action a continuous flow there. When we do this in the future, we shall speak of the segment in-

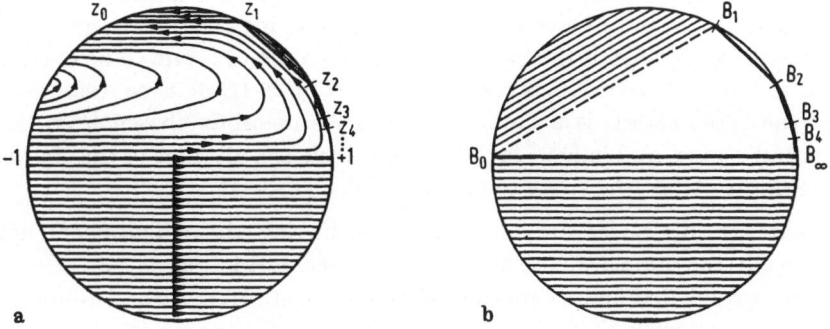

Fig. 11.3

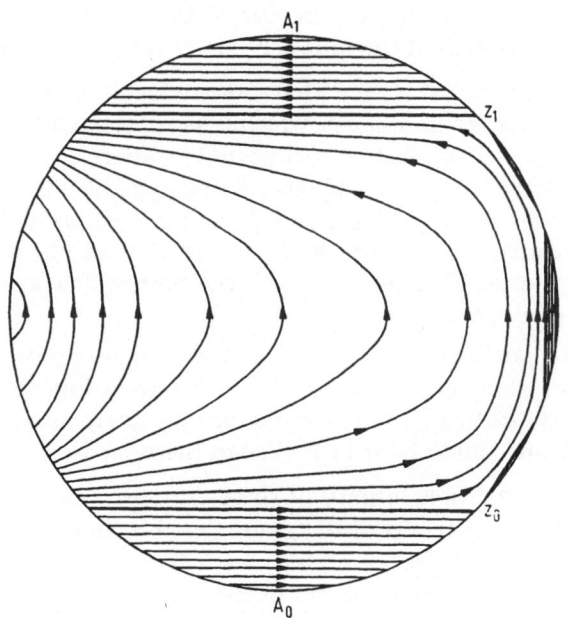

Fig. 11.4

volved as an *abutment*, and the flow as an *abutting flow* on the given chord. In the center of the disc, we have the flow follow the indicated orbits, in such a way as to be continuous to the abutting flows. We can make a topological space of the orbits by defining a set of orbits to be open if the union of those orbits is open as a subset of Ω. If we do so, then each of the three chords is a limit of a sequence of orbits chosen from the central region, and in fact, for each such sequence, all three of the chords are limits. Thus, the space of orbits is not Hausdorff.

Whatever intricacies are introduced by having three abutments are basically the same if we use any finite number. But in Fig. 11.3a we have gone beyond that and created a flow with one central region and infinitely many abutting flows. Ω is here the unit disc in $\mathbb{C}$, and we let A_0 be the lower half disc. We choose a sequence in the upper half disc converging to $+1$ monotonically along the unit circle, as shown. For each i, we let A_i be the abutment of the chord $[z_{i-1}, z_i]$ and define an abutting flow with the indicated orientation. Then we fill the center in the same way as before, making the central flow continuous to the abutting flows. We could do this, for example, by making all of these flows smooth and canonical with respect to the unit circle. We observe that for any sequence of orbits converging to any one of the chords from the center, that sequence has each of the infinitely many chords as a limit.

The example in Fig. 11.4 is somewhat more refined. Let a nowhere dense closed set F (such as a Cantor set, which is shown), be chosen in the part of the unit circle with positive real part. Let abutments be cut off at the top and bottom of this set by chords drawn parallel to the real axis. For each arc in the complement of F in the positive half-plane not already accounted for, let the chord of that arc cut off an abutment, and then define the abutting flows and the central flow as indicated in Fig. 11.4 to give a new flow.

Now that we can build these sorts of flows, let us combine them. In Fig. 11.5a, we combine a structure like the one in Fig. 11.1b and one like that in 11.2a. In Fig. 11.6, we combine infinitely many of the type in Fig. 11.1b and similarly in 11.7, though there we do it differently.

There is a large multiple infinity of these examples, and these are only shown to indicate what sort of things are possible. In order to catalogue these according to conjugacy classes, we will construct, for each flow, a *Kaplan diagram*, and this will aid in this task.

11.4. Kaplan Diagrams. Let Ω be the unit disc, and let φ be a flow in Ω which has no fixed points. Thus, for each orbit in Ω, the endpoints of that orbit lie in the unit circle. We want to define a specific invariant for this flow, which will be invariant under homeomorphisms of the disc. In

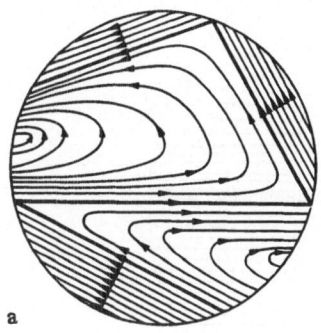

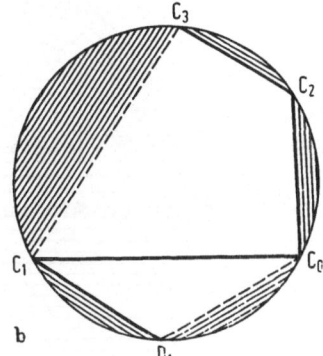

Fig. 11.5

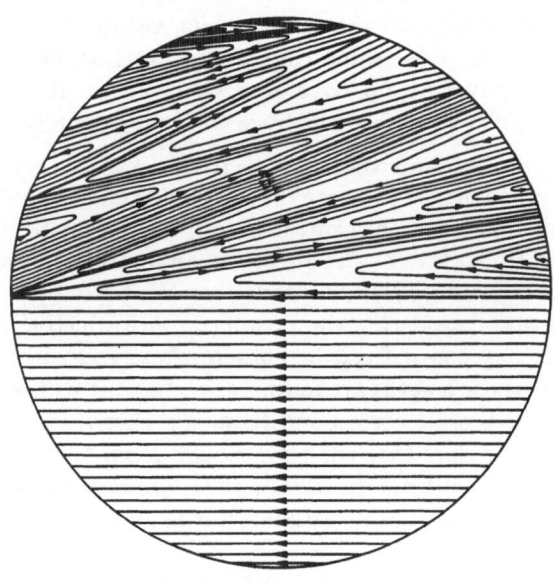

Fig. 11.6

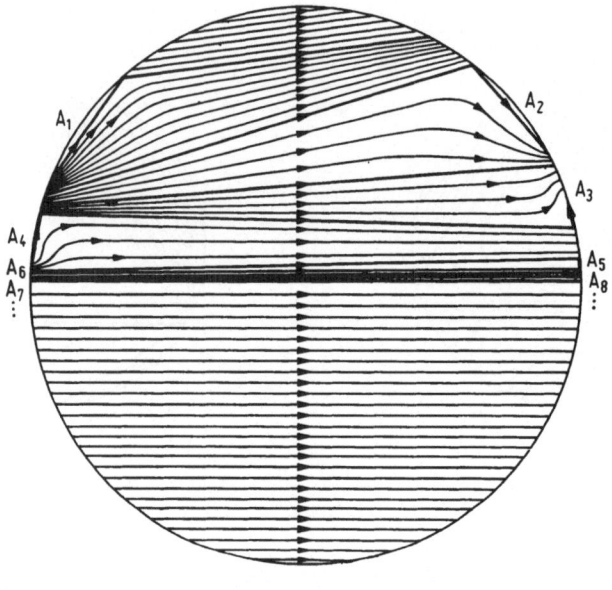

Fig. 11.7

fact, we will do somewhat better, since the Kaplan diagram depends not on the flow, but merely on the set of orbits. Given two flows with the same orbits, they have the same Kaplan diagrams. If, instead, the orbit diagrams are equivalent up to a homeomorphism, then the Kaplan diagrams are similar in a way which we will make definite later.

We begin this study with the additional assumption that each of the orbits under φ has planar measure 0.[1] We could remove this assumption later, but since that is not the central direction of this book, we leave that for the reader to work out. The theory is essentially the same in these cases, and the differences are not instructive. We begin by choosing some point $x_0 \in \Omega$ and considering the ratio $r(x_0)$ into which $\mathcal{O}_\varphi(x_0)$ cuts the area of Ω. Let any chord $L(x_0)$ be chosen whose endpoints divide $\partial(\Omega)$ into two parts whose lengths are in the ratio $r(x_0)$.[2] It will be our purpose to make each such orbit correspond to such a chord, and we must now look for a way to assure that we have enough room for each such chord. Thus, if x is some point not on $\mathcal{O}_\varphi(x_0)$, then we want to choose for $L(x)$ some chord which divides the perimeter of Ω as $\mathcal{O}_\varphi(x)$ divides its area,

---

[1] It seems no longer common knowledge that there exist simple arcs with positive planar measure. Cf. Osgood [1] or Beck, Bleicher, and Crowe [1] pp 201 – 208. It is not hard to show that flows exist which have such arcs among their orbits.

[2] *i.e.* if $\mathcal{O}_\varphi(x_0)$ divides Ω into pieces of areas a and b, then the endpoints of $L(x_0)$ divide $\partial(\Omega)$ into arcs of lengths $2a$ and $2b$.

and we must somehow choose the right one. The next few paragraphs indicate how to do this.

The orbits $\mathcal{O}_\varphi(x_0)$ and $\mathcal{O}_\varphi(x)$ divide Ω into three regions, one of which contains both $\mathcal{O}_\varphi(x_0)$ and $\mathcal{O}_\varphi(x)$ in its boundary. Since x_0 is fixed in this discussion, we will denote this region as Ω_x. Let γ be an arc in Ω_x joining x_0 to x. γ divides Ω_x into two subregions, one of which contains in its boundary the positive semi-orbit $\mathcal{O}_\varphi^+(x_0)$ and the other the negative semi-orbit. We call the first subregion Ω_x^+ and the second Ω_x^-. We note that for each $y \in \Omega_x$, $\varphi(t, y)$ lies in either Ω_x^+ or Ω_x^- for values of t large enough, and similarly for values of t small enough (large enough with negative sign). If, for a given y, these are the two sets Ω_x^+ and Ω_x^-, then we say that $\mathcal{O}_\varphi(y)$ *crosses* Ω_x. Otherwise, we say that $\mathcal{O}_\varphi(y)$ *is rooted in* Ω_x^+ or Ω_x^- depending on which set includes both ends of the orbit.

Let U^+ and U^- be the sets of points from Ω_x whose orbits are rooted in Ω_x^+ and Ω_x^- respectively, and let U be the region cut off from $\mathcal{O}_\varphi(x_0)$ by $\mathcal{O}_\varphi(x)$. Then it is easily seen that U^+, U^-, and U are all measurable, and we denote their measures as m^+, m^-, and m respectively. We see at once that U^+, U^-, and U are disjoint in pairs, and all lie in $[x \text{ side} : \mathcal{O}_\varphi(x_0)]$. Let $\bar{m}$ denote the area of $[x \text{ side} : \mathcal{O}_\varphi(x_0)]$. Then we have $\bar{m} - (m^+ + m^- + m) \geq 0$, since this number is the measure of

$$[x \text{ side} : \mathcal{O}_\varphi(x_0)] \setminus (U^+ \cup U^- \cup U).$$

We denote this number as $\tilde{m}$, and now consider the arc of length $2\tilde{m}$ cut off by $L(x_0)$. We mark off at the appropriate ends of this arc the arcs of length $2m^+ + \tilde{m}$ and $2m^- + \tilde{m}$. This leaves an arc of length $2m$, and the chord of this arc we denote as $L(x)$.

We now do the same construction for each $y \in [x \text{ side} : \mathcal{O}_\varphi(x_0)]$, assigning to the symbols $m(y)$, $m^+(y)$, $U^+(y)$, $\tilde{m}(y)$, etc., the meanings corresponding to the ones given these same symbols in the construction of $L(x)$. We wish to show that if $y \notin \mathcal{O}_\varphi(x)$, and y lies in $[x \text{ side} : \mathcal{O}_\varphi(x_0)]$, then $L(x)$ and $L(y)$ do not intersect, except possibly on the boundary, and then only in special cases.

Assume first that $x \in U(y)$. Then $U^+(y) \subseteq U^+(x)$, $U^-(y) \subseteq U^-(x)$, and $U(y) \supseteq U(x)$. We see that

$$[x \text{ side} : \mathcal{O}(x_0)] \setminus \big(U^+(x) \cup U^-(x) \cup U(x)\big) \supseteq$$

$$\supseteq [y \text{ side} : \mathcal{O}_\varphi(x_0)] \setminus \big(U^+(y) \cup U^-(y) \cup U(y)\big)$$

so that $\tilde{m}(y) \leq \tilde{m}(x)$. It follows that $2m^+(y) + \tilde{m}(y) \leq 2m^+(x) + \tilde{m}(x)$, and $2m^-(y) + \tilde{m}(y) \leq 2m^-(x) + \tilde{m}(x)$. Since $m(y) > m(x)$, we cannot have $m^-(x) = m^-(y)$, $m^+(x) = m^+(y)$, and $\tilde{m}(x) = \tilde{m}(y)$.

Thus at least one of the previous inequalities is strict, and $L(x)$ does not intersect $L(y)$, except maybe at an endpoint.

If $y \epsilon U(x)$, the result is the same.

If we have neither $x \epsilon U(y)$ nor $y \epsilon U(x)$, assume WoLOG that $y \epsilon U_x^+$. Then we have $U(y) \cup U^+(y) \subseteq U^+(x)$ and also $U(x) \cup U^-(x) \subseteq U^-(y)$. Thus, if $\tilde{m}(x) \geq \tilde{m}(y)$, then we have $2m^+(x) + \tilde{m}(x) \geq 2m^+(y) + \tilde{m}(y) + 2m(y)$, so that $L(x)$ and $L(y)$ cannot meet, except possibly at one endpoint. If $\tilde{m}(y) \geq \tilde{m}(x)$, then $2m^-(x) + \tilde{m}(x) + 2m(x) \leq 2m^-(y) + \tilde{m}(y)$, so that $L(x)$ and $L(y)$ again cannot meet, except possibly at one endpoint. Thus, if $y \epsilon U_x^+$, then $L(x)$ and $L(y)$ cannot meet, except possibly at a common endpoint, and similarly if $y \epsilon U_x^-$.

Thus, we have shown that there is an assignment of a chord to each of the orbits under φ which has no two chords intersecting in Ω. The collection of all the chords $L(x)$ constitutes a configuration in the unit disc which we call the *Kaplan diagram of the flow φ based at the point x_0*. We now note that the choice of another base point would not make any difference, up to rotations of the disc, so that we can consider each of these as the *Kaplan diagram* of φ.

Let $\mathcal{K}_1$ and $\mathcal{K}_2$ be two Kaplan diagrams in Ω. Suppose that there is a homeomorphism h of $\partial(\Omega)$ onto itself with these properties:

(i) If x_1 and x_2 are the endpoints of a chord in $\mathcal{K}_1$, then $h(x_1)$ and $h(x_2)$ are the endpoints of a chord in $\mathcal{K}_2$,

and

(ii) If y_1 and y_2 are the endpoints of a chord in $\mathcal{K}_2$, then $h^{-1}(y_1)$ and $h^{-1}(y_2)$ are the endpoints of a chord in $\mathcal{K}_1$.

Then we say the Kaplan diagrams $\mathcal{K}_1$ and $\mathcal{K}_2$ are *similar*, and we call h a *similarity* between them.

11.5. Polygonoids in Kaplan Diagrams. The chords which make up a Kaplan diagram for a flow in the disc do not, in general, exhaust the disc. In fact, the only case where they do so is the case when the Kaplan diagram consists of a collection of parallel chords. In general, however, there are points omitted in the diagram, and these are our next object of study. Let us first consider the Kaplan diagram for the family of orbits shown in Fig. 11.1 b. This can be seen in Fig. 11.1 c. The flows in Figs. 11.2a, 11.3a and 11.5a have their Kaplan diagrams exhibited in Figs. 11.2b, 11.3b, and 11.5b respectively. Note that in Fig. 11.1c, the chord AC does not represent any orbit, and the same is true of

$B_0 B_3$ in Fig. 11.2b, of $B_0 B_1$ in Fig. 11.3b, and of $C_1 C_3$ and $C_0 D_1$ in Fig. 11.5b. The triangle in Fig. 11.1c, the quadrangle in 11.2b, the infinitely-many-sided Fig. in 11.3b, and the triangle and quadrangle in 11.5b are called *polygonoids*, and they are characteristic of the anomalies described in Section 11.2.

Each polygonoid is a convex figure whose boundary is made up of non-intersecting chords together with a closed, nowhere dense set of measure 0 lying in the unit circle. Some of these chords are the images of φ-orbits; others are not. For those that are, we consider the closed set made up of the corresponding orbit, together with all the points on the orbits whose corresponding chords lie in the segment cut off by the given chord. For those that are not, we consider the open set made up of all the points on the orbits whose corresponding chords lie in the sector cut off by the chord. The union of these closed and open sets must exhaust the area of the disc. Thus the corresponding chords must exhaust the perimeter of the disc, showing that the remaining set must have measure 0. Let us now decipher the meaning of these polygonoids.

Let us begin with the case in which the polygonoid is actually a polygon. Then we have a finite number of sides, say n, and each one represents an open or closed portion of the disc. They cannot all be open, nor all closed, since in that case, it would be possible to separate the disc, contrary to its known connectedness. Similarly, the removal of the portions represented by the closed segments leaves an open connected region. By virtue of this connectedness, there is only one open segment. Thus, there is exactly one open side in the case where the polygonoid is a polygon. The same argument shows that a polygonoid with only finitely many closed sides is in fact a polygon and has only one open side.

In the case where there are infinitely many closed sides, let us examine the orbits they represent. By the lemma of the next section, if K is any compact set lying in Ω, K can meet at most finitely many of these orbits. Thus, employing this fact with a sequence of Jordan curves whose insides exhaust Ω, we see that the removal of these orbits and the closed sets they cut off leaves in Ω an open connected set. If there is more than one open set defined by the polygonoid, then any decomposition of the open sets into two collections gives a disconnection of the open set which was just seen to be connected. Consequently, we see that each polygonoid has at most one open side. On the other hand, we must have at least one open side, since otherwise, we would have the disc decomposed as the countable union of closed sets, which is known to be impossible in an arcwise connected space (cf. Sierpinski [1], also Theorem A.11).

Thus, each polygonoid has exactly one side which does not represent a φ-orbit, which we designate as *the open side* of the polygonoid.

11.6. Lemma. *Let φ be a flow in $\mathbb{R}^2$ with $\mathscr{F}(\varphi) = \square$. Let $\{x_i\}$ be a sequence of points of $\mathbb{R}^2$ so chosen that $\mathcal{O}_\varphi(x_i)$ can never separate $\mathcal{O}_\varphi(x_j)$ and $\mathcal{O}_\varphi(x_k)$, for any i, j, k.*

Then the sequence $\{x_i\}$ cannot be convergent.

PROOF: Suppose, contrariwise, that $x_i \to x_0$. Then $\mathcal{O}_\varphi(x_0)$ separates the plane, and we can assume, WoLOG, that all the x_i lie in one side of $\mathcal{O}_\varphi(x_0)$. We know that φ has no fixed points and thus no periodic orbits. It follows that there must be a number $\varepsilon_1 > 0$ small enough so that

$$d(x, x_0) < \varepsilon_1 \Rightarrow d(\varphi(t, x), x_0) > \varepsilon_1, \qquad \forall\, 1 \leq t \leq 2;$$

if there were no such ε_1, then we would have $x_0 = \varphi(t, x_0)$, $\exists\, 1 \leq t \leq 2$, by continuity and compactness. We choose $\varepsilon_1 > 0$ small enough to assure this implication and also satisfying $\varepsilon_1 < \frac{1}{2} d(x_0, \varphi(1, x_0))$.

We now choose an $\varepsilon_2 > 0$ so small that whenever $d(x, x_0) < \varepsilon_2$ and $0 \leq t \leq 2$, we have $d(\varphi(t, x), \varphi(t, x_0)) < \varepsilon_1$. Let i_0 be chosen so that $d(x_0, x_{i_0}) < \varepsilon_2$. We can assume, WoLOG, that x_{i_0} is the nearest point to x_0 on $\mathcal{O}_\varphi(x_{i_0})$.

We denote as L_0 the shortest subinterval of $[x_0, x_{i_0}]$ joining x_{i_0} to a point of $\mathcal{O}_\varphi(x_0)$, and denote as Ω_0 the portion of $\mathbb{R}^2$ cut off by $\mathcal{O}_\varphi(x_0)$ and $\mathcal{O}_\varphi(x_{i_0})$ which contains both of these orbits in its boundary (i.e., $\Omega_0 = [x_0 \text{ side} : \mathcal{O}_\varphi(x_{i_0})] \cap [x_{i_0} \text{ side} : \mathcal{O}_\varphi(x_0)]$). Define L_n for all values of n as $L_n = \varphi(n, L_0)$. Observe that $L_0 \cap L_1 = \square$, by the conditions on ε_1 and ε_2, and also that for any $x \in L_0$, $\varphi(t, x) \in L_0$ if $1 \leq t \leq 2$. For each n, L_n divides Ω_0 into two pieces, one of which contains in its boundary all the points $\varphi(t, x_0)$ with t large enough. We denote this portion as Ω_n^+, and the other as Ω_n^-. We note that $L_0 \cap L_1 = \square$, so that $\Omega_0^+ \supseteq \Omega_1^+$, and thus, $\Omega_m^+ \supseteq \Omega_n^+$ for all $m < n$. Also, L_1 is contained in Ω_0^+ except for its endpoints, so that for any $x \in L_0$ other than at its endpoints, $\varphi(t, x) \in \Omega_0^+$ for all $1 \leq t \leq 2$. Thus, $\varphi(t, x) \in \Omega_n^+ \subseteq \Omega_0^+$, $\forall\, n + 1 \leq t \leq n + 2$, $\forall\, n \geq 0$. It follows that $\varphi(t, x) \in \Omega_0^+$, $\forall\, t \geq 1$.

We now observe that if $x \in L_1$, then $\varphi(t, x) \in L_0$ for any $0 \leq t \leq 1$, since in that case, the point $y = \varphi(-1, x)$ would violate the basic condition on ε_1. It follows that for $-2 \leq t \leq -1$ and $x \in L_0$, other than at its endpoints, we have $\varphi(t, x) \in \Omega_0^-$. Thus, for $t \leq -1$, we have, anologously to the positive case, $\varphi(t, x) \in \Omega_0^-$.

We now observe that for $n < m$, $\Omega_n^+ \setminus \Omega_m^+$ is a set which has for its boundary a Jordan curve consisting of L_n, L_m and the arcs of the orbits of x_0 and x_{i_0} joining their endpoints. Thus, $\Omega_0 = \bigcup\limits_{n=-\infty}^{+\infty} (\Omega_n^+ \setminus \Omega_{n+1}^+)$.

Let $i_1 > i_0$ be chosen so that $d(x_{i_1}, x_0) < d(x_{i_0}, x_0) = d(x_0, \mathcal{O}_\varphi(x_{i_0}))$. Then $x_{i_1} \in \Omega_n^+ \setminus \Omega_{n+1}^+$ for appropriately chosen n. Therefore $\varphi(t, x_{i_1}) \in \Omega_0^+$ for t sufficiently large and $\varphi(t, x_{i_1}) \in \Omega_0^-$ for t sufficiently small (negatively large). Thus, $\mathcal{O}_\varphi(x_{i_1})$ separates $\mathcal{O}_\varphi(x_{i_0})$ and $\mathcal{O}_\varphi(x_0)$. Repeating the construction, we obtain x_{i_2} lying closer to x_0 than $\mathcal{O}_\varphi(x_{i_1})$, we see that $\mathcal{O}_\varphi(x_{i_1})$ separates $\mathcal{O}_\varphi(x_{i_0})$ from $\mathcal{O}_\varphi(x_{i_2})$, contradicting the hypothesis. Thus, $\{x_i\}$ cannot have a limit.   **QED**

11.7. Theorem. *Let φ and ψ be flows in Ω with $\mathscr{F}(\varphi) = \mathscr{F}(\psi) = \square$. Suppose further that φ and ψ are conjugate.*

Then the Kaplan Diagrams of the flows are similar.

PROOF: The proof of this theorem is to be found in Kaplan [1] and [2] and proceeds basically from the fact that the entire structure of the Kaplan diagram up to similarity is known when the order relationships of every triple is known. These order relationships for three orbits $\mathcal{O}_1$, $\mathcal{O}_2$, and $\mathcal{O}_3$ are one of these: either $\mathcal{O}_i$ separates $\mathcal{O}_j$ and $\mathcal{O}_k$ for some choice of indices, or the triple $\mathcal{O}_1 \mathcal{O}_2 \mathcal{O}_3$ run in clockwise order, or counter-clockwise. When these relations are known for every triple, the Kaplan diagram is fully known, up to similarity.   **QED**

11.8. Intrinsic Structure of Kaplan Diagrams. If we look at the chords constituting a Kaplan diagram, we see that except for the open sides of the polygonoids and diagonals of the polygonoids, there is no chord of the disc which fails to intersect one of the chords of the Kaplan diagram. Thus, the diagram is a maximal collection of chords, up to this special consideration. Put differently, if we look at a maximal collection of chords *and polygonoids*, then a Kaplan diagram is one of these; there is no chord in the disc which fails to intersect one of the chords or polygonoids of the diagram. Indeed, these chords and polygonoids exhaust the disc, if the polygonoids are taken as closed. Note that not every such maximal collection is a Kaplan diagram for some flow. To take the simplest counter-example, we can fill the disc with non-intersecting chords which are not parallel, *e.g.* the chords whose extensions pass through a given point lying outside Ω. We know from the theory that any flow in which, given any three orbits, one separates the other two, must have the property that its Kaplan diagram consists of parallel chords. However, this is not an essential point, and we could just as well widen our concept of a Kaplan diagram to allow invariance under similarities.

However, there is a more serious objection. Look at Fig. 11.8. In that figure, the point 0 is the limit of a sequence of triangles whose vertices are a_i, b_i, c_i, with $[a_i, c_i]$ in each case the shortest, and open, side. It follows that in that picture, $[a_i,\ b_i]$ is sensed in the direction from left to right if and only if $[b_i,\ c_i]$ is sensed in the opposite direction. Thus, there is no way to assign a direction to each orbit in such a way that the resulting configuration will be continuous for sense, much less actual motion. In fact, a close investigation of the corresponding collection of lines will show that the point corresponding to 0 cannot have any neighborhood in which the lines are "locally parallelizable". (We will not make this idea precise either. It means what it seems to mean.) If we could do this, then by Kaplan's theory, there would be a flow whose orbits would give the desired diagram, up to similarity. We will give a necessary and sufficient condition in the next sections.

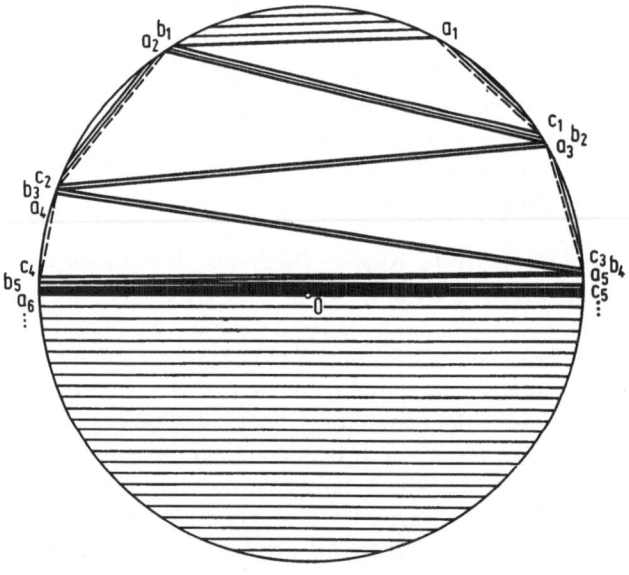

Fig. 11.8

Before we do this, however, let us note that whenever two polygonoids share a side, that side must be a *closed* side of each of the two polygonoids. If we satisfy this condition, and the local parallelizability condition, then, by Kaplan [1], [2], we can find a flow which gives us the desired configuration as its Kaplan diagram, up to similarity.

11.9. Definition. Let P be a polygonoid, and let its sides, in order of size, be denoted as $S_1, S_2, \ldots$. If the open side is either S_1 or S_2, *i.e.* if the open side is either the longest or second-longest side, then we call P *obtuse*; otherwise P is *acute*.[3]

11.10. Theorem. *Let $\mathcal{K}$ be a maximal disjoint collection of chords and polygonoids lying in the disc Ω. (i.e. any chord in the disc which does not intersect any chord in the collection lies in the interior of one of the polygonoids or on its boundary.) Let each polygonoid be assigned one of its sides as the open side with the proviso that any side which lies in two polygonoids is never chosen as the open side.*

Then $\mathcal{K}$ is the Kaplan diagram of a flow, up to similarity, if and only if every sequence $\{P_i\}$ of distinct acute polygonoids satisfies diam $(P_i) \to 0$.

PROOF: Let us suppose, first, that the condition is violated, *i.e.* that a sequence of acute polygonoids exists for which the diameters are bounded away from 0, say by $\varepsilon > 0$. Since the polygonoids are distinct, and thus have disjoint interiors, their areas form a convergent series, and are thus themselves convergent to 0. The polygonoids are convex, and thus for each, the area is no less than half the length of the longest side multiplied by the distance of the furthest point of the polygonoid from the line containing the longest side. At most one of the polygonoids contains the center of Ω, and for each of the others, the longest side is equal to its diameter, since the polygonoid is inscribed in the unit circle. Thus, $\{P_i\}$ lies in a strip of width w_i, where $w_i \to 0$, since the length of the longest side is bounded below. For w_i small enough, we see that the strip must intersect the unit circle in two arcs, and that all but one of the remaining sides must join two points lying in the same arc, while one side joins the two arcs. For w_i small enough, the distance between the two arcs is greater than either of them, so that the sides of the polygonoid which join the two arcs are the longest and the second-longest, and each of the other sides of the polygonoid joins two points of the same arc. We see that the open side must be one of these short sides, since the P_i are all acute. Now let us consider a flow φ for which $\mathcal{K}$ might be the Kaplan diagram and let us look at the direction of flow along the φ-orbits which correspond to the closed sides of P_i. We see that for chords lying in the open segment cut off by the open side, and very close to the open side, the corresponding orbits lie close to the orbits we get from the

---

[3] We must make some special comment about the case in which it is not clear which is the longest and/or second longest side of a polygonoid. These will all be called acute. Every polygonoid which does not contain the center of the disc Ω is inscribed in a semicircle and thus has a longest side. Any polygonoid whose longest side has length s and whose second and third longest sides are equal can be shown by elementary geometry to have an area of at least $\dfrac{s^3}{16}$. Thus, the diameters of these polygonoids always converge to 0.

closed side, so that the direction of flow on each of the closed side orbits must correspond to their order as sides of the polygonoid.

These considerations impose on each chord of the Kaplan diagram a sense, and these senses are consistent, in the obvious meaning, that every point has a neighborhood in which all the chords move in the same direction, more or less.

Now for the polygonoids P_i, we see that the longest and second-longest sides are oriented in the same direction *around the boundary of P_i*. This means that they must be oreinted in the opposite directions along the long axis of P_i, in a sense which will become obvious just below. In fact, let x_i, for each i, denote the center of the longest side of P_i. Since this side has length at least ε, the distance of x_i from 0 is no more than $1 - \dfrac{\varepsilon^2}{8}$, so that $\{x_i\}$ must have a subsequence which is convergent, say to a point x_0. If x_0 lies on a chord of $\mathcal{K}$, or on a closed side of one of the polygonoids, then we can see at once that for any point on that chord or side, we have a difficulty of the kind illustrated in Fig. 11.8. If x_0 lies on an open side, then the same sort of difficulty occurs at any point of the orbit corresponding to any closed side of the same polygonoid.

Conversely, when such a difficulty does exist, the sides of the polygonoids approaching the point of difficulty must be oppositely oriented. The corresponding polygonoids approach the given point, and so the longest sides of the polygonoids approach the length of the chord through that point (call it x_0), and are thus bounded away from 0. Also, the reversal of directions assures us that the open side is not one of the long sides, and thus, whenever we have a difficulty of the type shown, the condition of the theorem is violated.

Now assume we have a maximal collection of chords and polygonoids which satisfy the conditions. The remaining part of the proof rests heavily on the Moore Decomposition Theorem (cf. Moore [1], Theorem 22), which we shall state but not prove: Let $\mathcal{M}$ be a collection of compact sets in $\mathbb{R}^2$, none of which separates $\mathbb{R}^2$, which form an upper semi-continuous family (A family of sets is called upper semi-continuous if every neighborhood U of one of the sets contains a neighborhood V of the same set with the property that any set from the family which intersects V lies entirely in U.) and which exhaust $\mathbb{R}^2$. Let the space $\mathbb{R}^2/\mathcal{M}$ be defined with the elements of $\mathcal{M}$ as its points, where a set E of members of $\mathcal{M}$ is called open in $\mathbb{R}^2/\mathcal{M}$ if the union of those members is open in $\mathbb{R}^2$. Then the space $\mathbb{R}^2/\mathcal{M}$ is homeomorphic with $\mathbb{R}^2$.

We will begin by introducing into each of the polygonoids of $\mathcal{K}$ a structure which we will call a *striation*. Let the points of the open side of each polygonoid be paired with the points in the remainder of the boundary of the polygonoid by mapping that remaining boundary linearly onto the open side with respect to arc length. Note that most of the polygonoids are long and thin, and that in the case of obtuse polygonoids, the open side is about the same length as the remaining perimeter, while in the acute polygonoids, it is less than half, and often much less than that. We now join each pair of corresponding points by a closed line interval, and this collection of line intervals is called a *striation*. Note that for long, thin polygonoids, the elements of the striation are about the size of the width of the polygonoid if the polygonoid is obtuse, but can approximate its length if it is acute.

For each of the striation intervals which reaches to the boundary of Ω, we eliminate it from Ω. The conditions of the theorem assure us that the remaining set in Ω is still homeomorphic with $\mathbb{R}^2$. Now, we make up the elements of $\mathcal{M}$ as follows: Each point which lies on a chord of $\mathcal{K}$ which is not in the closure of any polygonoid is an element of $\mathcal{M}$. For each point which lies in just one polygonoid and is not on one of the removed segments, the striation interval in which it lies is an element of $\mathcal{M}$, *unless some point of that interval lies in two polygonoids*. For every point which lies in two polygonoids, the union of the two striation intervals on which it lies is an element of $\mathcal{M}$. It is clear that each of the elements of $\mathcal{M}$ is compact and does not separate Ω', the part of Ω which is left after the removal of the segments mentioned before. To see that the collection is upper semi-continuous, we note that in each polygonoid it is upper semi-continuous, and that for every sequence of distinct polygonoids, the acute ones must go to 0 in diameter, while the obtuse ones must go to 0 in width. Thus, the elements of $\mathcal{M}$ in these polygonoids become small uniformly as we pass down the sequence of polygonoids.

We will now define the flow φ by defining a flow $\hat{\varphi}$ along the chords and closed sides of polygonoids in $\mathcal{K}$. Let h be a homeomorphism of $\Omega'/\mathcal{M}$ onto $\mathbb{R}^2$, and for each pair (x, y) of points of Ω' which lie on chords or closed sides of $\mathcal{K}$, we define $d^*(x, y) = d\big(h(x), h(y)\big)$.[4] Then let a function f be defined by

$$f(x) = \inf_z \big(d^*(x, z) + d(z, \partial(\Omega'))\big).$$

We see immediately that $|f(x) - f(y)| \leq d^*(x, y)$, so that $f\big(h^{-1}(x)\big)$ is a continuous (indeed, Lipschitzian) function of $x \in \mathbb{R}^2$. Now we

---

[4] The mapping h is a homeomorphism of $\Omega'/\mathcal{M}$ onto $\mathbb{R}^2$. As such, we can also consider it, ambiguously, as a continuous mapping of Ω' onto $\mathbb{R}^2$. We use the symbol in both contexts. The reader should be able to follow the differences.

orient each of the chords and closed sides in Ω' in a consistent way, and define the algebraic flow $\hat{\varphi}$ to move along each of these paths in the indicated direction with speed $f(x)\,g(x)$, where $g(x)$ is a function which we will define below in such a way that the resulting flow will generate a continuous flow in $\mathbb{R}^2$.

The function g will be neither continuous nor bounded, but will have the property that it is constant on each chord of $\mathscr{K}$ and bounded on each compact set in Ω'. Note that f has the property that for all $x \in \Omega'$, $f(x) \leq d\big(x, \partial(\Omega)\big)$. Thus, since $g(x)$ is constant on each chord, the algebraic flow $\hat{\varphi}$ on that chord moves with a speed no faster than a constant multiple of canonical flow, and therefore that $\hat{\varphi}$ is defined as an *algebraic flow* is not in question. It is clear that if the resulting flow is to be continuous, then the value taken by g on any closed side of a polygonoid must bear approximately the same ratio to the values taken on chords near the open side that the sum of the lengths of the closed sides bears to the length of the open side. In fact, this will give us the definition of g. For each polygonoid P_i in the Kaplan diagram $\mathscr{K}$, let r_i be the ratio between the sum of the lengths of the closed sides and the length of the open side. Now we select a chord C_0 from $\mathscr{K}$ which is not a side of any polygonoid, and define g to be 1 on that chord. On any other chord C, we define g to be $\prod_i r_i^{\varepsilon_i}$, where ε_i is defined in the following way for each i:

(a) if C_0 and C lie in the same component of $\Omega \smallsetminus P_i$, then $\varepsilon_i = 0$.

(b) if C and C_0 lie in different components of $\Omega \smallsetminus P_i$, and a line segment joining an interior point of P_i to a point of C_0 intersects the open side of P_i, then $\varepsilon_i = 1$.

(c) if C and C_0 lie in different components of $\Omega \smallsetminus P_i$ and a line segment joining an interior point of P_i to a point of C intersects the open side of P_i, then $\varepsilon_i = -1$.

(d) If the line segments mentioned in (b) and (c) above both intersect closed sides of P_i, then $\varepsilon_i = 0$.

Clearly, the first order of business with such a definition is to show that the product converges. We observe that the chords in $\mathscr{K}$ which separate C from C_0 are bounded below in length, say by the number $c > 0$. For each polygonoid P_i making ε_i either 1 or -1, we define c_i to be the length of the open side, and b_i the length of the closed side which is intersected by one of the line segments in the definition. These two sides cut off one or two arcs which lie between them. Given any two polygonoids P_i and P_j both of which separate C from C_0, the arcs corresponding to them are disjoint. We denote the sum of the lengths of these arcs for each P_i as a_i, and observe that the perimeter of P_i is less than $a_i + b_i + c_i$, $\vee\, i$. It follows that $r_i < \dfrac{a_i + b_i}{c_i}$. Since $b_i < a_i + c_i$, we have

$$1 \le r_i < \frac{c_i + 2a_i}{c_i} = 1 + 2\,\frac{a_i}{c_i} \le 1 + 2\,\frac{a_i}{c},$$

so that $\prod\limits_i r_i^{\varepsilon_i}$ converges, since $\sum\limits_i 2\,\dfrac{a_i}{c} < \dfrac{4\pi}{c} < \infty$. In the same way, we see that g is bounded on compact sets.

We see from the definition that the values of g on the closed sides of any polygonoid and near the open side must be in the proper ratio. Furthermore, if $x_i \in \Omega'$ and $x_i \to x_0$, where each x_i lies on a chord of $\mathcal{K}$, and x_0 lies on a chord of $\mathcal{K}$, then $g(x_i) \to g(x_0)$. Thus, we need only show that $f(x) \neq 0$, $\vee\, x \in \Omega'$, to conclude that the movement in the assigned direction along each chord of $\mathcal{K}$ with a speed $f(x)\,g(x)$ at the point $x \in \Omega'$ induces in $\mathbb{R}^2$ a continuous flow which will have no fixed points. Suppose that $f(x_0) = 0$. Then there must be a sequence $\{x_n\}$ in Ω' such that $d^*(x_0, x_n) \to 0$ and $d(x_n, \partial(\Omega')) \to 0$. Let $[x_0]$ be the set of $\mathcal{M}$ containing x_0 and let the open set U be defined in Ω' by

$$U = \left\{ x \in \Omega' \,\middle|\, d(x, [x_0]) < \tfrac{1}{2}\, d([x_0], \partial(\Omega')) \right\}.$$

Then the set of elements of $\mathcal{M}$ lying wholly in U is an open set in $\mathbb{R}^2/\mathcal{M}$, and so all but finitely many of the x_n lie in it. But for all of these, $d(x_n, \partial(\Omega')) > \tfrac{1}{2}\, d([x_0], \partial(\Omega'))$. Thus, $f(x)$ is never 0 in Ω'.

If we now define $\varphi = h[\hat{\varphi}]$, then φ has the required collection $\mathcal{K}$ for its Kaplan diagram, up to similarity.   **QED**

11.11. Theorem. *Let φ and ψ be two flows in Ω with empty fixed sets and the same orbits.*

Then φ and ψ are conjugate.

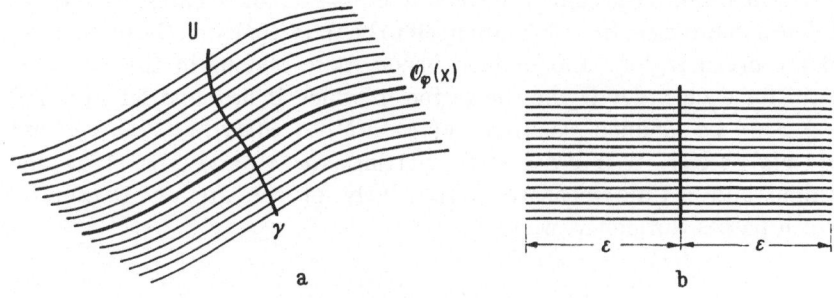

Fig. 11.9

PROOF: Again, we only sketch the proof, which relies heavily on the theory of local cross-sections (cf. Nemytskii and Stepanov [1] pp. 333 ff.). This theory has been explicitly avoided in this book up to now, and we will only quote the portion we need in this sketch. For each $x \in \mathscr{G}(\varphi)$, we can find a neighborhood U, a number $\varepsilon > 0$, and an arc γ so that every point $y \in U$ can be uniquely represented as $\varphi(t, z)$, where $z \in \gamma$ and $|t| < \varepsilon$ (see Fig. 11.9). Now we can modify the neighborhood and the representation in such a way that the local intersection of any orbit with the boundary of the neighborhood is at most two points, as shown in Fig. 11.10.

Let $\{J_n\}$ be a sequence of Jordan curves in Ω with the properties that $J_n \subseteq \text{ins}(J_{n+1})$, $\forall n$, and $\Omega = \bigcup_n \text{ins}(J_n)$. For each n, we can cover $J_n \cup \text{ins}(J_n)$ with finitely many of the modified neighborhoods in such a way that the covering for J_{n+1} always includes the covering for J_n. Let the sets in this covering be so numbered that the covering for J_n consists of the modified sets numbered from 1 to k_n, $\forall n$, and let these modified neighborhoods be denoted as V_i, $\forall i \geq 1$. For each number j, we can look at the set $W_j = V_j \setminus \bigcup_{i<j} V_i$, and map it back onto the representation of V_j, as shown in Fig. 11.11. We see that a finite number of "horns" intrude into V_j from the V_i, $i < j$. These wedge-shaped areas intrude from above, below, left, and right, but it is only those which intrude from above and below which are of special concern here. At the points of *these "horns"*, we add to the figure the "vertical" lines shown in Fig. 11.12, and these will divide each of the areas $V_j \setminus \bigcup_{i<j} V_i$ into a finite collection of regions with the property that each orbit which meets one of these regions crosses it locally in a simple arc.

As a first step, let us consider these simple arcs, and for each such arc, let us look at the quotient of its φ-time by its ψ-time. Let a function u_0 be defined which takes this value on each such arc, and the value 0 on all the boundaries of the regions. We see that the quasi-flow $u_0\varphi$ is a flow which takes the same time as ψ to cross each of the arcs. Now let us define a mapping h_0 from $\mathbb{R}^2$ onto itself to be the identity on the boundaries of the given regions and defined inside those regions in the following way: for each x, let U_x be the region in which it lies, and let $y(x)$ and $t(x) > 0$ be defined so that $y(x) \in \partial(U_x)$, $\psi(t(x), y(x)) = x$, and $\psi(t, y(x)) \in U_x$, $\forall 0 < t < t(x)$. Setting $s(\varphi, x)$ and $s(\psi, x)$ as the φ-time and ψ-time measures respectively of the arc of $\mathcal{O}_\psi(x) \cap U_x$ which passes through x, we define

$$h_0(x) = \varphi\left(t(x)\, \frac{s(\varphi, x)}{s(\psi, x)},\ y(x)\right).$$

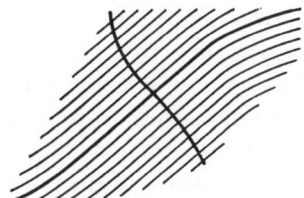
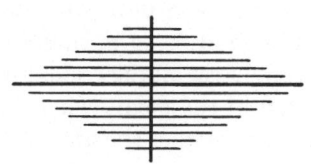

Fig. 11.10

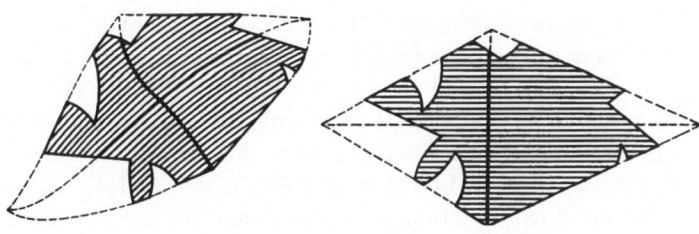

Fig 11.11

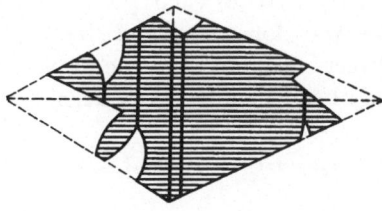

Fig. 11.12

It is clear that h_0 is a homeomorphism of $\mathbb{R}^2$ onto itself, and is so chosen that $h_0[\psi] = u_0\varphi$.

We note, however, that u_0 is not continuous on the boundaries of the regions, though it is continuous in the interiors of the regions, and constant on each orbit arc contained in a region. Our next step is to modify u_0 and h_0 to u_1 and h_1 so that $h_1[\psi] = u_1\varphi$ and such that u_1 is continuous in all the regions and on their boundaries, with the possible exception of the points where more than two regions come together. To do this, we number the arcs which make up the pieces of the boundaries, not including the excepted points, which we shall call *nodes*. We embed each open arc in an open set so that no two such sets intersect. We note that for each point on one of these arcs, u_0 takes the value 0 at that point and a constant value on each of the two subarcs of the orbit of the point lying on either side of the point. We now employ an analysis like that in Theorem 10.12 to alter u_0 to give us u_1, and alter h_0 to h_1 accordingly. All we need do now is take care of the nodes.

Note that we know very little about the behavior of u_1 at the node points; it need not even be bounded there. But we do know the behavior in the rest of the plane, and that u_1 is continuous there. The nodes do form a discrete set, so that we can form a modified (*i.e.* diamond-shaped) neighborhood about each node point such that no two intersect. Now, let us modify u_1 in each of these neighborhoods, and h_1 accordingly, so that the modified function u is continuous in the neighborhoods. Since these neighborhoods form a discrete set, and since each modification affects only the behavior *inside the neighborhood*, it is clear that we need only exhibit the modification in a single neighborhood.

Let x_0 be the node point, and let γ be the local cross-section through x_0. Recall that we formed the modified neighborhood, which we will call U, in the following way. We have a neighborhood V of x_0 and a number $\varepsilon_0 > 0$ with the property that for each $y \in V$, y can be represented as $\varphi(t, x)$ in exactly one way for $x \in \gamma$ and $|t| < \varepsilon_0$. We then index γ by the numbers between $-\varepsilon_0$ and ε_0 in such a way that x_0 corresponds to 0 with the image of s being denoted as x_s. Then for each $-\varepsilon_0 < s < \varepsilon_0$, we call

$$\{\varphi(t, x_s) \mid -\varepsilon_0 + |s| < t < \varepsilon_0 - |s|\}$$

the *major arc through* x_s, and denote it by A_s, and the neighborhood U is the union of all these major arcs.

We now define three functions on the interval $(-\varepsilon_0, \varepsilon_0)$, which will all turn out to be continuous. We define c and d by

$$c(s) = u_1\big(\varphi(-\varepsilon_0 + |s|, x_s)\big)$$

and

$$d(s) = u_1\big(\varphi(\varepsilon_0 - |s|, x_s)\big).$$

Note that c and d are in fact continuous in the interval $[-\varepsilon_0, \varepsilon_0]$ and take the same values at the endpoints, namely $u_1(x_{-\varepsilon_0})$ and $u_1(x_{\varepsilon_0})$. We define $M(s)$ to be the quotient of $2\varepsilon_0 - 2|s|$ by $\mu_{u_1\varphi}(A_s)$, and note that this is the quotient of continuous functions which are nowhere 0 and is thus continuous, and also that the limit of the quotient exists at $-\varepsilon_0$ and ε_0 and is the same as the functions c and d there.

Define $e(s) = \dfrac{1}{2M(s)} - \dfrac{1}{4c(s)} - \dfrac{1}{4d(s)}$. Then e is continuous in the interval $[-\varepsilon_0, \varepsilon_0]$, taking the value 0 at the endpoints. Let a constant $0 < b < 1$ be chosen so that $4|b\,e(s)\,M(s)| < 1$ for all values of s. We then define the function a by

$$\frac{1}{a(s)} - \frac{1}{M(s)} = 2b\,e(s).$$

Note that $a(s) > 0$.

Define a function f of five variables by

$$f(\alpha, a, c, d, t) = \begin{cases} c + \big(2(t+1)\big)^{\alpha^{\pm 1}}(a - c), & \forall\, -1 \le t \le -\dfrac{1}{2}, \\[2mm] a & , \quad \forall\, -\dfrac{1}{2} \le t \le \dfrac{1}{2}, \\[2mm] d + \big(2(1-t)\big)^{\alpha^{\pm 1}}(a - d), & \forall\, \dfrac{1}{2} \le t \le 1, \end{cases}$$

where the exponent of α is taken to be the sign of $a - c$ (resp. $a - d$). Note that f is continuous in the five variables, and is (weakly) decreasing in α for every choice of $a, c, d,$ and t, while the integral $\displaystyle\int_{-1}^{1} \frac{dt}{f(\alpha, a, c, d, t)}$ is strictly increasing in α when $\alpha > 0$ except when $a = c = d$. We now define $\alpha(s) > 0$ to satisfy the equation

$$\int_{-1}^{1} \frac{dt}{f\big(\alpha(s), a(s), c(s), d(s), t\big)} = \frac{2}{M(s)}, \quad \forall\, -\varepsilon_0 < s < \varepsilon_0, \text{ with } \alpha(s) = 1$$

whenever $e(s) = 0$, and define the function g in U by

$$g\big(\varphi(t(\varepsilon_0 - |s|), x_s)\big) = f\big(\alpha(s), a(s), c(s), d(s), t\big), \qquad \forall\, |t| \le 1.$$

In order to do this, we need to know that there is such an $\alpha(s)$ for each s. We see that in every case,

$$\frac{1}{4c(s)} + \frac{1}{2a(s)} + \frac{1}{4d(s)} = \frac{1}{2M(s)} - e(s) + \frac{1}{2M(s)} + be(s)$$

$$= \frac{1}{M(s)} - (1-b)\,e(s)$$

and

$$\frac{1}{M(s)} = \frac{1}{a(s)} - 2be(s).$$

Thus, when $e(s) > 0$, we have

$$\frac{1}{4c(s)} + \frac{1}{2a(s)} + \frac{1}{4d(s)} < \frac{1}{M(s)} < \frac{1}{a(s)}$$

with the reverse inequalities holding when $e(s) < 0$. When $a(s) > c(s)$ and $a(s) > d(s)$, then for very large values of α

$$\int_{-1}^{1} \frac{dt}{f(\alpha, a(s), c(s), d(s), t)}$$

is very close to $\dfrac{1}{2c(s)} + \dfrac{1}{a(s)} + \dfrac{1}{2d(s)}$ while for small values of α, it is close to $\dfrac{2}{a(s)}$, so that there is always a value where it satisfies the condition. When $a(s) < c(s)$ and $a(s) < d(s)$, we have the opposite inequalities and the same result. If $c(s) \leq a(s) \leq d(s)$, then for large values of α, the integral approximates $\dfrac{1}{2c(s)} + \dfrac{3}{2a(s)}$, while for small values it approximates $\dfrac{3}{2a(s)} + \dfrac{1}{2d(s)}$.

Since

$$\frac{1}{d(s)} \leq \frac{1}{a(s)} \leq \frac{1}{c(s)},$$

we have

$$\frac{3}{2a(s)} + \frac{1}{2d(s)} \leq \frac{1}{2c(s)} + \frac{1}{a(s)} + \frac{1}{2d(s)} \leq \frac{1}{2c(s)} + \frac{3}{2a(s)}$$

and

$$\frac{3}{2a(s)} + \frac{1}{2d(s)} \leq \frac{2}{a(s)} \leq \frac{1}{2c(s)} + \frac{3}{2a(s)}.$$

But $\dfrac{2}{M(s)}$ always lies between

$$\frac{1}{2c(s)} + \frac{1}{a(s)} + \frac{1}{2d(s)} \quad \text{and} \quad \frac{2}{a(s)},$$

and thus,

$$\frac{3}{2a(s)} + \frac{1}{2d(s)} \leq \frac{2}{M(s)} \leq \frac{1}{2c(s)} + \frac{3}{2a(s)},$$

so that in this case also, $\alpha(s)$ is defined and continuous. The same analysis holds when $d(s) \leq a(s) \leq c(s)$. We thus conclude that $\alpha(s)$ is everywhere well-defined, and is continuous by Theorem 10.11.

Now we have the function g continuous in U, and in fact continuous in $\bar{U}$, with $g\big(\varphi(t(\varepsilon_0 - |s|), x_s)\big) = u_1\big(\varphi(t(\varepsilon_0 - |s|), x_s)\big)$ for $t = \pm 1$, and for each $-\varepsilon_0 < s < \varepsilon_0$, we have

$$\int_{-\varepsilon_0 + |s|}^{\varepsilon_0 - |s|} \frac{dt}{g\big(\varphi(t, x_s)\big)} = \frac{2(\varepsilon_0 - |s|)}{M(s)}$$

$$= \mu_{u_1\varphi}(A_s).$$

Thus, the replacement of u_1 by g inside U will change u_1 to a function u meeting our requirements, and the corresponding change of h_1 to h will yield the theorem.   **QED**

11.12. Theorem. *Let φ and ψ be flows in $\mathbb{R}^2$ with $\mathscr{F}(\varphi) = \mathscr{F}(\psi) = \square$.*

Then φ and ψ are conjugate if and only if their Kaplan diagrams are similar.

PROOF: In case φ and ψ are conjugate, there is a homeomorphism of $\mathbb{R}^2$ onto itself which carries the orbits of φ onto the orbits of ψ. Thus, the order relations of any triple of orbits is preserved, and the work of Kaplan shows that the Kaplan diagram (up to similarity) is dependent only on these order relationships.

Let h_0 be a mapping of $\partial(\Omega)$ onto itself which carries the Kaplan diagram $\mathscr{K}_\varphi$ onto $\mathscr{K}_\psi$. We will consider the flow φ' defined in $\mathbb{R}^2$ from $\mathscr{K}_\varphi$ as in Theorem 11.10, and we will show that it is conjugate with both the flow ψ' defined in a similar way from K_ψ, and with the flow φ. We take up the former conjugacy first.

It will be sufficient to show that there is a homeomorphism of $\mathbb{R}^2$ onto itself which carries each orbit of φ' onto an orbit of ψ'. Reaching back to Ω, we see that this means the existence of a mapping of Ω onto itself which takes each chord of $\mathscr{K}_\varphi$ onto a chord of $\mathscr{K}_\psi$ and each polygonoid of

$\mathscr{K}_\varphi$ onto a polygonoid of $\mathscr{K}_\psi$ so as to preserve the striation. This visually "obvious" fact requires some proving, however, and for this, we reach again into the work of Kaplan.

Kaplan shows the existence of certain curves in Ω, which we will call *major transversals*, and we will rework his theory a little at this point. A *major transversal* of $\mathscr{K}_\varphi$ is an arc running from $\partial(\Omega)$ to $\partial(\Omega)$ which cuts each chord of $\mathscr{K}_\varphi$ at most once and traverses each polygonoid of $\mathscr{K}_\varphi$ in a striation interval. It is an easy, but tedious, exercise to show that one can construct countably many of these so that they are dense in Ω. The important thing is that since $\mathscr{K}_\varphi$ and $\mathscr{K}_\psi$ are similar, if we choose a major transversal T_φ^1 in $\mathscr{K}_\varphi$, we can find a major transveral T_ψ^1 in $\mathscr{K}_\psi$ which intersects the corresponding chords. If we then choose a major transversal T_ψ^2 in $\mathscr{K}_\psi$ which does not intersect T_ψ^1, then we can find a major transversal T_φ^2 in $\mathscr{K}_\varphi$ which intersects the corresponding chords but does not intersect T_φ^1. Next, we choose T_φ^3 not to intersect T_φ^1 and T_φ^2, continuing so that the odd-numbered transversals, T_φ^{2n+1}, are dense in $\mathscr{K}_\varphi$, while the even-numbered ones, T_φ^{2n}, are dense in $\mathscr{K}_\psi$.

For each point or striation interval on a major transversal, we identify it by the chord or polygonoid(s) on which it lies, and make correspond to it the point or striation interval on the corresponding transversal and on the corresponding chord or polygonoid. Mapping corresponding striation intervals linearly, we see easily that this gives a homeomorphism $\bar{h}$ of Ω onto itself which is an extension of the homeomorphism h and which carries each set of the decomposition $\mathscr{M}_\varphi$ corresponding to $\mathscr{K}_\varphi$ onto a set of $\mathscr{M}_\psi$, similarly defined, with each chord of $\mathscr{K}_\varphi$ mapped onto a chord of $\mathscr{K}_\psi$ and each polygonoid of $\mathscr{K}_\varphi$ mapped onto the corresponding polygonoid. Thus we have φ' conjugate with ψ'.

To show that φ' is conjugate with φ (and thus also that ψ' is conjugate with ψ), we make use of the major transversals once more. Let h be the mapping in Theorem 11.10 which carries $\Omega'/\mathscr{M}$ onto $\mathbb{R}^2$. We choose a sequence $\{J_i\}$ of Jordan curves such that $J_i \subseteq \mathrm{ins}\,(J_{i+1})$, $\forall\, i$, and $\bigcup_{i=1}^{\infty} \mathrm{ins}\,(J_i) = \mathbb{R}^2$. Let x_1' be any point of $h(T_\varphi^1)$, and let x_1 be any point of the φ-orbit whose corresponding chord in $\mathscr{K}_\varphi$ is the pre-image of $\mathcal{O}_{\varphi'}(x_1')$. Using theory of local cross-sections, we can construct an arc running from ∞ to ∞ which intersects in a single point each of the orbits whose corresponding chord is cut by T_φ^1.

We do this in the following way: for each chord of $\mathscr{K}_\varphi$ which intersects T_φ^1, we choose a point of the corresponding φ-orbit, and obtain a local cross-section at that point. For each such cross-section, we consider the portion of T_φ^1 corresponding to the orbits which meet the cross-section, including for each chord whose intersection with T_φ^1 belongs to a striation interval the entire interval. It is clear that each of these sets is an arc in T_φ^1, and by the local compactness of T_φ^1, we can select countably many of these open arcs so that they cover T_φ^1 and so that only finitely many intersect any compact sub-interval of T_φ^1. Denoting the countable family of cross-sections as $\{\gamma_i\}$, we see that for each γ_n, the set of points of γ_n which intersect orbits which meet any of the γ_i, $1 \leq i < n$, is a finite union of arcs. Assume that the cross-sections γ_i, $\forall\, 1 \leq i < n$, have been changed so that each φ-orbit which meets two of them meets them in the same point. Then on the set C_n of points of γ_n which intersect φ-orbits which also intersect $\bigcup\limits_{i=1}^{n-1} \gamma_i$, we can define a bounded continuous function f_n so that $f_n(x) \in \bigcup\limits_{i=1}^{n-1} \gamma_i$, $\forall\, x \in C_n$. Extend f_n to a continuous bounded function on γ_n in any way, and now replace γ_n in the sequence by $\{f_n(x) \mid x \in \gamma_n\}$. The cross-sections $\gamma_1, \ldots, \gamma_n$ will now have the additional property assumed for the cross-sections $\gamma_1, \ldots, \gamma_{n-1}$. Then $\bigcup\limits_{i=1}^{\infty} \gamma_i$ will be an arc of the type we require.

We will make the points of this arc correspond to the points of $h(T_\varphi^1)$ in the obvious way, so that the orbits corresponding to the same chord of $\mathcal{K}_\varphi$ will correspond. Next, we do the same thing for T_φ^2, in such a way that the two arcs indicated do not intersect, and so that the second arc lies on the "proper" side of the first. We will call these arcs A_1 and A_2 respectively, and we further require that for each arc of a φ'-orbit which lies wholly in ins (J_1) and joins a point of $h(T_\varphi^1)$ to a point of $h(T_\varphi^2)$, the corresponding arc between the corresponding points of A_1 and A_2 should have the same time measure, each with respect to its own flow.

Next, we build an image for $h(T_\varphi^3)$. We require that the image A_3 must cut the proper φ-orbits, each in one point, and something more. If a point of $h(T_\varphi^3)$ lies on one of the φ'-orbits between a point of $h(T_\varphi^1)$ and $h(T_\varphi^2)$, and if that arc lies wholly in ins (J_2), then the chosen point on A_3 must divide the arc between A_1 and A_2 on the appropriate φ-orbit so that the ratio of φ-time measures is the same as the corresponding ratio for the φ'-times on the corresponding φ'-orbit. If the point on $h(T_\varphi^3)$ does not lie between two such points, but if the arc between that point and the nearer of the two intersections with $h(T_\varphi^1)$ and $h(T_\varphi^2)$ along the orbit lies wholly in ins (J_2), then again we require that the time measures be equal.

Now we come to $h(T_\varphi^4)$. For each point $x' \in h(T_\varphi^4)$, we wish to choose a point of A_4 which comes from the appropriately chosen φ-orbit, and which has certain other properties. It may be that two or three of the arcs $h(T_\varphi^1)$, $h(T_\varphi^2)$, and $h(T_\varphi^3)$ intersect $\mathcal{O}_{\varphi'}(x')$. If so, then x' might lie between two of these intersections on $\mathcal{O}_{\varphi'}(x')$. If it does, and if the entire orbit arc between the two nearest intersections lies in ins (J_3), then the corresponding arcs must meet the corresponding φ-orbit, and we require that the point chosen on the φ-orbit divide the φ-time of the arc of the φ-orbit in the same ratio as x' divides the φ'-time of the arc of the φ'-orbit. If x' does not lie between two such intersections, there remains the possibility that the arc of the φ'-orbit joining x' to the nearest intersection point lies wholly in ins (J_3). In that case, we require that the point on A_4 be located so as to make the corresponding arcs have the same time measure, each in its own flow time.

We continue in this way, choosing an image point for each point lying on each $h(T_\varphi^n)$ in accordance with the way that the orbit of that point interacts with $h(T_\varphi^1), \ldots, h(T_\varphi^{n-1})$ and ins (J_{n-1}). It can be shown that this mapping is a homeomorphism, which we will call $\tilde{h}$, and thus $\tilde{h}[\varphi']$ and φ are flows with the same orbits. Thus, they are conjugate, so that φ' and φ are conjugate.   **QED**

To show sharpness in Theorem 11.12, and to further establish the importance of conjugacy as an invariant, we now exhibit two flows in the plane with the same orbits which are not homeomorphic.

11.13. Example. (See Fig. 11.13.) Let φ be a flow which is defined in the unit disc Ω in the following way: Let two chords C_1 and C_2 be chosen in Ω, of equal length and meeting at a point, and let D be the

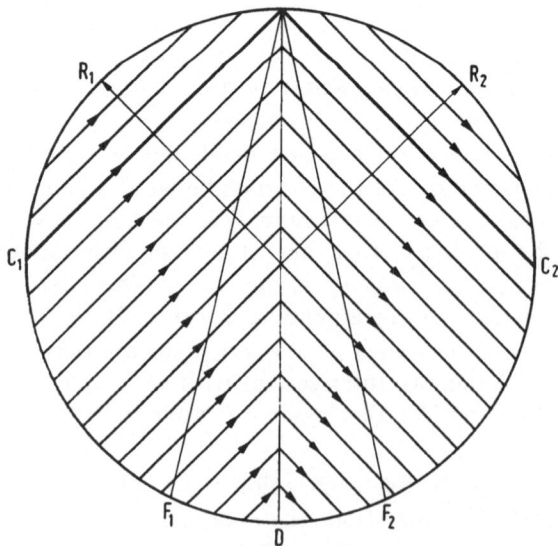

Fig. 11.13

diameter of Ω arising in that point. D cuts Ω into two semidiscs Ω_1 and Ω_2, respectively containing C_1 and C_2. A flow φ is defined in Ω which is canonical with respect to $\partial(\Omega)$ and in each Ω_i moves parallel to C_i.

Let R_1 and R_2 be the radii through the centers of C_1 and C_2 respectively, and let R be the radius to $C_1 \cap C_2$. For each point $x \in R$, let $\tau(x)$ be the φ-time from R_1 to R_2. Then $\tau(x)$ increases monotonically along R from 0 at the center of Ω to ∞ at the boundary. For $t > 0$, let $x(t) \in R$ be defined so that $\tau(x(t)) = t$.

Let F_1 and F_2 be two chords of equal length arising from $C_1 \cap C_2$ with $\mathscr{L}(C_1) < \mathscr{L}(F_1) < \mathscr{L}(D)$. We call the region of Ω between them the *inner wedge*. The flow ψ will be made from the flow φ by modifying its speed only inside this inner wedge. For each $x \in R$, we want the ψ-time from R_1 to R_2 to be $\sigma(x) = \tau(x) + (\sin(\tau(x)^2))^2$.

ψ is specifically constructed so that for any $0 < r < 1$, we can always find $x_1, x_2 \in R$ with $r < |x_1| < |x_2| < 1$ and $\sigma(x_1) - \sigma(x_2)$ as close to 1 as we wish. Using just this property, and the fact that $\varphi = \psi$ outside the inner wedge, we will show that φ and ψ are not homeomorphic, though they have the same orbits. Please note that φ and ψ have the "simplest" Kaplan diagram possible with the exception of parallel flow, and that all parallel flows are homeomorphic.

To show that φ and ψ are not homeomorphic assume that $\varphi = h[\psi]$, where h is a homeomorphism of Ω onto Ω. Clearly the images of C_1 and C_2 will be C_1 and C_2 in some order. WoLOG assume $h(C_i) = C_i$, $\forall i = 1, 2$. Let $\gamma_i = h(R_i)$, and $\{y_i\} = \gamma_i \cap C_i$, $\forall i = 1, 2$. Then for each $t > 0$, the φ-time along $\mathcal{O}_\varphi(h(x(t)))$ from γ_1 to γ_2 will be $t + (\sin(t^2))^2$.

For $i = 1, 2$, let s_i be chosen so that $R_i' = \varphi(s_i, R_i)$ passes through y_i. Then for $x \in R$ and $|x|$ close enough to 1, we have the φ-time of the arc of $\mathcal{O}_\varphi(x)$ between R_1 and R_2 differing from the φ-time between R_1' and R_2' by exactly $s_1 - s_2$, and thus the φ-time of $\mathcal{O}_\varphi(x(t))$ between R_1' and R_2' is exactly $t - s_1 + s_2$.

Select any δ satisfying $0 < \delta < \frac{1}{4}$ and let $\varepsilon > 0$ be chosen so small that $t \geq \delta$ implies either $d(y_1, x) > \varepsilon$ or $d(y_1, \varphi(t, x)) > \varepsilon$, and similarly for y_2. For $x \in R$ and $|x|$ large enough, $\mathcal{O}_\varphi(x)$ intersects R_i' and γ_i within ε of y_i, $\forall i = 1, 2$. It follows that the φ-time of $\mathcal{O}_\varphi(x(t))$ between γ_1 and γ_2 lies between $t - s_1 + s_2 - \delta$ and $t - s_1 + s_2 + \delta$. Thus, if $|x_1|$ and $|x_2|$ are large enough, say $r < |x_1| < |x_2| < 1$, then the φ-time of $\mathcal{O}_\varphi(x_2)$ between γ_1 and γ_2 exceeds the φ-time of $\mathcal{O}_\varphi(x_1)$, less 4δ. I.e., for all t_i large enough, with $t_1 < t_2$, $t_2 + (\sin(t_2^2))^2 > t_1 + (\sin(t_1^2))^2 - 4\delta$, which is false, since we can make $(t_1 + (\sin(t_1^2))^2) - (t_2 + (\sin(t_2^2))^2)$ as close to 1 as we like.

This same example is available with two flows whose common orbits give any Kaplan diagram which includes a polygonoid, *i.e.* any non-parallelizeable Kaplan diagram.

Kaplan-Markus Flows

11.14. Definition. Let Ω be a subset of $\mathbf{S}^2$. We say that φ is a *Kaplan-Markus flow* in Ω if

i) φ is a flow in $\bar{\Omega}$,

ii) $\mathrm{p}_\varphi(x) = \infty$, $\forall\, x \in \Omega$,

iii) $\alpha_\varphi(x) \cup \omega_\varphi(x) \subseteq \partial(\Omega)$, $\forall\, x \in \Omega$.

The work of Kaplan and Markus concerns Kaplan flows in the plane; we shall now take up Kaplan-Markus flows in finitely-connected regions. We begin with the case of the annulus, which we will reduce to the case of the disc by "cutting the annulus open" along an orbit. This requires a lemma, which is our next order of business.

11.15. Lemma. *Let $\Omega \subseteq \mathbf{S}^2$ be an annulus, and let $\acute{\varphi}$ be a Kaplan-Markus flow in Ω.*

Then there is a $y \in \Omega$ such that $\alpha_\varphi(y)$ and $\omega_\varphi(y)$ lie in the opposite components of $\partial(\Omega)$.

PROOF: Denote the boundaries of Ω as B^- and B^+. Assume the lemma fails. Then for each $x \in \Omega$, either $\alpha_\varphi(x) \cup \omega_\varphi(x) \subseteq B^-$ or $\alpha_\varphi(x) \cup \omega_\varphi(x) \subseteq B^+$, and we write $x \in \Omega^-$ or $x \in \Omega^+$ accordingly. Since $\bar{\Omega}$ is connected, we cannot have both $\Omega^- \cup B^-$ and $\Omega^+ \cup B^+$ closed in $\bar{\Omega}$. Thus, one of these sets has a limit point in the other, and WoLOG, we assume that there is a convergent sequence $\{x_i\} \subseteq \Omega^- \cup B^-$ whose limit is $x_0 \in \Omega^+ \cup B^+$.

For some of the possible subsequences $\{x_i\}$ and some sequences $\{t_j\}$ of real numbers, the sequence $\{\varphi(t_j, x_{i_j})\}$ has a limit. Let the set F consist of all such limits. We see by the method of Lemma 3.22 that F is compact and connected, and that $F \cap B^-$ and $F \cap B^+$ are both non-empty. We denote x_0 ambiguously as z_0 and choose an element $z_1 \in F$ so as to maximize the distance $\varrho(z_1, z_0)$. We can represent z_1 as the limit of a sequence $\{\varphi(t_{1,i}, x_{1,i})\}$, where $\{t_{1,i}\}$ is a sequence of real numbers and $\{x_{1,i}\}$ is a subsequence of $\{x_i\}$. Now let us consider the set $\bar{F}_1$ which consists of all limits of sequences of the form $\{\varphi(t_j, x_{1,i_j})\}$. We can select from $\bar{F}_1$ an element z_2 so as to maximize the distance $\varrho(z_2, \{z_0, z_1\})$. Then we represent z_2 as the limit of the sequence $\varphi(t_{2,i}, x_{2,i})$, where $\{t_{2,i}\}$ is a sequence of real numbers and $\{x_{2,i}\}$ is a subsequence of $\{x_{1,i}\}$. We continue with this process, always choosing z_i to maximize $\varrho(z_i, \{z_0, \ldots, z_{i-1}\})$.

Since $\mathbf{S}^2$ is compact in the metric ϱ, this distance must converge to 0 as $i \to \infty$. Each $\tilde{F}_i$ is a sub-continuum of $\tilde{F}_{i-1}$, and each must meet B^- and B^+. Let $\tilde{F} = \bigcap_{i=1}^{\infty} \tilde{F}_i$. We see at once that $z_i \in \tilde{F}$, $\forall\, i$, and in fact $\{z_i\}$ is dense in $\tilde{F}$. Furthermore, $\tilde{F}$ consists of all those points which can be represented as cluster-points of sequences of the form $\varphi(t_j, x_{j,j})$, and F meets both B^- and B^+. We can assume WoLOG that $\tilde{F} = F$, and that $x_{i,i} = x_i$, $\forall\, i$.

Let $z \in F \cap \Omega$. We wish to see that each of the orbits $\mathcal{O}_\varphi(x_i)$ only comes close to z "once". We will show that for any $\delta > 0$, there is an $\varepsilon > 0$ so that whenever $d(x_i, z) < \varepsilon$ and $|t| \geq \delta$, we must have $\big(d\varphi(t, x_i), z\big) \geq \varepsilon$. If that were not true, then for each $\varepsilon > 0$, no matter how small, we could always find a $y \in \Omega$ and a $t \geq \delta$ such that $d(y, z)$ and $d\big(\varphi(t, y), z\big)$ would both be less than ε. If ε is small enough (and we could take it this small), then the methods of the Gate Theorem assure us that we could make a φ-gate construction for the pair $\langle y, z \rangle$ so that there would be a Jordan curve in Ω with $\varphi(t, y)$ on one side of it for all t large, and on the other side for t small (large negative).

It follows that if y were of the form $\varphi(\bar{t}, x_i)$ then $\mathcal{O}_\varphi(x_i)$ could not have both ends in B^-, as we assumed. Thus, the $\varphi(\bar{t}, x_i)$ lying within some given ε of z must all have $d\big(z, \varphi(t, x_i)\big) > \varepsilon$ if $|t - \bar{t}| \geq \delta$. It follows from this that if $z_0 \in (F \cap \Omega) \setminus \mathcal{O}_\varphi(z)$, then we cannot have $d(z, z_0) < \varepsilon$, i.e., $\mathcal{O}_\varphi(z)$ is open in F.

For each $z \in F \cap \Omega$, we see from the foregoing that $\mathcal{O}_\varphi(z)$ is open in F. Thus, since F is separable, there are only countably many orbits in $F \cap \Omega$. Let these be denoted as $\mathcal{O}_\varphi(y_i)$, indexed over a subset of the integers, finite or infinite. We therefore have a compact connected set F_0 consisting of the two components B^- and B^+ of $\partial(\Omega)$ and all the orbits $\mathcal{O}_\varphi(y_i)$. We define $F_i = F_0 \setminus \bigcup_{j \leq i} \mathcal{O}_\varphi(y_j)$, $\forall\, i$, and we note that each F_i is compact. Thus, if all of these compacta were connected, then their intersection would be connected, by Theorem A.5. However, the intersection is exactly $B^- \cup B^+$, which is disconnected, so that F_i is disconnected, for some i. Let k be the least integer for which F_k is disconnected. Then $k > 0$, and F_{k-1} is connected. We will show that F_{k-1} is not connected. Let F_k be disconnected by the open sets U^- and U^+. Clearly, the set which contains B^- also contains the orbits which have their endpoints there, and the same is true for B^+. Since B^-, B^+, and the orbits $\mathcal{O}_\varphi(y_i)$ in F_k exhaust F_k, B^- and B^+ must lie in different ones of U^- and U^+, and WoLOG, let $B^- \subseteq U^-$, $B^+ \subseteq U^+$. $\mathcal{O}_\varphi(y_k)$ must have its endpoints in B^- or B^+; assume WoLOG it is B^-. Then $(F_k \cap U^-) \cup \mathcal{O}_\varphi(y_k)$ and $F_k \cap U^+$ disconnect F_{k-1}, contrary to the defining property of k.

Thus this contradiction gives us the lemma.   **QED**

We now employ the same method to obtain the next two theorems.

11.16. Theorem. *Let Ω be a finitely-connected region in $\mathbf{S}^2$ which is not simply-connected, and let φ be a Kaplan-Markus flow in $\bar{\Omega}$.*

Then there exists $y \in \Omega$ such that $\alpha_\varphi(y)$ and $\omega_\varphi(y)$ are contained in different components of $\partial(\Omega)$.

PROOF: Let the components of $\partial(\Omega)$ be denoted as $B_i, 1 \leq i \leq k$. If the theorem fails, then for every $z \in \Omega$, $\alpha_\varphi(z) \cup \omega_\varphi(z) \subseteq B_i$, for some i, and we then write $z \in \Omega_i$. We cannot have all the $\Omega_i \cup B_i$ closed in $\bar{\Omega}$, since that would violate the connectedness of $\bar{\Omega}$. Assume that $\Omega_n \cup B_n$ is not closed, and that some point $z_0 \in \partial(\Omega_n)$ lies in another $\Omega_i \cup B_i$, say for $i = m$. Then using the argument for Lemma 11.15 with B_n for B^- and B_m for B^+, we obtain a contradiction.   **QED**

11.17. Theorem. *Let Ω be a countably-connected region in $\mathbf{S}^2$ which is not simply connected, and let φ be a Kaplan-Markus flow in Ω.*

Then $\exists\, y \in \Omega$ such that $\alpha_\varphi(y)$ and $\omega_\varphi(y)$ are contained in different components of $\partial(\Omega)$.

PROOF: Let the components of $\partial(\Omega)$ be counted as $\{B_i\}$, $1 \leq i$. If the theorem fails, let Ω_i be defined as before. Since $\bar{\Omega}$ is connected, not all $\Omega_i \cup B_i$ can be closed in $\bar{\Omega}$. Suppose that $\Omega_n \cup B_n$ is not closed and that $z_0 \in \Omega_m \cup B_m$ lies in $\partial(\Omega_n)$. Then once again we use the argument of Lemma 11.15 with B_n for B^- and B_m for B^+ to obtain a contradiction.
                                                              QED

11.18. Definition. Let $\mathscr{K}$ be a Kaplan diagram in the unit disc Ω satisfying the condition of Theorem 11.10. The assignment of an orientation to any chord of $\mathscr{K}$ naturally assigns to each other chord a unique orientation consistent with it, in the sense that there can exist a flow with that Kaplan diagram and those orientations.

The orientation of each chord defines a corresponding orientation on the subtended arcs. For each chord, select the arc which contains no endpoint of the other chord.

If the orientations of these two arcs are opposite, then the chords are called *compatible* in $\mathscr{K}$; if they are the same, then the chords are *incompatible* in $\mathscr{K}$.

If two pairs (C_1, C_2) and (D_1, D_2) from $\mathscr{K}$ have the properties that

(i)   each pair is compatible in $\mathscr{K}$,

(ii) none of the four chords separates two of the others in Ω,

(iii) the line segment joining the midpoints of C_1 and C_2 does not intersect the line segment joining the midpoints of D_1 and D_2, then we call the *pairs* (C_1, C_2) and (D_1, D_2) *compatible* in $\mathcal{K}$.

If a sequence $\{(C_{i,1}, C_{i,2})\}$ (finite or infinite) of pairs of chords from $\mathcal{K}$ has the property that every two pairs of chords is compatible in $\mathcal{K}$, then the *sequence* is said to be *compatible* in $\mathcal{K}$.

11.19. Kaplan-Markus Flows in an Annulus. Let φ be a Kaplan-Markus flow in an annulus Ω. Let y be chosen so that $\alpha_\varphi(y)$ and $\omega_\varphi(y)$ lie in opposite boundaries of Ω. The removal of $\mathcal{O}_\varphi(y)$ from Ω leaves $\Omega \setminus \mathcal{O}_\varphi(y)$ a disc, and we can map $\Omega \setminus \mathcal{O}_\varphi(y)$ onto the unit disc U by a conformal mapping h. We extend h to $\mathcal{O}_\varphi(y)$, to give two images γ' and γ'' in $\partial(U)$, and in this extended disc, we define a flow ψ with $\varphi = h^{-1}[\psi]$.

For each point z on the radius from 0 to the midpoint z_1 of γ', let γ_z be the circular arc centered at z passing through the end points of γ', and lying outside U. We extend ψ to γ_z in the following way: For each point $x \in \gamma_z$, and each $t \in \mathbb{R}$, let $y(x)$ be the point of γ' satisfying $\arg(y(x)) = \arg(x)$. Then $\psi(t, x)$ is that point of γ_z satisfying

$$\arg(\psi(t, x)) = \arg(\psi(t, y(x))).$$

This process extends ψ continuously into the lune consisting of all the γ_z. In the same way, we can extend ψ into the lune formed at γ''. It is an elementary geometric fact that these two lunes do not intersect, so that adjoining them to U gives another disc U', and the extension of ψ into U' is a Kaplan-Markus flow in U'. Let $\mathcal{K}$ be its Kaplan diagram.

We see easily that the images of γ' and γ'' in $\mathcal{K}$ are compatible in $\mathcal{K}$.

Let $\bar{\gamma}'$ and $\bar{\gamma}''$ be the images of γ' and γ'' in $\mathcal{K}$. Then $\bar{\gamma}'$ and $\bar{\gamma}''$ divide the unit disc into three regions, one of which has both chords in its boundary. The Kaplan diagram in that region, together with $\bar{\gamma}'$ and $\bar{\gamma}''$, is called a *truncated Kaplan diagram*, where a part of the expanded concept is the assignment of which cords are paired.

Suppose now that we were dealing instead with a finitely-connected region Ω. Then we could find an orbit which joined two boundaries of Ω, by Theorem 11.16. Opening the region along this orbit, and adding abutting flows along the opened orbit, we would then have a region of connectivity one less, and could apply the theorem and the method again, continuing in this way until we had made Ω into a disc. Then we would produce a Kaplan diagram, remove the chords corresponding to the abutments, and label the orbits in the boundary as to pairs. The resulting configuration would also be called a *truncated Kaplan diagram*.

If, for any region, we can cut it open by using countably many orbits, in such a way that what remains is a disc, then we can apply the same process, and again we will use the same name.

It is not the case, however, that the truncated Kaplan diagram plays the same role for annular flows that the Kaplan diagram plays for disc flows. For one thing, the truncated Kaplan diagram is not an invariant, even up to similarity. To see that, consider the Kaplan-Markus flow in Fig. 11.14a. We can cut the annulus open along either of the two orbits shown in heavy lines. Depending on which we choose, we obtain one of the two truncated Kaplan diagrams in Fig. 11.14. These are not similar. However, this situation leads us to a concept of *equivalence* for truncated Kaplan diagrams.

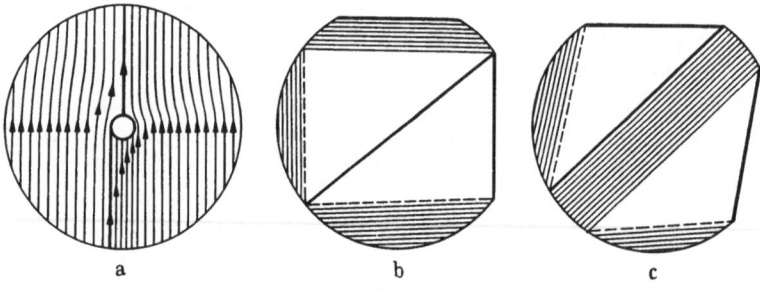

a   b   c

Fig. 11.14

Intuitively, we can obtain from any truncated Kaplan diagram another one by choosing a pair of corresponding chords in the boundary, orienting them with compatible orientations, cutting the diagram with a chord which separates these two, and reassembling the diagram with the two chords "sewn together" in the correct orientation, and the new chord now appearing as a pair in the boundary. It is easy to show that the new configuration is a truncated Kaplan diagram just if the first was. The two diagrams are said to be *semi-similar*, and the relation is now extended by iteration, so that it is transitive.

11.20. Definition. Let $\mathscr{K}_1$ and $\mathscr{K}_2$ be truncated Kaplan diagrams, with $\mathscr{K}_1$ containing in its boundary the pair of chords (C_1, C_2) and in its interior the chord D, while $\mathscr{K}_2$ contains in its boundary the pair of chords (D_1, D_2) and in its interior the chord C. Assume that the line segment joining the centers of C_1 and C_2 intersects D, and the segment joining the centers of D_1 and D_2 intersects C.

Let D divide $\mathscr{K}_1$ into two pieces, called Ω_1 and Ω_2, where $\partial(\Omega_1) \supseteq C_1 \cup D$ and $\partial(\Omega_2) \supseteq C_2 \cup D$. Let C divide $\mathscr{K}_2$ into two pieces, called Ω_3 and Ω_4, where $\partial(\Omega_3) \supseteq D_1 \cup C$ and $\partial(\Omega_4) \supseteq D_2 \cup C$. Let C_1 and C_2 be assigned compatible orientations, and D an orientation. Let D_1 and D_2 be assigned compatible orientations, and C an orientation.

Suppose now that there exist homeomorphisms h_1 of $\partial(\Omega_1)$ onto $\partial(\Omega_3)$ and h_2 of $\partial(\Omega_2)$ onto $\partial(\Omega_4)$ such that

(i) if x and y are the endpoints of a chord of $\mathcal{K}_1$ in Ω_i, then $h_i(x)$ and $h_i(y)$ are the endpoints of a chord of $\mathcal{K}_2$ in Ω_{i+2}, $\exists\, i = 1, 2$.

(ii) $h_i(C_i) = C$; $h_i^{-1}(D_i) = D$, $\forall\, i = 1, 2$.

(iii) the homeomorphism h_i carries the orientation of C_i onto the orientation of C, $\forall\, i = 1, 2$, and h_i^{-1} carries the orientation of D_i onto the orientation of D.

Then the pair (h_1, h_2) is called a *semi-similarity* of $\mathcal{K}_1$ with $\mathcal{K}_2$, and $\mathcal{K}_1$ and $\mathcal{K}_2$ are called *semi-similar* if there is a semi-similarity between them.

The truncated Kaplan diagrams $\mathcal{K}_1$ and $\mathcal{K}_2$ are *equivalent* if either

(i) $\mathcal{K}_1$ and $\mathcal{K}_2$ are similar,

or

(ii) there exists a finite sequence $\{\mathcal{K}^{(j)} \mid j = 0, 1, \ldots, n\}$ so that $\mathcal{K}^{(0)} = \mathcal{K}_1$, $\mathcal{K}^{(n)} = \mathcal{K}_2$, and $\mathcal{K}^{(i-1)}$ is semi-similar with $\mathcal{K}^{(i)}$, $\forall\, i = 1, \ldots, n$.

Clearly, the defined relation is an equivalence relation.

11.21. Theorem. *Let φ and ψ each be a Kaplan-Markus flow in an annulus.*

Then φ and ψ are conjugate if and only if their truncated Kaplan diagrams are equivalent.

PROOF: Assume that φ and ψ are conjugate, and let $u\varphi = h[\psi]$. Assume that in making the truncated Kaplan diagrams $\mathcal{K}_\varphi$ and $\mathcal{K}_\psi$ of φ and ψ, we cut the annuli in question by $\mathcal{O}_\varphi(y_\varphi)$ and $\mathcal{O}_\psi(y_\psi)$ respectively. If $h(y_\psi) \in \mathcal{O}_\varphi(y_\varphi)$, then we see at once that $\mathcal{K}_\varphi$ and $\mathcal{K}_\psi$ are similar. If not, then let the chords of $\mathcal{K}_\varphi$ corresponding to $\mathcal{O}_\varphi(y_\varphi)$ be designated as C_1 and C_2, the chord corresponding to $\mathcal{O}_\varphi(h[y_\psi])$ as D, the chords corresponding to $\mathcal{O}_\psi(y_\psi)$ in $\mathcal{K}_\psi$ as D_1 and D_2, and the chord corresponding to $\mathcal{O}_\psi(h^{-1}(y_\varphi))$ as C. Now, if the indices have been assigned in the proper order, we easily see that $\mathcal{K}_\varphi$ and $\mathcal{K}_\psi$ are semi-similar.

Conversely, assume that $\mathcal{K}_\varphi$ and $\mathcal{K}_\psi$ are equivalent. If they are similar, then an analysis just like that in Theorems 11.10—11.12 will show that φ and ψ are conjugate. If $\mathcal{K}_\varphi$ and $\mathcal{K}_\psi$ are semi-similar, then we can modify this approach to take account of the semi-similarity. Let $C, C_1, C_2, D, D_1, D_2, \Omega_1, \Omega_2, \Omega_3$, and Ω_4 be as in Definition 11.20. Let the portion of the Kaplan diagram $\mathcal{K}_\varphi$ lying in Ω_1 be extended to all of

Ω by putting abutments onto C_1 and D. Call the resulting Kaplan diagram $\mathscr{K}_1$. Let the portion of $\mathscr{K}_\psi$ lying in Ω_3 be extended to all of Ω by putting abutments onto C and D_1. Call the resulting Kaplan diagram $\mathscr{K}_3$. Then $\mathscr{K}_1$ and $\mathscr{K}_3$ are similar. We do the same for Ω_2 and Ω_4, giving the similar diagrams $\mathscr{K}_2$ and $\mathscr{K}_4$. Corresponding to the four Kaplan diagrams $\mathscr{K}_1 \dots \mathscr{K}_4$, we have four flows $\varphi_1 \dots \varphi_4$ in $\mathbb{R}^2$, and we see that φ_1 is conjugate with φ_3 and φ_2 with φ_4. Let us now discard all the points of $\mathbb{R}^2$ corresponding to the abutments we just added. This will leave us with four closed sets, numbered $F_1 \dots F_4$, with F_i being the remaining set corresponding to φ_i, $\forall\, i = 1, 2, 3, 4$. We will now "sew together" F_1 and F_2 to make an annulus. To do this, we index each point x of F_1 as the point $(1, x)$, and each y of F_2 as $(2, y)$, and consider these separate except for an identification made along $\partial(F_1)$ and $\partial(F_2)$ which makes $\varphi_1 = \varphi_2$ in the identified portion. This identification makes $F_1 \cup F_2$ into a topological space F_φ which is homeomorphic with an annulus, and in F_φ we define the flow $\varphi_3 = \varphi_1 \cup \varphi_2$. It is clear that φ and φ_3 are conjugate.

We now do the same for ψ, giving a topological annulus F_ψ on which the flow ψ_3 is defined which is conjugate with ψ. However, the mode of defining φ_3 and ψ_3 assures that these two are conjugate. Thus, φ and ψ are conjugate.

Our proof will now be complete when we show that two flows in an annulus are equivalent just if they are similar or semi-similar. We hesitate to show this in too great detail, since it is so clear. In essence it involves the following: Let φ be any Kaplan flow in an annulus Ω, and let $\mathscr{G}_0(\varphi)$ be the collection of all orbits of φ which have their α and ω sets in the two components of $\partial(\Omega)$. Then the orbits of $\mathscr{G}_0(\varphi)$ have a cyclic order which is not changed by cutting at any point. $\mathscr{G}_0(\varphi)$ is closed in Ω, as can be seen from an argument like that in Lemma 11.15 and for each component U of $\Omega \setminus \mathscr{G}_0(\varphi)$, U is a disc whose boundary in Ω is either two orbits from $\mathscr{G}_0(\varphi)$ or one orbit counted twice in case $\mathscr{G}_0(\varphi)$ contains just one orbit. The part of the truncated Kaplan diagram corresponding to U moves unchanged in the process of successive semi-similarities.

In these circumstances the relation of semi-similarity is like a cut in a deck of cards. The result of a sequence of cuts is either to leave the deck unchanged or else it is equivalent to a single cut. So is it in this case, and for the same reason, namely, that the cyclic order in $\mathscr{G}_0(\varphi)$ is never changed, and the other orbits retain their places inside this cyclic order.

QED

We now produce an example to show that in regions of connectivity greater than two, we can have two truncated Kaplan diagrams which are equivalent, but neither similar nor semi-similar.

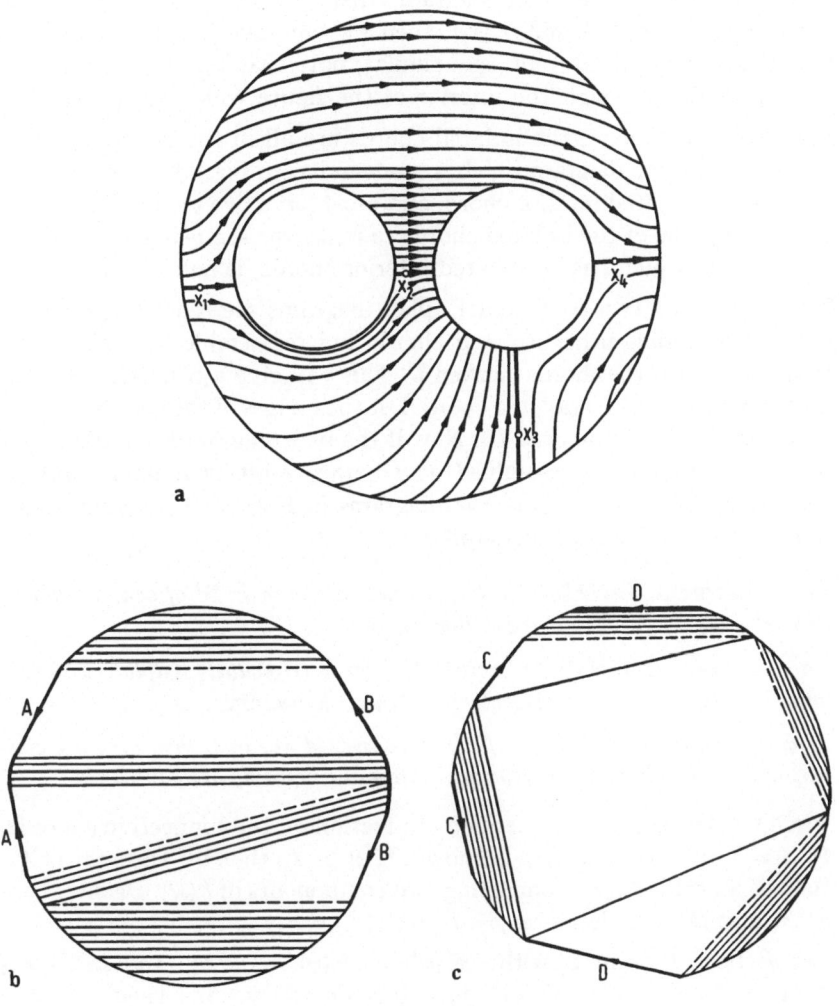

Fig. 11.15

11.22. Example. Let Ω be the triply-connected region shown in Fig. 11.15 a. and let φ be a flow in Ω which is canonical with respect to $\partial(\Omega)$ and has the orbits shown in the figure. If we cut Ω along $\mathcal{O}_\varphi(x_1)$ and $\mathcal{O}_\varphi(x_4)$ to obtain a truncated Kaplan diagram, then we will have Fig. 11.15 b. If we cut along $\mathcal{O}_\varphi(x_2)$ and $\mathcal{O}_\varphi(x_3)$, then we will have Fig. 11.15 c. To see that these diagrams are not semi-similar, note that Fig. 11.15 c has two chords passing through the interior of the figure which are isolated.

Fig. 11.15 b has none; the isolated chords are all in the boundary. Any truncated Kaplan diagram which is semi-similar to Fig. 11.15 c must have at least one isolated interior chord, and must have two unless Fig. 11.15 c is cut along one of the isolated chords to make the semi-similarity. Thus, Fig. 11.15 b, which has no isolated interior chords, is not one of these.

To see that the two truncated Kaplan diagrams are indeed equivalent, cut 11. 15 c along either isolated chord, and reassemble by making the identification at the chords marked C. This will give a truncated Kaplan diagram with just one isolated chord. Let the images of the chords marked D in Fig. 11.15 c be marked D'. Now if the new diagram is cut along the isolated chord and reassembled at D', the resulting diagram will be similar to Fig. 11.15 b. Thus, the diagrams in Fig. 11.15 are equivalent, but neither similar nor semi-similar.

11.23. Theorem. *Let Ω be a finitely-connected region in $\mathbf{S}^2$ of connectivity n, and let φ be a Kaplan-Markus flow in Ω.*

Then we can represent the orbits of φ by a truncated Kaplan diagram with exactly $n-1$ compatible pairs in its boundary.

If φ and ψ are two flows in finitely-connected regions, then they are conjugate if and only if their truncated Kaplan diagrams are equivalent.

PROOF: The theorem proceeds by induction on the connectivity n of Ω. If $n = 1$ or 2, the theorem is known. If $n > 2$, then by Theorem 11.17, there is an orbit $\mathcal{O}_\varphi(y_1)$ connecting two components of $\partial(\Omega)$. Denote these as B_1 and B_2, and define $B_0 = B_1 \cup \mathcal{O}_\varphi(y_1) \cup B_2$.

Let φ' be a flow in Ω with $\mathscr{F}(\varphi') = \mathscr{F}(\varphi) \cup \overline{\mathcal{O}_\varphi(y_1)}$, and satisfying $\mathcal{O}_\varphi(x) = \mathcal{O}_{\varphi'}(x)$, $\forall\, x \in \Omega \setminus \mathcal{O}_\varphi(y_1)$ (cf. Corollary 5.4). Then we can consider φ' as a Kaplan-Markus flow in $\Omega' = \Omega \setminus \mathcal{O}_\varphi(y_1)$, which is a region of connectivity $n-1$. Using the hypothesis of the induction, we select $n-2$ orbits, which we denote as $\mathcal{O}_{\varphi'}(y_2), \ldots, \mathcal{O}_{\varphi'}(y_{n-1})$ such that $\Omega' \setminus \bigcup_{i=2}^{n-1} \mathcal{O}_{\varphi'}(y_i)$ is a disc. However, $\mathcal{O}_{\varphi'}(y_i) = \mathcal{O}_\varphi(y_i)$, $\forall\, i = 2, \ldots, n-1$, so that by the definition of Ω', the disc is just $\Omega \setminus \bigcup_{i=1}^{n-1} \mathcal{O}_\varphi(y_i)$.

A similar induction will show that if $m < n - 1$, then for any $y_1, \ldots, y_m$, $\Omega \setminus \bigcup_{i=1}^{m} \mathcal{O}_\sigma(y_i)$ is not simply connected, while if $m > n - 1$, then $\Omega \setminus \bigcup_{i=1}^{m} \mathcal{O}_\varphi(y_i)$ is not connected. This proves the first part of the theorem.

The second part is shown by a proof essentially the same as the proof of Theorem 11.20.   **QED**

CHAPTER TWELVE

FINE STRUCTURE IN $\mathscr{G}_s(\varphi)$ II

12.1. Introduction. Having established some structure for Kaplan-Markus flows, we now turn to other singular flows. We let φ be a flow in $\mathbb{R}^2$ and Ω a component of $\mathscr{G}_s(\varphi)$. We note that Ω lies in a component of $\mathbb{R}^2 \setminus \mathscr{F}_s(\varphi)$, which we denote as Ω_0. We see at once by Corollary 3.27, that Ω is closed in Ω_0. The boundaries of Ω in Ω_0 lie entirely in $\mathscr{G}_s(\varphi)$, and by Theorem 3.12(c), each boundary point has either a periodic singular orbit or an aperiodic orbit which does not spiral at either end, *i.e.* whose ends lie in $\partial(\Omega_0)$. If $\mathcal{O}_\varphi(x)$ is periodic and lies in the interior of Ω, then $\mathcal{O}_\varphi(x)$ is singular on both sides, so that it separates the boundaries of Ω_0. If, for some reason, we know that $\partial(\Omega_0)$ has only finitely many components, then we know at once that there can be at most finitely many periodic orbits in the interior of Ω.

Note that for the purpose of examining the fine structure of $\mathscr{G}_s(\varphi)$, the character of the actions on the two sides of any periodic orbit are completely independent, except for the fact that they must be continuous to the periodic orbit. Thus, if we have two regions, each of which is a component of the singular moving set of a flow in $\mathbb{R}^2$, then we can "sew them together" to form another such region if they contain in their boundaries two orbits with the same period. *i.e.* if Ω_1 and Ω_2 are two such regions for the flows φ_1 and φ_2 respectively, and if $x_1 \in \partial(\Omega_1)$ and $x_2 \in \partial(\Omega_2)$, with $p_\varphi(x_1) = p_\varphi(x_2) < \infty$, then we can find a homeomorphism h of Ω_2 into $\mathbb{R}^2$ such that $\overline{h(\Omega_2)} \cap \overline{\Omega_1} = \mathcal{O}_\varphi(x_1)$, so that φ_1 agrees with $h[\varphi_2]$ on $\mathcal{O}_\varphi(x_1)$, and so that $\varphi_1 \cup h[\varphi_2]$ is continuous. Furthermore, there is a flow φ such that $\Omega_1 \cup h(\Omega_2)$ is a component of $\mathscr{G}_s(\varphi)$. Conversely, if Ω is a component of $\mathscr{G}_s(\varphi)$ for a given flow φ in $\mathbb{R}^2$, and if $\mathcal{O}_\varphi(x)$ lies in the interior of Ω, then for each of the two parts Ω_1 and Ω_2 of Ω, as divided by $\mathcal{O}_\varphi(x)$, there is a flow for which the designated region, together with $\mathcal{O}_\varphi(x)$, constitutes a component of the singular moving set.

It follows that we can restrict our attention to regions Ω which are assumed to have no periodic orbits in the interior, and we make that assumption throughout the chapter.

We begin by examining the structure of Ω. To do this, we shall first require the following lemma.

12.2. Lemma. *Let φ be a continuous flow in $\mathbb{R}^2$. Let $\{x_i\}$ be a sequence converging to x_0 with the property that $p_\varphi(x_i) < \infty$, $\forall\, i$, and*

$$x_i \in [x_0 \text{ side} : \mathcal{O}_\varphi(x_j)], \qquad \forall\, i \neq j.$$

Then $x_0 \in \mathscr{F}(\varphi)$.

PROOF: Assume, contrary to the lemma, that $x_0 \in \mathscr{G}(\varphi)$. We shall obtain a contradiction. Let δ_1 be chosen with $0 < \delta_1 < \frac{1}{2}\, p_\varphi(x_0)$. Then $B = \{\varphi(t, x_0) \mid |t| \le \delta\}$ is a simple curve. By the Schoenflies Theorem (Theorem C. 22), there is a homeomorphism of $\mathbb{R}^2$ onto itself which carries B onto $[0, 1]$. Let h be such a homeomorphism and let $\psi = h[\varphi]$. The method of the Gate Theorem assures us that for some $\varepsilon_1 > 0$, and for all i large enough, the only points of $\mathcal{O}_\varphi(x_i)$ which can lie within ε_1 of x_0 must lie on the closed arc $B_i == \{\varphi(t, x_i) \mid |t| \le \delta_1\}$. Let $\varepsilon_2 > 0$ be chosen so that $N_d\big(h(x_0), \varepsilon_2\big) \subseteq h\big(N_d(x_0, \varepsilon_1)\big)$,[1] and so that $\varepsilon_2 < \frac{1}{4} \min\big(|1 - h(x_0)|,\ |h(x_0)|\big)$. Define $z_i = h(x_i)$, $\forall\, i \ge 0$. Then for all but finitely many values of i, $z_i \in N_d(z_0, \varepsilon_2)$, which is bisected by $\mathcal{O}_\psi(z_0)$. Either infinitely many of the z_i lie in the upper half, or else infinitely many lie in the lower half. WoLOG, assume the former case holds, and in fact that *all* the z_i lie in the indicated half-disc, which we will now call U_0. For all i large enough, we have

$$d\big(\psi(t, z_i), \psi(t, z_0)\big) < \varepsilon_2, \qquad \forall\, |t| \le \delta_1,$$

and again we assume WoLOG that this inequality holds for all i.

Let $\delta_2 > 0$ be chosen so that $d\big(\psi(t, z_i), z_0\big) < \varepsilon_2$, $\forall\, |t| \le \delta_2$, and then define $\varepsilon_3 > 0$ so that $d\big(\psi(t, z_i), z_0\big) > \varepsilon_3$, $\forall\, \delta_2 \le |t| \le p_\psi(z_i) - \delta_2$, $\forall\, i$. The existence of such an ε_3 is assured by the simplicity of the arc $h(B)$, since the non-existence of ε_3 would mean that there would be points $\psi(t_i, z_i)$ converging to z_0 with $\delta_2 \le |t_i| \le \delta_1\ \forall\, i$, and we would have $z_0 = \psi(t, z_0)$ for some t with $\delta_2 \le |t| \le \delta_1$. For z_j close enough to z_0, say $d(z_j, z_0) < \varepsilon_4$, we have $d\big(\psi(t, z_j), \psi(t, z_0)\big) < \varepsilon_3$, $\forall\, |t| \le \delta_2$. Define C_i, $\forall\, i > 0$, by

---

[1] Note that this assures that $N_d\big(h(x_0), \varepsilon_2\big)$ does not meet $\mathcal{O}_\psi\big(h(x_0)\big)$ except in $[0, 1]$, since the assumption that $\psi(t, h(x_0))$ lies within ε_2 of $h(x_0)$, with $\delta_1 < |t| < p_\psi(x_0) - \delta_1$ must imply that $d\big(\varphi(t, x_i), x_0\big) < \varepsilon_1$ for that value of t and almost all i, contrary to the definition of ε_1.

$C_i = \{\psi(t, z_i) \,|\, |t| \leq \delta_2\}$. Now we choose any z_j with $d(z_j, z_0) < \varepsilon_4$, and define J to be a Jordan curve consisting of C_0, C_j, and two arcs lying in U_0, one within ε_3 of $\psi(\delta_2, z_0)$ and the other within ε_3 of $\psi(-\delta_2, z_0)$ and joining the corresponding endpoints of C_0 and C_j (Note that $\psi(\delta_2, z_0) \geq z_0 + \varepsilon_3$ and $\psi(-\delta_2, z_0) \leq z_0 - \varepsilon_3$). Clearly, all the points in the lower half-plane lie *outside* J, so that all the points in the upper half-plane which are close enough to z_0 will lie *inside* J. Choose any z_k which is close enough, and let γ_k be the longest arc of $\mathcal{O}_\psi(z_k)$ which lies inside J and contains z_k. Clearly, γ_k separates ins(J) into two pieces, one of which has C_0 in its boundary, and the other C_j. Thus, the line segment $[z_0, z_j]$ must intersect γ_k. We see that the minimal subsegment of $[z_0, z_j]$ which joins z_0 to $\mathcal{O}_\psi(z_k)$, together with a subarc of γ_k, is a simple arc joining z_0 to z_k, and lies in ins(J). A small change in this arc will give us an arc joining z_0 to z_k lying in ins(J) and disjoint from $\mathcal{O}_\psi(z_0)$, $\mathcal{O}_\psi(z_j)$ and $\mathcal{O}_\psi(z_k)$ except for its endpoints. Call this arc γ_0. Similarly, there is a subsegment of $[z_0, z_j]$ which, together with arcs of $\mathcal{O}_\psi(z_j)$ and $\mathcal{O}_\psi(z_k)$, forms an arc joining z_k to z_j, and this can be similarly modified to an arc γ_j joining z_k to z_j, lying in ins(J) and disjoint from $\mathcal{O}_\psi(z_0) \cup \mathcal{O}_\psi(z_k) \cup \mathcal{O}_\psi(z_j)$ except for z_k and z_j. Clearly the two arcs lie in the two parts of ins(J) formed by the exclusion of γ_k. Thus, for any homeomorphism g of $\mathbb{R}^2$ onto itself which carries γ_k onto $[0, 1]$, we can find a neighborhood N of $g(z_k)$ such that $g(\gamma_0) \cap N$ and $g(\gamma_j) \cap N$ lie in opposite half-planes (upper and lower). It follows that γ_0 and γ_j lie in opposite sides of $\mathcal{O}_\psi(z_k)$, so that $\mathcal{O}_\varphi(x_k)$ separates x_j from x_0, contrary to hypothesis. Thus, we have obtained our contradiction, and the lemma is proved.   **QED**

12.3. Corollary. *Let φ be a flow in $\mathbb{R}^2$, and let Ω be a component of $\mathscr{G}_s(\varphi)$. Let Ω_0 be the component containing Ω of $\mathbb{R}^2 \setminus \mathscr{F}_s(\varphi)$ and let K be a compact subset of Ω_0.*

Then K meets at most finitely many periodic orbits lving in the boundary of Ω.

12.4. Corollary. *Let φ be a flow in $\mathbb{R}^2$, and let Ω be a component of $\mathscr{G}_s(\varphi)$. Let $\mathcal{O}_\varphi(y)$ be a periodic orbit in $\partial(\Omega)$.*

Then $\mathcal{O}_\varphi(y)$ is open in $\partial(\Omega)$.

12.5. Theorem. *Let φ be a flow in $\mathbb{R}^2$, and let Ω be a component of $\mathscr{G}_s(\varphi)$. Let $\mathcal{O}_\varphi(y)$ be a periodic orbit in $\partial(\Omega)$.*

Then there exists an open annulus $U \subseteq \Omega$ such that

(i) $p_\varphi(x) = \infty$, $\forall \, x \in U$;

(ii) $\mathcal{O}_\varphi(y)$ *is one of the boundaries of U;*

(iii) *the other boundary of U contains a stagnation point;*

(iv) *every orbit in U has $\mathcal{O}_\varphi(y)$ as its α or ω set;*

(v) *the set of points in Ω having $\mathcal{O}_\varphi(y)$ as endpoints is U.*

PROOF: Earlier results (cf. Theorems 2.18, 2.23 and 3.9) show that $\{x \in [\Omega \text{ side} : \mathcal{O}_\varphi(y)] \mid y \text{ is an endpoint of } \mathcal{O}_\varphi(x)\}$ is an annulus U which has $\mathcal{O}_\varphi(y)$ as one of its boundaries, and in fact, for which $\mathcal{O}_\varphi(y)$ is the set of α (or ω)-points for each $x \in U$. Since $\mathcal{O}_\varphi(y)$ is singular on its Ω-side, some $x \in U$ must have a stagnation point, which clearly lies in the other boundary of U. Thus, the theorem is proved.   **QED**

12.6. Definition. Let Ω, φ and y be as in Theorem 12.5. Then if $\mathcal{O}_\varphi(y)$ serves as the α set for the elements of U, we call $\mathcal{O}_\varphi(y)$ a *periodic φ-Ω source*. If it serves as the ω set, we call it a *periodic φ-Ω sink*. When φ and Ω are understood, we say *periodic source* or *periodic sink*. The set U will be denoted as the *fief* of $\mathcal{O}_\varphi(y)$, and also of y, with respect to φ and Ω, and will be symbolized by $\mathrm{ff}\big(\mathcal{O}_\varphi(y), \varphi, \Omega\big)$ or $\mathrm{ff}(y, \varphi, \Omega)$, with $\mathrm{ff}\big(\mathcal{O}_\varphi(y)\big)$ or $\mathrm{ff}(y)$ being used when φ and Ω are understood.

12.7. Definition. The set $\mathrm{ff}(x) \cap \mathrm{ff}(y)$ is denoted as $\mathrm{ff}(x, y) = \mathrm{ff}(\varphi, \Omega, x, y)$. For each x and y, $\mathrm{ff}(x)$ and $\mathrm{ff}(y)$ are both open, so that $\mathrm{ff}(x, y)$ is open, though usually empty. If $\mathrm{ff}(x, y) \neq \square$, then one of $\mathcal{O}_\varphi(x)$, $\mathcal{O}_\varphi(y)$ is a periodic φ-Ω-source, and the other a periodic φ-Ω-sink. Each component of $\mathrm{ff}(x, y)$ is called a *φ-Ω-cell*, or merely a *cell*, and each cell has a *source* and a *sink*, defined in the obvious way.

12.8. Lemma. *Let D_1 be the points of $\mathrm{int}(\Omega)$ whose orbits have stagnation points.*

Then D_1 is closed in $\mathrm{int}(\Omega)$.

PROOF: The complement of D_1 in $\mathrm{int}(\Omega)$ is the union of all the cells, which is open.   **QED**

12.9. Lemma. *Let D_2 be the points of Ω which have both α-stagnation points and ω-stagnation points.*

Then D_2 is closed in Ω.

PROOF: For each $x \in \partial(\Omega)$ with $\mathrm{p}_\varphi(x) < \infty$, $\mathrm{ff}(x) \cup \mathcal{O}_\varphi(x)$ is open in Ω. Thus, the union of these fiefs and orbits is open in Ω. The complement of this union is D_2.   **QED**

Structure of Cells

12.10. Theorem. *Let U be a component of* $\mathrm{ff}(y^-, y^+)$, *where* $\mathcal{O}_\varphi(y^-)$ *is a periodic source, and* $\mathcal{O}_\varphi(y^+)$ *is a periodic sink. Assume that* $z \in \partial(U)$.
Then exactly one of these occurs:

(i) $z \in \mathcal{O}_\varphi(y^-)$,

(ii) $z \in \mathcal{O}_\varphi(y^+)$,

(iii) z α-*spirals to* y^-,

(iv) z ω-*spirals to* y^+,

(v) $\mathrm{p}_\varphi(z) = \infty$, *and z does not spiral*,

(vi) $z \in \mathcal{F}(\varphi)$.

PROOF: For every $x \in U$, $\mathrm{p}_\varphi(x) = \infty$, and x α-spirals to y^- and ω-spirals to y^+. If $0 < \mathrm{p}_\varphi(z) < \infty$, then for all $x \in U$ close enough to z, x spirals to z. Thus, if $0 < \mathrm{p}_\varphi(z) < \infty$, then $z \in \mathcal{O}_\varphi(y^+)$ or $z \in \mathcal{O}_\varphi(y^-)$.

If $\mathrm{p}_\varphi(z) = \infty$, and z spirals to w, then we must show that $w \in \mathcal{O}_\varphi(y^-) \cup \mathcal{O}_\varphi(y^+)$. Suppose not. Assume WoLOG that z ω-spirals to w. Note that $\omega_\varphi(z)$ meets neither $\mathcal{O}_\varphi(y^-)$ nor $\mathcal{O}_\varphi(y^+)$. (Theorem 2.18). Nor can $\omega_\varphi(z)$ separate these two orbits, since no members of $\omega_\varphi(z)$ can spiral. Thus, we can find a φ-gate construction $\langle J_0, L_0, T \rangle$ for the pair $\langle z, w \rangle$ such that J_0 separates $\omega_\varphi(z)$ from $\mathcal{O}_\varphi(y^+)$. For all x close enough to z, J_0 separates $\omega_\varphi(x)$ from $\mathcal{O}_\varphi(y^+)$. Since we can choose $x \in U$ as close as we like to z, and $\omega_\varphi(x) = \mathcal{O}_\varphi(y^+)$ for all $x \in U$, we have a contradiction. Thus, if z ω-spirals, it ω-spirals to y^+. Similarly, if z α-spirals, it α-spirals to y^-. If z did both, then we would have $z \in \mathrm{ff}(y^-, y^+)$, which violates $z \in \partial(U)$.

Thus, we see that $0 < \mathrm{p}_\varphi(z) < \infty \Rightarrow$ (i) or (ii), but not both.

$\mathrm{p}_\varphi(z) = \infty \Rightarrow$ (iii), (iv), or (v), but not two of these, and $\mathrm{p}_\varphi(z) = 0 \Rightarrow$ (vi).
                                                                          QED

12.11. Theorem. *Let U be a component of* $\mathrm{ff}(y^-, y^+)$, *where* $\mathcal{O}_\varphi(y^-)$ *is a periodic source and* $\mathcal{O}_\varphi(y^+)$ *is a periodic sink.*
Then U is a disc.

PROOF: U is open and connected, and thus arcwise connected. Choose two points x_1 and x_{-1} from U which do not lie on the same orbit, and join them by an arc lying in U. By the method of Theorem 10.2, we can replace that arc by an arc γ_0 which intersects each φ-orbit in at most one point. Set $V_0 = \{\varphi(t, x) \mid t \in \mathbb{R} \text{ and } x \in \gamma_0\}$. Then V_0 is clearly a disc.

We wish to show that there is no arc γ lying in $U \setminus \overline{V_0}$ joining x_1 to x_{-1}. If there were, then we could assume WoLOG that γ also intersects each φ-orbit in at most one point. The union of γ_0 and γ, with their endpoints, would then be a Jordan curve J lying in U, which intersects each φ-orbit in at most one point. Since $\{\varphi(t, x) \mid t \in \mathbb{R} \text{ and } x \in J\}$ is an annulus contained in U which separates $\mathcal{O}_\varphi(y^-)$ from $\mathcal{O}_\varphi(y^+)$, we see that J separates $\mathcal{O}_\varphi(y^-)$ from $\mathcal{O}_\varphi(y^+)$. Now the fief $\mathrm{ff}(y^-)$ is a singular annulus whose one boundary is $\mathcal{O}_\varphi(y^-)$ and whose other properly contains $\mathcal{O}_\varphi(y^+)$. Since no point of J can be in $\partial(\mathrm{ff}(y^-))$, we see that J separates the two boundaries of $\mathrm{ff}(y^-)$. This implies that every orbit in $\mathrm{ff}(y^-)$ meets J, and therefore ω-spirals to $\mathcal{O}_\varphi(y^+)$. But since $\mathrm{ff}(y^-)$ contains at least one orbit which does not ω-spiral, we have a contradiction.

Thus $U \setminus \overline{V_0}$ has at least two components, one of which has x_1 in its boundary, the other x_{-1}. Call these two U_1 and U_{-1} respectively. Choose $x_2 \in U_1$ so that $\varrho(x_2, V_0) > \frac{1}{2} \sup \{\varrho(x, V_0) \mid x \in U_1\}$ and $x_{-2} \in U_{-1}$ similarly. Then join x_1 to x_2 by an arc in U_1 which intersects each φ-orbit in at most one point, and similarly for x_{-1} and x_{-2}. Let the union of these two arcs with γ_0 and its endpoints be denoted by γ_1, and define $V_1 = \{\varphi(t, x) \mid t \in \mathbb{R} \text{ and } x \in \gamma_1\}$. Then V_1 is also a disc.

Continuing in this manner, we construct an expanding sequence $\{V_n\}$ of discs. Their union V is clearly a disc, and $V \subseteq U$. However, the distances $\varrho(x_i, V_{|i|-2})$ must converge to zero as $i \to \pm\infty$, since $\mathbb{S}^2$ is compact in the metric ϱ. Thus, no point of U can lie outside of V. It follows that U is the disc V.   **QED**

12.12. Theorem. *Let U be a component of $\mathrm{ff}(y^-, y^+)$, where $\mathcal{O}_\varphi(y^-)$ is a periodic source and $\mathcal{O}_\varphi(y^+)$ is a periodic sink. Let h be a conformal map of U onto the unit disc $U_0 = \{z \mid |z| < 1\}$.*

Then

(i) *There exist points z^- and z^+ with norm 1 such that the neighborhoods of z^- in U_0 are the images of the neighborhoods of $\mathcal{O}_\varphi(y^-)$ in U, and similarly for z^+ and $\mathcal{O}_\varphi(y^+)$.*

(ii) *The flow $h[\varphi]$ extends continuously to a flow in Γ_1 which moves in the direction from z^- toward z^+ (including some fixed points) on each of the arcs of Γ_1 between z^- and z^+.*

PROOF: Let V^- be a neighborhood of $\mathcal{O}_\varphi(y^-)$ which contains no stagnation point of φ. Then since the boundary of U is connected, we must have a point x^- of V^- which lies in the boundary. If V^- is chosen small enough, that will assure that $\alpha_\varphi(x^-) = \mathcal{O}_\varphi(y^-)$, so that $\mathcal{O}_\varphi(x^-)$ will be a genuine spiral. We know (cf. Collingwood and Lohwater [1], pp. 167 ff.) that such a spiral in the boundary assures the existence of a prime end, and

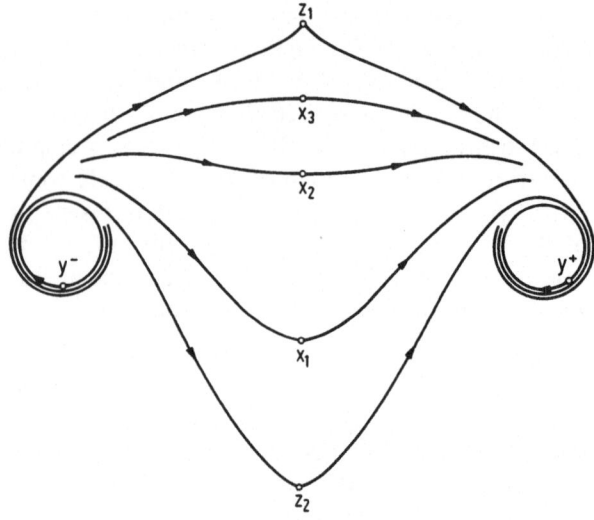

Fig. 12.1

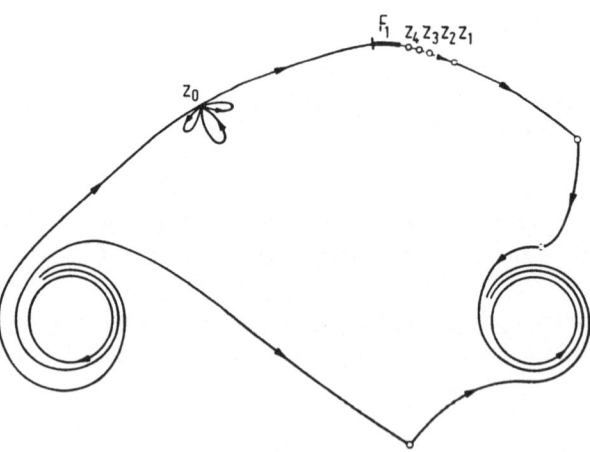

Fig. 12.2

is reflected in the conformal mapping as a single point of $\partial(U_0)$. Thus, $h[\varphi]$ is a flow in U_0, with every orbit running from z^- to z^+. We know that h extends to the boundary, within limits, and that the portions of $\partial(U)$ which lie in $\mathscr{G}(\varphi)$ are represented by arcs of Γ_1, and the extension of $h[\varphi]$ to these arcs gives an image for φ there. By continuity, the extension of $h[\varphi]$ must move in the direction from z^- to z^+ along Γ_1. Assigning all the remaining points of Γ_1 to be fixed points of the extension of $h[\varphi]$ gives us the required flow in $\partial(U_0)$.   **QED**

12.13. Remark. In Figs. 12.1, 12.2, and 12.3, we show some examples of cells. The simplest is 12.1, where we have the two periodic orbits, $\mathcal{O}_\varphi(y^-)$ and $\mathcal{O}_\varphi(y^+)$, and some of the points (x_1, x_2, x_3) which run from the one to the other. Since no orbit in the boundary can run between these orbits, we must have some fixed points z_1 and z_2 as shown in this figure. In Fig. 12.2, we see more complications. Several orbits lying in $\partial(U)$ begin and end at z_0, which is a fixed point. There could, in fact, be infinitely many of these, though the number is always countable at most. The boundary includes a whole arc F_1 of fixed points, and a sequence $\{z_1, z_2, z_3, \ldots\}$ of fixed points converging to a point of F_1. The role of F_1 might be taken by a more complicated continuum, and there might be infinitely many of these. In Fig. 12.3, we see that the orbits in $\partial(U)$

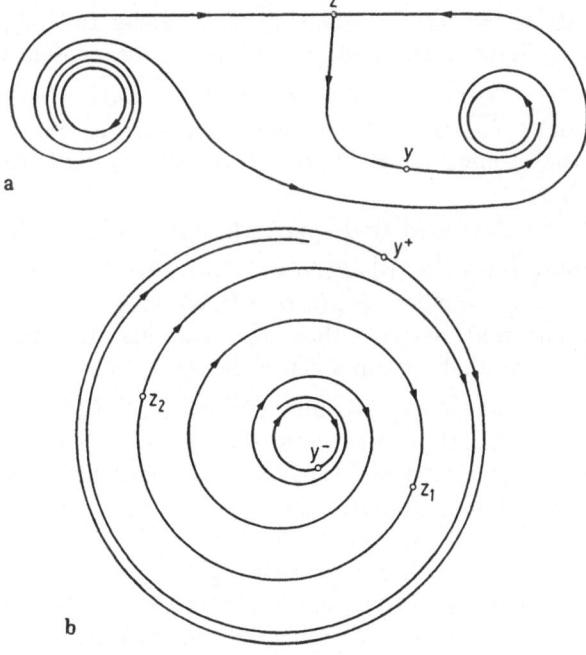

Fig. 12.3

can be type 2 in $\partial(U)$, as is $\mathcal{O}_\varphi(y)$ in Fig. 12.3 (a). Carrying this further, we see that all the aperiodic orbits in $\partial(U)$ can be of type 2, as in Fig. 12.3 (b).

Combinations of Cells

12.14. Theorem. *Let C_1 and C_2 be two cells, and let z_1, z_2 and z_3 be three distinct simple stagnation points in $\partial(C_1) \cap \partial(C_2)$.*

Then there cannot exist a third cell C_3 with $\{z_1, z_2, z_3\} \subseteq \partial(C_3)$.

PROOF: The proof of this theorem turns basically on the fact that z_1, z_2 and z_3 are arcwise accessible in C_1 and C_2 by disjoint arcs, a fact which we shall prove below. Let $x_1 \in C_1$ and $x_2 \in C_2$, and for each $i = 1, 2$ and $j = 1, 2, 3$, let $\gamma_{i,j}$ be an arc in C_i joining x_i to z_j so that no two intersect, except at their endpoints. For each j, let $\gamma_j = \gamma_{1,j} \cup \{z_j\} \cup \gamma_{2,j}$. Then the arcs γ_j would satisfy the Theta Curve Theorem (Theorem C. 26), and C_3 would have to lie in one of the discs bounded by two of these. Assume WoLOG that C_3 lies in the disc bounded by γ_2 and γ_3. Then C_3 would be bounded away from z_1, contrary to the assumption that it has all the z_j in its boundary.

To complete the proof, we must show that the z_j are arcwise accessible in the C_i. We shall show that z_1 is arcwise accessible in C_1. Assume WoLOG that $\{z_1\} = \omega_\varphi(y_1)$, where y_1 must belong to $\partial(C_1)$, since no elements of C_1 have stagnation points. Let ε_0 be chosen so that no element of $N_0 = N_\varrho(y_1, \varepsilon_0)$ belongs to $\mathscr{F}(\varphi) \cup \overline{\mathcal{O}_\varphi(x_1)}$, which is possible since $\mathscr{F}(\varphi)$ and $\overline{\mathcal{O}_\varphi(x_1)}$ are both compact and neither contains y_1. We will show first that some subneighborhood N_1 of N_0 contains no point of $\partial(C_1) \setminus \mathcal{O}_\varphi(y_1)$.

Let $\delta_1 > 0$ be chosen so that $\varrho(y_1, \varphi(t, y_1)) < \frac{1}{2} \varepsilon_0$, $\forall |t| \leq \delta_1$, and set $\varepsilon_1 = \min\left(\varrho(y_1, \varphi(\delta_1, y_1)), \varrho(y_1, \varphi(-\delta_1, y_1))\right)$. Choose $\varepsilon_2 > 0$ so that $\varrho(x, y_1) < \varepsilon_2$ implies $\varrho(\varphi(t, x), \varphi(t, y_1)) < \frac{1}{2} \varepsilon_1$, $\forall |t| \leq \delta_1$. Set $N_1 = N_\varrho(y_1, \varepsilon_2)$ and assume that N_1 contains an element y_2 of $\partial(C_1) \setminus \mathcal{O}_\varphi(y_1)$; we shall obtain a contradiction.

Let L_0 be the interval $[y_1, y_2]$, and define $K_0 = L_0 \cap \partial(C_1)$. Define $F_0 = \{\varphi(t, x) \mid x \in K_0, |t| \leq \delta_1\}$, and denote each component of F_0, which is compact, as a *filament*. Clearly, y_1 and y_2 lie in different filaments, and thus we can find a Jordan curve J in $\mathbf{S}^2 \setminus F_0$ with y_1 and y_2 in opposite sides of J, by Theorem C.30. Since each interval of $L_0 \setminus K_0$ which meets J must have length at least $2\varrho(J, F_0)$, there can be at most finitely many of these, and thus, since y_1 and y_2 lie in opposite sides of J, one of these intervals (call it L) joins two points (call them y_3 and y_4) which lie in different filaments.

Let $L^+ = \varphi(\delta_1, L)$, and let the three components of $C_1 \setminus (L \cup L^+)$ be denoted as C^-, C^0, and C^+, where C^0 is the component bounded by L and L^+, together with $\{\varphi(t, y_3) \mid 0 \leq t \leq \delta_1\}$ and $\{\varphi(t, y_4) \mid 0 \leq t \leq \delta_1\}$, and C^- is the other component with L in its boundary, and C^+ is the other component with L^+ in its boundary. For every point x which lies in C^0, we see at once that $\mathcal{O}_\varphi(x)$ must meet $\partial(C^0)$, since C^0 is bounded away from the source and sink of C_1. It follows that $\mathcal{O}_\varphi(x)$ must meet either L or L^+, and thus must meet both. Thus, for every point whose orbit does not meet L, that orbit must lie entirely in C^- or C^+. For each such point, we can apply the Gate Theorem to its orbit in the vicinity of the source or sink of C_1, to conclude that the set of points whose orbits lie in C^- is open, as is the set of those points whose orbits lie in C^+. By our hypothesis, the point x_1 lies in one of these two open sets. Assume WoLOG that it is C^-, which is thus non-empty. Then we shall show that the set of points whose orbits meet L or lie in C^+ is also open, thus leading to a contradiction, since C_1 is known to be connected.

We know that the points whose orbits lie in C^+ form an open set. We shall show that for every point $x' \in L$, there is a neighborhood of x' which contains only points from this open set or points whose orbits meet L, and that will show that the indicated set is open. But $\varphi(\delta_1, x') \in L^+$, and thus does not belong to $\overline{\mathrm{cl}(C^-)}$. Thus, for some neighborhood of x', every point x in that neighborhood has $\varphi(\delta_1, x) \notin C^-$, so that $\mathcal{O}_\varphi(x)$ must meet L or lie entirely in C^+. Thus, the assumption that N_1 meets $\partial(C_1) \setminus \mathcal{O}_\varphi(y_1)$ leads to a contradiction, and we have $N_1 \cap \partial(C_1) \subseteq \mathcal{O}_\varphi(y_1)$.

Let $x_0 \in N_1 \cap C_1$, and let y_0 be the first point of $[x_0, y_1]$ which belongs to $\partial(C_1)$. Then the interval $[x_0, y_0]$ lies wholly in C_1. Let a continuous function f be defined in $[x_0, y_0)$ which is 0 at x_0 and approaches $+\infty$ at y_0. If f grows slowly enough, then the arc $\{\varphi(f(x), x) \mid x \in (x_0, y_0)\}$ joins x_0 to z_1.

Thus, we have shown that the z_j are arcwise accessible in the C_i, and this completes our proof. **QED**

12.15. Corollary. *Let C_1 and C_2 be two cells lying in $\mathrm{ff}(y)$, and let z_1 and z_2 be two distinct simple stagnation points in $\partial(C_1) \cap \partial(C_2)$.*

Then there cannot exist a third cell $C_3 \subseteq \mathrm{ff}(y)$ with $\{z_1, z_2\} \subseteq \partial(C_3)$.

PROOF: Let $K = [y \text{ side}: \Omega]$ and let h be any continuous mapping of $\mathbb{S}^2$ onto itself which is one-to-one except that $h(K)$ is a single point. Now apply the argument of Theorem 12.14 to the regions $h(C_1)$, $h(C_2)$, and $h(C_3)$ and the points $h(z_1)$, $h(z_2)$, and $h(K)$. **QED**

12.16. Corollary. *Let C_1 and C_2 be two cells lying in* ff(x, y), *and let z be a simple stagnation point in* $\partial(C_1) \cap \partial(C_2)$.

Then there cannot exist a third cell $C_3 \subseteq$ ff(x, y) with $z \in \partial(C_3)$.

PROOF: Same proof, *mutatis mutandis*.   **QED**

12.17. Examples. These examples act as counterpoise to the previous three results. We will show flows in subsets of the plane in which infinitely many cells share two stagnation points, or a source and a sink, or a sink and a stagnation point.

(a) Let U be the unit disc, and let $F = \{i, -i\}$. Let K^+ and K^- be disjoint circles in U tangent to Γ_1 at i and $-i$ respectively. Consider all circular arcs from i to $-i$ in U. The portions of these arcs lying outside K^+ and K^- will be arcs of orbits of φ. Let $\{x_n\}$ be a sequence of real numbers indexed from $-\infty$ to $+\infty$ so as to be monotonically increasing and to have -1 and 1 as limits. Let the circular arc passing through x_n be extended to i or $-i$ according as n is even or odd (see Fig. 12.4(a)). The arcs passing through x_{n-1} and x_{n+1} bound in K^+ (or in K^-) a set which we will denote as U_n, and in each U_n, we select a circle, which we shall call K_n, $\forall -\infty < n < +\infty$. (See Fig. 12.4(b)).

We now define $\Omega = U \setminus \bigcup_{n=-\infty}^{\infty}$ ins (K_n). Our flow φ will be defined in $\bar{\Omega}$. Each of the K_n is a φ-orbit, and φ will be defined to be canonical with respect to F and to move in the positive direction on each K_n. In each set U_n, the flow φ will have orbits like those shown in Fig. 12.4(c). The region shown lies in K^+; the modifications for K^- are obvious. When all these have been defined and combined (see Fig. 12.4(d)), the resulting flow is canonical with respect to F, and the boundaries of the cells are shown in Fig. 12.4(e). If we had wished, we could have made φ orbit-analytic as well.

(b) Let A be the annulus $\{z \mid 1 \leq |z| \leq 2\}$, and let φ_0 be the standard orientation-preserving spiral flow defined in A. Let F be a compact infinite subset of int (A) with the properties that $\varphi_0(\mathbb{R}, F)$ is nowhere dense in A and no two points of F lie on the same φ_0-orbit. Let φ be that element of $\mathrm{Rep}^+(\varphi_0)$ which is canonical with respect to F.[2] Then we will have infinitely many cells with Γ_1 and Γ_2 as their source and sink respectively, and their closures comprise A.

(c) We begin with the flow in (a) above. Our new flow will act the same in the upper half of U. We will make modifications below. We introduce a new circle, $\check{K}$, lying in U. $\check{K}$ will be the universal sink for this example.

---

[2] Actually $\varphi \in \mathrm{Prod}^+(\varphi_0)$.

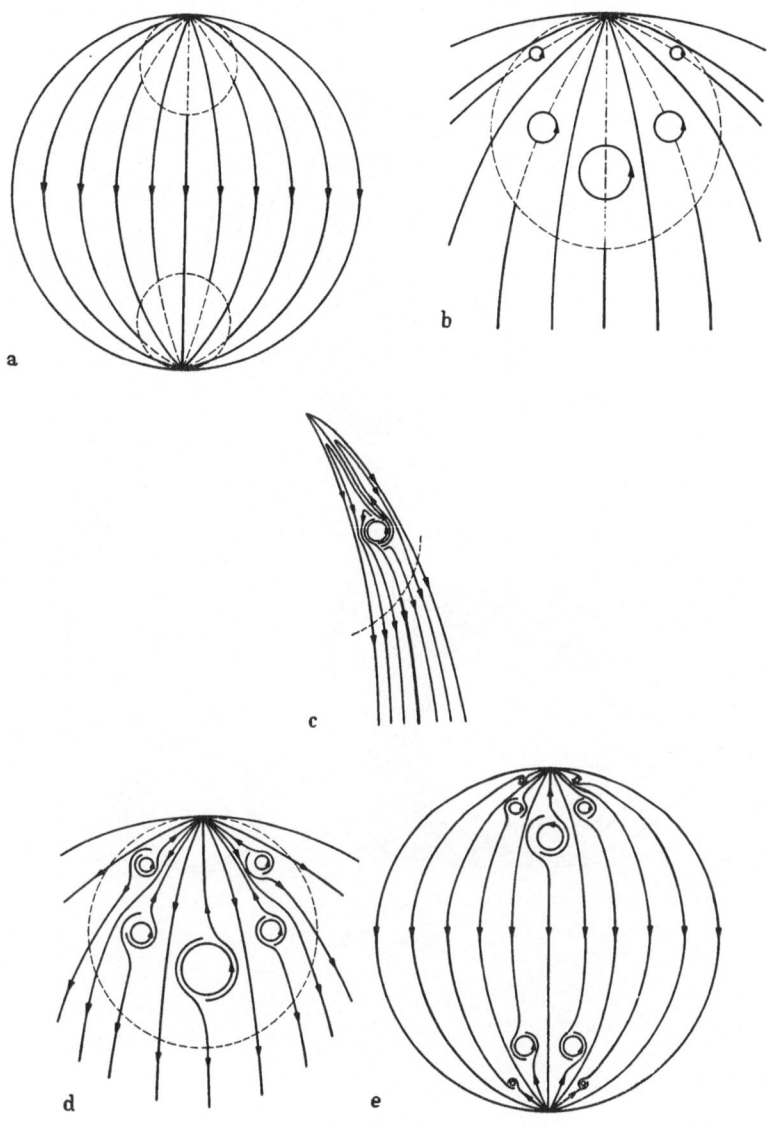

Fig. 12.4

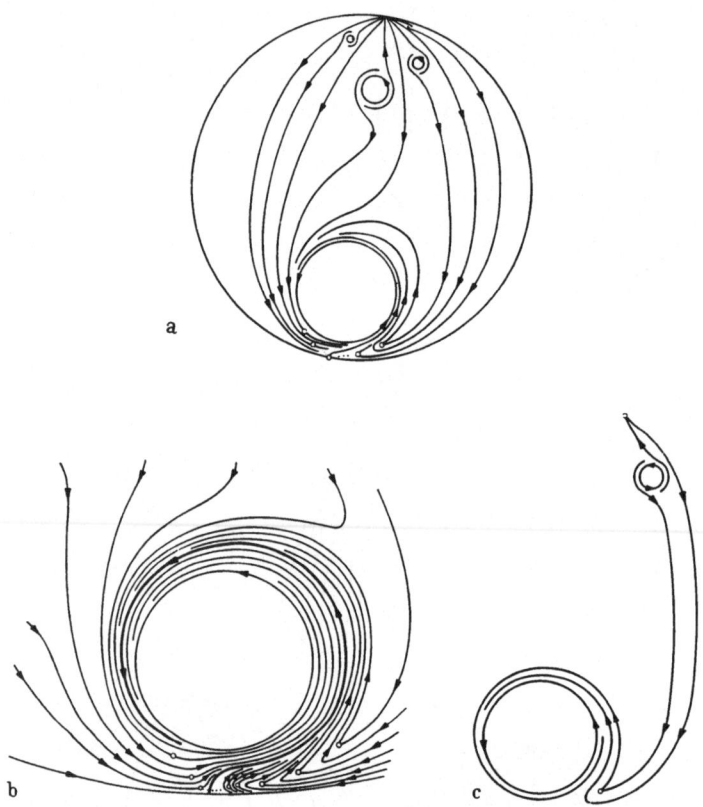

Fig. 12.5

We choose an orbit beginning at $-i$ and spiraling to $\hat{K}$. Then the orbits which are boundary orbits in Example (a) will be extended. On each side of the orbit joining $-i$ to $\hat{K}$, we choose a sequence of points converging to $-i$. These will become stagnation points. Choose any boundary orbit which begins at i and let it spiral to $\hat{K}$ (see Fig. 12.5 for this argument). Then let the adjacent boundary orbits stagnate at the first stagnation points on each side, and let an orbit begin at each such stagnation point and spiral to $\hat{K}$. The next boundary orbits adjacent to these will also spiral to $\hat{K}$, and the boundary orbits adjacent to *those* will stagnate in the next stagnation points, *etc.* The boundary orbits can now be extended in such a way that they encompass all of U in the obvious way. The flow can be taken to be canonical with respect to the set consisting of all the indicated stagnation points. A typical cell is seen in Fig. 12.5 (c), and has $\hat{K}$ for a periodic sink and i for a stagnation point.

12.18. Definition. Let Ω be a component of $\mathscr{G}_s(\varphi)$, and again assume that Ω contains no periodic orbit in its interior. For each periodic orbit $\mathcal{O}_\varphi(y) \in \partial(\Omega)$, we say $\mathcal{O}_\varphi(y)$ is *bilateral* if ff(y) consists of exactly two cells and two aperiodic orbits. In case ff(y) is bilateral, the cells in its fief are called *contiguous*. A *tissue* is a collection of cells which is closed under the property that for any cell included, every contiguous cell is included, and is minimal with respect to inclusion among all collections which are closed under the property. A cell which is contiguous with exactly one cell is called an *end cell*.

Given any tissue, we shall use the word *tissue* ambiguously to refer also to the set of points lying in the union of the fiefs of all the bilateral sources and sinks of the cells in the tissue. This set is the union of all the points in the cells, together with two aperiodic orbits for each bilateral source or sink.

12.19. Examples. The following are examples of tissues:

(a) A cell whose source and sink are both not bilateral (*i.e.* the class of contiguous cells is empty).

(b) Example 12.17 (a) shows an example of a tissue with infinitely many cells.

(c) The removal from Example 12.17 (a) of two non-contiguous cells leaves three tissues. Two of these have only one end cell each; the third has finitely many cells, two of which are end cells.

(d) Assume the tissue in (c) above has an even number of cells, so that it has either two sources or two sinks in its boundary. We can then identify the points of these two periodic orbits, and also the aperiodic orbits lying in the boundary, so as to create a bilateral source or sink (see Fig. 12.6). Such a tissue comprises an annulus.

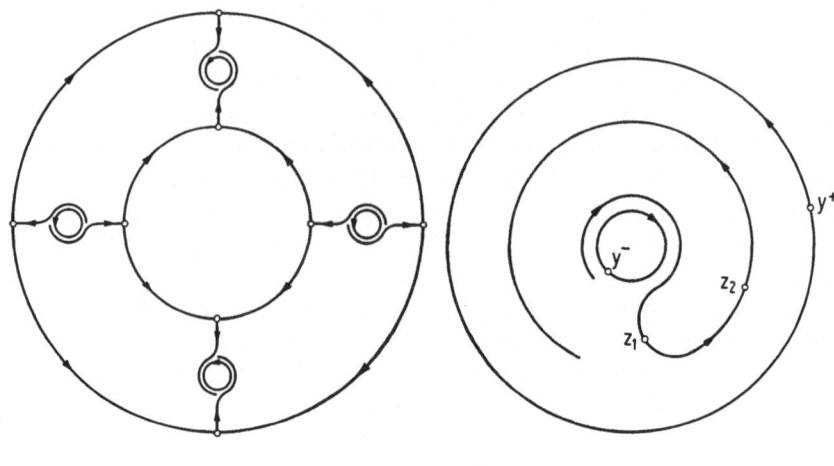

Fig. 12.6   Fig. 12.7

In a certain sense, these examples exhaust the possible tissues. The sense is the following: If a tissue consists of just one cell, then we can see the sense in which it is like Example (a). If there is more than one cell, then we can introduce an order relation into the cells in the obvious way: the cells contiguous to a given cell have the given cell between them in the indicated order, and all these ordered triples must match up. This ordering is either circular or linear. If circular, then it has only finitely many cells, and is similar to example (d). If linear, then it has two end cells, or else one, or else none. If it has two end cells, then the number of cells is finite, and the example is like the finite example in (c). If one end cell, then it is like an infinite example in (c), and if it has no end cells, then it like Example (b). The likeness under discussion is little more than the recognition of the order relationships. It is not even as strong as any form of conjugacy. To illustrate this point, we see that even the flows in two single cells might not be conjugate.

For example, the flow in Fig. 12.3 (b) is not conjugate to that in Fig. 12.7. The proof of this is essentially the same as that of Theorem 10.15. The same sort of argument, somewhat refined, shows that the flow shown in the cell in Fig. 12.1 is essentially different from that in Fig. 12.8, when the behavior near the boundary is taken into account. These comments require very careful wording, for there are great differences in the truth values of statements which seem to be nearly identical in meaning. First of all, we see at once that the flows shown in the two cells in Fig. 12.1 and Fig. 12.8 are both homeomorphic with translation flow in the plane, and thus with each other. If we look instead at the flows in the closures of the cells, then the two flows might or might not be homeomorphic depending on the periods of the sources and sinks, and the ways in which the spiral flows converge to the periodic ones (cf. *e.g.* Example 10.6.) It is the case, however, that these two flows are conjugate, as we shall indicate below. On the other hand, there do not exist two flows in the plane which are conjugate, with the homeomorphism of the conjugacy carrying a cell of the type of Fig. 12.1 onto the one of the type of Fig. 12.8.

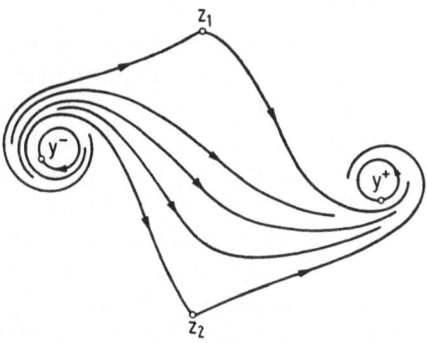

Fig. 12.8

12.20. Theorem. *Let φ be a flow in the closure of the cell shown in Fig. 12.1 which has the orbits indicated there, and let ψ be a flow in the closure of the cell in Fig. 12.8 which has the orbits indicated there.*

Then $\varphi \equiv^{+} \psi$.

PROOF: Let the figures be labelled in the following way. In Fig. 12.1, we call the source $\mathcal{O}_{\varphi}(y_{\varphi}^{-})$, the sink $\mathcal{O}_{\varphi}(y_{\varphi}^{+})$, the stagnation points $z_{1,\varphi}$ and $z_{2,\varphi}$, the orbit from $\mathcal{O}_{\varphi}(y_{\varphi}^{-})$ to $z_{1,\varphi}$ will be called $\mathcal{O}_{\varphi}(z_{1,\varphi}^{-})$, with $z_{1,\varphi}^{-}$ being a point appropriately chosen to make that accurate, and $z_{2,\varphi}^{-}$, $z_{1,\varphi}^{+}$, and $z_{2,\varphi}^{+}$ being defined accordingly. In Fig. 12.8, we make the corresponding labelling. Let $\varepsilon > 0$ be chosen so that the ε-neighborhoods of the source and sink for φ are disjoint, and similarly for ψ.

Using the techniques of Theorem 10.2, we can find a simple curve $\gamma_{0,\varphi}^{-}$ which lies within a distance ε of y_{φ}^{-}, which has for its endpoints points on $\mathcal{O}_{\varphi}(z_{1,\varphi}^{-})$ and $\mathcal{O}_{\varphi}(z_{2,\varphi}^{-})$, and which intersects each orbit in the indicated cell in exactly one point. WoLOG, we assume the endpoints are $z_{1,\varphi}^{-}$ and $z_{2,\varphi}^{-}$ respectively, and that these are the only points of the curve on the boundary of the cell. We make a similar construction of a curve $\gamma_{0,\varphi}^{+}$ near the sink, and similar construction for the flow ψ.

Now, following the methods of Theorem 10.7, we construct infinitely many curves $\gamma_{n,\varphi}^{-}$, and a function g_{φ}^{-} defined on their union γ_{φ}^{-} so that for each $x \in \gamma_{n,\varphi}^{-}$, $\varphi(g_{\varphi}^{-}(x), x) \in \gamma_{n+1,\varphi}^{-}$, while $\varphi(t, x)$ does not belong to

γ_φ^- for any $0 < t < g_\varphi^-(x)$, every sequence $\{x_n\}$ with $x_n \in \gamma_{n,\varphi}^-$, $\forall\, n$, converges to y_φ^-, and $g_\varphi^-(x_n) \to p_\varphi(y_\varphi^-)$ for each such sequence. We make similar constructions at the φ-sink and for the flow ψ.

For each point $x \in \gamma_{0,\varphi}^- \setminus \{z_{1,\varphi}^-, z_{2,\varphi}^-\}$, we see that there is a positive number, which we will call $f_\varphi(x)$, such that $\varphi(f_\varphi(x), x) \in \gamma_{0,\varphi}^+$. It is clear that $f_\varphi(x) \to \infty$ as $x \to z_{1,\varphi}^-$ or $z_{2,\varphi}^-$, and that f_φ is continuous. We define a function f_ψ similarly for the flow ψ.

Our next step is to construct a homeomorphism h_0 and a positive function u_0 such that $h_0[\varphi] = u_0\psi$, without yet requiring that u_0 be continuous. Let a homeomorphism h_0 be defined on the set of all the curves we have mentioned in the following way: we start with any homeomorphism h_0 of $\gamma_{0,\varphi}^-$ onto $\gamma_{0,\psi}^-$ which takes $z_{1,\varphi}^-$ onto $z_{1,\psi}^-$ and $z_{2,\varphi}^-$ onto $z_{2,\psi}^-$. We extend h_0 to all of γ_φ^- by requiring that $h_0\big(\varphi(g_\varphi^-(x), x)\big)$ should be $\psi\big(g_\psi^-(h_0(x)), h_0(x)\big)$, and to $\gamma_{0,\varphi}^+$ by $h_0\big(\varphi(f_\varphi(x), x)\big) = \psi\big(f_\psi(h_0(x)), h_0(x)\big)$, $\forall\, x \in \gamma_{0,\varphi}^- \setminus \{z_{1,\varphi}^-, z_{2,\varphi}^-\}$; $h_0(z_{i,\varphi}^+) = z_{i,\psi}^+$, $i = 1, 2$. We extend h_0 to all of γ_φ^+ by requiring that $h_0\big(\varphi(g_\varphi^+(x), x)\big)$ be $\psi\big(g_\psi^+(h_0(x)), h_0(x)\big)$ for all $x \in \gamma_\varphi^+$. We next define $u_0\big(\psi(t, x)\big)$ for each $x \in \gamma_{0,\psi}^-$. If $t < 0$, then we can write $\psi(t, x) = \psi(s, \tilde{x})$, where $\tilde{x} \in \gamma_\psi^-$ and $g_\psi^-(\tilde{x}) \leq s < 0$. At each such point, we define $u_0\big(\psi(t, x)\big) = g_\psi^-(\tilde{x})/g_\varphi^-(h_0^{-1}(\tilde{x}))$. For $t > f_\psi(x)$, we can write $\psi(t, x) = \psi(s, \tilde{x})$, where $\tilde{x} \in \gamma_\psi^+$ and $0 < s \leq g_\psi^+(\tilde{x})$. At that point we write $u_0\big(\psi(t, x)\big) = g_\psi^+(\tilde{x})/g_\varphi^+(h_0^{-1}(\tilde{x}))$. Now we set $\bar{l}(x) = \tfrac{1}{4}\min\big(f_\varphi(x), f_\psi(h_0(x))\big)$, $\forall\, x \in \gamma_{0,\varphi}^-$. For $0 \leq t \leq \bar{l}(x)$ and $f_\psi(h_0(x)) - \bar{l}(x) \leq t \leq f_\psi(h_0(x))$, we set $u_0\big(\psi(t, h_0(x))\big) = 1$, $\forall\, x \in \gamma_{0,\varphi}^-$, while for $\bar{l}(x) < t < f_\psi(h_0(x)) - \bar{l}(x)$, we set

$$u_0\big(\psi(t, h_0(x))\big) = \frac{f_\psi(h_0(x)) - 2\bar{l}(x)}{f_\varphi(x) - 2\bar{l}(x)}.$$

Finally we define u_0 to be $p_\psi(y_\psi^-)/p_\varphi(y_\varphi^-)$ on $\mathscr{O}_\psi(y_\psi^-)$ and $p_\psi(y_\psi^+)/p_\varphi(y_\varphi^+)$ on $\mathscr{O}_\psi(y_\psi^+)$. We now see that there is exactly one homeomorphism h_0 for which we will have $h_0[\varphi] = u_0\psi$.

We now use the methods of Theorem 10.12 to change from u_0 to a continuous ψ-multiplier u, and correspondingly change h_0 to h. By using care in this change, we can preserve the property that $u\big(\psi(t, h_0(x))\big) = 1$, $\forall\, x \in \gamma_{0,\varphi}^-$, whenever $0 \leq t \leq \bar{l}(x)$ or $f_\psi(h_0(x)) - \bar{l}(x) \leq t \leq f_\psi(h_0(x))$.

QED

12.21. Theorem. *Let φ and ψ be continuous flows in the regions Ω_φ and Ω_ψ respectively. Assume that one of the cells of the flow φ is of the form of Fig. 12.1, and that one of the cells of ψ is of the form of Fig. 12.8.*

Then there is no conjugacy of φ with ψ which carries any point of the first cell onto a point of the second.

PROOF: Let the two cells be labelled in the same way as in Theorem 12.20. Assume that h is a homeomorphism of Ω_φ onto Ω_ψ and u a positive continuous function such that $h[\varphi] = u\psi$. (The case where u is negative is exactly the same.) Let x_φ be a point of Ω_φ with $h(x_\varphi) \in \Omega_\psi$. Then clearly $h(\alpha_\varphi(x_\varphi)) = \alpha_\psi(h(x_\varphi))$, so that $h(\mathcal{O}_\varphi(y_\varphi^-)) = \mathcal{O}_\psi(y_\psi^-)$, and we can assume WoLOG that $h(y_\varphi^-) = y_\psi^-$. Similarly, we assume $h(y_\varphi^+) = y_\psi^+$. Every point of the boundary of either cell is either periodic, or fixed, or has a fixed endpoint for its orbit. Thus, h carries one boundary onto the other, and it is immediate that the images of the fixed points are fixed. We choose two non-intersecting Jordan curves J_φ^- and J_φ^+ in the neighborhoods of $\mathcal{O}_\varphi(y_\varphi^-)$ and $\mathcal{O}_\varphi(y_\varphi^+)$ respectively so that the region between J_φ^- and $\mathcal{O}_\varphi(y_\varphi^-)$ is an annulus in Ω_φ without any fixed points of φ, and a similar thing happens at the sink. Choose J_φ^- and J_φ^+ in such a way that each orbit in the first cell intersects each of them in a single point. Let $\tilde{\Omega}_\varphi$ be the union of these two annuli with the first cell, and let $h(\tilde{\Omega}_\varphi)$ be denoted as $\tilde{\Omega}_\psi$. It is clear that each of these regions is triply-connected.

We now see at once that the homeomorphism h cannot take the boundary of $\tilde{\Omega}_\varphi$ into the boundary of $\tilde{\Omega}_\psi$ in the proper orientation. Just as in Theorem 10.15, let us consider the identification map made by identifying $\varphi(t\mathrm{p}_\varphi(y_\varphi^-), y_\varphi^-)$ with $\varphi(t\mathrm{p}_\varphi(y_\varphi^+), y_\varphi^+)$, $\forall t \in \mathbb{R}$. This manifold would have to be homeomorphic with the manifold obtained by identifying the images of these points. However, the first of these manifolds is a Klein bottle with a hole in it, while the second is a torus with a hole in it, and these are not homeomorphic. **QED**

12.22. Definition. Let φ be a flow in a region of $\mathbf{S}^2$, and let Ω be a cell of φ. Let one of the fixed points of $\partial(\Omega)$ be chosen to play the role of ∞. Then the cell Ω is called *orientable* if the source and sink rotate in opposite directions (as in Fig. 12.8) and *non-orientable* if they rotate in the same direction (as in Fig. 12.1). These designations conform to the appearance of the punctured torus and punctured Klein bottle in Theorem 12.21. Note that the cell in Fig. 12.3 (b) is orientable, while that in Fig. 12.7 is non-orientable.

12.23. Theorem. *Let φ be a flow in a region Ω_φ of $\mathbf{S}^2$, and ψ a flow in a region Ω_ψ of $\mathbf{S}^2$. Assume that φ is conjugate with ψ.*

Then the homeomorphism h of the conjugacy takes each orientable cell of φ onto an orientable cell of ψ, and each non-orientable cell of φ onto a non-orientable cell of ψ.

PROOF: This is just Theorem 12.21 reworded in the notation of Definition 12.22. **QED**

12.24. Theorem. *Let φ and ψ be flows in regions in $\mathbf{S}^2$, and let Ω_φ and Ω_ψ be tissues of φ and ψ respectively. Let the restrictions of φ and ψ to Ω_φ and Ω_ψ respectively be denoted as φ_Ω and ψ_Ω respectively. Assume that there is a pairing of φ-cells of Ω_φ with ψ-cells of Ω_ψ such that the pairs of contiguous cells are contiguous, the pair of each orientable cell is orientable, and the pair of each non-orientable cell is non-orientable.*

Then the flows φ_Ω and ψ_Ω are conjugate.

PROOF: Let the cells of Ω_φ be denoted as $\{\Omega_{i,\varphi}\}$, where these latter are indexed over some interval, finite or infinite, of the integers, in such a way that $\Omega_{i,\varphi}$ is contiguous with $\Omega_{i-1,\varphi}$ and $\Omega_{i+1,\varphi}$, if there are cells for these indices, and where the cell of least index is contiguous with that of greatest index if the tissue has a circular ordering. For each source and sink in the boundary of the given tissue, we select a point, which we denote as $y_{i,\varphi}$, in such a way that the source and sink of $\Omega_{i,\varphi}$ are $\mathcal{O}_\varphi(y_{i,\varphi})$ and $\mathcal{O}_\varphi(y_{i+1,\varphi})$, in some order, and where the obvious identification is made between two of these in the case of a circular ordering of cells. For each $y_{i,\varphi}$, we select a Jordan curve J_i lying in the region of definition of φ, in such a way that the annulus lying between $\mathcal{O}_\varphi(y_{i,\varphi})$ and J_i contains no fixed point of φ, and so that these annuli are disjoint in pairs. Following the work of Theorem 10.2 and Corollary 10.3, we can find in each of these annuli a Jordan curve with the same properties and also with the property that it cuts each orbit in $\mathrm{ff}(y_{i,\varphi})$ in a single point. We can assume WoLOG that J_i has this last property also.

What we have done for the flow φ, we can do for the flow ψ as well, and let the Jordan curves for ψ be denoted as $\bar{J}_i$, where the corresponding values of i are chosen so that the paired cells of Ω_φ and Ω_ψ have the same numbers, and so that the sources and sinks of Ω_ψ are denoted as $\mathcal{O}_\psi(y_{i,\psi})$, where the same numbering scheme is used with respect to the numbers of the cells and the numbering of their sources and sinks. This arrangement gives a natural pairing of the sources and sinks of Ω_φ with those of Ω_ψ, and these must be consistently paired as sources with sources and sinks with sinks, or else consistently as sources with sinks. In the latter case, we could look at the flows φ and $-\psi$, instead of φ and ψ, where the question of conjugacy is equivalent, and where the sources are paired with sources, sinks with sinks. Thus, WoLOG, we assume this is true of φ and ψ. Now, in the mode of Theorems 10.9 and 10.12, we can find a conjugacy between the flow φ_i which is the restriction of φ to $\mathrm{ff}(y_{i,\varphi}) \cup \mathcal{O}_\varphi(y_{i,\varphi})$ with the flow of Example 3.14(b) in the annulus $\{z \mid 1 < |z| \leq 3\}$ in such a way that the image of J_i under the homeomorphism of the conjugacy is just Γ_2. Clearly, the conjugacy is positive just if $\mathcal{O}_\varphi(y_{i,\varphi})$ is a sink. We can perform the same analysis for the flow ψ, and the two conjugacies between the fiefs and the fixed annulus yield a conjugacy between the flows on the fiefs, with the source or sink included. Thus φ_i is conjugate with ψ_i, and the conjugacy is, in fact, positive. We denote the homeomorphism of the two sets by h_i and the multiplier by u_i, so that $h_i[\varphi_i] = u_i \psi_i$.

Next, we pay attention to the cells $\Omega_{i,\varphi}$ and $\Omega_{i,\psi}$, where in each cell the intersection with J_i and J_{i+1} on the one hand and $\bar{J}_i$ and $\bar{J}_{i+1}$ on the other will play the role given to $\gamma_{0,\varphi}^-$, $\gamma_{0,\varphi}^+$, $\gamma_{0,\psi}^-$ and $\gamma_{0,\psi}^+$ in Theorem 12.20. The method used there will give us a homeomorphism g_i of $\Omega_{i,\varphi}$ onto $\Omega_{i,\psi}$ and a multiplier v_i such that $g_i[\varphi] = v_i \psi$ when these flows are restricted to the cells under consideration, and such that v_i takes the value 1 in a neighborhood of the boundary orbits of ψ_Ω which spiral to either a source or a sink, in the region between the Jordan curves J_i and J_{i+1}.

We now construct a homeomorphism $\bar{h}$ made by applying h_i in the annulus between J_i and $\mathcal{O}_\varphi(y_{i,\varphi})$, $\forall\, i$, and g_i in the part of $\Omega_{i,\varphi}$ not accounted for in this way, including the boundary orbits which spiral. We construct a positive function $\bar{u}$ which is u_i on the annulus between $\bar{J}_i$ and $\mathcal{O}_\psi(y_{i,\psi})$, $\forall\, i$, and v_i in all of $\Omega_{i,\psi}$ which is not accounted for in this way. We now have $\bar{h}[\varphi_\Omega] = \bar{u}\psi_\Omega$, and our job is finished except for the continuity of $\bar{u}$. However, this is easily taken care of by the methods of Theorem 10.12.   **QED**

12.25. Conjugacy Classes and Likeness Classes. We have not actually defined formally what we mean by likeness, nor are we going to, since the concept does not appear to be particularly fruitful for analysis of structure. It is ample to observe that there are infinitely many likeness classes, that these are the ones set forth in Example 12.19, and that each likeness class (except the first) contains several conjugacy classes. Note that in Fig. 12.4, we could assign the orientation of the flow on each periodic orbit in either direction, making local alterations to accommodate this change. Thus, we easily see that there are $\aleph$ different flows of this same likeness class. The same is true for the infinite examples in (c). For tissues with finite numbers of cells, there are only finitely many different ways of assigning the orientations on the periodic orbits. When the orientations are properly matched, then we can obtain a conjugacy for the flows on the tissues.

Flows with Finitely Many Stagnation Points

12.26. Introduction. Let φ be a flow in $\bar{\Omega}$ for which the assumptions of the previous sections hold, *i.e.*, there are no fixed points and no periodic orbits in Ω, and every point in $\bar{\Omega}$ is singular. Let x be any point of Ω, and let us examine $\omega_\varphi(x)$. If $\omega_\varphi(x)$ is a periodic orbit, then we can examine its fief. The opposite boundary of the fief cannot be a periodic orbit, since in that case, the fief would be an annulus of regular points, contrary to hypothesis. Thus, we understand the behavior of φ in the neighborhood of $\omega_\varphi(x)$. If, on the other hand, $\omega_\varphi(x)$ is not a periodic orbit, then let us assume further that $\mathcal{O}_\varphi(x)$ has a moving ω-point, and that there is a neighborhood N of $\omega_\varphi(x)$ such that $N \cap [x$ side : $\omega_\varphi(x)]$ contains no stagnation points. We can build a φ-gate construction so that every point on the gate has the same ω-set, namely, $\omega_\varphi(x)$. The next results will help us to use the set $\omega_\varphi(x)$ in somewhat the same way that we used periodic orbits in the work just preceding.

12.27. Definition. Let Ω be a region in $\mathbb{R}^2$, and let φ be a flow in $\bar{\Omega}$ which has no periodic orbits or fixed points in Ω and no regular points in $\bar{\Omega}$. Let x be a point which has a moving α-point, and also has the property that there is a neighborhood N of $\alpha_\varphi(x)$ such that $N \cap [x$ side : $\alpha_\varphi(x)]$ contains no stagnation points. Then the annulus consisting of those points of Ω which have $\alpha_\varphi(x)$ as their α-set will be called the α-φ-Ω-fief of x, written $\mathrm{ff}_\alpha(x, \varphi, \Omega)$, or $\mathrm{ff}_\alpha(x)$ when the other symbols are understood. The same definition is made with the symbol ω replaced for α throughout.

12.28. Theorem. *Let* $x \in \Omega$, *and let* $\mathrm{ff}_\alpha(x)$ *exist.*

Then $\mathrm{ff}_\alpha(x)$ *is open, and a similar result holds for* $\mathrm{ff}_\omega(x)$.

PROOF: $\mathrm{ff}_\alpha(x)$, if it exists, is an open annulus.   **QED**

12.29. Theorem. *Let* x_1, $x_2 \in \Omega$, *and assume that* $\mathrm{ff}_\alpha(x_1)$ *and* $\mathrm{ff}_\alpha(x_2)$ *are non-empty.*

Then $\mathrm{ff}_\alpha(x_1)$ *and* $\mathrm{ff}_\alpha(x_2)$ *are either disjoint or co-incident, and a similar result holds for* $\mathrm{ff}_\omega(x_1)$ *and* $\mathrm{ff}_\omega(x_2)$.

PROOF: Assume that $x_3 \in \mathrm{ff}_\alpha(x_1) \cap \mathrm{ff}_\alpha(x_2)$. Then by Corollary 2.25, we see at once that $\mathrm{ff}_\alpha(x_1) = \mathrm{ff}_\alpha(x_3) = \mathrm{ff}_\alpha(x_2)$.   **QED**

12.30. Theorem. *Let* Ω *be a region in* $\mathbb{R}^2$, *and let* φ *be a flow in* $\bar{\Omega}$ *which has no periodic orbits or fixed points in* Ω *and no regular points in* $\bar{\Omega}$. *Assume that* $\mathscr{S}(\varphi)$ *is finite.*

Then $\{x \mid \alpha_\varphi(x)$ *and* $\omega_\varphi(x)$ *are each a single point}* *is closed.*

PROOF: For each point x which has a moving endpoint, its fief at that end exists and is open, and consists only of points which spiral. If x has only fixed endpoints, then since there are only finitely many stagnation points, $\mathcal{O}_\varphi(x)$ has only one endpoint at each end (Theorem 1.46).   **QED**

12.31. Organs. If Ω and φ satisfy the hypotheses of Theorem 12.30, then the set of all points in $\bar{\Omega}$ having a simple endpoint at each end is closed. Let us imagine that we were to enlarge Ω to a region Ω_0 by adding all the periodic orbits in the boundary of Ω, together with the sides of those orbits away from Ω. Then Ω_0 would be a finitely-connected region with at least one stagnation point in each boundary. The orbits which have simple stagnation points form a closed set, and this set of curves is a subset of a regular curve-family filling Ω_0 in the sense of Kaplan [1] and [2]. It can no doubt be easily shown (though we have not yet done so) that the flow φ restricted to this set can be extended to a Kaplan-Markus flow on all of Ω_0 without introducing any new stagnation points. Thus, we will, in some sense, know all there is to know if we can describe the action of φ in each of the remaining open sets and describe how they hang together. The first of these tasks is easier than the second, and we will take up that problem next.

The open sets in question are made up mostly of cells, together with all the orbits which arise in or terminate in a periodic orbit. For any cell, the entire tissue to which that cell belongs lies in a single one of these open sets. Pursuing the metaphor of anatomical nomenclature, we call each of these an *organ* of the flow φ in Ω. The remaining points of the organ are made up of sets in which the flow has a source or sink at one end and stagnation points at the other. These are like cells, except that they are sort of half-cells. Having gone this far into the anatomical metaphor, we call them *gametes*.

12.32. Definition. Let Ω be a region in $\mathbb{R}^2$ and let φ be a flow in $\bar{\Omega}$. Assume that $\partial(\Omega)$ has a stagnation point of φ in each component, and that Ω contains no fixed point, no periodic orbit, and no orbit whose endpoints are all fixed, but that every point of Ω is singular with respect to φ. Then we say that Ω is an *organ* of φ, and that φ is an *organ flow* in $\bar{\Omega}$.

12.33. Definition. Let φ be an organ flow in $\bar{\Omega}$ and let y be a periodic point in $\partial(\Omega)$. Let Ω_0 be an open, connected subset of Ω such that every point of Ω_0 has $\mathcal{O}_\varphi(y)$ for its endpoints at one end, while all the endpoints at the other end are stagnation points. Then Ω_0 is a φ-*gamete*, or just a *gamete* when φ is understood, and φ is a *gamete flow* in Ω_0. Examples of gametes are seen in Fig. 12.9.

12.34. Theorem. *Let Ω be a finitely-connected region in $\mathbb{R}^2$ and let φ be a flow in $\bar{\Omega}$. Assume that Ω contains no periodic orbit or fixed point, and that every point of $\bar{\Omega}$ is singular. Let F be the set of points in Ω with only fixed points for endpoints.*

Then every component of $\Omega \setminus F$ is an organ of φ.

PROOF: Immediate from Definition 12.32 and the foregoing discussion.
QED

12.35. Theorem. *Let Ω be a region in $\mathbf{S}^2$ and let φ be an organ flow in Ω. Let Ω_0 be a tissue of φ.*

Then $\Omega = \Omega_0$ precisely if Ω_0 lies in the likeness class of either Example 12.17 (a) or Example 12.19 (d).

PROOF: Clearly, no boundary point of any tissue of either of these types has a spiral orbit. Thus, each such tissue is a whole organ. Conversely, every other tissue has an end cell, and thus, is only part of an organ, since each end cell shares a source or a sink with another cell from a different tissue or a gamete.   **QED**

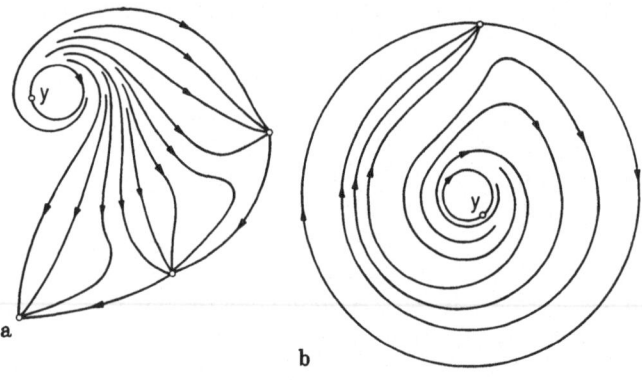

Fig. 12.9

12.36. Definition. Let Ω be an organ of the flow φ, and let Ω_0 be a tissue of φ. For each cell or gamete of Ω, let a point of that set be chosen to represent it. For each source or sink in $\partial(\Omega)$, let a point of its fief be chosen to represent it. For each cell or gamete, let an arc be drawn joining the representative point of the set to that of its source and/or sink. These arcs should be pairwise disjoint. At the representative point, arcs should be added, in the manner of Fig. 12.10, which will indicate whether the set is a gamete, an orientable cell, or a non-orientable cell. The resulting one-dimensional configuration will be called the *skeleton of φ in Ω*, and the portion corresponding to the cells of Ω_0 we will call the *skeleton of φ in Ω_0*.

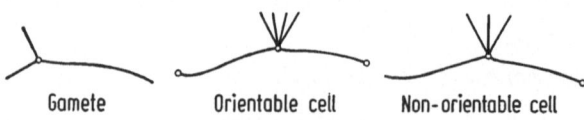

Gamete   Orientable cell   Non-orientable cell

Fig. 12.10

12.37. Theorem. *Let φ and ψ be flows in regions of $\mathbf{S}^2$, and let Ω_φ and Ω_ψ be tissues of φ and ψ respectively. Let φ_Ω and ψ_Ω be the flows φ and ψ respectively, restricted to the union of the fiefs of the sources and sinks of the cells in Ω_φ and Ω_ψ.*

Then φ_Ω and ψ_Ω are conjugate if and only if the skeletons of φ and ψ in Ω_φ and Ω_ψ respectively are homeomorphic.

PROOF: The skeleton of each tissue is a tree-like continuum unless the tissue is of the form of Example 12.19(d). In that case, it is a Jordan curve with markings on it which will indicate the order of orientable and non-orientable cells. In the former case, it is an arc with markings on it to indicate the order of orientable and non-orientable cells. If there is a conjugacy of the sort indicated, then the markings must occur in the same order, and the skeletons are homeomorphic. If the skeletons are homeomorphic, then the homeomorphism gives a pairing of the cells of Ω_φ with those of Ω_ψ which meets the hypotheses of Theorem 12.23, and thus, by that theorem, the required conjugacy exists.   **QED**

12.38. Example. The same theorem does not hold for organs, rather surprisingly. The difference is that there are two "orientations" at any source or sink which is not bilateral: the order of tissues around the source or sink, and the direction of motion of the flow. These can agree or disagree, and these circumstances are basically different. Consider the organ flows in Figs. 12.11 and 12.13. They both have for their skeletons the continuum in Fig. 12.12.

Now let us look at the flow in Fig. 12.11. Call it φ. Let y^+ be a point in the outer boundary, and let a Jordan curve J be constructed, in the manner of Corollary 10.3, such that there is no stagnation point or boundary point of the organ in the annulus between $\mathcal{O}_\varphi(y^+)$ and J, and so that J intersects the orbit of every point in that annulus in exactly one point. Now we see that J intersects each of the tissues of this organ in an arc. Let us choose a point of the arc which passes through the tissue with only one cell, and call it x_1. Then if we look at the images $\varphi(n, x_1)$, we see that some subsequence $\{\varphi(n_i, x_1)\}$ converges as $n \to +\infty$. Choosing another point, we find a sub-subsequence for which there is convergence there also. We continue, and finally find a subsequence of the positive integers with the property that every point converges over this subsequence. With a small additional quantum of care, we assure that every point of $\mathcal{O}_\varphi(y^+)$ is included in the limit set. We now see that the flow φ on $\mathcal{O}_\varphi(y^+)$ passes in the positive time direction through the points which are limits of points from the tissue with one cell, followed by those from the tissue with three cells, followed by those from the tissue with five cells, and then those from the tissue with one cell again.

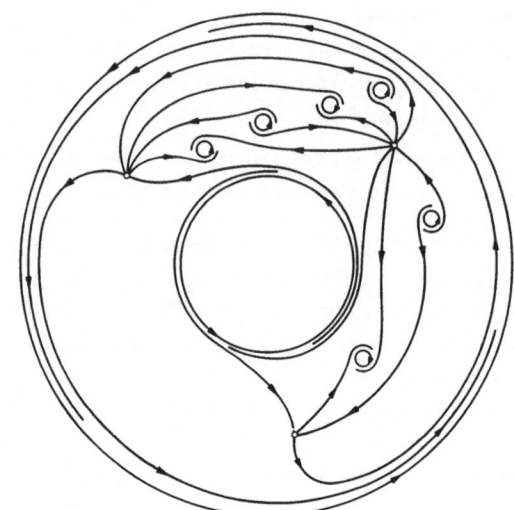

Fig. 12.11

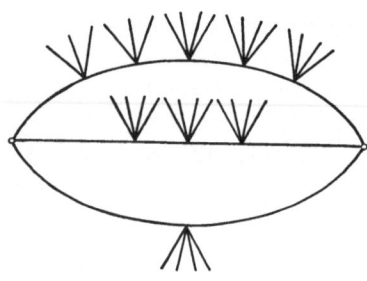

Fig. 12.12

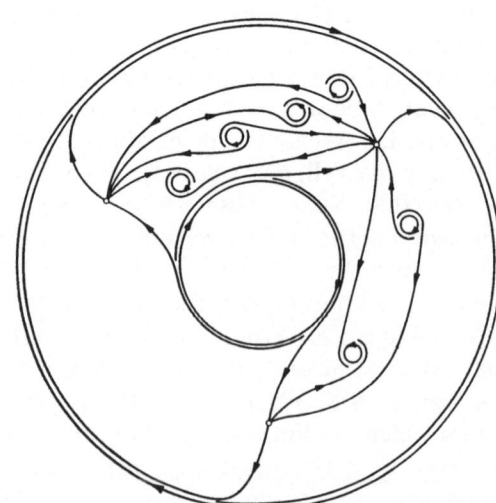

Fig. 12.13

Now let us turn our attention to the flow in Fig. 12.13, which we will call ψ. Because of the location of the two non-orientable cells, any conjugacy must take the inner boundary of the φ-organ onto the inner boundary of the ψ-organ. Thus, if we assume the two flows are conjugate, they must be positively conjugate, with the inner boundary being a source and the outer boundary a sink. Looking at the images of the Jordan curves $\varphi(n_i, J)$, we see that each of these curves passes through the ψ-tissues in the order: first the tissue with one cell, then the tissue with five, and then the tissue with three, and then the tissue with one again. That is, the limit points have that order, as in the case of the φ-organ. But this is a contradiction. Thus, there is no such conjugacy.

12.39. Definition. Let Ω be an organ of a flow φ in a region of $\mathbf{S}^2$. For each source and sink in $\partial(\Omega)$, let a Jordan curve be drawn in Ω so that the annulus between that curve and the source or sink is an annulus of points from Ω (and thus contains no points of $\partial(\Omega)$). Choose these Jordan curves so that all the closures of these annuli are disjoint. For each cell or gamete of Ω, let a point of that set be chosen to represent it which does not lie in any closed annulus. For each representative point, let an arc lying in the represented set join the point to the Jordan curve at the source and/or sink for that set. At the representative point, arcs should be added, as for the skeleton (Definition 12.36), to show whether the set is a gamete, or an orientable cell, or a non-orientable cell. Let each Jordan curve be sensed so that the orientation on that curve is the same as the direction of flow on the corresponding source or sink. On each Jordan curve, for each cell or gamete in the fief of the corresponding source or sink, let two points be chosen so that the arc of the Jordan curve which begins at the intersection of the Jordan curve and the arc to the representative point and moves in the indicated sense through both points does not leave that cell or gamete. At the first of these points, let one of the symbols in Fig. 12.14 be added, according as the Jordan

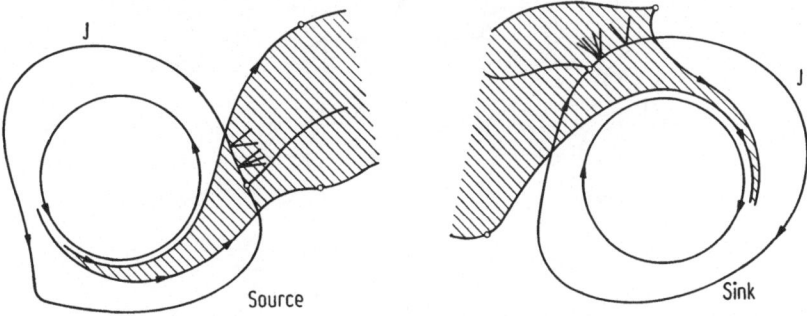

Fig. 12.14

curve corresponds to a source or a sink, and at the second, add the other symbol shown in Fig. 12.14 so that the location of this symbol will show the direction of motion.

The union of all these markings is called the *oriented skeleton* of φ in Ω.

12.40. Theorem. *Let φ and ψ be flows in regions of $\mathbf{S}^2$. Let Ω_φ and Ω_ψ be organs of φ and ψ respectively. Let the flows φ and ψ restricted respectively to Ω_φ and Ω_ψ be denoted as φ_Ω and ψ_Ω.*

Then φ_Ω and ψ_Ω are positively conjugate if and only if their oriented skeletons are homeomorphic.

PROOF: The nature of the construction is such that if the flows are positively conjugate, then the construction puts the same markings in the same places, except for small changes, which come within the area of a homeomorphism.

Going the other way, we see that the oriented skeleton identifies the sources and the sinks, and the directions of motion on each of these, and automatically co-ordinates these motions with the order of the cells and

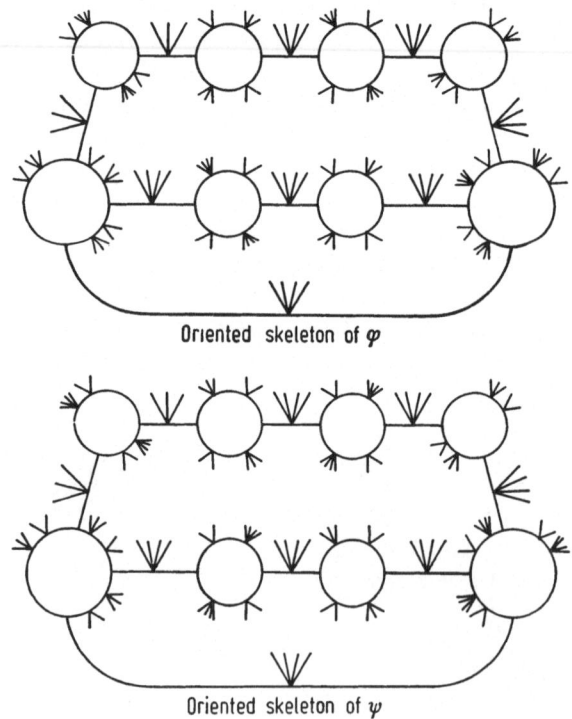

Oriented skeleton of φ

Oriented skeleton of ψ

Fig. 12.15

gametes in the fief of each periodic orbit. Thus, the method of Theorem 12.37 works here to give a conjugacy between the two organ flows φ_Ω and ψ_Ω. **QED**

12.41. Example. (Footnote to Example 12.38). The oriented skeletons of the flows φ and ψ of Example 12.38 are displayed in Fig. 12.15. The reader is invited to notice that they are not homeomorphic.

12.42. Combinations of Organs. The major difficulty in combining organs lies in the continuity of the flows in the case of convergence of a sequence of points, all from different organs and converging to a moving point with only stagnation points for endpoints. When the limit points could only be fixed points, we used the technique of bounded flows to assure the continuity across the regions being pieced together to a point not in any of them. However, to paste organs together, where the limit points might be moving points, is much harder. Even the use of canonical flows would not assure continuity; there exist quasi-flows which are canonical but not continuous. This whole question is more or less open.

12.43. Conclusion. Chapters 11 and 12 have indicated the current state of work on characterizing the fine structure of the singular moving points of a continuous flow in the plane. To some extent, it is as complete as the analysis of the regular moving points. However, it is not as easy to build up from topological invariants to standard models for each conjugacy class. This is perhaps the major area for work in this subject.

APPENDIX A

TOPOLOGY

A.1. Definitions and Notation. If A is a subset of a topological space X, then the closure of A is denoted by $\bar{A}$ or $\mathrm{cl}(A)$, the boundary of A by $\partial(A)$, and the interior of A by $\mathrm{int}(A)$.

If A and B are subsets of a topological space X, then the union of those components of A which meet B, is denoted by $\mathscr{C}(A, B)$.

If A is a subset of a topological space X, then the components of $X \setminus A$ are sometimes called the *sides of A in X*, or simply, the *sides of A*, if X is understood.[1] If a subset B of X is contained in a single component of $X \setminus A$, then this component of $X \setminus A$ is called the B *side of A*, and is denoted by $[B \text{ side} : A]$. (Again, X is understood.) A subset of $[B \text{ side} : A]$ is said to *lie in the B side of A*.

The family of *Borel* subsets of a topological space X is defined to be the smallest σ-algebra (σ-field) of subsets of X which contains all open sets.

If f and g are continuous functions from a topological space X into a topological space Y, then f and g are said to be *homotopic* if there exists a continuous function h from $[0, 1] \times X$ into Y, such that

$$h(0, x) = f(x) \text{ and } h(1, x) = g(x), \qquad \forall\, x \in X.$$

If f and g are homeomorphisms from a topological space X into a topological space Y, then f and g are said to be *isotopic* if there exists a continuous function h from $[0, 1] \times X$ into Y, such that

(a) $h(0, x) = f(x)$ and $h(1, x) = g(x), \qquad \forall\, x \in X,$
and

(b) For each $t \in [0, 1]$, the function $x \to h(t, x)$ is a homeomorphism of X into Y.

---

[1] This terminology is motivated by the fact that if A is a Jordan curve in $\mathbb{R}^2$, then the two components of $\mathbb{R}^2 \setminus A$ are called the inside of A and the outside of A.

Connectedness in Compact Hausdorff Spaces

A.2. Definition. A compact, connected Hausdorff space which contains more than one point, is called a *continuum*.

A.3. Definition. Let H and K be mutually disjoint subsets of a Hausdorff space X. A compact connected subset C of X is called a *continuum from H to K* if $C \cap H \neq \square$ and $C \cap K \neq \square$.

A.4. Lemma. *Let H and K be closed subsets of a Hausdorff space X,[2] let $\{C_\alpha \mid \alpha \in A\}$ be a chain of compact subsets of X, and suppose*

(i) *For each $\alpha \in A$, $C_\alpha \cap H \neq \square$ and $C_\alpha \cap K \neq \square$,*

and

(ii) *For each $\alpha \in A$, C_α is not the union of two mutually disjoint compact subsets, one meeting H and the other meeting K.*

Let $C = \bigcap_\alpha C_\alpha$.

Then C meets H and C meets K, but C is not the union of two mutually disjoint compact subsets, one meeting H and the other meeting K.

PROOF: We see at once that C is compact. Since $\{C_\alpha \cap H \mid \alpha \in A\}$ and $\{C_\alpha \cap K \mid \alpha \in A\}$ are chains of non-empty, compact sets, it is clear that C meets H and C meets K.

To obtain a contradiction, suppose $C = A \cup B$ where A and B are compact, $A \cap B = \square$, A meets H, and B meets K. Choose mutually disjoint open sets U and V such that $A \subseteq U$ and $B \subseteq V$. It is easy to see that, for each $\alpha \in A$, $C_\alpha \setminus (U \cup V)$ is compact and non-empty, and it follows that

$$C \setminus (U \cup V) = \bigcap_\alpha \left(C_\alpha \setminus (U \cup V) \right) \neq \square,$$

contradicting the choice of U and V.

This completes the proof.   **QED**

A.5. Theorem. *Let $\{C_\alpha \mid \alpha \in A\}$ be a chain of compact, connected subsets of a Hausdorff space X.*
Then $\bigcap_\alpha C_\alpha$ is compact and connected.

PROOF: If $C_\alpha = \square$ for some α, the theorem is trivial. If $C_\alpha \neq \square$ for every $\alpha \in A$, the theorem follows from Lemma A.4 with $H = K = X$.
                                                            QED

---

[2] H and K are not necessarily disjoint.

A.6. Lemma. *Let H and K be mutually disjoint closed subsets of a Hausdorff space X. Suppose X contains a compact subset C, such that C meets H, C meets K, and C is not the union of two mutually disjoint compact subsets, one meeting H and the other meeting K.*

Then X contains a subset C_0 of C which is minimal with respect to the properties of being compact, meeting H, meeting K, and not being the union of two mutually disjoint compact subsets of which one meets H and the other meets K. Furthermore, C_0 is a minimal continuum from H to K.

PROOF: In view of Lemma A.4, the existence of C_0 follows from an easy application of Zorn's lemma.

We claim that C_0 is connected: If C_0 is not connected, write $C_0 = A \cup B$, where A and B are non-empty, compact, and mutually disjoint. Because of the properties of C_0, we may clearly assume that $C_0 \cap H$ and $C_0 \cap K$ are both included in A. Therefore the minimality of C_0 implies that A can be written in the form $A = A_1 \cup A_2$, where A_1 and A_2 are mutually disjoint and compact, and A_1 meets H and A_2 meets K. But now, C_0 is the union of A_1 and $A_2 \cup B$, which is impossible, since A_1 and $A_2 \cup B$ are mutually disjoint and compact, and A_1 meets H, and $A_2 \cup B$ meets K. This contradiction shows that C_0 is connected, as claimed.

It now follows easily that C_0 is a minimal continuum from H to K, and this completes the proof.   **QED**

A.7. Theorem. *Let H and K be mutually disjoint closed subsets of a Hausdorff space X, and suppose X contains a continuum C from H to K.*

Then X contains a minimal continuum C_0 from H to K, such that $C_0 \subseteq C$.

PROOF: The result follows at once from Lemma A.6.   **QED**

A.8. Theorem. *Let H and K be mutually disjoint closed subsets of a compact Hausdorff space X, and suppose X contains no continuum from H to K.*

Then X can be written in the form $A \cup B$, where A and B are mutually disjoint and compact, and $H \subseteq A$ and $K \subseteq B$.

PROOF: Since the theorem is trivial when either H or K is empty, we shall assume that H and K are non-empty.

We assert that if $h \in H$ and $k \in K$, then X can be written as a union of two mutually disjoint compact sets, of which one contains h, and the other contains k, for otherwise, by Lemma A.6, X would have to contain a continuum from $\{h\}$ to $\{k\}$, and this would be a continuum from H to K.

Temporarily, fix $h \in H$. For each $k \in K$, let A_{hk} and B_{hk} be mutually disjoint compact subsets of X, such that $X = A_{hk} \cup B_{hk}$, $h \in A_{hk}$, and $k \in B_{hk}$. We note that, for every $k \in K$, A_{hk} and B_{hk} are open. Therefore, since K is compact, we can find finitely many members $k_1, k_2, \ldots, k_n$ of K such that

$$K \subseteq \bigcup_{i=1}^{n} B_{hk_i}.$$

Let $B_h = \bigcup_{i=1}^{n} B_{hk_i}$, and $A_h = \bigcap_{i=1}^{n} A_{hk_i}$.

Then A_h and B_h are mutually disjoint compact sets, $A_h \cup B_h = X$, $h \in A_h$, and $K \subseteq B_h$.

Now choose A_h and B_h in this way, for every $h \in H$. We note that A_h and B_h are open, for each $h \in H$. Therefore, since H is compact, we can find finitely many members $h_1, \ldots, h_m$ of H such that

$$H \subseteq \bigcup_{i=1}^{m} A_{h_i}.$$

Let $A = \bigcup_{i=1}^{m} A_{h_i}$, and $B = \bigcap_{i=1}^{m} B_{h_i}$.

It is clear that A and B satisfy the requirements of the theorem. **QED**

A.9. Theorem. *Let H and K be mutually disjoint closed subsets of a Hausdorff space X, and let C be a minimal continuum from H to K.*
Then

(a) $C \setminus (H \cup K)$, $C \setminus H$, and $C \setminus K$ *are all connected.*

(b) *Every point belonging to either $C \cap H$ or $C \cap K$ is a limit-point of $C \setminus (H \cup K)$, i.e., the closure of $C \setminus (H \cup K)$ is C.*

PROOF: Since $H \cap C$ and $K \cap C$ are closed, there is no loss of generality in assuming, for convenience of notation, that $C = X$.

First, we shall prove that $X \setminus (H \cup K)$ is connected: if not, we can write $X \setminus (H \cup K) = A \cup B$, where A and B are separated (*i.e.*, $\bar{A} \cap B = \bar{B} \cap A = \square$), and A and B are non-empty.

Now $\bar{A}$, being a proper, compact subset of X, cannot contain a continuum from H to K. Therefore, by Theorem A.8, we can write $\bar{A} = A_H \cup A_K$, where A_H and A_K are mutually disjoint compact sets, $\bar{A} \cap H \subseteq A_H$, and $\bar{A} \cap K \subseteq A_K$. Similarly, we can write $\bar{B} = B_H \cup B_K$, where B_H and B_K are mutually disjoint compact sets, $\bar{B} \cap H \subseteq B_H$, and $\bar{B} \cap K \subseteq B_K$.

It is easy to see that $H \cup A_H \cup B_H$ and $K \cup A_K \cup B_K$ are non-empty, mutually disjoint compact sets whose union is X. But this is impossible, since X is connected.

We have therefore shown that $X \setminus (H \cup K)$ is connected. It follows that $\text{cl}\,(X \setminus (H \cup K))$ is compact and connected.

We claim that $\text{cl}\,(X \setminus (H \cup K))$ meets both H and K. If $\text{cl}\,(X \setminus (H \cup K))$ does not meet H, then we can write X as the union of the mutually disjoint, non-empty, compact subsets H, and $\text{cl}\,(X \setminus (H \cup K)) \cup K$, which is impossible, since X is connected. Therefore $\text{cl}\,(X \setminus (H \cup K))$ meets H. A similar argument shows that $\text{cl}\,(X \setminus (H \cup K))$ meets K.

Therefore, $\text{cl}\,(X \setminus (H \cup K))$ is a continuum from H to K, and it follows that

$$\text{cl}\,(X \setminus (H \cup K)) = X.$$

(b) follows at once from this, and (a) follows from the trivial observation that

$$X \setminus (H \cup K) \subseteq X \setminus H \subseteq \text{cl}\,(X \setminus (H \cup K)),$$

and

$$X \setminus (H \cup K) \subseteq X \setminus K \subseteq \text{cl}\,(X \setminus (H \cup K)). \qquad \textbf{QED}$$

A.10. Theorem. *Let X be a compact Hausdorff space, and let H be a closed subset of X.*
Then $\mathscr{C}(X, H)$ is closed.

PROOF: Suppose $\mathscr{C}(X, H)$ is not closed, and choose

$$x \in \overline{\mathscr{C}(X, H)} \setminus \mathscr{C}(X, H).$$

Then X contains no continuum from x to H, and therefore, by Theorem A.8, we can find an open, closed subset U of X such that

$$x \in U \quad \text{and} \quad H \subseteq X \setminus U.$$

Since no component of X can meet U and $X \setminus U$, we clearly have

$$\mathscr{C}(X, H) \subseteq X \setminus U,$$

and therefore

$$\overline{\mathscr{C}(X, H)} \subseteq X \setminus U,$$

contradicting the fact that $x \in \overline{\mathscr{C}(X, H)}$.

Therefore $\mathscr{C}(X, H)$ must be closed. $\qquad \textbf{QED}$

A.11. Theorem. *If a compact, connected Hausdorff space is a union of a countable, pairwise disjoint family of closed subsets, then only one member of the family can be non-empty.*

PROOF: It is easy to see that if the theorem is false, we can find a compact, connected Hausdorff space X, which is a union $\bigcup_{n=1}^{\infty} A_n$, where

$\{A_n \mid n = 1, 2, \ldots\}$ is a pairwise disjoint, countably infinite family of closed, non-empty subsets.

Suppose X and $\{A_n \mid n = 1, 2, \ldots\}$ satisfy these conditions. We shall construct a decreasing sequence $\{X_m \mid m = 1, 2, \ldots\}$ of compact, connected subspaces of X, such that, for each $m = 1, 2, \ldots, X_m$ meets more than one member of $\{A_n \mid n = 1, 2, \ldots\}$, but

$$X_m \cap A_m = \square.$$

Such a sequence $\{X_m \mid m = 1, 2, \ldots\}$ will at once yield a contradiction, since

$$\bigcap_{m=1}^{\infty} X_m = \square,$$

in spite of the fact that $\{X_m \mid m = 1, 2, \ldots\}$ is a chain of non-empty, compact sets.

Since X is a continuum from A_1 to A_2, X contains a minimal continuum Y from A_1 to A_2. Now

$$Y = \bigcup_{n=1}^{\infty} (Y \cap A_n),$$

and it follows from the Baire Category Theorem (Kelley [1], page 200) that there exists n such that $Y \cap A_n$ has interior in the relative topology on Y. From Theorem A.9, we see that $n \geq 3$, and that Y is not a minimal continuum from A_2 to A_n. Let $X_1 \subseteq Y$ be a minimal continuum from A_2 to A_n. Then since $X_1 \neq Y$, X_1 is not a continuum from A_1 to A_2. We therefore have

$$X_1 \cap A_1 = \square.$$

If m is a positive integer, and $X_1, \ldots, X_m$ have been defined in such a way that $X_1 \supseteq X_2 \supseteq \ldots \supseteq X_m$, X_m is compact, connected, and disjoint from A_m, and X_m meets more than one member of $\{A_n \mid n = 1, 2, \ldots\}$; we construct X_{m+1} as follows:

If $X_m \cap A_{m+1} = \square$, let $X_{m+1} = X_m$.

If X_m meets A_{m+1}, then since X_m meets more than one member of $\{A_n \mid n = 1, 2, \ldots\}$ and does not meet $\bigcup_{n=1}^{m} A_n$, we can choose $m' > m + 1$ such that $X_m \cap A_{m'} \neq \square$. Now since X_m is a continuum from A_{m+1} to $A_{m'}$, X_m contains a minimal continuum Y_m from A_{m+1} to $A_{m'}$. Since

$$Y_m = \bigcup_{n=m+1}^{\infty} (Y_m \cap A_n),$$

it follows from the Baire Category Theorem (Kelley [1]) that there exists
$m'' \geq m + 1$ such that $Y_m \cap A_{m''}$ has interior, in the relative topology
on Y_m. From Theorem A.9, we see that $m'' \neq m + 1$, $m'' \neq m'$,
and Y_m it not a minimal continuum from $A_{m'}$ to $A_{m''}$. Let $X_{m+1} \subseteq Y_m$
be a minimal continuum from $A_{m'}$ to $A_{m''}$. Then since $X_{m+1} \neq Y_m$,
X_{m+1} is therefore not a continuum from A_{m+1} to $A_{m'}$. We must therefore
have

$$X_{m+1} \cap A_{m+1} = \square.$$

This completes the proof of the theorem.   **QED**

Cantor Sets

A.12. Definition. For each number δ satisfying $0 < \delta \leq \frac{1}{3}$, the
δ-*Cantor subset of* $[0, 1]$ is defined in the following manner: Let $A_1 = [0, 1]$.
If we have already defined $A_1, \ldots, A_n$, where n is a positive integer,
in such a way that A_n is a finite, disjoint union of compact subintervals of
$[0, 1]$, let A_{n+1} be obtained from A_n by removing from each component of
A_n the centrally located open subinterval of that component of length

$$\delta \left(\frac{1}{3} \right)^{n-1}.$$

Having defined $A_1, A_2, \ldots$, in this way, we define the δ-Cantor subset of
$[0, 1]$ to be $\bigcap_{n=1}^{\infty} A_n$.

A.13. Definition. If a and b are real numbers and $a < b$, then the
δ-*Cantor subset of* $[a, b]$ is defined to be the image of the δ-Cantor subset
of $[0, 1]$, under the function $t \to a(1 - t) + bt$.

A.14. Theorem. *Let a and b be real numbers, and suppose $a < b$.
Then for each $\delta \in (0, \frac{1}{3}]$, we have*

(a) *The δ-Cantor subset of $[a, b]$ is homeomorphic to the usual Cantor
middle third set (which is, of course, the $\frac{1}{3}$ -Cantor subset of $[0, 1]$).*

(b) *The δ-Cantor subset of $[a, b]$ has Lebesgue measure $(b - a)(1 - 3\delta)$.*

PROOF: This proof is elementary, and can be omitted.   **QED**

THE KURZWEIL INTEGRAL

In this appendix, we describe a special case of an integral developed by J. Kurzweil (cf. Kurzweil [1]). Kurzweil refers of this integral as a "Generalized Perron integral", and it is thus, *a fortiori*, a generalization of the Lebesgue integral. It is this last fact in which we are interested here, and we shall prove in this appendix that if $[a, b]$ is any compact subinterval of $\mathbb{R}$, and $f: [a, b] \to [0, \infty)$ is Lebesgue measurable, then the Kurzweil integral of f over $[a, b]$ exists, and coincides with the Lebesgue integral $\int_a^b f \, dm$.

B.1. Definition. Let $[a, b]$ be a compact subinterval of $\mathbb{R}$, and let δ be a function from $[a, b]$ into $(0, \infty)$. A finite, ordered family $\{\tau_0, t_1, \tau_1, t_2, \tau_2, \ldots, t_n, \tau_n\}$ of points in $[a, b]$ is said to $\mathscr{K}$-*conform to* δ if

(i) $a = \tau_0 \leq t_1 \leq \tau_1 \leq t_2 \leq \tau_2 \leq \ldots \leq t_n \leq \tau_n = b$,

(ii) $a = \tau_0 < \tau_1 < \tau_2 < \ldots < \tau_n = b$,

and

(iii) For each $j = 1, \ldots, n$, both τ_{j-1} and τ_j lie in $\left(t_j - \delta(t_j), t_j + \delta(t_j)\right)$.

B.2. Definition. Let $[a, b]$ be a compact subinterval of $\mathbb{R}$, and let $f: [a, b] \to [0, \infty)$.

(a) If $0 \leq l < \infty$, we say that the Kurzweil integral of f over $[a, b]$ is l, and write

$$\mathscr{K} \int_a^b f = l,$$

provided that, for each $\varepsilon > 0$, there exists a function δ from $[a, b]$ into $(0, \infty)$ such that, whenever a family $\{\tau_0, t_1, \tau_1, \ldots, t_n, \tau_n\}$ $\mathscr{K}$-conforms to δ, we have

$$\left| \sum_{j=1}^{n} f(t_j) (\tau_j - \tau_{j-1}) - l \right| < \varepsilon.$$

(b) We say that the Kurzweil integral of f over $[a, b]$ is ∞, and write

$$\mathcal{K} \int_a^b f = \infty,$$

provided that, for each $\lambda \in \mathbb{R}$, there exists a function δ from $[a, b]$ into $(0, \infty)$ such that, whenever a family $\{\tau_0, t_1, \tau_1, \ldots, t_n, \tau_n\}$ $\mathcal{K}$-conforms to δ, we have

$$\sum_{j=1}^{n} f(t_j) (\tau_j - \tau_{j-1}) > \lambda.$$

B.3. Lemma. *Let $[a, b]$ be a compact subinterval of $\mathbb{R}$, and δ_1, δ_2 be functions from $[a, b]$ into $(0, \infty)$. Suppose that, for each $t \in [a, b]$, $\delta_1(t) \leq \delta_2(t)$. Then any family $\{\tau_0, t_1, \tau_1, \ldots, t_n, \tau_n\}$ of points in $[a, b]$ which $\mathcal{K}$-conforms to δ_1, must also $\mathcal{K}$-conform to δ_2.*

PROOF: Trivial.   **QED**

B.4. Lemma. *Let $[a, b]$ be a compact subinterval of $\mathbb{R}$, and let δ be a function from $[a, b]$ into $(0, \infty)$. Then there exists a family $\{\tau_0, t_1, \tau_1, \ldots, t_n, \tau_n\}$ which $\mathcal{K}$-conforms to δ.*

PROOF: The family $\left\{ \left(t - \delta(t), t + \delta(t) \right) \mid t \in [a, b] \right\}$, being an open cover of the compact set $[a, b]$, contains a finite, irredundant subcover of $[a, b]$. Choose $t_1, \ldots t_n$ in $[a, b]$ such that

$$a \leq t_1 < t_2 < \ldots < t_n \leq b,$$

and such that $\left\{ \left(t_j - \delta(t_j), t_j + \delta(t_j) \right) \mid j = 1, \ldots, n \right\}$ is an irredundant cover of $[a, b]$. It is easy to see that, for each $j = 1, \ldots, n - 1$, we have

$$t_j + \delta(t_j) < t_{j+1} + \delta(t_{j+1}),$$

and

$$t_j - \delta(t_j) < t_{j+1} - \delta(t_{j+1})$$

and that, consequently, for each $j = 1, \ldots, n - 1$, there exists a number τ_j such that

$$t_j < \tau_j < t_{j+1},$$

and

$$\tau_j \in \left(t_j - \delta(t_j), t_j + \delta(t_j) \right) \cap \left(t_{j+1} - \delta(t_{j+1}), t_{j+1} + \delta(t_{j+1}) \right).$$

Let $\tau_0 = a$ and $\tau_n = b$. It is easy to see that $\{\tau_0, t_1, \tau_1, \ldots, t_n, \tau_n\}$ $\mathcal{K}$-conforms to δ.   **QED**

B.5. Theorem. *Let* $[a, b]$ *be a compact subinterval of* $\mathbb{R}$, f *and* g *functions from* $[a, b]$ *into* $[0, \infty)$, *such that*

$$\mathcal{K}\int_a^b f \quad \text{and} \quad \mathcal{K}\int_a^b g$$

both exist, and let c be a nonnegative real number.

(a) $\mathcal{K}\int_a^b (f + g)$ *exists, and we have*

$$\mathcal{K}\int_a^b (f + g) = \mathcal{K}\int_a^b f + \mathcal{K}\int_a^b g \ .$$

(b) $\mathcal{K}\int_a^b cf$ *exists, and we have* $\mathcal{K}\int_a^b cf = c \cdot \mathcal{K}\int_a^b f \ .$

(c) *if* $f \leq g$, *then* $\mathcal{K}\int_a^b f \leq \mathcal{K}\int_a^b g \ .$

PROOF: In view of Lemma B.3, the proof of this theorem is analogous to the proof of the corresponding theorem for the Riemann integral, and will therefore be omitted. **QED**

B.6. Lemma. *Let* $[a, b]$ *be a compact subinterval of* $\mathbb{R}$, f *a bounded function from* $[a, b]$ *into* $[0, \infty)$, *and suppose* $\{f_n\}$ *is a sequence of functions from* $[b]$ a, *into* $[0, \infty)$, *which converges uniformly to* f *on* $[a, b]$. *Suppose further, that for each* $n = 1, 2, \ldots$, $\mathcal{K}\int_a^b f_n$ *exists.*
Then $\mathcal{K}\int_a^b f$ *exists, and we have*

$$\mathcal{K}\int_a^b f_n \to \mathcal{K}\int_a^b f.$$

PROOF: Since, for any positive integers n_1 and n_2, and any family $\{\tau_0, t_1, \tau_1, \ldots, t_n, \tau_n\}$, (where $a = \tau_0 \leq t_1 \leq \tau_1 \leq \ldots \leq t_n \leq \tau_n = b$) we have

$$\left| \sum_{j=1}^n f_{n_1}(t_j)(\tau_j - \tau_{j-1}) - \sum_{j=1}^n f_{n_2}(t_j)(\tau_j - \tau_{j-1}) \right|$$
$$\leq (b - a) \cdot \sup\{|f_{n_1}(t) - f_{n_2}(t)| \mid t \in [a, b]\},$$

it is clear that for any positive integers n_1 and n_2, we have

$$\left| \mathcal{K}\int_a^b f_{n_1} - \mathcal{K}\int_a^b f_{n_2} \right| \leq (b - a) \cdot \sup\{|f_{n_1}(t) - f_{n_2}(t)| \mid t \in [a, b]\}.$$

It follows that $\left\{\mathscr{K}\int_a^b f_n\right\}$ is a Cauchy sequence of real numbers. Let l be its limit.

Let $\varepsilon > 0$, and choose n_0 such that

(i) $\left|\mathscr{K}\int_a^b f_{n_0} - l\right| < \varepsilon$,

and

(ii) $|f_{n_0}(t) - f(t)| < \varepsilon$, $\qquad \forall \, t \in [a, b]$.

Now choose a function δ from $[a, b]$ into $(0, \infty)$ such that for any family $\{\tau_0, t_1, \tau_1, \ldots, t_n, \tau_n\}$ which $\mathscr{K}$-conforms to δ, we have

$$\left|\sum_{j=1}^n f_{n_0}(t_j)\,(\tau_j - \tau_{j-1}) - \mathscr{K}\int_a^b f_{n_0}\right| < \varepsilon.$$

Then for any family $\{\tau_0, t_1, \tau_1, \ldots, t_n, \tau_n\}$ which $\mathscr{K}$-conforms to δ, we have

$$\left|\sum_{j=1}^n f(t_j)\,(\tau_j - \tau_{j-1}) - l\right| < \varepsilon\,(b - a) + 2\varepsilon.$$

This shows that $\mathscr{K}\int_a^b f = l$, and the proof is complete.   **QED**

B.7. Theorem. *Let $[a, b]$ be a compact subinterval of $\mathbb{R}$, and let f be a Lebesgue measurable function from $[a, b]$ into $[0, \infty)$.*

Then $\mathscr{K}\int_a^b f$ exists, and we have

$$\mathscr{K}\int_a^b f = \int_a^b f\,dm.$$

PROOF: We prove this theorem in a number of steps.

CASE 1. Suppose A is a Lebesgue measurable subset of $[a, b]$, and $f = \chi_A$. In this case, we wish to show that

$$\mathscr{K}\int_a^b f = m(A).$$

Let $\varepsilon > 0$. Let C be a compact subset of A, and U a neighborhood of A, such that $m(U \setminus C) < \varepsilon$, and let δ be the function from $[a, b]$ into $(0, \infty)$ defined by:

$$\delta(t) = d(t, \mathbb{R} \setminus U), \qquad \forall \, t \in A$$
$$\delta(t) = d(t, C), \qquad \forall \, t \in [a, b] \setminus A.$$

Now if $\{\tau_0, t_1, \tau_1, \ldots, t_n, \tau_n\}$ is a family which $\mathscr{K}$-conforms to δ, then for any $j = 1, \ldots, n$, we have: if $t_j \in A$, then $(\tau_{j-1}, \tau_j) \subseteq U$, and consequently,

$$\sum_{j=1}^{n} f(t_j)\,(\tau_j - \tau_{j-1}) = \sum_{t_j \in A} (\tau_j - \tau_{j-1}) \leq m(U);$$

and also, for any $j = 1, \ldots, n$, we have: if $t_j \notin A$, then $(\tau_{j-1}, \tau_j) \cap C = \square$, and consequently

$$\sum_{j=1}^{n} f(t_j)\,(\tau_j - \tau_{j-1}) = \sum_{t_j \in A} (\tau_j - \tau_{j-1})$$

$$= m\left(\bigcup_{t_j \in A} [\tau_{j-1}, \tau_j]\right) \geq m(C).$$

It follows that

$$\left| \sum_{j=1}^{n} f(t_j)\,(\tau_j - \tau_{j-1}) - m(A) \right| < \varepsilon,$$

and this shows that $\mathscr{K}\int_a^b f = m(A)$.

CASE 2. Suppose f is a (Lebesgue measurable) simple function. The fact that

$$\mathscr{K}\int_a^b f = \int_a^b f\,dm$$

follows at once from Case 1 and Theorem B.5.

CASE 3. Suppose f is bounded. Since f is the uniform limit of a sequence of simple functions, the required result follows at once from Lemma B.6.

CASE 4. $\int_a^b f\,dm = \infty$. For each positive integer n, let $f_n = f \wedge n$, i.e.,

$$f_n(t) = f(t) \quad \text{whenever } f(t) \leq n$$

and

$$f_n(t) = n \quad \text{whenever } f(t) > n.$$

Then by the Lebesgue monotone convergence theorem, $\int_a^b f_n\,dm \to \infty$, and therefore, by Case 3, we have that $\mathscr{K}\int_a^b f_n\,dm \to \infty$. Now to prove that $\mathscr{K}\int_a^b f = \infty$, let $\lambda \in \mathbb{R}$. Choose n_0 such that $\mathscr{K}\int_a^b f_{n_0} > \lambda + 1$, and choose a function $\delta : [a, b] \to (0, \infty)$, such that, whenever a family $\{\tau_0, t_1, \tau_1, \ldots, t_n, \tau_n\}$ $\mathscr{K}$-conforms to δ, we have

$$\left| \sum_{j=1}^{n} f_{n_0}(t_j) (\tau_j - \tau_{j-1}) - \mathscr{K}\int_a^b f_{n_0} \right| < 1.$$

Then for any such family, we have

$$\sum_{j=1}^{n} f(t_j) (\tau_j - \tau_{j-1}) \geq \sum_{j=1}^{n} f_{n_0}(t_j) (\tau_j - \tau_{j-1}) > \lambda.$$

This shows that $\mathscr{K}\int_a^b f = \infty = \int_a^b f \, dm$.

CASE 5. $\int_a^b f \, dm < \infty$. For each $n = 0, 1, 2, \ldots$, let

$A_n = \{t \in [a, b] \mid n \leq f(t) < n+1\}$. Clearly $\sum_{n=0}^{\infty} n \cdot m(A_n) \leq \int_a^b f \, dm < \infty$,

and consequently,

$$\sum_{n=0}^{\infty} (n + 1) \cdot m(A_n) < \infty.$$

For each $n = 0, 1, \ldots$, choose a neighborhood U_n of A_n such that $m(U_n) < m(A_n) + 2^{-n}$. Then clearly,

$$\sum_{n=0}^{\infty} (n + 1) \cdot m(U_n) < \infty.$$

Let $\varepsilon > 0$, and choose an integer N so large that

$$\sum_{n=N}^{\infty} (n + 1) \cdot m(U_n) < \varepsilon.$$

It is easy to see that, if g is the function from $[a, b]$ into $[0, N)$ defined by:

$g(t) = f(t), \quad \forall \, t \in \bigcup_{n < N} A_n$ and $g(t) = 0, \quad \forall \, t \in \bigcup_{n=N}^{\infty} A_n,$

then

$$\int_a^b g \, dm \leq \int_a^b f \, dm < \int_a^b g \, dm + \varepsilon.$$

By Case 3, we can choose a function δ_1 from $[a, b]$ into $(0, \infty)$ such that, whenever a family $\{\tau_0, t_1, \tau_1, \ldots, t_n, \tau_n\}$ $\mathscr{K}$-conforms to δ, we have

$$\left| \sum_{j=1}^{n} g(t_j) (\tau_j - \tau_{j-1}) - \mathscr{K}\int_a^b g \right| < \varepsilon.$$

Let δ_2 be the function from $[a, b]$ into $(0, \infty)$ defined as follows:

$$\delta_2(t) = d(t, \mathbb{R} \setminus U_n), \qquad \forall\, t \in A_n, \qquad \forall\, n = 0, 1, 2, \dots .$$

Let $\delta = \delta_1 \wedge \delta_2$. Now if $\{\tau_0, t_1, \tau_1, \dots, t_n, \tau_n\}$ is any family which $\mathscr{K}$-conforms to δ, we have, firstly,

$$\sum_{j=1}^{n} f(t_j)(\tau_j - \tau_{j-1}) \geq \sum_{j=1}^{n} g(t_j)(\tau_j - \tau_{j-1})$$

$$> \mathscr{K}\int_a^b g - \varepsilon$$

$$= \int_a^b g\, dm - \varepsilon$$

$$> \int_a^b f\, dm - 2\varepsilon,$$

and secondly,

$$\sum_{j=1}^{n} f(t_j)(\tau_j - \tau_{j-1}) = \sum_{j=1}^{n} g(t_j)(\tau_j - \tau_{j-1}) + \sum_{\substack{t_j \in \bigcup A_n \\ n \geq N}} f(t_j)(\tau_j - \tau_{j-1})$$

$$\leq \mathscr{K}\int_a^b g + \varepsilon + \sum_{n \geq N}\left(\sum_{t_j \in A_n}(n+1)(\tau_j - \tau_{j-1})\right)$$

$$\leq \int_a^b f\, dm + \varepsilon + \sum_{n \geq N}(n+1) \cdot m(U_n)$$

$$\leq \int_a^b f\, dm + 2\varepsilon.$$

Therefore, whenever the family $\{\tau_0, t_1, \tau_1, \dots, t_n, \tau_n\}$ $\mathscr{K}$-conforms to δ, we have

$$\left| \sum_{j=1}^{n} f(t_j)(\tau_j - \tau_{j-1}) - \int_a^b f\, dm \right| \leq 2\varepsilon,$$

and this completes the proof of the theorem.   **QED**

B.8. Remark. Our description of Kurzweil's integration process in this appendix is much narrower than the actual integral that Kurzweil developed. From the point of view of integration theory the most interesting property of the Kurzweil Integral is that like the Perron Integral it is not absolutely convergent, so that in addition to being able to integrate all nonnegative real Lebesgue measurable functions, as we have shown in B.7, it can also handle such integrals as

$$\int_{-1}^{1} f' \quad \text{where} \quad f(x) = x^2 \sin\left(\frac{1}{x^2}\right)$$

which the Lebesgue integral cannot handle.

For nonnegative real functions it can be shown that the Kurzweil Integral is not broader than the Lebesgue and that consequently the two integrals are exactly the same. It is perhaps a little peculiar that it is precisely under these circumstances where the two integrals coincide that we need the Kurzweil Integral in this book. Instead of using the Kurzweil Integral to integrate functions deemed not *kosher* by the Lebesgue process (we could also have used the Perron Integral for that), we look at the Kurzweil Integral as another way to arrive at the Lebesgue Integral of a nonnegative measurable function. Now because the construction of the Kurzweil Integral is so much like that of the Riemann Integral (of course the Riemann Integral is simply the Kurzweil Integral with the constraint that the functions δ are constant), looking at a Lebesgue Integral as a Kurzweil Integral gives us a much closer comparison between the Riemann and Lebesgue Integrals than would otherwise have been possible. We are therefore able to apply techniques of proof to Lebesgue Integrals that one might have thought would be applicable only to Riemann Integrals.

The Kurzweil Integral first appeared in Kurzweil [1] as an integral in $\mathbb{R}^n$ to be used as a tool for the theory of differential equations. In Kurzweil [2] the theory is extended and in Kurzweil [3] the Integral is again put to work in differential equation theory. An essentially similar process was developed simultaneously by R. Henstock and is described in Henstock [1] and Henstock [2]. In the latter paper Henstock compares the Lebesgue, Perron and Henstock processes. More recently W. F. Pfeffer has developed a Perron-like integration process for functions defined on a topological space, and this integral includes all the others as special cases. This work may be found in Pfeffer [1] and Pfeffer [2], and a comparison of the various integration processes may be found in Pfeffer and Wilbur [1].

APPENDIX C

SOME PROPERTIES OF THE PLANE

C.1. Introduction. There are several ways of looking at the Euclidean plane $\mathbb{R}^2$.

Firstly, we can regard $\mathbb{R}^2$ as a topological space, whose topology is induced by the Euclidean metric d, where, for any points $\langle x_1, y_1 \rangle$ and $\langle x_2, y_2 \rangle$ in $\mathbb{R}^2$,

$$d(\langle x_1, y_1 \rangle, \langle x_2, y_2 \rangle) = \sqrt{(x_1 - x_2)^2 + (y_1 - y_2)^2}.$$

Secondly, we can regard $\mathbb{R}^2$ as a real, normed linear space, whose norm induces the metric d.

Thirdly, we can regard $\mathbb{R}^2$ as the field $\mathbb{C}$ of complex numbers (where we identify the point $\langle x, y \rangle$ of $\mathbb{R}^2$ with the complex number $x + iy$).

Finally, we can regard $\mathbb{R}^2$ as a topological subspace of the unit sphere $\mathbb{S}^2$ in $\mathbb{R}^3$, and at the same time, regard $\mathbb{S}^2$ as the one-point compactification of $\mathbb{R}^2$. This we do as follows:

We define a homeomorphism h from $\mathbb{R}^2$ onto $\mathbb{S}^2 \setminus \{\langle 0, 0, 1 \rangle\}$ by letting $h(x, y)$ be that point of $\mathbb{S}^2 \setminus \{\langle 0, 0, 1 \rangle\}$ which lies on the straight line joining $\langle x, y, 0 \rangle$ to $\langle 0, 0, 1 \rangle$.[1] Explicitly, h is given by

$$h(x, y) = \left(\frac{2x}{x^2 + y^2 + 1}, \frac{2y}{x^2 + y^2 + 1}, \frac{x^2 + y^2 - 1}{x^2 + y^2 + 1} \right), \qquad \forall \langle x, y \rangle \in \mathbb{R}^2.$$

or in polar coordinates,

$$h(r e^{i\vartheta}) = \left(\frac{2r \cos \vartheta}{r^2 + 1}, \frac{2r \sin \vartheta}{r^2 + 1}, \frac{r^2 - 1}{r^2 + 1} \right), \qquad \forall r e^{i\vartheta} \in \mathbb{C}.$$

---

[1] We could just as well have used certain other homeomorphisms from $\mathbb{R}^2$ onto a punctured 2-sphere. We could, for example, have used the stereographic projection (cf. Hille, Analytic Function Theory, Vol. 1, pages 38ff.).

If we now adjoin a new point ∞ to the plane, and topologize $\mathbb{R}^2 \cup \{\infty\}$ as the one-point compactification of $\mathbb{R}^2$, (see Kelley [1], p. 150), then the extension of h obtained by letting $h(\infty) = \langle 0, 0, 1 \rangle$, is a homeomorphism of $\mathbb{R}^2 \cup \{\infty\}$ onto $\mathbf{S}^2$.

The sphere $\mathbf{S}^2$ is a metric space, whose topology is induced by the chordal metric ϱ, defined simply to be the Euclidean metric in $\mathbb{R}^3$, i.e.,

$$\varrho(\langle x_1, y_1, z_1 \rangle, \langle x_2, y_2, z_2 \rangle) = \sqrt{(x_1 - x_2)^2 + (y_1 - y_2)^2 + (z_1 - z_2)^2}.$$

Regarding $\mathbb{R}^2$ as a subspace of $\mathbf{S}^2 = \mathbb{R}^2 \cup \{\infty\}$, we obtain a new metric ϱ which induces the topology of $\mathbb{R}^2$. Explicitly,

$$\varrho(r_1 e^{i\vartheta_1}, r_2 e^{i\vartheta_2}) =$$

$$\sqrt{\left(\frac{2r_1 \cos\vartheta_1}{r_1^2 + 1} - \frac{2r_2 \cos\vartheta_2}{r_2^2 + 1}\right)^2 + \left(\frac{2r_1 \sin\vartheta_1}{r_1^2 + 1} - \frac{2r_2 \sin\vartheta_2}{r_2^2 + 1}\right)^2 + \left(\frac{r_1^2 - 1}{r_1^2 + 1} - \frac{r_2^2 - 1}{r_2^2 + 1}\right)^2}.$$

It can be shown that if $r_0 e^{i\vartheta_0} \in \mathbb{R}^2$, then

$$\frac{\varrho(r e^{i\vartheta}, r_0 e^{i\vartheta_0})}{d(r e^{i\vartheta}, r_0 e^{i\vartheta_0})} \to \frac{2}{1 + r_0^2} \quad \text{as} \quad r e^{i\vartheta} \to r_0 e^{i\vartheta_0}.$$

C.2. Some Notation. Let $a \in \mathbb{R}^2$, and $\delta > 0$. With the Euclidean metric, the open ball with center at a and radius δ is denoted by $N_d(a, \delta)$, i.e.,

$$N_d(a, \delta) = \{x \in \mathbb{R}^2 \mid d(a, x) < \delta\}.$$

Let $a \in \mathbb{R}^2 \cup \{\infty\}$, and $\delta > 0$. With the chordal metric, the open ball with center at a and radius δ is denoted by $N_\varrho(a, \delta)$, i.e.,

$$N_\varrho(a, \delta) = \left\{x \in \mathbb{R}^2 \cup \{\infty\} \mid \varrho(a, x) < \delta\right\}.$$

For each number $\delta > 0$, the circle with center at 0, radius δ, (with respect to d) is denoted by Γ_δ, i.e.,

$$\Gamma_\delta = \{\delta e^{i\vartheta} \mid 0 \le \vartheta \le 2\pi\}.$$

C.3. Choosing a New ∞. We recall, that if a function f from $\mathbb{R}^2$ into $\mathbb{R}^2$ is holomorphic in $\mathbb{R}^2 \setminus K$, where K is some compact subset of $\mathbb{R}^2$, and if we define $\tilde{f}(z) = f(z^{-1})$ for $z \neq 0$, then ∞ is said to be a removable singularity of f if 0 is a removable singularity of $\tilde{f}$. This happens precisely when $\tilde{f}$ is bounded near 0, i.e., when f is bounded near ∞, and in this case, defining $\tilde{f}(0)$ suitably, we can make $\tilde{f}$ holomorphic in a neighborhood of 0. The extension of f to $\mathbf{S}^2$ defined by letting $f(\infty) = \tilde{f}(0)$, is said to be holomorphic in $\mathbf{S}^2 \setminus K$.

Now suppose α and $\beta \in \mathbb{R}^2$. Let g be the homeomorphism of $\mathbf{S}^2$ onto itself, defined by

$$g(z) = \frac{1}{z - \alpha} + \beta, \qquad \forall \, z \in \mathbb{R}^2 \setminus \{\alpha\},$$

$$g(\alpha) = \infty, \text{ and } g(\infty) = \beta.$$

It is clear that the restriction of g to $\mathbf{S}^2 \setminus \{\alpha\}$ is a conformal equivalence of $\mathbf{S}^2 \setminus \{\alpha\}$ onto $\mathbb{R}^2$. This allows us to identify $\mathbf{S}^2 \setminus \{\alpha\}$ with $\mathbb{R}^2$, and whenever we do this, we shall say we are *letting α play the role of ∞*.

C.4. Definition. A non-empty, connected, proper, open subset of $\mathbf{S}^2$ will be called a *region*. A region Ω is said to be a *disc* if Ω is homeomorphic with the open unit disc $\{z \in \mathbb{R}^2 \mid |z| < 1\}$. It is a simple consequence of the Riemann Mapping Theorem (cf. for example, Rudin [1], Theorem 14.8) that a region Ω is a disc iff $\mathbf{S}^2 \setminus \Omega$ is connected.

C.5. Lemma. *Let Ω be a region, and let F be the union of a family of components of $\mathbf{S}^2 \setminus \Omega$.*
Then $\Omega \cup F$ is connected, and the components of $\mathbf{S}^2 \setminus (\Omega \cup F)$ are precisely those components of $\mathbf{S}^2 \setminus \Omega$ which are not contained in F.

PROOF: Since every component of $\mathbf{S}^2 \setminus \Omega$ contains a limit-point of Ω, we see at once that $\Omega \cup F$ is connected. The lemma now follows easily.
QED

C.6. Lemma. *If Ω is a disc, then $\partial(\Omega)$ is connected.*

PROOF: Let h be a homeomorphism from Ω onto $\{z \in \mathbb{R}^2 \mid |z| < 1\}$, and for each $n = 1, 2, \ldots$, let

$$K_n = \left\{ x \in \Omega \,\middle|\, |h(x)| \geq 1 - \frac{1}{n} \right\}.$$

It is easy to see that for each $n = 1, 2, \ldots$, we have

$$\bar{K}_n = K_n \cup \partial(\Omega),$$

and it follows that

$$\partial(\Omega) = \bigcap_{n=1}^{\infty} \bar{K}_n.$$

It therefore follows from Theorem A. 5 that $\partial(\Omega)$ is connected.   **QED**

C.7. Theorem. *Let Ω be a region.*

(a) *If A is a side of Ω (i.e., a component of $\mathbf{S}^2 \setminus \Omega$), then $\partial(A)$ is a component of $\partial(\Omega)$.*

(b) *If B is a component of $\partial(\Omega)$, then*

$$B = \partial([B \text{ side} : \Omega]).$$

PROOF: To prove (a), suppose A is a side of Ω. It follows from Lemma C.5 that $\mathbf{S}^2 \setminus A$ is connected, and therefore $\mathbf{S}^2 \setminus A$ is a disc. Therefore, since $\partial(A) = \partial(\mathbf{S}^2 \setminus A)$, Lemma C.6 implies that $\partial(A)$ is connected. Thus $\partial(A)$ is contained in a component B of $\partial(\Omega)$, and the assertion $B = \partial(A)$ follows from the observation that $A = [B \text{ side} : \Omega]$.

We omit the proof of (b).   **QED**

C.8. Corollary. *Let Ω be a region.*

Then the mapping $A \to \partial(A)$ gives a one-to-one correspondence from the sides of Ω onto the components of $\partial(\Omega)$.

C.9. Theorem. *Let Ω be a region, and let Ω_0 be a subregion of Ω. Suppose a component B_0 of $\partial(\Omega_0)$ is contained in a component B of $\partial(\Omega)$.*

Then $B_0 = B$, and $[B \text{ side} : \Omega_0] = [B \text{ side} : \Omega]$.

PROOF: Since $\mathbf{S}^2 \setminus \Omega \subseteq \mathbf{S}^2 \setminus \Omega_0$, $[B \text{ side} : \Omega]$ must be contained in a side of Ω_0. Therefore, since $B_0 \subseteq B$, we have $[B \text{ side} : \Omega] \subseteq [B_0 \text{ side} : \Omega_0]$, and it follows that

$$\mathbf{S}^2 \setminus [B_0 \text{ side} : \Omega_0] \subseteq \mathbf{S}^2 \setminus [B \text{ side} : \Omega].$$

But $\mathbf{S}^2 \setminus [B_0 \text{ side} : \Omega_0]$ and $\mathbf{S}^2 \setminus [B \text{ side} : \Omega]$ are discs whose boundaries are, respectively, B_0 and B. Therefore $\mathbf{S}^2 \setminus [B_0 \text{ side} : \Omega_0]$, being a relatively open and closed subset of the connected set $\mathbf{S}^2 \setminus [B \text{ side} : \Omega]$, must coincide with $\mathbf{S}^2 \setminus [B \text{ side} : \Omega]$. The result now follows directly.

QED

C.10. Definition. Let A be any subset of $\mathbf{S}^2$.

The *connectivity* of A is the number of components of $\mathbf{S}^2 \setminus A$ (*i.e.*, the cardinality of the family of sides of A).

If the connectivity of A is 1, A is said to be *simply connected*.

If the connectivity of A is finite, A is said to be *finitely connected*.

If the connectivity of A does not exceed $\aleph_0$, A is said to be *countably connected*.

C.11. Theorem. *Let K be a compact subset of $\mathbf{S}^2$.*

Then K is simply connected iff every component of K is simply connected.

PROOF: The theorem is trivial if either $K = \square$ or $K = \mathbf{S}^2$. We shall therefore assume that $\square \neq K \neq \mathbf{S}^2$. It follows from Lemma C.5 that, if K is simply connected, then every component of K is simply connected.

Now suppose that every component of K is simply connected, and let $\Omega = \mathbf{S}^2 \setminus K$. Ω is a non-empty, open, proper subset of $\mathbf{S}^2$. To obtain a contradiction, assume that Ω is not connected, and let Ω_1 and Ω_2 be two distinct components of Ω. Ω_1 and Ω_2 are regions, whose boundaries are contained in K. Let $H = [\Omega_1 \text{ side} : \Omega_2]$, and let $B = \partial(H)$. Then since B is a component of $\partial(\Omega_2)$, B is contained in a component C of K. Clearly,

$$\mathbf{S}^2 \setminus C \subseteq \text{int}(H) \cup (\mathbf{S}^2 \setminus H),$$

and therefore, since $\mathbf{S}^2 \setminus C$ is connected, and $\text{int}(H) \cap (\mathbf{S}^2 \setminus H) = \Box$, $\mathbf{S}^2 \setminus C$ cannot meet both the sets $\text{int}(H)$ and $\mathbf{S}^2 \setminus H$. But

$$(\mathbf{S}^2 \setminus C) \cap \text{int}(H) \supseteq \Omega_1 \neq \Box, \text{ and } (\mathbf{S}^2 \setminus C) \cap (\mathbf{S}^2 \setminus H) \supseteq \Omega_2 \neq \Box.$$

This contradiction shows that Ω is connected, *i.e.*, that K is simply connected. **QED**

C.12. Corollary. *Let Ω be a region, let K be a relatively closed subset of Ω, and suppose every component of K is simply connected and compact.*

Then $\Omega \setminus K$ is connected.

PROOF: Let $H = (\mathbf{S}^2 \setminus \Omega) \cup K$. Since $\mathbf{S}^2 \setminus H = \Omega \setminus K$, the result will follow from Theorem C.11 when we have shown that every component of H is simply connected.

Suppose A is any component of H. If $A \subseteq \mathbf{S}^2 \setminus \Omega$, then A, being a component of $\mathbf{S}^2 \setminus \Omega$, is simply connected. If $A \subseteq K$, then A, being a component of K, is simply connected. We claim that A cannot meet both the sets $\mathbf{S}^2 \setminus \Omega$ and K: if A meets both $\mathbf{S}^2 \setminus \Omega$ and K, then A meets a component C of K, and A contains a minimal continuum B from $\mathbf{S}^2 \setminus \Omega$ to C. (See Theorem A.7.) By Theorem A.9, $B \cap \Omega$ is connected, and has limit-points in $\mathbf{S}^2 \setminus \Omega$, and it follows easily from this that C has limit-points in $\mathbf{S}^2 \setminus \Omega$, contradicting the fact that C is a compact subset of Ω.

This completes the proof. **QED**

C.13. Theorem. *Let K be a compact subset of $\mathbf{S}^2$, and let Ω be a component of $\mathbf{S}^2 \setminus K$.*

Then the connectivity of Ω does not exceed the number of components of K.

PROOF: We prove the theorem by showing that each side of Ω contains a component of K. Let C be any side of Ω. Then $\partial(C)$, being a connected subset of K, is included in a component of K. It is easy to see that this component of K is included in C. **QED**

Curves and Arcs

C.14. Definitions. *A* (*continuous*) *curve* γ *in a Hausdorff space X is a* continuous function from $[0, 1]$ into X.

A continuous curve γ is said to be a *closed curve* if $\gamma(0) = \gamma(1)$.
If a and b are any points in $\mathbb{R}^2$, $[a, b]$ denotes the curve γ defined by

$$\gamma(t) = (1 - t)\, a + tb, \qquad \forall\, t \in [0, 1].$$

C.15. Remark and Definition. If γ is a closed curve, we shall identify γ with the function defined on the unit circle that carries $e^{2\pi i t}$ to $\gamma(t)$, for every $t \in [0, 1]$. This function is clearly well-defined and continuous, and if it is one-one, it is a homeomorphism. When this function is a homeomorphism, γ is called a *Jordan curve*.

C.16. Definition.[2] An *arc* Λ in a Hausdorff space X is a homeomorphism from $(0, 1)$ into X. If a and b are distinct points in $\mathbb{R}^2$, then (a, b) denotes the arc Λ defined by

$$\Lambda(t) = (1 - t)a + tb, \qquad \forall\, t \in (0, 1).$$

An arc Λ in a Hausdorff space X is said to *join a set A to a set B if*

$$A = \bigcap_{t \in (0,1)} \mathrm{cl}\, \{\Lambda(s) \mid 0 < s \le t\}$$

and

$$B = \bigcap_{t \in (0,1)} \mathrm{cl}\, \{\Lambda(s) \mid t \le s < 1\}.$$

C.17. Lemma. *Let Λ be an arc in a Hausdorff space X, and suppose Λ joins A to B.*
Then neither A nor B can meet Λ. Furthermore, if X is compact, both A and B are non-empty, compact, and connected.

PROOF: The first part of the lemma is easy, and the second part is an immediate consequence of Theorem A.5.   **QED**

C.18. Definition. Given a curve or arc γ in a Hausdorff space X, we can identify γ with its range, upon which is placed an appropriate orientation. Sometimes the orientation does not concern us, and therefore, in addition to the above definitions of *curve* and *arc*, we shall define a *curve* to be a continuous image of $[0, 1]$, a *Jordan curve* to be a homeomorphic image of a circle, and an *arc* to be a homeomorphic image of $(0, 1)$. It is hoped

---

[2] Some authors use the word *arc* to denote a homeomorphism from $[0, 1]$ into a space X.

that even when these definitions are freely interchanged, the context will always make it clear exactly which one we are using.

C.19. Definition. Let γ be a curve or arc in a Hausdorff space X, and let $A \subseteq X$.

The *first (last) time γ meets A* is defined to be the least (greatest) t in the domain of γ, such that $\gamma(t) \in A$.

The *first (last) point of $\gamma \cap A$* is defined to be $\gamma(t)$, where t is the first (last) time γ meets A.

C.20. Definition.

(a) Let Ω be a region which contains $(0, 1)$, and let f be a conformal equivalence from Ω onto a region Δ.

The restriction of f to $(0, 1)$ (and also the range of this restriction) is called an *analytic arc.*

(b) Let Ω be a region which contains the unit circle $\Gamma_1 = \{e^{i\vartheta} \mid 0 \leq \vartheta < 2\pi\}$, and let f be a conformal equivalence from Ω onto a region Δ.

The restriction of f to Γ_1 (and also the range of this restriction) is called an *analytic Jordan curve.*

C.21. Note. It is easy to show that every open finite, half-infinite, or infinite straight line in $\mathbb{R}^2$ is an analytic arc.

C.22. The Schoenflies Theorem. We need to make use of the following two results from the Schoenflies Theorem, which most mathematicians accept on faith:

(a) *Every homeomorphism from the unit circle into $\mathbb{R}^2$ can be extended to homeomorphism of $\mathbb{R}^2$ onto itself.*

(b) *Every homeomorphism from $[0, 1]$ into $\mathbb{R}^2$ can be extended to a a homeomorphism of $\mathbb{R}^2$ onto itself.*

C.23. Corollary, the Jordan Curve Theorem.

(a) *Let J be a Jordan curve in $\mathbb{S}^2$. Then $\mathbb{S}^2 \setminus J$ has two components (i.e., J has two sides), and each of these is a disc whose boundary is J.*

(b) *Let J be a Jordan curve in $\mathbb{R}^2$. Then $\mathbb{R}^2 \setminus J$ has two components, of which one is a bounded disc whose boundary is J, and the other is an unbounded open subset of $\mathbb{R}^2$ whose boundary is J, and whose union with $\{\infty\}$ is a disc.*

PROOF: To prove (a), we note first that since $\mathbb{S}^2$ is not homeomorphic to a circle, we have $\mathbb{S}^2 \setminus J \neq \square$. Choose a point $\alpha \in \mathbb{S}^2 \setminus J$ to play the role of ∞, (and identify $\mathbb{S}^2 \setminus \{\alpha\}$ with $\mathbb{R}^2$ in the usual way). Choose a homeomorphism h from the unit circle Γ_1 onto J, and using Theorem C.22(a), let h be extended to a homeomorphism from $\mathbb{R}^2$ onto itself. Extend h further to a homeomorphism from $\mathbf{S}^2$ onto itself, by defining $h(\infty) = \infty$. It is now easy to see that (a) holds.

(b) is an elementary consequence of (a).   **QED**

C.24. Definition. Let J be a Jordan curve in $\mathbb{R}^2$.

The bounded side of J, i.e., the bounded component of $\mathbb{R}^2 \setminus J$, is called the *inside of* J, and is denoted by $\mathrm{ins}(J)$.

The unbounded side of J, i.e., the unbounded component of $\mathbb{R}^2 \setminus J$, is called the *outside of* J, and is denoted by $\mathrm{outs}(J)$.

C.25. Corollary. *Let L be a homeomorphic image of $[0, 1]$ in $\mathbb{R}^2$.*

Then $\mathbb{R}^2 \setminus L$ is connected, and there exists an arc Λ in $\mathbb{R}^2 \setminus L$, which joins the endpoints of L.

PROOF: It follows from Theorem C.22(b) that it is sufficient to prove this result for the case $L = [0, 1]$; but this proof is clear.   **QED**

C. 26. The Theta-Curve Theorem. *Let a and b be distinct points in $\mathbb{R}^2$, let Λ_1, Λ_2 and Λ_3 be arcs joining a to b, and suppose no two of these three arcs intersect. For each choice of two distinct integers i and j in $\{1, 2, 3\}$, let*

$$J_{ij} = \Lambda_i \cup \Lambda_j \cup \{a, b\}.$$

Then

(a) *For each choice of i and j, J_{ij} is a Jordan curve.*

(b) *Exactly one of the following inclusions holds:*

$$\Lambda_1 \subseteq \mathrm{ins}(J_{23}), \quad \Lambda_2 \subseteq \mathrm{ins}(J_{13}), \quad \Lambda_3 \subseteq \mathrm{ins}(J_{12}).$$

(c) *If i, j and k are 1, 2 and 3 in some order, and $\Lambda_j \subseteq \mathrm{ins}(J_{ik})$, we have*

   (i) $\Lambda_i \subseteq \mathrm{outs}(J_{jk})$ *and* $\Lambda_k \subseteq \mathrm{outs}(J_{ij})$,

   (ii) $\mathrm{diam}(\Lambda_j) \leq \mathrm{diam}(\Lambda_i \cup \Lambda_k) \leq \mathrm{diam}(\Lambda_i) + \mathrm{diam}(\Lambda_k)$,

   and

   (iii) $\mathrm{ins}(J_{ik})$ *is the disjoint union of Λ_j, $\mathrm{ins}(J_{ij})$, and $\mathrm{ins}(J_{jk})$.*

PROOF: (a) is obvious.

To prove (b), we shall first show that at least one of the three arcs Λ_1, Λ_2, Λ_3 must lie in the inside of the Jordan curve formed by the other two.

We claim that if $\text{ins}(J_{ij}) \cap \text{ins}(J_{ik}) \neq \square$, then either $\Lambda_j \subseteq \text{ins}(J_{ik})$ or $\Lambda_k \subseteq \text{ins}(J_{ij})$, for suppose $x \in \text{ins}(J_{ij}) \cap \text{ins}(J_{ik})$, and $\Lambda_j \subseteq \text{outs}(J_{ik})$. Using Corollary C.25, choose an arc γ which joins x to ∞, but does not meet $\overline{\Lambda_i}$. Let t_0 be the first time γ meets Λ_k. Then since

$$\gamma(t) \in \text{ins}(J_{ik}) \quad \text{whenever} \quad 0 < t < t_0,$$

it is clear that

$$\gamma(t) \notin \Lambda_j \quad \text{whenever} \quad 0 < t < t_0,$$

and it follows easily that $\gamma(t_0) \in \text{ins}(J_{ij})$. From this we deduce at once that $\Lambda_k \subseteq \text{ins}(J_{ij})$.

Now to obtain a contradiction, suppose that each of the three arcs Λ_1, Λ_2, Λ_3, lies outside the Jordan curve formed by the other two. Then no two of the three sets $\text{ins}(J_{12})$, $\text{ins}(J_{13})$, $\text{ins}(J_{23})$ can meet. In view of Theorem C.22(b), it is clearly sufficient to obtain the contradiction for the special case in which $a = 0$, $b = 1$, and $\Lambda_3 = (0, 1)$. Let c be any point in Λ_3, and let γ be an arc which joins c to ∞, and which does not meet J_{12}. Since γ meets Λ_3 for a last time, we may clearly assume (by changing the point c if necessary) that $\gamma \cap \Lambda_3 = \square$. Choose $\delta > 0$ such that $N_d(c, \delta)$ does not meet J_{12}. Since both $\text{ins}(J_{13})$ and $\text{ins}(J_{23})$ contain points arbitrarily close to c, and since $\text{ins}(J_{13})$ and $\text{ins}(J_{23})$ do not intersect, the (open) upper half of $N_d(c, \delta)$ must lie in one of the two sets $\text{ins}(J_{13})$, $\text{ins}(J_{23})$, and the lower half must lie in the other. It follows that, for small t, $\gamma(t)$ must lie in $\text{ins}(J_{13}) \cup \text{ins}(J_{23})$. But this is clearly impossible, for since γ does not meet either J_{13} or J_{23}, we clearly have

$$\gamma \subseteq \text{outs}(J_{13}) \cap \text{outs}(J_{23}).$$

We have therefore proved that at least one of the three arcs Λ_1, Λ_2, Λ_3, must lie in the inside of the Jordan curve formed by the other two. We now assume, for once and for all, that $\Lambda_2 \subseteq \text{ins}(J_{13})$.

The proof of (b) will be complete when we have shown that $\Lambda_1 \subseteq \text{outs}(J_{23})$ and $\Lambda_3 \subseteq \text{outs}(J_{12})$. By Theorem C.22(a), it is clearly sufficient to do this for the case in which J_{13} is a circle. But in this case, the inclusions $\Lambda_1 \subseteq \text{outs}(J_{23})$ and $\Lambda_3 \subseteq \text{outs}(J_{12})$ can be seen at once by drawing a straight line that joins a point of Λ_1 (or Λ_3) radially to ∞. Thus (b) is proved.

The assertion (c)(i) follows immediately from (b).

To prove the assertion (c)(iii) we note first that since $\Lambda_1 \subseteq \text{outs}(J_{23})$ and $\Lambda_3 \subseteq \text{outs}(J_{12})$, the observation made at the beginning of the proof implies that the sets $\text{ins}(J_{12})$, $\text{ins}(J_{23})$ are disjoint. It is clear that

$$\Lambda_2 \cup \text{ins}(J_{12}) \cup \text{ins}(J_{23}) \subseteq \text{ins}(J_{13}).$$

To obtain the reverse inclusion, let x be any point of ins (J_{13}), and suppose, without loss of generality, that $\Lambda_2 = (0, 1)$. If $x \notin \Lambda_2$, choose an arc $\gamma \subseteq \text{ins}(J_{13})$, that joins x to a point $y \in \Lambda_2$. We may clearly assume (by changing the point y if necessary), that $\gamma \cap \Lambda_2 = \square$. Choose $\delta > 0$ such that $N_d(y, \delta)$ does not meet J_{13}. Then arguing as before, we see that $N_d(y, \delta)$ has its upper half in one of the sets ins (J_{12}), ins (J_{23}), and its lower half in the other. Thus γ contains points in ins $(J_{12}) \cup \text{ins}(J_{23})$ and therefore lies entirely in one of the sets ins (J_{12}) or ins (J_{23}). This shows that

$$x \in \Lambda_2 \cup \text{ins}(J_{12}) \cup \text{ins}(J_{23}),$$

thus completing the proof of (c) (iii).

The second inequality in the assertion (c) (ii), is clear, and the first is an immediate consequence of the following lemma.   **QED**

C.27. Lemma. *Let J be a Jordan curve in $\mathbb{R}^2$. Then*

$$\text{diam}(J) = \text{diam}(J \cup \text{ins}(J)) = \text{diam}(\text{ins}(J)).$$

PROOF: We need only show that

$$\text{diam}(\text{ins}(J)) \le \text{diam}(J).$$

Suppose a and b are any two points of ins (J). Then since each of the sets $\{a(1 - t) + bt \mid t > 1\}$ and $\{a(1 - t) + bt \mid t < 0\}$ must meet J, we see that a and b belong to a straight line segment which joins two points of J. Therefore

$$d(a, b) \le \text{diam}(J),$$

and the proof is complete.   **QED**

Jordan Curves near the Boundary of a Region

C.28. Lemma. *Let K be a compact subset of $\mathbb{R}^2$, and suppose K is simply connected. Let $\varepsilon > 0$.*

Then there exists $\delta > 0$ such that, for every compact set H satisfying

$$K \subseteq H \subseteq N_d(K, \delta),$$

the set $N_d(K, \varepsilon)$ contains all bounded components of $\mathbb{R}^2 \setminus H$.

PROOF: If the theorem is false, then for each $n = 1, 2, \ldots$, we can choose a compact set H_n and a point x_n in a bounded component of $\mathbb{R}^2 \setminus H_n$, such that

$$K \subseteq H_n \subseteq N_d\left(K, \frac{1}{n}\right) \quad \text{and} \quad d(x_n, K) \ge \varepsilon.$$

Since $\{x_n\}$ is bounded, there is no loss of generality in assuming that $x_n \to x$ as $n \to \infty$. We may further assume that $x_n \in N_d\left(x, \dfrac{\varepsilon}{2}\right)$ for every $n = 1, 2, \ldots$. Now since $x \notin K$, and K is simply connected, we can find an arc $\Lambda \subseteq \mathbb{R}^2 \setminus K$, which joins x to ∞. Let $\eta = d(\Lambda, K)$. Then $\eta > 0$. Choose an integer n_0 such that

$$\frac{1}{n_0} < \eta \quad \text{and} \quad \frac{1}{n_0} < \frac{\varepsilon}{2}.$$

Then $[x_{n_0}, x] \cup \Lambda$ connects x_{n_0} to ∞, and does not meet H_{n_0}. But this contradicts the fact that x_{n_0} lies in a bounded component of $\mathbb{R}^2 \setminus H_{n_0}$.

QED

C. 29. Theorem. *Let P and Q be mutually disjoint compact sets in $\mathbb{R}^2$ and suppose that P and Q are simply connected.*

Then there exists an analytic Jordan curve J in $\mathbb{R}^2 \setminus (P \cup Q)$ such that

$$P \subseteq \text{ins}(J) \quad \text{and} \quad Q \subseteq \text{outs}(J).$$

PROOF:

CASE 1: P is connected.

If $P = \square$, the theorem is trivial, and if P consists of a single point, J can be chosen to be a small circle around P. If P contains more than one point, then choose a conformal equivalence h from $S^2 \setminus P$ onto the unit disc ins(Γ_1). Choose $r < 1$, so large that

$$|h(\infty)| < r, \text{ and } |h(x)| < r \text{ for all } x \in Q,$$

and let $J = h^{-1}(\Gamma_r)$. It is clear that J has the prescribed properties.

CASE 2: P has two components.

Let C_1 and C_2 be the components of P. Choose a homeomorphism f_1 from $S^2 \setminus C_1$ onto $\mathbb{R}^2$, and extend f_1 to a continuous function from S^2 onto itself, by defining

$$f_1(x) = \infty, \qquad \forall\, x \in C_1.$$

Now choose a homeomorhism f_2 from $S^2 \setminus f_1(C_2)$ onto $\mathbb{R}^2$, and extend f_2 to a continuous function from S^2 onto itself, by defining

$$f_2(x) = \infty, \qquad \forall\, x \in f_1(C_2).$$

Define $f(x) = f_2(f_1(x))$, $\forall\, x \in S^2$. Since $S^2 \setminus f(Q)$ is connected, we can choose an arc Λ in $S^2 \setminus f(Q)$ which joins the two points $f(C_1)$,

$f(C_2)$. By adjusting Λ, if necessary, we may require that Λ does not contain the point $f(\infty)$. Since Theorem C. 22(b) implies that $\Lambda \cup f(C_1) \cup f(C_2)$ is simply connected, the proof of Case 2 may be completed by applying Case 1 to the set $P \cup f^{-1}(\Lambda)$.

CASE 3: P has finitely many components.

We prove this case by induction on the number of components of P. Suppose, for some integer n, that the theorem is true when P has at most n components; and suppose that P has $n + 1$ components: $C_1, C_2, ..., C_{n+1}$.

Since $\bigcup\limits_{j=1}^{n} C_j$ and $C_{n+1} \cup Q$ are mutually disjoint and simply connected (see Theorem C.11), our induction hypothesis enables us to choose a Jordan curve J_1 in $\mathbf{S}^2 \setminus (P \cup Q)$ such that

$$\bigcup_{j=1}^{n} C_j \subseteq \text{ins}\,(J_1) \text{ and } C_{n+1} \cup Q \subseteq \text{outs}\,(J_1).$$

We can now apply Case 2 to find an analytic Jordan curve J such that

$$C_{n+1} \cup \overline{\text{ins}\,(J_1)} \subseteq \text{ins}\,(J) \text{ and } Q \subseteq \text{outs}\,(J).$$

J clearly has the prescribed properties.

GENERAL CASE: Let $\varepsilon = d(P, Q)$. Then $\varepsilon > 0$. Choose δ between 0 and $\dfrac{\varepsilon}{4}$, so small that given any compact set H satisfying

$$P \subseteq H \subseteq N_d(P, \delta),$$

all bounded components of $\mathbb{R}^2 \setminus H$ lie in $N_d\left(P, \dfrac{\varepsilon}{4}\right)$.

Construct a grid of horizontal and vertical lines which intersect in squares of side $\dfrac{\delta}{2}$. Let P^* be the union of the squares which meet P. Now let $\tilde{P}$ be the union of P^* and the bounded components of $\mathbb{R}^2 \setminus P^*$.

It is easy to see that $\tilde{P}$ and Q are mutually disjoint and simply connected, and that $\tilde{P}$ has only finitely many components. Using Case 3, choose an analytic Jordan curve J such that

$$\tilde{P} \subseteq \text{ins}\,(J) \quad \text{and} \quad Q \subseteq \text{outs}\,(J).$$

J clearly has the required properties.   **QED**

C.30. Theorem. *Let Ω be a region, let B be a component of $\mathbf{S}^2 \setminus \Omega$, and let $\varepsilon > 0$.*

Then there exists an analytic Jordan curve $J \subseteq \Omega$ *such that*

$$[B \text{ side} : J] \subseteq N_\varrho(B, \varepsilon).$$

PROOF: Let $K_1 = \{x \in \mathbf{S}^2 \mid \varrho(x, B) \geq \varepsilon\}$, and let

$$K = K_1 \cup \mathscr{C}(\mathbf{S}^2 \setminus \Omega, K_1).$$

It follows from Theorem A.10 that K is compact.

We claim that $K \cup (\mathbf{S}^2 \setminus \Omega)$ contains no continuum from K to B, for otherwise, $K \cup (\mathbf{S}^2 \setminus \Omega)$ contains a minimal continuum C from K to B. (See Theorem A.7.) By Theorem A.9, $C \setminus (K \cup B)$ is connected, and has limit-points in B. But this is impossible, since $C \setminus (K \cup B)$ is contained in $\mathbf{S}^2 \setminus \Omega$.

Thus $K \cup (\mathbf{S}^2 \setminus \Omega)$ contains no continuum from K to B, and it follows from Theorem A.8 that we can write $K \cup (\mathbf{S}^2 \setminus \Omega)$ as the disjoint union of compact sets P_1 and Q_1, where $K \subseteq P_1$ and $B \subseteq Q_1$.

Let $P = \mathbf{S}^2 \setminus [B \text{ side} : P_1]$, and $Q = Q_1 \setminus P$. It is easy to see that P and Q are mutually disjoint and compact, and that $P \supseteq P_1$. It is clear that P is simply connected, and since Q is the union of a family of components of $\mathbf{S}^2 \setminus \Omega$, it follows from Lemma C.5 that Q is also simply connected.

Choose a point z of $\mathbf{S}^2 \setminus (P \cup Q)$ to play the role of ∞, and using Theorem C.29, choose an analytic Jordan curve $J \subseteq \mathbf{S}^2 \setminus \{z\}$ which has P and Q in opposite sides. J clearly satisfies the requirements of the theorem.   **QED**

Mappings between Regions

C.31. Cluster Sets. Let f be a continuous function from a region Ω into $\mathbf{S}^2$, and suppose A is a compact subset of $\partial(\Omega)$.

The *Ω-cluster set of f at A* is denoted by $C_\Omega(f, A)$, and is defined by

$$C_\Omega(f, A) = \bigcap_{n=1}^{\infty} \overline{f(V_n)},$$

where, for each $n = 1, 2, \ldots,$

$$V_n = \left\{ x \in \Omega \,\middle|\, \varrho(x, A) < \frac{1}{n} \right\}.$$

C.32. Theorem. *Let f be a continuous function from a region Ω into $\mathbf{S}^2$, and suppose A is a non-empty, compact subset of $\partial(\Omega)$.*

(a) $C_{\Omega}(f, A)$ is a non-empty, compact subset of $\overline{f(\Omega)}$.

(b) $C_{\Omega}(f, A)$ is the set of all cluster points of all sequences of the form $\{f(x_n)\}$, where $\{x_n\}$ is a sequence of points of Ω which converges to a point of A.

PROOF: This proof is elementary, and is omitted. **QED**

C.33. Theorem. *Let f be a continuous function from a region Ω into $\mathbf{S}^2$, and let A be a compact subset of $\partial(\Omega)$. Suppose there exists a family $\{V_n \mid n = 1, 2, \ldots\}$ neighborhoods of A, with the properties*

(i) *$\Omega \cap V_n$ is connected, for every $n = 1, 2, \ldots$,*

(ii) *$V_1 \supseteq V_2 \supseteq \cdots$,*

(iii) *For every neighborhood V of A, there exists n such that $\Omega \cap V_n \subseteq \Omega \cap V$.*

Then $C_{\Omega}(f, A)$ is connected.

PROOF: This result is an elementary consequence of Theorem A.5, and the fact that

$$C_{\Omega}(f, A) = \bigcap_{n=1}^{\infty} \overline{f(\Omega \cap V_n)}.$$ **QED**

C.34. Remark. It is clear that a set A satisfying the hypotheses of the above theorem must itself be connected.

C.35. Lemma. *Let f be a homeomorphism from a region Ω onto a region Δ, and let A be a compact subset of $\partial(\Omega)$.*

Then $C_{\Omega}(f, A) \subseteq \partial(\Delta)$.

PROOF: Clear. **QED**

C.36. Theorem. *Let f be a homeomorphism from a region Ω onto a region Δ.*

Then the map

$$B \to C_{\Omega}(f, B)$$

is a one-one correspondence from the family of components of $\partial(\Omega)$ onto the family of components of $\partial(\Delta)$. Furthermore, for each component B of $\partial(\Omega)$, we have

$$B = C_{\Delta}\big(f^{-1}, C_{\Omega}(f, B)\big).$$

PROOF: Let B any component of $\partial(\Omega)$. To see that $C_{\Omega}(f, B)$ is connected, choose Jordan curves $J_n \subseteq \Omega$, for each $n = 1, 2, \ldots$, in such a way that

$$\Omega \cap [B \text{ side} : J_n] \subseteq N_{\varrho}\left(B, \frac{1}{n}\right), \qquad \forall\, n = 1, 2, \ldots.$$

(This is possible by Theorem C.30.) For each $n = 1, 2, \ldots,$ let $V_n = [B$ side $: J_n]$. Then for each n, since each component of $\mathbf{S}^2 \setminus (\Omega \cap V_n)$ is either a side of Ω or the set $\mathbf{S}^2 \setminus V_n$, and is thus simply connected, it follows from Theorem C.11 that $\Omega \cap V_n$ is connected. It therefore follows from Theorems C.33 and C.35 that $C_\Omega(f, B)$ is a connected subset of $\partial(\Delta)$.

Let B' be the component of $\partial(\Delta)$ which contains $C_\Omega(f, B)$. Then since $C_\Delta(f^{-1}, B')$ is clearly a connected subset of $\partial(\Omega)$, we see that

$$C_\Delta(f^{-1}, B') \subseteq B,$$

and the theorem follows easily from Theorem C.32 (b).   **QED**

C.37. Corollary. *If two regions are homeomorphic, they have the same connectivity.*

For finitely connected regions, this corollary has a converse:

C.38. Theorem. *Two finitely connected regions are homeomorphic iff they have the same connectivity.*

PROOF: The "only if" assertion has already been proved.

Suppose Ω is a finitely connected region, and let $n > 1$ be its connectivity. Using techniques similar to those used in Case 2 of the proof of Theorem C.29, choose a continuous function f from $\mathbf{S}^2$ onto itself, whose restriction to Ω is a homeomorphism, such that the image under f of each component of $\mathbf{S}^2 \setminus \Omega$ is a single point. Denote these n points by $x_1, x_2, \ldots, x_n$, and let Λ be an arc in $\mathbf{S}^2$ which joins x_1 to x_n, and contains x_j for each $j = 2, 3, \ldots, n-1$. Choose a homeomorphism h from $\overline{\Lambda}$ onto $[0, 1]$ which carries $\{x_1, x_2, \ldots, x_n\}$ onto $\left\{0, \dfrac{1}{n-1}, \dfrac{2}{n-1}, \ldots, 1\right\}$, and using Theorem C.22 (b) extend h to a homeomorphism of $\mathbf{S}^2$ onto itself.

It is now clear that Ω is homeomorphic to $\mathbf{S}^2 \setminus \left\{0, \dfrac{1}{n-1}, \dfrac{2}{n-1}, \ldots, 1\right\}$, and we have therefore shown that any two finitely connected regions with the same connectivity $n > 1$, must be homeomorphic.

We complete the proof by noting that any two simply connected regions must also be homeomorphic.   **QED**

C.39. Theorem. *Let J be a Jordan curve in a region Ω, let B_1 and B_2 be components of $\partial(\Omega)$, and let f be a homeomorphism from Ω onto a region Δ.*

Then B_1 and B_2 lie in the same side of J iff $C_\Omega(f, B_1)$ and $C_\Omega(f, B_2)$ lie in the same side of $f(J)$.

PROOF: By Theorem C.11, $\Omega \setminus J$ has two components, and we denote these by Ω_1 and Ω_2. It is clear that Ω_1 and Ω_2 lie in opposite sides of J. Since Δ is the disjoint union of $f(J)$, $f(\Omega_1)$, and $f(\Omega_2)$, it is easy to see that $f(\Omega_1)$ and $f(\Omega_2)$ are the two components of $\Delta \setminus f(J)$, and that consequently, $f(\Omega_1)$ and $f(\Omega_2)$ lie in opposite sides of $f(J)$.

Now let B be any component of $\partial(\Omega)$, which is contained in $[\Omega_1 \text{ side}: J]$. Then it is easy to see that $C_\Omega(f, B) \subseteq [f(\Omega_1) \text{ side}: f(J)]$. From this observation, the theorem follows at once. **QED**

C.40. Annuli. An *annulus* is a region of connectivity 2. Thus a region is an annulus iff it is homeomorphic to $\{z \in \mathbb{R}^2 \mid 1 < |z| < 2\}$.

C.41. Definition. Let A and B be mutually disjoint compact subsets of $\mathbf{S}^2$, and suppose each of the sets A and B lies in a side of the other.

Then the region $[A \text{ side}: B] \cap [B \text{ side}: A]$ is called the *region between A and B*.

When A and B are non-empty, and connected, the region between them is an annulus, and is called the *annulus between A and B*.

C.42. Definition. If J is a Jordan curve in an annulus Ω, then we say that J *separates* $\partial(\Omega)$ if the two components of $\partial(\Omega)$ lie in opposite sides of J.

C.43. Theorem. *Let J_1 and J_2 be mutually disjoint Jordan curves in an annulus Ω, and suppose both J_1 and J_2 separate $\partial(\Omega)$.*

Then the annulus between J_1 and J_2 is a subset of Ω.

PROOF: Clear. **QED**

Mappings between Finitely Connected Regions

In this section, we state, without proof, those theorems from conformal mapping theory which we need.

C.44. Definition. Let Ω be a finitely connected region.

Then Ω is said to be an *elementary* region if every component of $\partial(\Omega)$ is either a circle or a single point.

C.45. Theorem.

(a) *Every finitely connected region is conformally equivalent to an elementary region.*

(b) *Every annulus is conformally equivalent to an annulus of the form $\{x \in \mathbb{R}^2 \mid r_1 < |x| < r_2\}$, where $0 \leq r_1 < r_2 \leq \infty$.*

PROOF: The proof of this theorem can be found in most texts on conformal mapping theory.   **QED**

C.46. Theorem. *Let h be a conformal equivalence from a finitely connected region Ω onto a region Δ. Suppose some component of $\partial(\Omega)$ consists of a single point x.*

Then

(a) *$C_\Omega(h, x)$ is a single point.*

(b) *Writing $y = C_\Omega(h, x)$, and defining $h(x) = y$, we may extend h to a conformal equivalence from $\Omega \cup \{x\}$ onto $\Delta \cup \{y\}$.*

PROOF: This theorem follows at once from the observation that the fact that h is one-one implies that x cannot be an essential singularity of h.   **QED**

C.47. Definition. Let Ω be a finitely connected region.

An arc Λ in Ω which joins a point of Ω to a point of $\partial(\Omega)$, will be called an *end-cut* of Ω.

If $y \in \partial(\Omega)$, and there exists an end-cut of Ω which joins a point of Ω to y, then y is said to be be an *arcwise-accessible point* of $\partial(\Omega)$.

An arc Λ in Ω which joins two points which lie in the same component of $\partial(\Omega)$, is called a *cross-cut* of Ω.

C.48. Theorem. *Let h be a conformal equivalence from a finitely connected region Ω onto an elementary region Δ, and let Λ be an end-cut [cross-cut] in Ω.*

Then $h(\Lambda)$ is an end-cut [cross-cut] in Δ.

PROOF: The proof of this theorem can be found in Collingwood and Lohwater [1].   **QED**

C.49. Definition. For the definition of the *prime ends* of a finitely connected region Ω, we refer the reader to page 169 of Collingwood and Lohwater [1]. The disc D there, should be replaced by Ω.

For the definition of the *impression $I(P)$* of a prime end P of Ω, we refer the reader to page 170 of Collingwood and Lohwater [1]. The reader should also take note of Theorem 9.2 on that page.

C.50. Theorem. *Let h be a conformal equivalence from a finitely connected region Ω onto an elementary region Δ.*

Then there is a one-one correspondence

$$x \to P(x)$$

from $\partial(\Delta)$ *onto the set of prime ends of* Ω. *Further, for every* $x \in \partial(\Delta)$, *we have*

$$C_\Delta(h^{-1}, x) = I(P(x)).$$

PROOF: We refer the reader to Theorem 9.4 on p. 173 of Collingwood and Lowhater [1], and to the final paragraph of Section 6 on p. 176.

QED

C.51. Principal and Subsidiary points. Let P a be prime end of a finitely connected region Ω. Then $I(P)$ can be partitioned into its *principal points* and its *subsidiary points*. The definition can be found on page 176 of Collingwood and Lohwater [1]. On the same page, the reader can find the definition of *convergence of an arc* Λ *to* P. A point $x \in I(P)$ is said to be *accessible relative to* P if there exists an end-cut Λ of Ω, which converges to P, and joins a point of Ω to x.

C.52. Theorem. *Let* P *be a prime end of a finitely connected region* Ω, *let* $x \in I(P)$, *and suppose* x *is accessible relative to* P.

Then x *is the only principal point of* $I(P)$.

PROOF: See Collingwood and Lohwater [1], Theorem 9.7, p. 177. **QED**

C.53. Definition. Let Ω be a finitely connected region, let B be a component of $\partial(\Omega)$, and suppose B is a circle with center b and radius r. Let $x \in B$ and let $z \in \Omega$ be so chosen that the three points z, x, and b, are collinear. The given any ϑ satisfying $0 < \vartheta < \frac{\pi}{2}$, the set of those points $y \in \Omega$ for which $|x - y| < \frac{1}{2}r$ and the angle between $[x, y]$ and $[x, z]$ does not exceed ϑ, is called a *Stolz angle of* Ω *at* x, and ϑ is called the *opening* of this Stolz angle.
If A is any subset of B and $0 < \vartheta < \frac{\pi}{2}$, then the *Stolz angle of* Ω *at* A, *of opening* ϑ, *is defined to be*

$$\bigcup_{x \in A} L_x,$$

where for each $x \in A$, L_x is the Stolz angle of Ω at x, of opening ϑ.

C.54. Definition. Let Ω be a finitely connected region, and let x be contained in a circular component of $\partial(\Omega)$. An end-cut Λ which joins a point of Ω to x, is said to be a *Stolz path of* Ω *at* x if there exists a Stolz angle of Ω at x which contains Λ.

C.55. Theorem. *Let* h *be a conformal equivalence from a finitely connected region* Ω *onto an elementary region* Δ. *Let* $x \in \partial(\Delta)$, *and let* $P(x)$ *be the prime end which corresponds to* x *under the correspondence of Theorem* $C.50$.

Then given a Stolz path Λ of Δ at x, the arc $h^{-1}(\Lambda)$ joins a point of Ω to the set of principal points of $I(P(x))$.

PROOF: See Collingwood and Lohwater [1], Theorem 9.8 on p. 178.

QED

Extension of a Conformal Equivalence to an Arc of Boundary Points

C.56. Lemma. *Let Λ be an arc in $\mathbf{S}^2$.*

Then there exists a homeomorphism f from a neighborhood V of $(0, 1)$ onto a neighborhood of Λ, such that $f((0,1)) = \Lambda$.

PROOF: Let Λ join A to B.

CASE 1: $A \cap B = \square$.

It is easy to see that $[\Lambda \text{ side} : A \cup B]$ is an annulus, and we can therefore choose a homeomorphism h from $[\Lambda \text{ side} : A \cup B]$ onto $\mathbf{S}^2 \smallsetminus \{0, 1\}$. It is clear that $h(\Lambda)$ joins 0 to 1. By Theorem C.22(b), we can choose a homeomorphism g from $\mathbf{S}^2$ onto itself that carries $h(\Lambda)$ onto $(0, 1)$. It is clear that the function f defined by

$$f(x) = h^{-1}(g^{-1}(x))$$

has the required properties.

CASE 2: $A \cap B \neq \square$.

Since $[\Lambda \text{ side} : A \cup B]$ is homeomorphic to $\mathbf{S}^2 \smallsetminus \{1\}$, there is no loss of generality in assuming that Λ is an arc in $\mathbb{R}^2$ joining 1 to 1. Now, in view of Theorem C.22(a), we may assume that $\Lambda = \Gamma_1 \smallsetminus \{1\}$, and for this special arc, the result is elementary.   **QED**

C.57. The "Type" of an Arc. Let Ω be a finitely connected region, and let Λ be an arc, which is a relatively open subset of $\partial(\Omega)$.

If $\Omega \cup \Lambda$ is open, then Λ is said to be *of type* 2, and if $\Omega \cup \Lambda$ is not open, Λ is said to be *of type* 1.

Let f be a homeomorphism from a neighborhood V of $(0, 1)$ onto a neighborhood of Λ, and suppose that $f((0, 1)) = \Lambda$. Using the fact that Λ is relatively open in $\partial(\Omega)$, adjust V if necessary so as to ensure that

$$f(V) \cap \partial(\Omega) = \Lambda.$$

Let V^+ and V^- respectively be the intersections of V with the open upper and lower half-planes, and let

$$W = f^{-1}(\Omega \cap f(V)).$$

It is easy to see that if $x \in (0, 1)$ and $N_d(x, \delta) \subseteq V$, then either $N_d(x, \delta) \cap V^+ \subseteq W$ or $N_d(x, \delta) \cap V^+$ does not meet W. Furthermore, either $N_d(x, \delta) \cap V^- \subseteq W$ or $N_d(x, \delta) \cap V^-$ does not meet W.

For each $x \in (0, 1)$, choose $\delta_x > 0$ such that $N_d(x, \delta_x) \subseteq V$. It is now clear that exactly one of the following occurs:

(a) For every $x \in (0, 1)$, $N_d(x, \delta_x) \cap V^+ \subseteq W$ and $N_d(x, \delta_x) \cap V^-$ does not meet W.

(b) For every $x \in (0, 1)$, $N_d(x, \delta_x) \cap V^- \subseteq W$ and $N_d(x, \delta_x) \cap V^+$ does not meet W.

(c) For every $x \in (0, 1)$, $N_d(x, \delta_x) \cap V^+ \subseteq W$ and $N_d(x, \delta_x) \cap V^- \subseteq W$.

It is clear that Λ is of type 1 iff   (c) does not occur.

C.58. Theorem. *Let h be a conformal equivalence from a finitely connected region Ω onto an elementary region Δ.*

(a) *If Λ is an analytic arc in $\partial(\Omega)$, and Λ is of type 1, then $C_\Omega(h, \Lambda)$ is a homeomorphic image γ of Λ, and there exists a conformal equivalence from a neighborhood of γ onto a neighborhood of Λ which carries γ onto Λ, and whose restriction to Δ coincides with h^{-1}.*

(b) *If Λ is an analytic arc in $\partial(\Omega)$, and Λ is of type 2, then $C_\Omega(h, \Lambda)$ is the union of two mutually disjoint homeomorphic images γ_1 and γ_2 of Λ, and for each $j = 1, 2$, there exists a conformal equivalence from a neighborhood of γ_j onto a neighborhood of Λ which carries γ_j onto Λ, and whose restriction to Δ coincides with h^{-1}.*

Furthermore, if Λ is given an orientation, then these conformal equivalences induce opposite orientations in γ_1 and γ_2.[3]

(c) *If J is an analytic Jordan curve, and J is a component of $\partial(\Omega)$, then $C_\Omega(h, J)$ is a circular component γ of $\partial(\Delta)$, and there exists a conformal equivalence from a neighborhood of J onto a neighborhood of γ which carries J onto γ and whose restriction to Ω coincides with h.*

(d) *if Λ_1 and Λ_2 are mutually disjoint analytic arcs in $\partial(\Omega)$, then $C_\Omega(h, \Lambda_1)$ and $C_\Omega(h, \Lambda_2)$ are mutually disjoint.*

PROOF: We shall not give a complete proof of this theorem, but we shall make some general remarks, from which a proof of this theorem can be constructed without difficulty.

Let Ω be a finitely connected region, and let Λ be an analytic arc, which is a relatively open subset of $\partial(\Omega)$. Let h be a conformal equivalence from Ω onto an elementary region Δ, suppose that Δ is a subregion of

---

[3] Since γ_1 and γ_2 lie in a circle, there is an obvious meaning to the statement that γ_1 and γ_2 are oppositely oriented.

the open unit disc, and suppose that the unit circle Γ_1 is the component of $\partial(\Delta)$ which contains $C_\Omega(h, \Lambda)$.

Choose a neighborhood V of $(0, 1)$ and a conformal equivalence f from V onto a neighborhood of Λ, such that f carries $(0, 1)$ onto Λ, and $f(V) \cap \partial(\Omega) = \Lambda$.

Define V^+, V^- and W, in the same way as was done in Section C.57, let $W^+ = W \cap V^+$, and let $W^- = W \cap V^-$. Suppose that either (a) or (c) of Section C.57 occurs, *i.e.*, suppose that for every point $x \in (0, 1)$, W^+ contains all points in the upper half plane which are sufficiently close to x. Let f^+ be the restriction of f to W^+, and let $[a, b]$ be a compact subinterval of $[0, 1]$. Let L be an arc in W^+ which joins a to b, and lies so close to $[a, b]$ that W^+ contains the inside of the Jordan curve $L \cup [a, b]$. $f^+(L)$ is a cross-cut of Ω, which joins a' to b', where $a' = f(a)$ and $b' = f(b)$. By Theorem C.48, $h(f^+(L))$ is a cross-cut in Δ which joins a point a'' of Γ_1 to a point b'' of Γ_1. It follows easily from Theorem C.52 that $a'' \neq b''$. Now it is clear that the boundary of $h(f^+(\text{ins}(L \cup [a, b])))$ consists of $h(f^+(L))$, the points a'' and b'', and one of the two subarcs of Γ_1 which join a'' to b''. Let this subarc of Γ_1 be denoted by γ.

We assert that there is a conformal equivalence from a neighborhood of γ onto a neighborhood of Λ_0 $\left(\text{where } \Lambda_0 = f((a, b))\right)$, which carries γ onto Λ_0, and whose restriction to Δ coincides with h^{-1}. To see this, choose any point z in $\Gamma_1 \setminus \bar{\gamma}$, and let g be a conformal equivalence from $S^2 \setminus \{z\}$ onto $\mathbb{R}^2$ which carries the open unit disc onto the upper half plane. Define the function k^+ by

$$k^+(x) = g(h(f^+(x))), \qquad \forall \, x \in W^+.$$

The required conformal equivalence can now be obtained by an application of the Schwartz reflection principle to k^+.

Having made these remarks, we leave the proof of Theorem C.58 to the reader.   **QED**

C.59. Theorem. *Let h be a conformal equivalence from a finitely connected region Ω onto an elementary region Δ, let Λ be an analytic arc which is a relatively open subset of $\partial(\Omega)$, and suppose Λ joins a subset A of $\partial(\Omega)$ to a subset B of $\partial(\Omega)$. Let γ be one of the at most two images of Λ under h, in the sense of Theorem C.58, and suppose γ joins p to q, (where p and q correspond, respectively, to A and B). Let L be an end-cut in Δ, which joins a point of Δ to p, and suppose that L lies in a Stolz angle of Δ at $\gamma \cup \{p\}$.*

Then $h^{-1}(L)$ joins a point of Ω to a subset of A.

PROOF: Let $P(p)$ be the prime end of Ω which is associated with p in the sense of Theorem C.50. It is not hard to see from the definition of a principal point that every principal point of $I\big(P\,(p)\big)$ must belong to A. Therefore, in the special case in which L is a Stolz path of $\varDelta$ at p, the result follows at once from Theorem C.55.

Now in general, choose a point $c \in \gamma$, a Stolz path L_1 of $\varDelta$ at p and a Stolz path L_2 of $\varDelta$ at c such that L_1, L_2 and the part of γ between p and c constitute a Jordan curve J, and let this be done in such a way that

$$L \subseteq \operatorname{ins}(J) \subseteq \varDelta.$$

Let $D = h^{-1}\big(\operatorname{ins}(J)\big)$. Then D is a disc, and $\partial(D)$ contains $A \cup h^{-1}(J \setminus \{p\})$. But from the above special case, we see that $A \cup h^{-1}(J \setminus \{p\})$ is compact, and therefore, if $A \cup h^{-1}(J \setminus \{p\})$ were not the whole of $\partial(D)$, we could deduce a contradiction as follows: Draw an arc in D which joins a point of D to a point $x \in \partial(D) \setminus [A \cup h^{-1}(J \setminus \{p\})]$. Then since $x \in I\big(P(p)\big)$ and is accessible relative to $P(p)$, it follows from Theorem C.52 that x is a principal point of $I\big(P(p)\big)$, and this contradicts the fact that $x \notin A$.

Therefore

$$\partial(D) = A \cup h^{-1}(J \setminus \{p\}),$$

and it follows easily that $h^{-1}(L)$ joins a point of $\varOmega$ to a subset of A. **QED**

Index of a Closed Curve

If γ is a closed curve in $\mathbb{R}^2$, and x belongs to $\mathbb{R}^2 \setminus \gamma$, the index of γ at x, which is denoted by $\operatorname{Ind}_\gamma(x)$, is the "number of times γ winds around x counterclockwise". For our purposes, however, it is simpler to use the following equivalent definition:

C.60. Definition. Let γ be a continuously differentiable (oriented) closed curve in $\mathbb{R}^2$. The *index* $\operatorname{Ind}_\gamma$ *of* γ is the function from $\mathbb{R}^2 \setminus \gamma$ defined by

$$\operatorname{Ind}_\gamma(x) = \frac{1}{2\pi i} \int_0^1 \frac{\gamma'(t)}{\gamma(t) - x}\, dt, \qquad \forall\, x \in \mathbb{R}^2 \setminus \gamma.$$

C.61. Theorem. *Let γ be a continuously differentiable closed curve in $\mathbb{R}^2$.*

(a) $\operatorname{Ind}_\gamma$ *is a continuous function from* $\mathbb{R}^2 \setminus \gamma$ *into the set of integers.*
(b) *For every x in the unbounded component of* $\mathbb{R}^2 \setminus \gamma$, *we have* $\operatorname{Ind}_\gamma(x) = 0$.

PROOF: See Rudin [1] Theorem 10.10.   **QED**

C.62. Theorem. *Let γ_1 and γ_2 be continuously differentiable closed curves in $\mathbb{R}^2$, let $z \in \mathbb{R}^2$, and suppose*

$$|\gamma_1(t) - \gamma_2(t)| < |\gamma_1(t) - z|, \qquad \forall \, t \in [0, 1].$$

Then $\mathrm{Ind}_{\gamma_1}(z) = \mathrm{Ind}_{\gamma_2}(z)$.

PROOF: See Rudin [1], Theorem 10.35.   **QED**

C.63. Definition. Let γ be a closed curve in $\mathbb{R}^2$, (which need not be continuously differentiable). We define $\mathrm{Ind}_\gamma(x)$, for each $x \in \mathbb{R}^2 \setminus \gamma$, by

$$\mathrm{Ind}_\gamma(x) = \lim_{n \to \infty} \mathrm{Ind}_{\gamma_n}(x)$$

where $\{\gamma_n\}$ is any chosen sequence of continuously differentiable closed curves, which converges uniformly to γ on $[0, 1]$.

C.64. Remark

(a) Since any closed curve γ, regarded as a function on the unit circle in the sense of Remark C.15, is the uniform limit of a sequence of trigonometric polynomials, it is clear that a sequence $\{\gamma_n\}$ of continuously differentiable closed curves can always be found, converging uniformly to γ.

(b) We see at once from Theorem C.62 that Definition C.63 is not ambiguous.

(c) It is clear that Theorems C.61 and C.62 hold without the assumption of continuous differentiability.

C.65. Theorem. *Let γ_0 and γ_1 be closed curves in a region $\Omega \subseteq \mathbb{R}^2$, and suppose γ_0 and γ_1 are Ω-homotopic, (i.e., γ_0 and γ_1 are homotopic, as functions from $[0, 1]$ into the topological space Ω).*

Then $\mathrm{Ind}_{\gamma_0}(x) = \mathrm{Ind}_{\gamma_1}(x)$, *for every* $x \in \mathbb{R}^2 \setminus \Omega$.

PROOF: Let $x \in \mathbb{R}^2 \setminus \Omega$. By Theorem C.62 and Remark C.64(c), it is clear that the function

$$t \to \mathrm{Ind}_{\gamma_t}(x) \quad (t \in [0, 1])$$

is continuous. Therefore, since $\mathrm{Ind}_{\gamma_t}(x)$ is always an integer, the theorem is proved.   **QED**

C.66. Theorem. *Let J be a (oriented) Jordan curve in $\mathbb{R}^2$. Then Ind_J is identically zero in* $\mathrm{outs}(J)$, *and either*

(i) $\mathrm{Ind}_J(x) = -1$ *for all* $x \in \mathrm{ins}(J)$,

or

(ii) $\mathrm{Ind}_J(x) = 1$ *for all* $x \in \mathrm{ins}(J)$.

PROOF: It is already known that Ind_J is identically zero in outs(J).

Let z be any point of ins(J), and let C be a circle whose center is z, and which contains J in its inside. Using Theorem C.45 (b), choose a conformal equivalence h from the annulus between J and C onto the annulus between two concentric circles. Extend h to a homeomorphism which carries J onto one of these circles and C onto the other. Now since $h(J)$ and $h(C)$ are concentric circles, $h(J)$ may be continuously deformed into $h(C)$, by continuously changing its radius. It follows that, defining $\Omega = \mathbb{R}^2 \setminus \{z\}$, the (oriented) Jordan curve J is Ω-homotopic to an (oriented) Jordan curve γ whose range is the circle C.

Therefore $\text{Ind}_J(z) = \text{Ind}_\gamma(z)$.

We complete the proof by showing that $\text{Ind}_\gamma(z) = \pm 1$, and to do this, it is sufficient to assume that C is the unit circle, $z = 0$, and $\gamma(0) = 1$. Let f be the function from $[0, 1]$ into $[0, 2\pi)$ defined by the equation

$$e^{if(t)} = \gamma(t), \qquad \forall\, t \in [0, 1].$$

It is clear that f is continuous and strictly monotone on $(0, 1)$.

If f is increasing on $(0, 1)$, define γ_s, for each $s \in [0, 1]$, by

$$\gamma_s(t) = e^{i(s \cdot f(t) + (1-s) \cdot 2\pi t)}, \qquad \forall\, t \in [0, 1].$$

If f is decreasing on $(0, 1)$, define γ_s for each $s \in [0, 1]$, by

$$\gamma_s(t) = e^{i(s \cdot f(t) + (s-1) \cdot 2\pi t)}, \qquad \forall\, t \in [0, 1].$$

This shows that γ is homotopic to a curve γ_0, where either

$$\gamma_0(t) = e^{2\pi it} \qquad \forall\, t \in [0, 1],$$

in which case $\text{Ind}_\gamma(z) = \text{Ind}_{\gamma_0}(z) = 1$, or

$$\gamma_0(t) = e^{-2\pi it} \qquad \forall\, t \in [0, 1],$$

in which case $\text{Ind}_\gamma(z) = \text{Ind}_{\gamma_0}(z) = -1$.

This completes the proof of the theorem.   **QED**

C.67. Theorem. *Let Ω be a subregion of $\mathbb{R}^2$, let h be a homeomorphism from Ω onto Ω, and suppose h is Ω-homotopic to the identity map 1_Ω from Ω onto itself. Let J be a Jordan curve in Ω.*

Then

$$h\big(\text{ins}(J) \cap \Omega\big) = \text{ins}\big(h(J)\big) \cap \Omega,$$

and

$$h\big(\text{outs}(J) \cap \Omega\big) = \text{outs}\big(h(J)\big) \cap \Omega.$$

PROOF: It is sufficient to show that for every $z \in \Omega \setminus J$,

$$\mathrm{ind}_J(z) = \mathrm{ind}_{h(J)}\big(h(z)\big).$$

Let H be a continuous function from $[0, 1] \times \Omega$ into Ω such that

$$H(0, x) = h(x) \ \text{ and } \ H(1, x) = x, \qquad \forall \, x \in \Omega.$$

For each $t \in [0, 1]$, let $J_t = H(t, J)$. Then for each $z \in \Omega \setminus J$, the function

$$t \to \mathrm{Ind}_{J_t}\big(H(t, z)\big)$$

is clearly continuous, and therefore constant.

This completes the proof of the theorem.   **QED**

C.68. Definition. A Jordan curve J in $\mathbb{R}^2$ is said to be *oriented clockwise* if

$$\mathrm{Ind}_J(x) = -1, \qquad \forall \, x \in \mathrm{ins}(J),$$

and is said to be *oriented counter-clockwise* if

$$\mathrm{Ind}_J(x) = 1, \qquad \forall \, x \in \mathrm{ins}(J).$$

Two Jordan curves J_1 and J_2 in $\mathbb{R}^2$ are said to be *similarly oriented* if they are either both oriented clockwise, or both oriented counter-clockwise. If two Jordan curves J_1 and J_2 in $\mathbb{R}^2$ are not similarly oriented, they are said to be *oppositely oriented*.

EPILOGUE

Now we have come to the end of this analysis of the behavior of continuous actions in the Euclidean plane. Using the terminology and theory generated in the foregoing pages, what can we now say of the qualitative theory of these actions? To begin with, the most obvious and prominent feature of the work is the discovery and development of the notion of *stagnation point* and the concept of singularity which is built around it, rather than the older idea that every fixed point, indeed every point of zero velocity, whether it is fixed or not, is a singularity. The perception, and the attendant analysis, leads us to categorize certain fixed points as singular, and others as regular, and we also then decompose the set of moving points into a singular and regular part.

In analyzing the construction of a flow, we first observe that the singular fixed points constitute a compact subset of the 2-sphere, and that any such compact set could serve in this capacity. The set of stagnation points is intimately associated with this singular set; not only must every component contain a point from the closure of the stagnation points, but there are requirements on accessibility which must be met. Once we have established our singular moving set, and identified the accessible stagnation points, we then know that the regular fixed points form a T_σ-set in the complement, which is weakly accessible with respect to the stagnation points, and this is as full a description of the regular fixed points as one can obtain, since every such set is so represented, under appropriate circumstances. The regular moving points are the union of disjoint open annuli, and the singular moving points are what they must be to make the above characterizations hold.

To describe the action of the flow, we note that it is fixed on the fixed points, which is easy enough, and that on the regular moving points, the actions in the annuli can be completely described in terms of certain very simple building blocks. In fact, we exhibit explicitly a whole class of standard annular flows, and show that every component of the regular moving points has an action which is conjugate to one of these models.

The models, in turn, are completely characterized by certain continua, one for each, in such a way that two models are positively conjugate if and only if their continua are homeomorphic. Thus, in some sense, we can lay claim to knowing all there is to be known about the regular moving points.[1]

With the singular moving points, things do not come so easily. To build theory, we first consider the very special case of flows in a multiply-connected region where every orbit is aperiodic and has all its endpoints in the boundary of the region. We call these the Kaplan-Markus flows, and we obtain a certain amount of theory for these, based on the work of those two forerunners of this theory. In the basic case, where the singular fixed set has finitely many components, we can claim a strong grasp on the structure of such a flow. But this is after all a very special case. Nonetheless, it is the basis for what comes next.

We can choose any Kaplan-Marcus flow and delete from its moving set an open invariant set, and embed into that open invariant set a collection of Jordan curves, these to be the non-trivial periodic orbits of the flow we will construct. These Jordan curves must behave according to the basic properties of periodic orbits as we have defined them and proven them, but every such collection will serve. Among other things, the union of the periodic orbits must be closed in the open set we have chosen, and have other properties. The collection of Jordan curves separates the remaining points into those which are thereby cut off from the stagnation points, and those which are not. Those which are cut off will be regular points: the moving points must move in the manner we have discussed, and the fixed points are collected into T-sets of their components of the regular points. The remaining moving singular points form themselves into *organs* of the flow, and thus the study of these organs is the next order of business. We have seen that the organs are decomposed as the union of certain *tissues* and *gametes*, where the tissues are further decomposed into *cells*. The structure and interrelation of these objects with the physiological names is explored, and certain topological invariants are obtained. To a very major degree, we can say that we know the structure of these cells, of the tissues and gametes, and of the organs themselves, though not the complete mastery which we have for the regular moving points. Still, we can say that to a very large degree, we have laid bare the fine structure of this portion of the flow as well.

Along the way, we have investigated new methods of dealing with flows and new ways of looking at them. The work of the Lewins, which has

---

[1] One must be careful about saying that one knows all there is to be known about anything. It was once thought that everything worth knowing was known about continuous flows in the line. The excellent work which the Lewins have created shows that that judgment was premature (cf. Chaps. 4, 5, 8 and 9).

developed and formalized the method of multiplication of flows, and investigated the new and exciting field of algebraic combinations of flows, has been presented here in the setting from which it arose, and we hope to see these areas develop into major areas in the field of dynamics. Their value to the work of this book is clear.

And so we take our leave of you, dear reader, and commend you to the interesting problems which remain in this area of work, and in the closely related areas on which we may have cast some light. And as you go, don't forget to take with you some of our little stagnation points. You may find them useful.

BIBLIOGRAPHY

ANATOLE BECK

[1] *On Invariant Sets*, Annals of Mathematics [2] **67** (1958), pp. 99–103.
[2] *Plane Flows with Closed Orbits*, Transactions of the American Mathematical Society **114** (1965), pp. 539–551.
[3] *Plane Flows with Finitely Many Stagnation Points*, mimeographed and circulated privately.
[4] *Plane Flows with Few Stagnation Points*, Bulletin of the American Mathematical Society **71** (1965), pp. 886–890.
[5] *Upper Semi-continuous Decompositions of an Arc*, Mathematical Systems Theory **5** (1971), pp. 292–294.

ANATOLE BECK, MICHAEL N. BLEICHER and DONALD W. CROWE

[1] *Excursions into Mathematics*, Worth, New York (1969).

ANATOLE BECK, JONATHAN LEWIN and MIRIT LEWIN

[1] *On Compact One-to-one Continuous Images of the Real Line*, Colloquium Mathematicum **23** (1971), pp. 251–256.

E. F. COLLINGWOOD and A. J. LOHWATER

[1] *The Theory of Cluster Sets*, Cambridge University Press, Cambridge, England, (1966).

LEONARD GILLMAN and MEYER JERISON

[1] *Rings of Continuous Functions*, D. Van Nostrand Company, Inc., Princeton, N.J. (1960).

WALTER H. GOTTSCHALK

[1] *Bibliography for Topological Dynamics, Fourth Edition*, Wesleyan University (Department of Mathematics) Middletown, Conn. (1969).

WALTER H. GOTTSCHALK and GUSTAV A. HEDLUND

[1] *Topological Dynamics*, American Mathematical Society Colloquium Publications. Vol. 36, American Mathematical Society, Providence, R.I. (1955).

RALPH HENSTOCK

[1] *On Ward's Perron Stieltjes Integral*, Canadian Journal of Mathematics **9** (1957), pp. 96–109.
[2] *Tauberian Theorems for Integrals*, Canadian Journal of Mathematics **15** (1963), pp. 433–439.

458   Bibliography

WILFRED KAPLAN

[1] *Regular Curve-families Filling the Plane I*, Duke Mathematical Journal 7 (1940), pp. 154—185.
[2] *Regular Curve-families Filling the Plane II*, Duke Mathematical Journal 8 (1941), pp. 11—46.

J. L. KELLEY

[1] *General Topology*, D. Van Nostrand Company, Inc., Princeton, N.J. (1963).

JAROSLAV KURZWEIL

[1] *Generalized Ordinary Differential Equations and Continuous Dependence on a Parameter*, Czech Mathematical Journal 7 (1957), pp. 418—449.
[2] *On Integration by Parts*, Czech Mathematical Journal 8 (1958), pp.356—359.
[3] *Generalized Ordinary Differential Equations*, Czech Mathematical Journal 8 (1958), pp. 360—385.

R. L. MOORE

[1] *Concerning Upper Semi-continuous Collections*, Transactions of the American Mathematical Society 27 (1925), pp. 416—428.

LAWRENCE MARKUS

[1] *Global Structure of Ordinary Differential Equations in the Plane*, Transactions of the American Mathematical Society 76 (1954), pp. 127—148.

SAM B. NADLER, JR. and J. QUINN

[1] *Embeddability and Structure Properties of Real Curves*, Memoirs of the American Mathematical Society, Vol. 125, Providence, R.I. (1972).

V. V. NEMYTSKII and V. V. STEPANOV

[1] *Qualitative Theory of Differential Equations*, Princeton University Press, Princeton, N.J. (1960).

W. F. OSGOOD

[1] *A Jordan Curve of Positive Area*, Transactions of the American Mathematical Society 4 (1903), pp. 107—112.

W. F. PFEFFER

[1] *An Integral in Topological Spaces I*, Journal of Mathematics and Mechanics. 18 (1968/69), pp. 953—972.
[2] *An Integral in Topological Spaces II*, Mathematica Scandinavica 27 (1970), pp. 77—104.

W. F. PFEFFER and W. J. WILBUR

[1] *On the Measurability of Perron Integrable Functions*, Pacific Journal of Mathematics 34 (1970), pp. 131—144.

WALTER RUDIN

[1] *Real and Complex Analysis*, McGraw-Hill Book Company, New York (1966).

WACLAW SIERPINSKI

[1] *Une théoreme sur les continus*, Tôhoku Mathematical Journal 13 (1918), pp. 300—303.

TARO URA

[1] *Isomorphism and Local Characterization of Local Dynamical Systems*, Funkcial Ekvac 12 (1969), pp. 99—122.

SUBJECT INDEX

Die Grundlehren der mathematischen Wissenschaften in Einzeldarstellungen mit besonderer Berücksichtigung der Anwendungsgebiete

Eine Auswahl